Land Surface Evaporation

Thomas J. Schmugge Jean-Claude André
Editors

Land Surface Evaporation

Measurement and Parameterization

With 249 Illustrations

Springer-Verlag
New York Berlin Heidelberg London
Paris Tokyo Hong Kong Barcelona

Thomas J. Schmugge
USDA Hydrology Laboratory
Building 265
Beltsville, Maryland 20705
USA

Jean-Claude André
Director
Centre National de Recherches Météorologiques
31057 Toulouse
France

Library of Congress Cataloging-in-Publication Data
Land surface evaporation: measurement and parameterization / Thomas
J. Schmugge, Jean-Claude André, editors.
p. cm.
Papers presented at a workshop held in Banyuls, France, 10–21
Oct., 1988.
Includes bibliographical references and index.
ISBN-13:978-1-4612-7769-9
1. Evaporation (Meteorology)—Measurement—Congresses.
I. Schmugge, Thomas J. II. André, J.-C. (Jean-Claude)
QC915.6.L36 1991
551.57′2—dc20 90-9938

Printed on acid-free paper.

Typeset by Publishers Service of Montana, Bozeman, Montana.

9 8 7 6 5 4 3 2 1

ISBN-13:978-1-4612-7769-9 e-ISBN-13:978-1-4612-3032-8
DOI:10.1007/978-1-4612-3032-8

This volume is dedicated to Dr. Peter Camillo of the NASA Goddard Space Flight Center who died after a four-year bout with cancer on October 30, 1988, one week after returning from the Banyuls workshop. Peter gave two excellent talks during the workshop, as evidenced by his papers in this volume. However, it was clear that his energy was fading as the workshop went on. His ceaseless good humor in the face of his adversity was an inspiration to all of us.

Preface

The objective of this volume is to present a review of land surface evaporation processes, both from the measurement and modeling points of view. The various chapters address the description, modeling, and parameterization of these processes for atmospheric, hydrological, and climatic studies, with particular emphasis on the up-scaling methods. They include a discussion of the role of land surface evaporation in atmospheric circulation, in hydrology, and in global climate dynamics. They also deal with its measurement at various spatial scales, from the very local, micrometeorological patch scale, up to a scale of a few tens of kilometers typical of the grid size of large-scale atmospheric, climate, and hydrological models. The described methods range from the local *in situ* techniques to the airborne and satellite remote sensing techniques. The volume is based on the papers presented at a workshop on these topics held at Banyuls, France, in October 1988.

Acknowledgments

The workshop was organized by Dr. Jean-Claude André, Centre National de Recherches Météorologiques, Toulouse, France, and Dr. Thomas J. Schmugge, US Department of Agriculture, Beltsville, Maryland, USA. It was sponsored/supported by the World Climate Research Program (WCRP) and the World Meteorological Organization (WMO), the European Center for Medium Range Weather Forecast (ECMWF) in Reading, UK; the National Center for Atmospheric Research (NCAR) in Boulder, Colorado, USA; the US Department of Agriculture (USDA) and the National Aeronautics and Space Administration (NASA), both in the USA; and the Centre National de la Recherche Scientifique (CNRS), the Centre National d'Etudes Spatiales (CNES), the Institut National de la Recherche Agronomique (INRA) and the Direction de la Météorologie Nationale (DMN), from France. We would also like to acknowledge the excellent editorial assistance provided by Ms. Leoma Kruzic of the Hydrology Laboratory; without her I am sure I would have lost a paper or two.

Thomas J. Schmugge
Jean-Claude André

Contents

Contributors

F. Becker, GSTS, Universite L. Pasteur ENSPS, 7 rue de l'Université, 67000 Strasbourg, France

N.J. Bink, Department of Physics and Meteorology, Wageningen Agricultural University, Duivendaal 2, 6701 AP, Wageningen, The Netherlands

Christian Blondin, Ministere des Transports et de la Mer Direction de la Météorologie Nationale, Service Central d'Exploitation de la Météorologie, 2, avenue Rapp, 75340 Paris, France

P. Bougeault, Centre National de Recherches Météorologiques, 42, avenue Coriolis, 31057 Toulouse, France

B. Bret, Centre National de Recherches Météorologiques, 42, avenue Coriolis, 31057 Toulouse, France

H.A.R. de Bruin, Department of Physics and Meteorology, Wageningen Agricultural University, Duivendaal 2, 6701 AP, Wageningen, The Netherlands

Peter J. Camillo, (deceased), Hydrological Sciences Branch, Laboratory for Terrestrial Physics, NASA/Goddard Space Flight Center, Greenbelt, Maryland 20771, USA

Toby N. Carlson, Department of Meteorology, Pennsylvania State University, 619 Walker Building, University Park, Pennsylvania 16802, USA

Richard H. Cuenca, Department of Bioresource Engineering, Gilmore Hall, Oregon State University, Corvallis, Oregon 97331-3906, USA

R.L. Desjardins, Land Resource Research Center, Central Experimental Farm, Building 74, Ottawa, Ontario K1A OC6 Canada

Judith E. Devaney, Hydrological Sciences Branch, Laboratory for Terrestrial Physics, NASA/Goddard Space Flight Center, Greenbelt, Maryland 20771, USA

J.-P. Goutorbe, Centre National de Recherches Météorologiques, 42, avenue Coriolis, 31057 Toulouse, France

Robert L. Grossman, Cooperative Institute for Research in the Environmental Sciences, Campus Box 449, University of Colorado INOAA, Boulder, Colorado 80309-0449, USA

Robert J. Gurney, Hydrological Sciences Branch, Laboratory for Terrestrial Physics, NASA/Goddard Space Flight Center, Greenbelt, Maryland 20771, USA

F.G. Hall, Earth Resources Branch, Laboratory for Terrestrial Physics, NASA/Goddard Space Flight Center, Greenbelt, Maryland 20771, USA

Peter H. Hildebrand, National Center for Atmospheric Research, P.O. Box 3000, Boulder, Colorado 80307, USA

L.J.M. Kroon, Department of Physics and Meteorology, Wageningen Agricultural University, Duivendaal 2, 6701 AP, Wageningen, The Netherlands

P. Lacarrere, Centre National de Recherches Météorologiques, 42, avenue Coriolis, 31057 Toulouse, France

J.I. MacPherson, Land Resource Research Center, Central Experimental Farm, Building 74, Ottawa, Ontario K1A OC6, Canada

L. Mahrt, Oregon State University, Department of Atmospheric Sciences, Corvallis, Oregon 97331, USA

Joël Noilhan, Centre National de Recherches Météorologiques, 42, avenue Coriolis, 31057 Toulouse, France

J. Philip O'Kane, Civil Engineering Department, University College Dublin, Earlsfort Terrace, Dublin 2, Ireland

C. Ottlé, CNET/CRPE, 38-40, rue de Général LeClerc, 92131 Issy-les-Moulineaux, France

Alain Perrier, INRA Bioclimatologie, 78850 Thiveral Grignon, France

P.R. Rowntree, Meteorological Office, London Road, Bracknell, Berkshire RG12 2SZ, Great Britain

Steven W. Running, School of Forestry, University of Montana, Missoula, Montana 59812, USA

Thomas J. Schmugge, USDA Hydrology Laboratory, Building 265, Beltsville, Maryland 20705, USA

P.J. Sellers, COLA, Department of Meteorology, University of Maryland, College Park, Maryland 20742, USA

W. James Shuttleworth, Institute of Hydrology, Maclean Building, Crowmarsh Gifford, Wallingford, Oxfordshire OX14 1DQ, United Kingdom

Donald E. Strebel, Versar, Inc., 9200 Rumsey Road, Columbia, Maryland 21045, USA

C. Tarrieu, Centre National de Recherches Météorologiques, 42, avenue Coriolis, 31057 Toulouse, France

James E. Tillman, Department of Atmospheric Sciences, AK 40, University of Washington, Seattle, Washington 98195, USA

Andrée Tuzet, INRA Bioclimatologie, 78850 Thiveral Grignon, France

D. Vidal-Madjar, CNET/CRPE, 38-40, rue de Général LeClerc, 92131 Issy-les-Moulineaux, France

A. Weill, CNET/CRPE, 38-40, rue de Général LeClerc, 92131 Issy-les-Moulineaux, France

John C. Wyngaard, NCAR, P.O. Box 3000, Boulder, Colorado 80307, USA

1
Introduction

The exchanges of momentum, heat, and moisture between the atmosphere and the earth's surface have a fundamental influence on the dynamics and thermodynamics of the atmosphere. The oceans and continents store a portion of the incoming solar energy and eventually return it to the atmosphere. Over land, this delay is short because most of the net radiation returns to the atmosphere through turbulent exchanges that have rather small characteristic time scales and are very efficient. Part of the returned energy is immediately usable in the form of the sensible heat flux. The other and larger part is the latent heat (or moisture) flux. The energy transfer from the surface to the atmosphere approximately compensates for the loss by radiative transfer. Thus on the global scale (typically the whole globe for a space scale and 1 year for a time scale) the atmosphere maintains a constant level of energy (internal + potential + kinetic). However, this global energy balance is not achieved on smaller space and time scales, and atmospheric motions act to transfer energy among the different regions of the globe. A workshop was held in Banyuls in 1988 with the objective of improving the understanding of how land surface processes affect this transfer and the local climates.

Evidence from atmospheric general circulation model (GCM) experiments suggests that the climate is sensitive to variations in evaporation at the land surface. The early experiments of Walker and Rowntree (1977) demonstrated the sensitivity of climate to the initial soil moisture and the representation of the evaporation process in the models. Thus it is important to estimate evaporation realistically in the climate modeling context. This requires an adequate parameterization scheme or representation for the evaporation process in the models. Accurate initial conditions are important in the context of short-term predictions from a few hours or days to a few weeks. Up to now, the GCM sensitivity experiments have used rather simple parameterizations of the evaporation process.

Recently, more complex representations of land surface processes have been developed. To test the effectiveness of these new approaches and to develop them further, the HAPEX-MOBILHY (Hydrologic Atmospheric Experiment–Modélisation du Bilan Hydrique) program was initiated in 1986 to study the hydrological budget and evaporation flux at the scale of a GCM square, i.e., 100 × 100 km. The experiment was conducted in the southwest of France. Some of the surface networks operated from April 1985 to January 1987 (André et al., 1986, 1988), and a special observing period was held from 7 May to 15 July 1986, when the surface networks were augmented by airborne measurements of the surface fluxes and remote sensing observations of surface properties. The results of the HAPEX-MOBILHY measurement campaign and modeling experiments were discussed at a Fall School and Workshop held in Banyuls, France, from 10 to 21 October, 1988.

The Banyuls Fall School and Workshop

The workshop was organized around the discussion of the following scientific issues:

- The strategy for the use of experimental data in the testing, validation, and development of the

parameterization schemes for the land surface fluxes in GCMs.
- The physical principles involved in the various measurement techniques and their relevance with respect to spatial and temporal scales of interest. Descriptions of and results from specific instruments were given.
- The spatial and temporal scales that can be studied using the various techniques. The measurements with different techniques and from different experiments, such as HAPEX-MOBILHY and FIFE were compared.

More than 20 invited speakers and 80 participants assembled to discuss these topics during the 10-day meeting. The meeting was organized with both general overview presentations and open discussion/workshop sessions. It was intended for both young scientists desiring to obtain an overview of land surface processes and for senior scientists interested in a thorough discussion of the current topics in this field. The invited speakers included modelers, both GCM and mesoscale, and experimentalists. They were asked to provide a written version of their talks for distribution to the attendees. Refereed versions of these papers are the basis for this volume.

The first section of this publication is concerned with the parameterization of land surface processes in large-scale meteorological and hydrological models. Peter Rowntree's chapter presents the basic concepts for parameterizing these processes based on energy, heat, and moisture budgets, and introduces important quantities such as the aerodynamic and stomatal resistances. After discussing some important climatic effects depending on the relative magnitude of these resistances, he reviews the various schemes that have been used for climate simulations, and describes the results of simulations for various climatic scenarios such as CO_2 doubling and Amazon deforestation experiments.

Christian Blondin's chapter addresses the question of how to implement such schemes within numerical weather prediction models, particularly, those that are used in the context of short to medium range weather forecasting. Of particular importance is the determination of realistic surface conditions, including not only the prescription of the changing surface and vegetation characteristics, but also the determination of initial water content of the upper soil layers. He also discusses the validation methods that are presently used to control these parameterization schemes.

Philippe Bougeault's chapter goes one step further in the refinement of the description of land surface processes in mesoscale atmospheric models, for which the grid size may be as small as a few kilometers. At this scale it is possible to use more advanced parameterization schemes, and several such schemes are described in the chapter. He also considers validation of the scheme.

The chapter by James Shuttleworth reviews the various concepts that have been developed for the description of evaporative processes within hydrological models, some of them being the same as those used within atmospheric models. However, more emphasis is put on the physiological processes at the scale of either the leaf, the plant, or the canopy. He presents some approaches for handling the upscaling problem. The chapter by Peter Camillo et al. describes the application of a particular hydrological model to a situation where movement of the soil water plays an important role, i.e., hillslope runoff, thus exemplifying some of the main problems that have to be faced in the hydrological modeling of realistic situations. The final chapter of this section by Catherine Ottlé and Daniel Vidal-Madjar presents an improved description of evaporative processes within hydrological models, and how such refined hydrological models could then be used for estimating the area averaged evaporation flux over a large region.

The second section is concerned with the physical principles that underlie the measurement techniques, and with the possible limitations of any of these measurement techniques. First, Alain Perrier and Andrée Tuzet's chapter reviews the basic micrometeorological equations that can be used for estimating evaporative fluxes from the measurement of vertical gradients of meteorological parameters (the so-called similarity theory). They discuss in detail how these relationships either break down or have to be augmented in the cases where local advection is present. This problem is discussed further in the chapter by Henk de Bruin et al. that deals with the measurement and simulation of a situation encountered where advection between irrigated and nonirrigated parcels was a significant process.

Jean Paul Goutorbe's chapter describes how the evaporation flux can be estimated as a residual of the surface energy balance, using measured net radiation and estimation of the sensible heat flux from a simplified aerodynamic method. This chapter also discusses the accuracy of such a method, based on results obtained during the HAPEX-MOBILHY experiment. Peter Camillo's second chapter describes a combined flux estimation method, in which a one-dimensional model and remotely sensed radiance data are used to estimate spatial and temporal variations of the evaporation fluxes. The results of this method are checked against more direct measurements taken during HAPEX-MOBILHY.

John Wyngaard's chapter surveys the behavior of scalar fluxes in the PBL, describes some of the fundamental aspects of the flux measurement process, and concludes with a discussion of the eddy correlation technique to measure turbulent fluxes, and particularly those of water vapor. Considerable attention is paid to the question of estimating the influence of the measurement system itself on the measurements, particularly considering flow distortion.

Peter Hildebrand's chapter is concerned with the application of the eddy correlation technique to airborne measurements, and discusses the basic principles that govern aircraft measurement systems. Examples are given in this chapter from the HAPEX-MOBILHY experiment. Raymond Desjardins and Ian MacPherson's chapter discusses complementary, and sometime alternative, approaches to these same questions, primarily based on data taken during the FIFE program. Larry Mahrt's chapter continues this series of papers on airborne eddy correlation measurements by discussing the influence of spatial averaging on computed turbulence fluxes in the presence of surface inhomogeneity. He also discusses sampling strategy and the elimination of longer wavelengths that are observed with inadequate sample size. In his chapter, Robert Grossman describes the application of airborne flux measurement techniques in the FIFE program, where the vertical divergence of the heat and moisture fluxes is estimated for some selected cases.

The chapter by Richard Cuenca and Joel Noilhan presents various methods that can be used to infer the evaporation flux from hydrological measurements. Of particular interest are the methods based on the monitoring of the upper soil moisture content from neutron soundings. It is shown that such a measurement technique does indeed provide a reliable estimate of the regional, i.e., area averaged, evaporation flux. Alain Weill's chapter is concerned with a quite different technique, as it reviews the possibility of estimating eddy fluxes of momentum and of virtual heat from Doppler Sodar measurements. Here again examples taken from the HAPEX-MOBILHY experiment are given, which show the degree of accuracy that can be expected from such indirect techniques. James Tillman's chapter gives a complete overview of the various spectroscopic techniques that allow for *in situ* measurements of water vapor content in the atmosphere, both in the Lyman-alpha and infrared bands, from the theoretical as well as practical points of view.

Turning now to remote sensing approaches, the chapter by Thomas Schmugge and François Becker presents a review of the remote sensing approaches that can be used in the estimation of the land surface fluxes. Toby Carlson's chapter describes a combined approach where remotely sensed temperature data are used together with physiological modeling to estimate the moisture content of the upper soil and of the rooting zone. Indications are also given that concern the diurnal and seasonal variations of the stomatal resistance and leaf water potential. The last chapter of this section, by Steve Running, presents a new approach, in which a terrestrial ecosystem model is used to simulate the land surface processes and their spatial inhomogeneity. This model includes a prescription of the leaf area index, simple soil properties and makes use of routine meteorological data. This approach provides a framework within which satellite-derived normalized difference vegetation indexes (NDVI) are used in conjunction with the model-derived surface resistance factor to provide an estimation of continental scale evapotranspiration fluxes on a weekly basis.

Philip O'Kane discusses the problem of observing highly nonlinear processes, e.g., threshold phenomena, which appear very often when dealing with hydrological processes. In particular he considers the problem of switching from atmosphere-controlled to soil-controlled evaporation and how this may be remotely sensed.

The third section addresses the central problem of aggregating, or scaling up, the smaller scale values of the evaporation fluxes. The chapter by Philippe Bougeault et al. uses examples from the HAPEX-MOBILHY program to illustrate spatial integration using atmospheric mesoscale modeling. It shows which kind of strategy can be used to form a realistic description of the surface boundary conditions in the mesoscale model based on satellite remotely sensed data (e.g., NDVI, albedo, etc.). This turns out to be a necessary step for obtaining simulation results that compare well with airborne *in situ* measurements of turbulent heat and moisture fluxes.

The last two chapters, by Jean Paul Goutorbe and Catherine Tarrieu and by Don Strebel et al., respectively, describe the only two data bases presently available to study the land surface processes and their associated spatial variability, at scales of about 100 km in the case of the HAPEX-MOBILHY program, primarily concerned with the development and testing of parameterization schemes for use in atmospheric general circulation models, or of about 10 km in the case of the FIFE program, mostly concerned with the development of algorithms to retrieve surface fluxes from remotely sensed data.

References

André JC, Goutorbe JP, Perrier A (1986). HAPEX-MOBILHY: A hydrologic atmospheric experiment for the study of water budget and evaporation flux at the climatic scale. Bull Am Meteorol Soc 67:138–144.

André JC, Goutorbe JP, Perrier A, Becker F, Bessemoulin P, Bougeault P, Brunet Y, Brutsaert W, Carlson T, Cuenca R, Gash J, Gelpe J, Hildebrand P, Lagouarde JP, Lloyd C, Mahrt L, Mascart P, Mazaudier C, Noilhan J, Ottlé C, Payen M, Phulpin T, Stull R, Shuttleworth J, Schmugge T, Taconet O, Tarrieu C, Thepenier RM, Valencogne C, Vidal-Madjar D, Weill A (1988). Evaporation over land surfaces: First results from HAPEX-MOBILHY special observing period. Ann Geophys 6:477–492.

Walker J, Rowntree P (1977). The effect of soil moisture on circulation and rainfall in a tropical model. Q J R Meteorol Soc 103:29–46.

2
Atmospheric Parameterization Schemes for Evaporation over Land: Basic Concepts and Climate Modeling Aspects

P.R. Rowntree

1. Introduction

For more than a decade, evidence has been accumulating from climate model experiments that the climate is sensitive to variations in evaporation at the land surface. The experiments of Walker and Rowntree (1977) and Charney et al. (1977) demonstrated considerable sensitivity to initial soil moisture and the parameterization of evaporation, respectively. These and some subsequent experiments were reviewed by Mintz (1984) and Rowntree (1988). Rowntree and Sangster (1986) reviewed experiments on the effects of changes in albedo and surface moisture availability, both of which affect evaporation, on climate in the Sahel region. Table 2.1, taken from that paper, summarizes these responses.

It is evidently important to estimate evaporation realistically; in the climate modeling context, this requires an adequate parameterization scheme; accurate initial conditions are mainly important in the contexts of shorter term predictions from a few hours or days (Rowntree and Bolton, 1983) up to a few months. The general circulation model (GCM) sensitivity experiments mentioned above have mostly been rather simple, involving only assessments of model sensitivities to changes in initial data, to total elimination of evaporation, or to changes in model parameterizations. This was inevitable because the models contained no parameters representing geographic variations of, say, vegetation cover, which could be varied to allow any other type of study. More complex schemes have, however, been developed in recent years; first, geographic variations of a few important parameters such as albedo, roughness length, and root depth were introduced (e.g., Hansen et al., 1983); then, more elaborate and hopefully more realistic parameterizations were developed (Dickinson, 1984; Sellers et al., 1986). These are only just beginning to be used in sensitivity experiments (e.g., Dickinson and Henderson-Sellers, 1988). Since schemes approaching these in complexity are likely to be the basis of future developments, I shall present the first of them in some detail in Section 4. As a necessary introduction to this, I shall first discuss, in Section 2, the basic concepts behind the parameterizations, and then, in Section 3, some aspects of the dependence of evaporation on surface characteristics.

The sensitivity of climate simulations to evaporation is, as discussed above, well established. It may be valuable to those working on the development and use of techniques of measurements that may be used in the validation and development of parameterizations of evaporation to be aware of some recent applications of such parameterizations in climate sensitivity studies. In Section 5, I shall present some examples in which evaporation parameterizations may affect the conclusions obtained from experiments designed to assess the response of climate to man's changes of the environment. Finally, some uncertainties and outstanding problems are reviewed in Section 6.

2. Basic Concepts for Parameterization of Evaporation

2.1 Equations for Surface and Subsurface Thermal and Moist Processes

For a layer of ground between depths z and $z + \delta z$, neglecting horizontal subsurface transfers of heat

TABLE 2.1a. Effects of albedo changes on evaporation and rainfall in the Sahel region.

Experiment	Change in albedo (δA^*)	Evaporation (δE) (mm/day)	Rainfall (δR) (mm/day)	$\delta R/R$ [a]	$\frac{\delta R/R}{\delta A^*}$
Charney et al. (1977) Sahel	0.21	−0.9	−3.4	−0.46	−2.2
Sud and Fennessy (1982) Sahel	0.12	−0.7	−1.5	−0.26	−2.1
Chervin (1979) Sahara	0.22		−2.5	−0.4	−1.8
Picon (1983) Sahel	0.19	−1.1	−0.8	−0.30	−1.6
Carson and Sangster (1981)					
0.1 → 0.3 Global	0.2	−0.95	−1.22	−0.27	−1.3
0.2 → 0.3 Global	0.1	−0.26	−0.4	−0.14	−1.4

[a] R is rainfall in the control experiment.

TABLE 2.1b. Changes in African rainfall at 15°N in experiments with large-scale reductions of evaporation.[a]

	Anomaly		15°N African rainfall (mm/day)		
Reference	Area	Change in β[b] (wet → dry)	Wet	Dry	Change
Manabe (1969)	Global	1 → [c]	5	1	−4
Kurbatkin et al. (1979)	Global	[c] → 0	3	2	−1
Shukla and Mintz (1982)	Global	1 → 0	2–8	1–5	−2
Suarez and Arakawa (see Mintz (1984))	Global	1 → 0	7	1–2	−5
Carson and Sangster (1981)	Global	1 → 0[d]	1–10	1–5	−3
Cunnington and Rowntree (1986)	Africa north of 10°N	1 → 0[d]	2–7	0–2	−3

[a] All the rainfall data given are estimates from maps.
[b] $\beta = E/E_p$ (see Equation (2.30)).
[c] β is time-dependent function of soil moisture.
[d] β is time-dependent function of soil moisture, the values given being the initial values.

and water, and defining z, the ground heat flux G, and the water flux M as positive downward

$$\rho_g C\, \partial T/\partial t = -\partial G/\partial z + Q_g \qquad (2.1)$$

$$\partial m/\partial t = -\partial M/\partial z + N \qquad (2.2)$$

where ρ_g and C are the soil density and heat capacity, m is the soil water content, and Q_g and N are source/sink terms. For the heat budget, the only significant subsurface heat source is due to moisture phase changes. In the moisture budget, provided we consider m (and M) to refer to the sum of water vapor and liquid water contents (and flux), there are no sources and sinks except those associated with melting and freezing. With this definition, the "surface" is strictly the interface between vegetation and air, and it is there that the surface boundary condition

$$M(0) = -E(0) + [P + M_s - Y(0)] \qquad (2.3)$$

applies. The terms in parentheses represent the net contribution of surface and atmospheric processes [rainfall P and snowmelt M_s less surface runoff $Y(0)$] to the downward water flux $M(0)$ at the soil surface, while $E(0)$ is the evapotranspiration that may be partly a sink at the soil surface, but in the presence of vegetation also takes water out of the soil throughout the root zone, and transfers it to the atmosphere throughout the canopy.

The surface boundary condition for heat fluxes is

$$G(0) = R_N(0) - H(0) - LE(0) \qquad (2.4)$$

(see the next section for definitions).

The subsurface heat flux $G(z)$ can be represented by a diffusive term of form

$$G(z) = -\lambda_g \, \partial T/\partial z \qquad (2.5)$$

Both heat capacity C and conductivity λ_g depend mostly on moisture content so that changes in m can affect $G(0)$.

The subsurface moisture flux $M(z)$ may be represented (following Dickinson, 1984) in terms of the soil water suction $\Phi = z - \Phi_w$, where z is the negative of the gravitational potential or, equivalently the depth in the soil, and Φ_w is the matric potential, which is determined by surface tension and other forces binding water to the soil. Then

$$M(z) = -K(z) \, \partial\Phi/\partial z \qquad (2.6)$$

Here K is the hydraulic conductivity of the soil, which is expressed by Dickinson after Clapp and Hornberger (1978) in terms of the fractional soil water content s (ratio of soil water volume to volume of voids in the soil) as

$$K = K_0 s^{2B+3} \qquad (2.7)$$

Similarly Dickinson assumes

$$\Phi_w = \Phi_0 s^{-B} \qquad (2.8)$$

In these expressions, B is a nondimensional parameter, which varies with soil type, and Φ_0 and K_0 are values at saturation.

In the rest of this section, we will discuss the role of surface characteristics in determining the individual terms of the heat budget Equation (2.4), that is, the surface net radiation, R_N divided into the solar and longwave components

$$R_N(0) = R_{SN} + R_{LN} \qquad (2.9)$$

the surface sensible heat flux H and latent heat flux LE and the surface soil heat flux G. [Note that $H(0)$ is henceforth abbreviated to H for compactness and similarly for other surface fluxes.] This will allow broad assessments of likely sources of errors in the estimation of evaporation.

2.2 Solar Radiation

The absorption of solar radiation at the earth's surface can be written as

$$R_{SN} = (1-\alpha)R_s \Downarrow \qquad (2.10)$$

Here, α is the surface albedo and $R_S \Downarrow$ is the downward solar flux. Both of these are spectral quantities, that is, they vary with the wavelength of the radiation. An example of why such variations are important is that, because of absorption by water vapor and droplets in the near infrared, the solar flux below clouds is relatively more intense than with clear skies at wavelengths of less than 0.7 μm, for which vegetation is particularly strongly absorbing ($\alpha < 0.1$), but snow is especially highly reflective ($\alpha > 0.9$).

Spectrally averaged surface albedo varies from near 0.1 over forests (Oguntoyinbo, 1970; Stewart, 1971) to as high as 0.4 over some sandy deserts (e.g., Rockwood and Cox, 1978; Kondratyev et al., 1982) and around 0.9 over fresh snow (Kondratyev et al. 1982). With mean clear sky daily solar radiation of 300 W m^{-2} or more in summer in middle and high latitudes, and all year in the tropics, the effects of changes of land surface cover in the absence of snow can approach 100 W m^{-2}—enough energy to warm the atmosphere below a tropopause at 200 mb by over 1 K/day. Even larger changes are possible with snow, as the albedo of forests is little affected by snow once the branches are snowfree, so that albedo changes of 0.6 are possible with deforestation of high latitudes or of middle latitude mountains. With a moist surface, a large part of such differences will be seen in the evaporation.

2.3 Longwave Radiation

At the temperatures characteristic of the earth's surface, radiation is emitted at wavelengths longer than the approximate upper limit (3 μm) of radiation received from the sun. The peak of the emission is at about 11 μm, but there is strong emission from about 5 to 50 μm. The emissivity of most natural surfaces is near to 1 over most of this range (i.e., the surface is almost a black body). The atmosphere emits over a similar range of wavelengths because it has a similar temperature; however, at some wavelengths, the atmosphere is far from being a black body, especially in the so-called water vapor window near 9–12 μm in which there is only weak water vapor absorption and emission, and some strong but narrow trace gas absorption bands, notably due to ozone.

The net absorbed longwave radiation

$$R_{LN} = \varepsilon \Downarrow R_L \Downarrow - \varepsilon \Uparrow \sigma T^4 \quad (2.11)$$

Here, there are two emissivities, one appropriate for the spectral composition of the downward radiation, and one for the upward, near black-body, radiation from the surface, whose intensity depends on the fourth power of the surface temperature. Outside the water vapor window, $R_L \Downarrow$ originates from levels close to the ground and so usually differs little from the emission from the earth's surface. It is therefore the emissivity in the window that is important for the surface radiation budget. Data collected by Kondratyev et al. (1982) show that this is nearer to unity than the integrated emissivity, being typically 0.96–0.98. Generally then, errors in the common GCM assumption that $\varepsilon = 1$ are not likely to exceed about 0.04. This implies an error (with a relatively high net radiation of 100 W m^{-2})

$$\delta R_N \approx 4 \text{ W m}^{-2}$$

Emissivities as low as 0.7 have been estimated for some Saharan surfaces with a large quartz content at 9 μm from satellite data by Prabhakara and Dalu (1976). Their data suggest that average 9–12 μm values probably exceed 0.8 even for these regions, so that R_N for such surfaces could be increased by as much as 20 W m^{-2}. However, evaporation is usually small in these conditions because of the aridity. The role of errors in the estimation of emissivity are thus usually much less serious than errors in albedo for the calculation of evaporation.

2.4 Latent and Sensible Heat Fluxes

The evaporation at the surface can be written in a number of ways:

$$E = \rho w'q' = \rho C_E V \delta q = \rho \delta q / r_E = \rho \delta q / (r_S + r_A) \quad (2.12)$$

Here w' and q' are small perturbations of vertical velocity and specific humidity, respectively, C_E is the surface transfer coefficient for moisture, δq is the atmospheric specific humidity deficit below the surface saturation value, and r_E is the resistance to moisture transfer. The formal definition of the flux of moisture is given first, then the formulation commonly used in the simpler GCM parameterizations, and then expressions in terms of resistances.

As shown in Equation (2.12), the resistance r_E can be separated into the surface resistance r_S and the atmospheric resistance r_A. For vegetation, the former normally represents the stomatal resistance to transfer from moist surfaces within the leaves. It decreases to zero when there is water (dew or intercepted rainfall or snow) covering the stomata and also, in the absence of vegetation, with a moist soil surface. The resistance r_A represents that between the leaf or soil surface and the atmosphere and is generally taken to be the same as that in the equation for the sensible heat flux:

$$H = \rho c_P w'T' = \rho c_P C_H V \delta T = \rho c_P \delta T / r_A \quad (2.13)$$

Here, T' is a temperature perturbation and δT is the difference between the surface temperature (T_*) and the potential temperature at a reference height (z_1), commonly 10 m.

2.5 Typical Values of r_s

As indicated above, the surface resistance, r_S, is zero in the presence of sufficient surface moisture such as when a vegetation canopy is wet. Otherwise, r_S depends on the vegetation type and a number of atmospheric and hydrological variables affecting the supply of and demand for moisture, including soil moisture distribution relative to root development, atmospheric vapor pressure deficit, wind speed, and temperature [see Dickinson (1984) and Stewart (1988) for detailed discussions of the hydrological and atmospheric aspects, respectively].

Even for the relatively simple case of freely transpiring vegetation, observed values of r_S from different sources sometimes conflict; there appears to be agreement on 40–60 s/m for growing crops and 80 to 130 s/m for forests in the absence of low temperatures or high vapor pressure deficits (see Perrier, 1982; Thompson et al., 1981; Buckley and Warrilow, unpublished Meteorological Office note, 1988). Grassland values are more variable—from 60 to 200 s/m can be found.

The transpiration cannot exceed the plant's ability to supply water through the roots and stems. This may be seriously limited by soil moisture deficits. Dickinson et al. (1986) assume that transpiration is restricted to the product of a maximum value (of order 1.5×10^{-7} m s^{-1}, or 13 mm/day) and a term $(1 - W_{LT})$ dependent on soil water potential for each model layer where

$$W_{LT} = (s^{-B} - 1)/(s_w^{-B} - 1) \quad (2.14)$$

Here s_w is the soil water content for which transpiration becomes zero—some water being unavailable; Dickinson et al. use $s_w = 0.125$. The dependence of W_{LT} on soil moisture (s) through Equation (2.14) is illustrated by the following values for a typical value $B = 5$:

s	1	0.8	0.6	0.4	0.3	
W_{LT}	0	0.0001	0.0004	0.003	0.013	
s	0.2	0.175	0.15	0.14	0.13	0.125
W_{LT}	0.095	0.19	0.40	0.57	0.82	1

Note the small effect of variations in s for $s > 0.3$. The abrupt decrease in water availability over a small range of s is consistent with the observations summarized by Priestley and Taylor (1972).

2.6 Atmospheric Resistance and Its Dependence on Roughness Length

The atmospheric resistance (r_A) to moisture transfer is a function of boundary layer structure and surface roughness through both the transfer coefficient C_E and the wind speed u. To relate r_A to roughness length z_0 and gradient wind speed V_G, it is necessary to use Rossby similarity theory to find u.

The atmospheric resistance r_A to moisture transfer is defined by Equation (2.12) with $r_S = 0$ as

$$r_A = 1/C_E \quad (2.15)$$

We wish to relate r_A to the surface roughness length z_0. We shall consider here only the near-neutral case, for which most observed data are relevant. The magnitude of the surface stress

$$\tau = \rho C_D u^2 = \rho u^{*2} \quad (2.16)$$

so

$$r_A = 1/C_E u = (u/u^{*2})(C_D/C_E) \quad (2.17)$$

where C_D is the drag coefficient for the same level as the wind, typically 10 m above the vegetation, and u_* is the friction velocity. Assuming a logarithmic profile

$$\frac{u}{u_*} = \frac{1}{k} \ln\left(\frac{z-d}{z_0}\right) \quad (2.18)$$

then, for $z-d = 10$ m, where d is the zero-plane displacement, since $C_E = k^2/[\ln(10/z_0)\ln(10/_M)]$:

$$r_A = \frac{1}{ku_*} \ln\left(\frac{10}{z_0}\right) \ln\left(\frac{10}{z_M}\right) \Big/ \ln\left(\frac{10}{z_0}\right) = \frac{1}{ku_*} \ln\left(\frac{10}{z_M}\right) \quad (2.19)$$

Here k is the von Karman constant, usually taken as 0.4. Because u_* is dependent both on z_0 and on (V_G), we must use Rossby similarity theory to calculate u_*/V_G from

$$\ln\frac{V_G}{fz_0} = A - \ln\frac{u_*}{V_G} + \left(\frac{k^2 V_G^2}{u_*^2} - B^2\right)^{1/2} \quad (2.20)$$

(see, e.g., Tennekes and Lumley, 1972). Here, f is the Coriolis parameter, and A and B are constants.

The values of A and B for this neutral case were taken from Arya (1975) as $A = 1.01$ and $B = 5.14$. u_*/V_G was calculated by interpolation between values of V_G/fz_0 calculated from Equation (2.20), for a range of values of u_*/V_G. Values of r_A for selected z_0 were then computed for selected V_G and f. Table 2.2 presents a selection of results for various z_0, assuming $z_M = z_0$. An obvious result is that the larger z_0 is, the more important a change by a given factor becomes.

The comparisons of results for different V_G show only a small (10%) deviation from an inverse linear dependence of r_A on V_G. There is a small decrease in r_A as f increases (not shown; see Rowntree, unpublished Meteorological Office paper, 1989) from 10^{-5} s^{-1} (4° latitude) to 10^{-4} s^{-1} (44° latitude), but this is only 12.5% for $z_0 = 10^{-4}$ m and 20% for $z_0 = 1$ m. The variation is linear with log f. Other values for A and B from Deacon (1973) (1.9, 4.7) and P.J. Mason (personal communication) (1.2, 2.3—derived theoretically) were also tried and for the results in Table 2.2 led to reductions in r_A of less than 10% except for $z_0 = 1$ m (up to 15%).

2.7 Typical Values of z_0 and r_A

Estimates of roughness length (z_0) for different surfaces have been reviewed by Garratt (1977) on

TABLE 2.2. Variations in r_A (s m^{-1}) for selected z_0 and geostrophic wind (V_G) (m s^{-1}) ($f = 10^{-4}$ s^{-1}).

	z_0 (m)				
	10^{-4}	10^{-3}	10^{-2}	10^{-1}	1
			(a) $z_0 = z_M$		
$V_G = 20$	62.6	44.2	28.8	16.4	6.9
$V_G = 10$	120.7	84.8	55.0	31.1	13.0
$V_G = 5$	232.5	162.5	104.8	59.0	24.4
$V_G = 2.5$	447.2	311.2	199.7	111.7	46.0
			(b) $\ln(z_0/z_M) = 2$		
$V_G = 20$	73.5	53.7	37.1	23.5	12.8
$V_G = 10$	141.7	103.2	70.9	44.6	24.2
$V_G = 5$	272.9	197.8	135.2	84.6	45.6
$V_G = 2.5$	524.9	387.8	257.5	160.2	86.0

the basis of the extensive literature on the subject. Garratt notes that values fall in the range

$$0.02 < z_0/h < 0.2$$

where h is the height of roughness elements. It is common in numerical modeling to assume the roughness lengths for temperature (z_T) and water vapor (z_M) to be equal to z_0; indeed, in some models z_0 is assumed to be smaller than z_M. However, Garratt suggests $z_0/z_M \simeq 7$ and some studies propose larger ratios. The sensitivity to this assumption is illustrated in Table 2.2 by comparison of the values of r_A for $z_0 = z_M$ and for $\ln(z_0/z_M) = 2$, close to Garratt's proposal. This shows that r_A is higher with the second assumption, with the fractional change a maximum, nearly a doubling, for the largest z_0 (1 m) quoted.

The value of z_0 varies considerably between different surfaces. We consider here forest, savanna, pasture, and semidesert. Typical values of z_0 for these surfaces are discussed below.

1. Forest. Tropical forest z_0 values of 0.9–5 m are estimated by Garratt (see appendix to his paper). Note that the largest z_0 values in his table (4–5 m) are for 30–70% coverage.
2. Savanna. For savanna with grass of $\simeq$50 cm and shrubs or small trees of 2–8 m, Garratt suggests $z_0 \simeq$ 12–48 cm.
3. Pasture. Because its height varies widely (from 1 to 65 cm in Garratt's Table 2 with values exceeding 1 m for some tropical types (Garratt, 1977), grassland has a large range of z_0. For extensive grasslands, we are not concerned with the lowest values of z_0, which are for mown grass. Grazed areas would be expected to have heights of order 10 cm and $z_0 \simeq$ 1 cm.
4. Sparse grassland or semidesert. The lowest z_0 for land surfaces, excluding snow, in Garratt's review is that for deserts. Values as low as 0.02–0.03 cm have been observed. These were for barchan dunes (Lettau, 1969) and sunbaked sandy alluvium (Calder quoted by Deacon, 1953). Garratt, however, suggests a value for typical Australian deserts of 0.1 cm, similar to the figure of Clarke et al. (1971) for very sparse grass, arguably an upper bound for nonstony deserts. Note that Sud and Smith (1985) used 0.02 cm in a GCM sensitivity experiment to investigate the effects of allowing for the smoothness of deserts.

The above discussion suggests that a range of z_0 from 10^{-4} to 1 m needs to be considered. Table 2.2 shows that this corresponds to a factor of 10 variation in r_A; there is a strong, roughly linear, inverse dependence on gradient wind speed. These calculations are for near-neutral conditions. This is generally appropriate as evaporation mostly occurs with upward energy flux when departures from neutrality are small. It is relevant that Van Zyl and De Jager (1987) found the Penman–Monteith equation, in which we shall be using these values of r_A, to give good accuracy when compared with lysimeter data, and that the accuracy was not improved by allowing for stability variations.

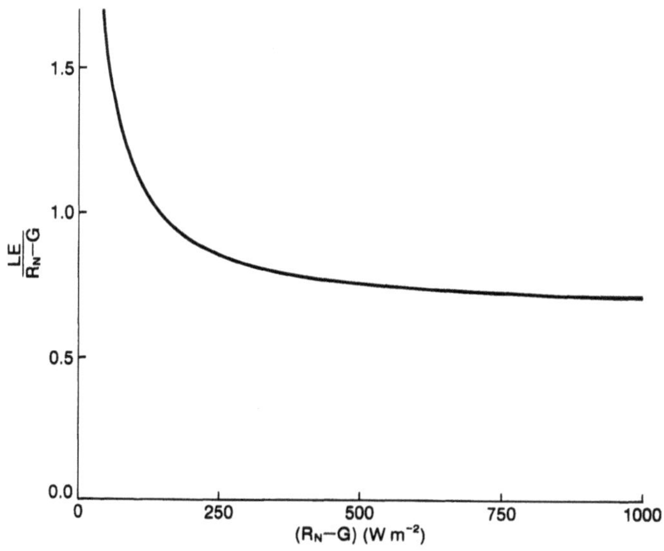

FIGURE 2.1. $LE/(R_N - G)$ as a function of $(R_N - G)$ for typical tropical conditions (near surface air temperature of 303.2 K, r_A of 50 s/m, r_S of 50 s/m, and $\delta'q$ of 5 g/kg).

3. Dependence of Evaporation on Surface Characteristics

3.1 Use of the Penman–Monteith Equation for Estimation of Surface Fluxes

Because of the difficulty of measuring the surface temperature, it is desirable, for estimating the evaporation, to eliminate it from Equation (2.12). In the Penman-Monteith approach (e.g., Monteith, 1973), this is achieved by using the surface energy balance:

$$LE = R_N - G - \rho c_P \delta T / r_A \qquad (2.21)$$

By writing:

$$\Delta = \partial q_s(T)/\partial T$$

$$\delta' q = q_s[T(z_1)] - q(z_1)$$

so that

$$\delta q = q_s(T_*) - q(z_1) = (\delta T)\Delta + \delta' q \qquad (2.22)$$

δT (and so T_*) can be eliminated between Equations (2.12), (2.21), and (2.22) to give the Penman–Monteith equation:

$$LE = \frac{\Delta(R_N - G) + \rho c_P \delta' q / r_A}{\Delta + (c_P/L)(1 + r_S/r_A)} \qquad (2.23)$$

For some purposes it is more useful to consider the ratio

$$\frac{LE}{(R_N - G)} = \frac{1 + \rho c_P \delta' q / [(R_N \equiv G) r_A \Delta]}{1 + (c_P/L\Delta)(1 + r_S/r_A)} \qquad (2.24)$$

In the rest of this section, Equations (2.23) and (2.24) are used to study the sensitivity of evaporation estimates to the surface characteristics discussed in Section 2: the surface albedo important for determining the available energy, and the atmospheric and surface resistance to turbulent transfers, r_A and r_S, which control the partitioning of the available energy between the sensible and latent energy fluxes.

3.2 Dependence on Available Energy and Vapor Pressure Deficit

We have noted in Section 2 the large variations in albedo and, consequently, net radiation that can exist. For a typical pair of values of r_A and r_S, Figure 2.1 shows the effect of varying $(R_N - G)$ with fixed $\delta' q$; for $(R_N - G) < 130$ W m^{-2}, $\mu \equiv LE/(R_N - G) > 1$ so that sensible heat flux would be negative in these conditions. In practice, with clear skies in middle latitude summer or tropical conditions, for which the $\delta' q$ is appropriate, such small $(R_N - G)$, though similar to the daily mean, occurs only within 2 to 3 hr of sunrise and sunset. Moreover, $\delta' q$ varies diurnally; examination of Equation (2.24) shows that μ is linearly related to

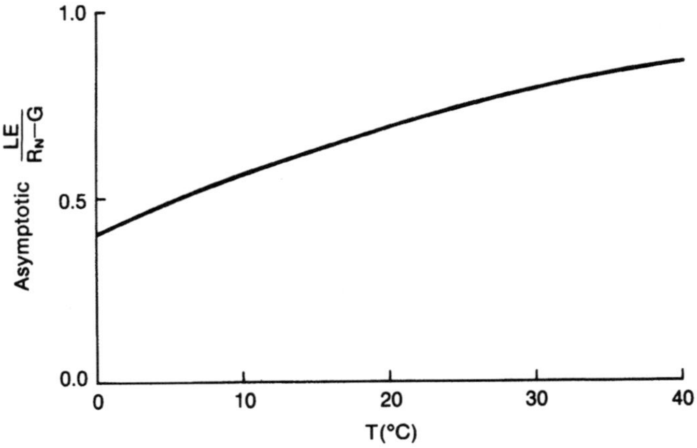

FIGURE 2.2. The asymptotic value $(1+\varepsilon)^{-1}$ of $LE/(R_N - G)$ as a function of temperature.

$\kappa \equiv \delta'q/(R_N - G)$. Consequently, the diurnal variation of μ with $(R_N - G)$ can be quite small, since there is a tendency for $\delta'q$, like temperature, to lag only a little behind $(R_N - G)$, with a maximum in the early afternoon (e.g., Shuttleworth et al., 1985). This diurnal asymmetry in $\delta'q$ makes the occurrence of $H < 0$ more likely toward sunset. With typical midday $(R_N - G)$ of 500 W m^{-2}, $LE/(R_N - G)$ is near 0.75, close to what is observed in a tropical forest for the daily mean (Shuttleworth et al., 1984). As $\delta'q \rightarrow 0$, with $r_S = 0$, Equation (2.24) shows that

$$\mu \rightarrow 1/[1 + (c_P/L\Delta)] \equiv 1/(1 + \varepsilon) \qquad (2.25)$$

where ε, so defined, is a function only of temperature. Figure 2.2 and Table 2.3 show that this asymptotic value of μ, which also occurs in a number of other contexts, is twice as large for high temperatures ($\approx$40°C) as for temperatures near freezing point.

TABLE 2.3. Variation with temperature of Δ $(=dq_S/dT)$, ε $(=c_P/L\Delta)$, and $1/(1+\varepsilon)$ [i.e., $LE/(R_N - G)$ for the value of r_S at which E is independent of r_A] and $\Delta(1+\varepsilon)$, the denominator in Equation (2.27).

	Temperature (°C)				
	0	10	20	30	40
Δ (10^{-3} K^{-1})	0.28	0.516	0.908	1.53	2.51
$\varepsilon = c_P/L\Delta$	1.439	0.783	0.445	0.264	0.161
$1/(1+\varepsilon)$	0.410	0.561	0.692	0.791	0.861
$\Delta(1+\varepsilon)$	0.68	0.92	1.31	1.93	2.91

Differentiation of Equation (2.23) with respect to $(R_N - G)$ reveals that the rate of change of LE with $(R_N - G)$

$$d(LE)/d(R_N - G) = 1 / [1 + \varepsilon\,(1 + r_S / r_A)] \qquad (2.26)$$

is a function only of temperature and r_S/r_A, as displayed in Figure 2.3. For zero r_S, the differential is the same as μ in Equation (2.25). Thus, an albedo change produces a much larger evaporation change with high than with low temperatures, even if $(R_N - G)$ is the same. The decrease in the effect of albedo on evaporation as surface resistance to evaporation increases is to be expected since E itself is similarly reduced. Equation (2.26) shows that LE always increases with increasing $(R_N - G)$, but that the rate of change as a fraction of that in $(R_N - G)$ is always less than unity. Thus both latent and sensible heat fluxes always increase with an increase in available energy $(R_N - G)$.

3.3 Dependence on r_A

As already discussed, the atmospheric resistance r_A is a function of wind speed and roughness length z_0, so that as wind speed is itself strongly dependent on z_0 as well as on gradient wind, the gradient wind speed V_G must be specified if sensitivities are to be estimated correctly. The variation of $LE/(R_N - G)$ with r_A is shown in Figure 2.4a for selected values of r_S with values of temperature and $\kappa = \delta'q/(R_N - G)$ appropriate for tropical forests. The r_S values

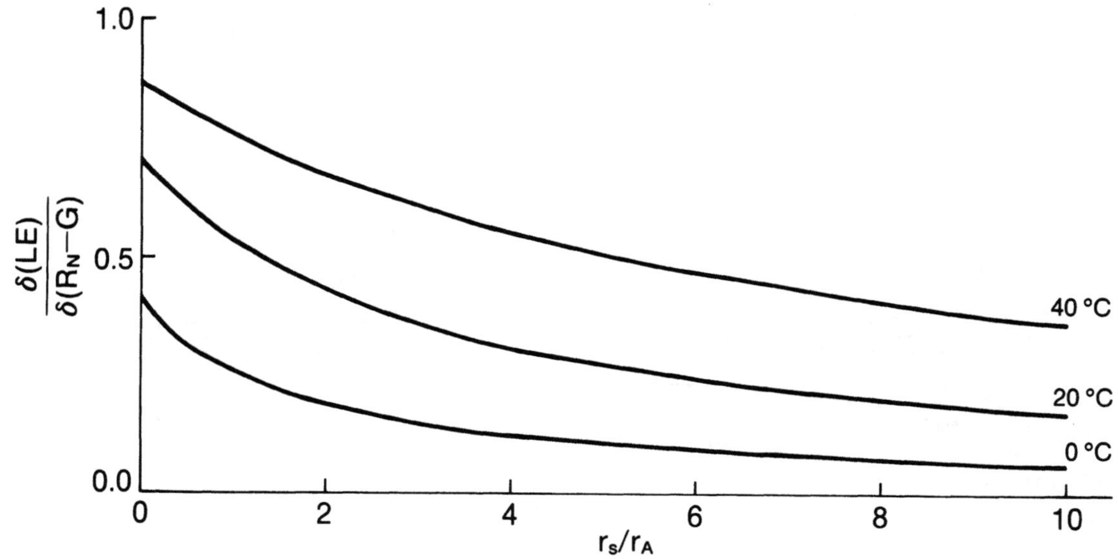

FIGURE 2.3. The change in latent heat flux as a function of the change in $(R_N - G)$, as a function of temperature and r_S/r_A.

are characteristic of forest vegetation in varying stages of foliage wetness—zero when wet, 100 s m^{-1} when almost dry but freely transpiring.

Of particular interest in Figure 2.4 is the different response to increasing roughness for low and high values of r_S. This is because if $\delta' q$ is large, E can be large with a small temperature lapse (and so also small H). If r_S is small, this effect is dominant and E decreases as r_A increases. On the other hand, if r_S is large (and E relatively small), the restriction of E by r_S dominates while r_A is small, but as r_A increases, the relative effect of r_S decreases so E increases. Note that as $r_A \rightarrow \infty$, $LE/(R_N - G)$ tends to the asymptotic value $(1+\varepsilon)^{-1}$ already shown in Figure 2.2 and Table 2.3. The critical value of r_S above which E increases with increasing r_A or decreasing z_0 is shown in Figure 2.5. Note that for typical values of κ of 0.01 to 0.03 and for freely transpiring vegetation, it is in the range typically observed for r_S (40–200 s m^{-1}). For this critical r_S, $LE/(R_N - G)$ again equals the asymptotic value.

The behavior for zero r_S can be examined by differentiating (2.23)

$$\frac{d(LE)}{d(r_A)} = \frac{-\rho c_P \delta' q / r_A}{\Delta(1 + \varepsilon)} \tag{2.27}$$

Table 2.3 shows the temperature dependence of the denominator; there is a factor of two decrease from 30 to 10°C, showing that the sensitivity to variations of r_A increases considerably with decreasing temperature. Even greater sensitivity is evident for 0°C, but 10°C is typical of large areas in middle and high latitudes in late spring and summer when $(R_N - G)$ is comparable to tropical values. Figure 2.4b shows that the larger sensitivity to r_A at 10°C is also evident for r_S = 50 s m^{-1}, which is typical of growing crops; this was to be expected from the temperature dependence of the critical r_S value in Figure 2.5.

3.4 Dependence on r_S

The previous diagrams gave an impression of the variation of $LE/(R_N - G)$ with r_S. The variation of this dependence with r_A is clearly shown by Figure 2.6. Note how much steeper is the cutoff in E with increasing r_S for small r_A, i.e., large z_0 or strong winds. It should be noted that r_S increases to very large values as soil moisture becomes limiting and plants approach wilting point. These values are not shown in Figure 2.6. The dependence of E on s is then perhaps better expressed by the W_{LT} parameter of Dickinson et al. (1986) discussed in Section 2.

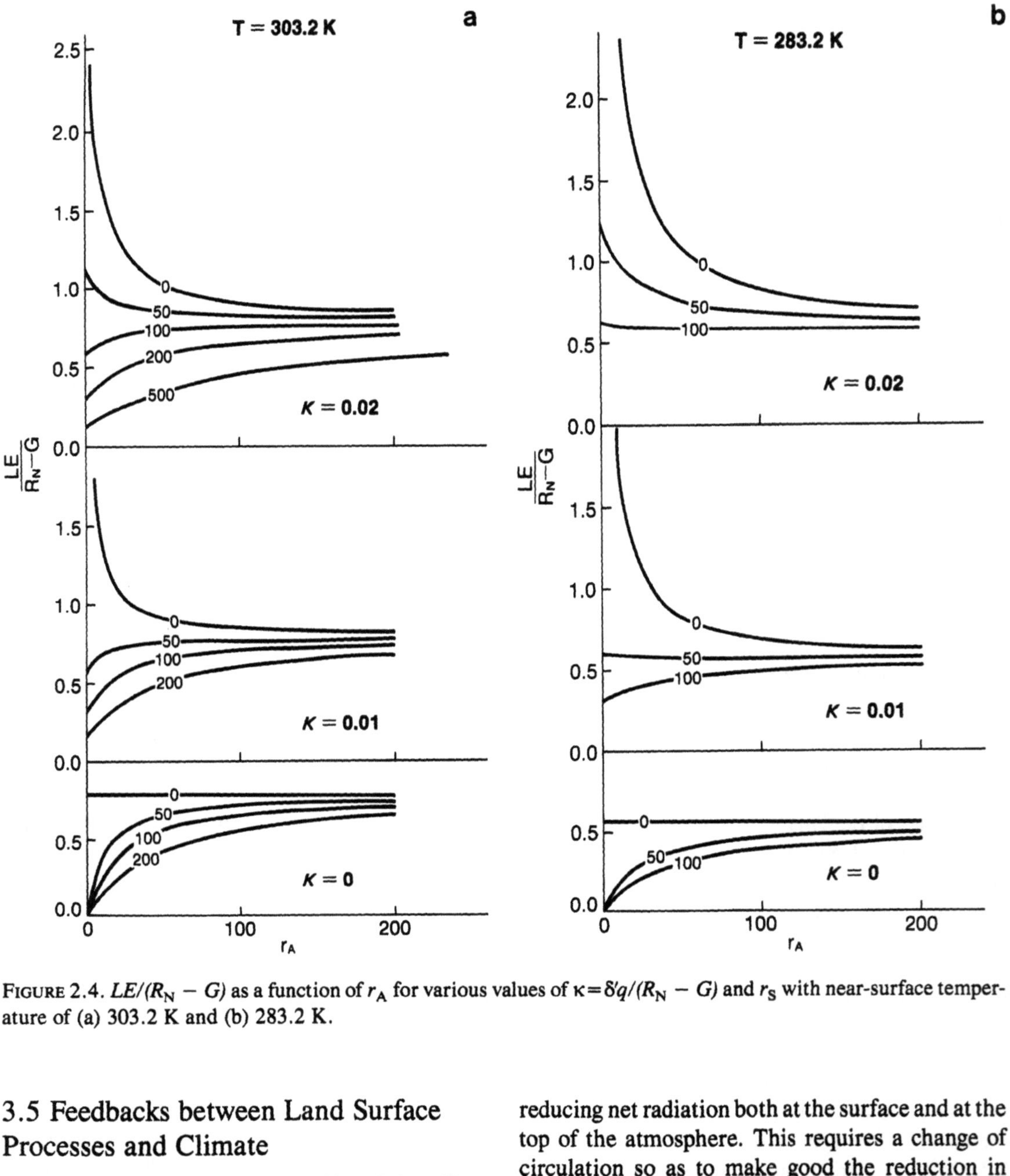

FIGURE 2.4. $LE/(R_N - G)$ as a function of r_A for various values of $\kappa = \delta' q/(R_N - G)$ and r_S with near-surface temperature of (a) 303.2 K and (b) 283.2 K.

3.5 Feedbacks between Land Surface Processes and Climate

In the previous sections we considered the effects of land surface characteristics on evaporation in terms of three parameters: albedo, surface resistance, and surface roughness. For completeness, we will now consider possible feedbacks between climate and these parameters.

Albedo

First, for albedo, we have the well-known feedback proposed by Charney (1975) with increasing albedo reducing net radiation both at the surface and at the top of the atmosphere. This requires a change of circulation so as to make good the reduction in heating. In the tropics this is normally achieved through increased subsidence, so drying the troposphere and tending to maintain arid conditions and so the higher albedos. There is also a secondary feedback: the decrease in evaporation that we have seen to accompany any decrease in $(R_N - G)$ also tends to reduce the rainfall. On a regional scale, the water balance requires this. GCM experiments support the existence of both feedbacks (Mintz, 1984; Rowntree, 1988).

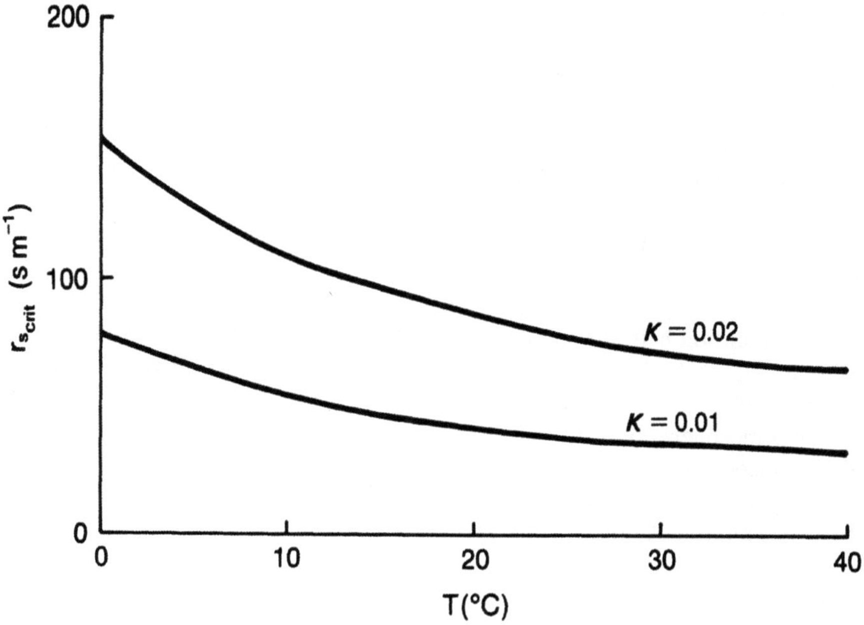

FIGURE 2.5. The critical r_S below which evaporation increases with roughness length for two values of $\kappa = \delta' q/(R_N - G)$.

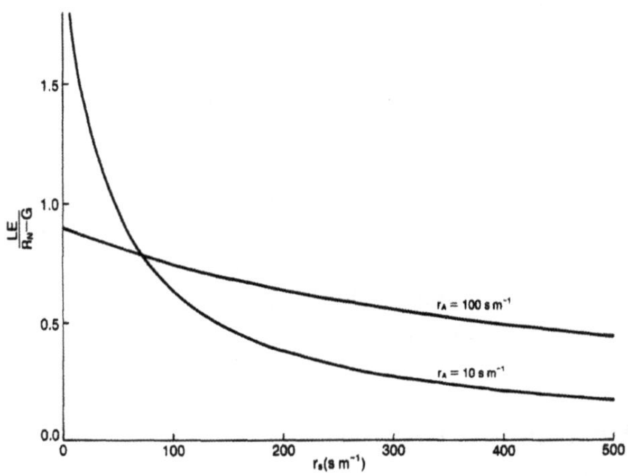

FIGURE 2.6. Dependence of $LE/(R_N - G)$ on r_S for r_A of 10 and 100 s/m, with near-surface temperature of 303.2 K and $\delta' q/(R_N - G) = 0.02$ (g/kg)/(W m^{-2}).

Surface Resistance to Evaporation

Where an increase in resistance to evaporation is due to a loss of vegetation, a positive feedback can be expected because the loss of evaporation can lead to a drier atmosphere and reduction of precipitation, and so, as with the albedo feedback, a maintenance of reduced vegetation. Here, the surface input of energy may not be reduced but, because the upper atmosphere is not heated by latent heat release, subsidence is required there to compensate for radiative cooling (Cunnington and Rowntree, 1986). There is an obvious negative feedback in that the reduction of evaporation helps to conserve soil moisture. There are also radiative feedbacks, positive due to the reduced greenhouse effect of water vapor in the drier atmosphere, positive due to the warming of dry surfaces in the absence of evaporative cooling, and negative if the drying of the atmosphere leads to a reduction in cloudiness.

Surface Roughness

There are two mechanisms involved here:

1. Through the effect on partitioning between evaporation and heat flux: we have seen that this has opposite signs for large and small surface resistance. Of particular interest may be the potential for more rapid recycling of water with rough wet surfaces, thus returning moisture to the atmosphere more rapidly from forests than semidesert or short grass surfaces.
2. Through the impact on frictional moisture convergence: increased roughness tends to strengthen cross-isobaric flow and frictional convergence, which by mass conservation leads to more ascent and a greater likelihood of precipitation (Sud and Smith, 1985).

4. Parameterizations of Evaporation Used in GCMs

4.1 Parameterizations in Use in the Early 1980s

Parameterizations of land surface processes used in general circulation models (GCMs) have been reviewed by Carson (1982) and updated where appropriate by Rowntree (1988). Most of the parameterizations of evaporation reviewed at that time were based on Equation (2.12) in the form:

$$E = \rho C_E V \delta q \quad (2.28)$$

with C_E approximated by βC_{EO}; C_{EO} is the transfer coefficient when $r_S = 0$, generally taken as a constant, with the same value for all land. β is a soil moisture-dependent parameter varying between 1 in moist ($r_S = 0$) and 0 in arid conditions. Comparison of Equations (2.28) and (2.12) shows that

$$\beta = r_A/(r_S + r_A) \quad (2.29)$$

Typically, however, β was taken as a linear function of the soil moisture m. A more realistic parameterization similar to Equation (2.29) was tested by Laval et al. (1984), giving substantial decreases in evaporation. Note that Equation (2.28) may also be written

$$E = \beta E_P \quad (2.30)$$

Here E_P is at first sight the evaporation for a moist surface, commonly called the potential evaporation. However, this is a misleading interpretation in that the saturation mixing ratio q_S in the term δq ($\equiv q_S - q$) in (2.28) is calculated for the surface temperature, which in dry conditions will be considerably enhanced due to the weakness of evaporative cooling; observationally, on the other hand, E_P is usually estimated for a moist surface as in the work of Priestley and Taylor (1972), which might otherwise justify the use of Equation (2.30). Mintz and Serafini (1981) proposed an alternative approach in which a separate wet surface energy balance was used to compute the surface temperature needed for the calculation of E_P.

The soil moisture required for use in Equation (2.30) was calculated by a simple "bucket" model (Manabe, 1969), based on Equation (2.3); the bucket was of finite depth m_{max} and the runoff Y was zero until the bucket was full, after which Y was sufficient to maintain $m = m_{max}$. More elaborate treatments designed to allow some runoff for $m < m_{max}$ were incorporated in some models.

4.2 The Introduction of Geographic Variations of Land Surface Characteristics

Apart from prescribed land ice extents and modeled variations of snowdepth and soil moisture, the models discussed above included no geographic variations of land surface type. A major step to including these was taken by Hansen et al. (1983). They included fields of surface visible and near-infrared albedo, roughness length, field capacity, and masking depth which modifies albedo in the presence of snow. Two soil moisture stores were included, with rapid upward moisture transfer between them during growing seasons, to represent the direct transfer of water from the lower layer to the atmosphere by transpiration. The land surface type definitions were based on the data sets compiled by Matthews (1983); subsequently, such geographic variations have been introduced in several other models, using either Matthews' data sets, or those of Wilson and Henderson-Sellers (1985).

4.3 More Elaborate Representations of Land Surface Processes

By introducing the explicit control of evaporation by soil and plant processes, as outlined in Section 2.1, and including the role of the foliage and canopy following the work of Deardorff (1978), Dickinson (1984) has placed the parameterization of evaporation on a more scientific basis. More recently, Sellers et al. (1986) developed a scheme based on similar principles, though differing in many detailed features. In this section, I shall describe Dickinson's scheme, the Biosphere–Atmosphere Transfer Scheme (BATS), since it is the earlier of the two, and then mention some of the differences between the two.

Figure 2.7 shows a schematic picture of BATS, and Table 2.4 gives values of the vegetation and soil parameters used in the scheme over South America. Dickinson uses (2.12) to calculate evaporation with the resistances r_S and r_A. Separate calculations are made for evaporation from the wet parts and transpiration from the dry parts of the canopy, and of evaporation from the soil. The total is calculated by weighting these estimates by the fractions of the gridbox occupied by each, based on the prescribed bare soil fraction for the surface type, and an estimate of the wet fraction of the canopy, based on the water calculated to be held by the canopy following dew or interception of rainfall. The maximum canopy water content is proportional to the leaf and stem area L_{SAI} defined below.

The surface resistance is expressed as

$$r_S = r_{S\,min}\, R_L\, S_L\, M_L/A_D < r_{S\,max} \quad (2.31)$$

Here $r_{S\,min}$ is the prescribed minimum stomatal resistance factor for the vegetation specified for the gridbox, 250 s/m, except 150 for mixed farming;

$r_{S\,max}$, the upper limit on r_S, is 5000 s/m;

R_L represents the dependence of r_S on solar radiation; it varies between about 1 for overhead sun and $r_{S\,max}/r_{S\,min}$ for nighttime where $r_{S\,max}$ is the cuticular resistance of the leaves; it is defined as $(1+f)/(f+r_{S\,min}/r_{S\,max})$, where f is the product of the visible solar flux and the light sensitivity factor (Table 2.4a);

S_L is a temperature dependence, equal to $[1-0.0016\,(298-T)^2]^{-1}$;

M_L is used to restrict transpiration to the maximum that the soil/root system can supply, having a minimum value of 1; the maximum transpiration has already been discussed in Section 2.5; the model has two soil layers;

$A_D = (\sigma_F\, L_{SAI}\, L_D)$ represents the area of the transpiring surface, with σ_F the fraction of the gridbox covered by vegetation, L_{SAI} the area of leaf and stem per unit area of gridbox, and L_D the unwetted fraction of this free to transpire; typical values of these quantities used by Dickinson lead to maximum values of A of about 5 for forests and savanna, 3.6 for tundra, 1.6 for short grass, 0.6 for semidesert, and 0 for desert.

The moisture and heat fluxes from the air within the canopy to the atmosphere above are calculated using bulk transfer formulae [e.g., Equation (2.28)] with transfer coefficients for heat, moisture, and momentum assumed to be equal. These are equated to the sum of the fluxes into the canopy from the soil and foliage, so allowing calculation of the within canopy values of temperature and humidity.

Finally, in this discussion of the BATS scheme, it is useful to note that there are 14 specified parameters for each of the 18 land cover/vegetation types (Table 2.4a). These are maximum vegetation cover and its variation with temperature, roughness length, depths of the total and upper soil layers, root distribution between the layers, albedos for < 0.7 μm and > 0.7 μm, minimum r_S, maximum and minimum leaf area indexes, a stem and dead matter area index, and measures of leaf dimension and

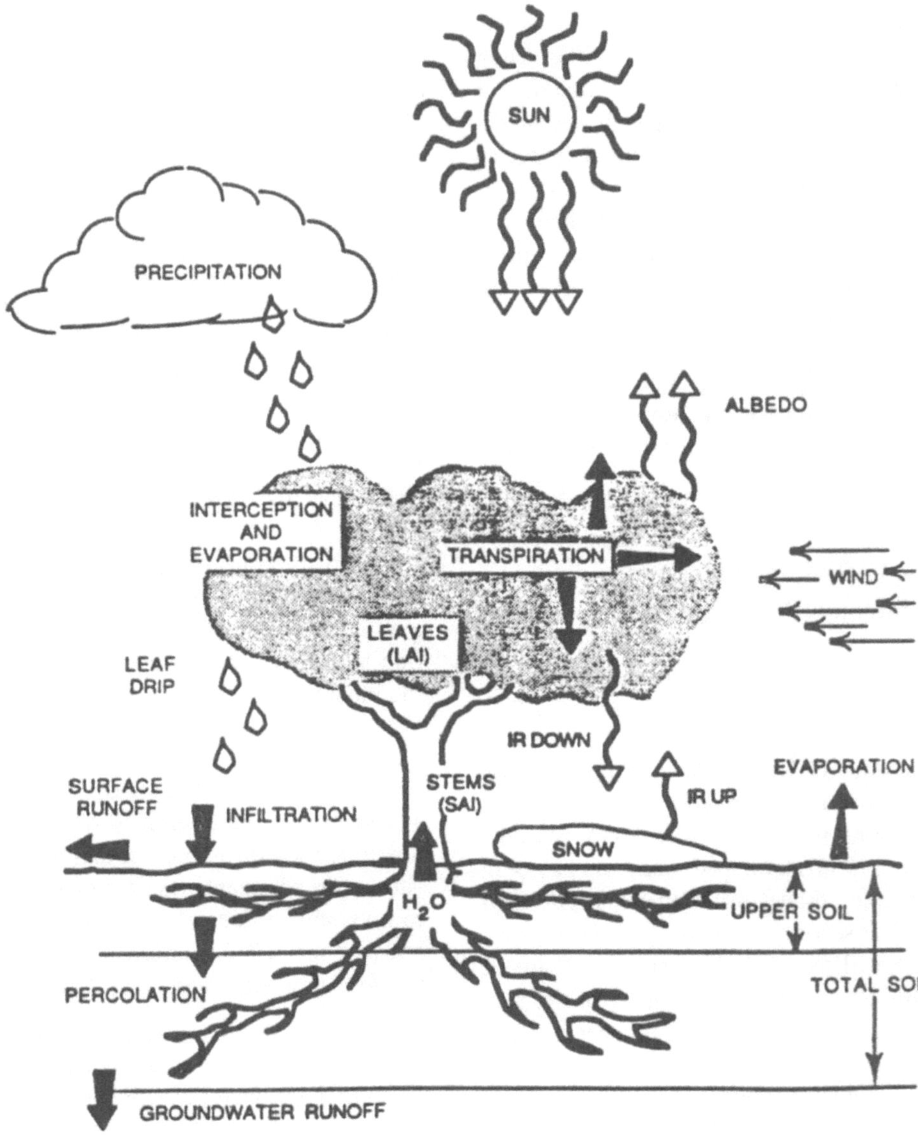

FIGURE 2.7. Schematic diagram illustrating the features included in the BATS land-surface parameterization. (From Dickinson et al., 1986.)

light sensitivity. For soil (Table 2.4b), for each of 12 texture classes, there are six parameters (porosity, maximum soil suction, saturated hydraulic and thermal conductivities, and B and s_w as defined in Sections 2.1 and 2.5, respectively); additionally, for each of eight color classes, there are dry and wet soil albedos for < 0.7 μm and > 0.7 μm [see Dickinson et al. (1986) for the actual values].

The SiB or simple biosphere model of Sellers et al. (1986) is similar in concept to the BATS scheme, while differing in detail. Some differences are the representation of both an upper and a lower vegetation layer each with an interception water

TABLE 2.4a. Vegetation/land cover parameters used for South America.[a]

Parameter	Land cover/vegetation type[b]								
	1	2	5	6	7	11	17	18	19[c]
Maximum fractional vegetation cover	0.85	0.80	0.80	0.90	0.80	0.10	0.80	0.80	0.80
Difference between maximum fractional vegetation cover and cover at temperature of 269 K	0.6	0.1	0.3	0.5	0.3	0.1	0.3	0.2	0.3
Roughness length (m) of vegetation	0.06	0.02	0.8	2.0	0.1	0.1	0.1	0.8	0.05
Depth of the total soil layer (m)	1.0	1.0	2.0	1.5	1.0	1.0	1.0	2.0	1.0
Depth of upper soil layer (m)	0.1	0.1	0.1	0.1	0.1	0.1	0.1	0.1	0.1
Rooting ratio (upper to total soil layers)	3	8	10	12	8	8	5	10	10
Vegetation albedo for wavelengths <0.7 μm	0.10	0.10	0.08	0.04	0.08	0.17	0.08	0.06	0.08
Vegetation albedo for wavelengths >0.7 μm	0.30	0.30	0.28	0.20	0.30	0.34	0.28	0.24	0.30
Minimum stomatal resistance ($s\ m^{-1}$)	150	250	250	250	250	250	250	250	250
Maximum leaf area index (LAI)	6	2	6	6	6	6	6	6	6
Minimum LAI	0.5	0.5	1.0	5.0	0.5	0.5	1.0	3.0	0.5
Stem (and dead matter) area index	0.5	4.0	2.0	2.0	2.0	2.0	2.0	2.0	2.0
Inverse square root of leaf dimension ($m^{-1/2}$)	10	5	5	5	5	5	5	5	5
Light sensitivity factor ($m^2\ W^{-1}$)	0.01	0.01	0.03	0.03	0.01	0.01	0.01	0.03	0.01

[a] From Dickinson and Henderson-Sellers (1988). By courtesy of the Royal Meteorological Society.

[b] Vegetation types are: crop (1); shortgrass (2); deciduous broadleaf tree (5); evergreen broadleaf tree (6); semidesert (11); deciduous shrub (17); mixed woodland (18).

[c] Vegetation type 19 has been defined especially for this study and is *not* part of the NCAR CCM land-type data set.

TABLE 2.4b. Soil parameters used for South America.[a]

Functions of texture parameter	Texture class [from sand (1) to clay (12)]								
	4	5	6	7	8	9	10	11	12
Porosity (volume of voids to volume of soil)	0.42	0.45	0.48	0.51	0.54	0.57	0.60	0.63	0.66
Maximum soil suction (m)	0.2	0.2	0.2	0.2	0.2	0.2	0.2	0.2	0.2
Saturated hydraulic conductivity (m s^{-1})	0.13×10^{-4}	0.89×10^{-5}	0.63×10^{-5}	0.45×10^{-5}	0.32×10^{-5}	0.22×10^{-5}	0.16×10^{-5}	0.11×10^{-5}	0.8×10^{-6}
Ratio of saturated thermal conductivity to that of loam	1.2	1.1	1.0	0.95	0.90	0.85	0.80	0.75	0.70
Exponent "B," Clapp and Hornberger (1978)	5.0	5.5	6.0	6.8	7.6	8.4	9.2	10.0	10.8
Moisture content (relative to saturation) at which transpiration ceases	0.266	0.300	0.332	0.378	0.419	0.455	0.487	0.516	0.542

Functions of color parameter	Color [from lightest (1) to darkest (8)]							
	1	2	3	4	5	6	7	8
Dry soil albedo								
<0.7 μm	0.23	0.22	0.20	0.18	0.16	0.14	0.12	0.10
≥0.7 μm	0.46	0.44	0.40	0.36	0.32	0.28	0.24	0.20
Saturated soil albedo								
<0.7 μm	0.12	0.11	0.10	0.09	0.08	0.07	0.06	0.05
≥0.7 μm	0.24	0.22	0.20	0.18	0.16	0.14	0.12	0.10

[a] From Dickinson and Henderson-Sellers (1988). By courtesy of the Royal Meteorological Society.

store, the use of three soil moisture stores, the lowest not being directly accessible to the vegetation, and the inclusion of a linear dependence of r_S on vapor pressure deficit.

The Meteorological Office (MO) model (Warrilow et al., 1986) occupies an intermediate position on the scale of complexity. Eight vegetation-dependent and eight soil-dependent parameters have geographic variations based on the Wilson and Henderson-Sellers data sets. The evaporation is calculated assuming $E = \beta E_w$, where E_w is the evaporation allowing for minimum stomatal resistance, but with the limitation by soil moisture represented by $\beta = m/(5 \text{ cm}) \leq 1$. Soil temperatures and heat fluxes are calculated with a four-layer model designed to give accurate representations of variations on time scales from a few hours to over a year. Moisture is held in (1) a single soil layer of capacity determined from the vegetation rooting depth, with quick (surface) and slow (percolation) runoff and (2) a canopy store which can also represent near-surface water in bare soil.

5. Examples of Climate Model Sensitivity to Land Surface Parameterizations

5.1 Amazon Deforestation Experiment Using the BATS Scheme

Dickinson and Henderson-Sellers (1988) used the BATS scheme, described in the last section, in the NCAR Community Climate Model (CCM), to investigate the model's response to deforestation of Amazonia. They changed both vegetation and soil specifications; all vegetation of type 6 (evergreen broadleaf tree) was replaced by a new vegetation type 19 (see Table 2.4), "supposed to possess the characteristics of impoverished scrub-grassland typical of deforested regions in Amazonia." The soil characteristics were also changed at the same points, texture being made finer by two classes up to a maximum of 12, and soil color lighter by two classes (Table 2.4). The major changes were decreases in z_0, total soil depth, vegetation cover and sensitivity of stomata to visible radiation, and increases in surface runoff and albedos. Note that although minimum leaf area index was decreased, this would have little impact in the tropics, because the seasonal variation, depending on the S_L factor defined in Section 4, is small. Consequently, the canopy interception capacity would also be little changed.

The experiment was run for 13 months, and most of the results shown were regional averages for most of the deforested area south of the equator. As expected from both theory and from previous experiments with increased albedo, evaporation was reduced (Fig. 2.8), and, as expected from the reduced water-holding capacity and decreased evapotranspiration—unless compensated by rainfall decreases—runoff was increased (Fig. 2.9). Note that evaporation of intercepted water also decreased, mainly due to the decreased roughness (see Fig. 2.1 for small r_S). The roughness changes were also responsible for a marked rise of near-surface air temperatures (Fig. 2.10), as expected from (2.12) and (2.13) if r_A increases with little change in $(H + LE)$. Somewhat surprisingly, in view of the increases in albedo and decreased water storage, rainfall was only slightly reduced (Fig. 2.11); the experiment was not long enough, given the noisy nature of the rainfall simulation, either to show that such relatively small mean changes were statistically significant, or on the other hand to prove that over a longer experiment substantial significant changes would not be obtained. Decreases of rainfall have been obtained in experiments with other models (Henderson-Sellers and Gornitz, 1984; Wilson, 1984; Lean and Warrilow, 1989), in which, probably mainly due to the effect of increased albedo in reducing energy input and so also evaporation and [by the Charney (1975) hypothesis (Section 3.5)] low level moisture convergence, there were clear decreases in rainfall over a deforested Amazonia (e.g., Fig. 2.12 from Lean and Warrilow, 1989).

5.2 Impacts of Land Surface Parameterizations on CO_2/Climate Change Experiments

There has been considerable interest in recent years in the modeling of climate changes due to increasing atmospheric contents of greenhouse gases. There is general agreement between models that global mean temperatures and precipitation

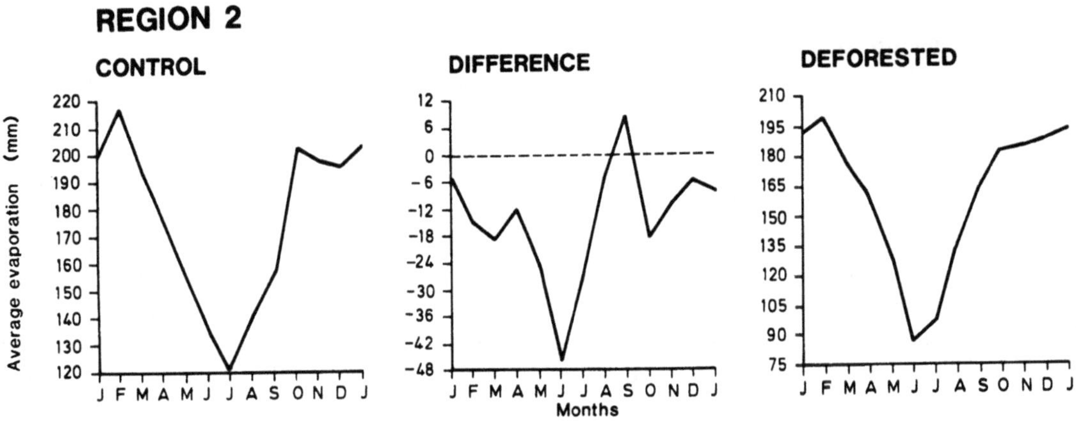

FIGURE 2.8. Monthly distribution of evaporation (mm/month) in control and deforested experiments by Dickinson and Henderson-Sellers (1988), and the difference between them. (Reprinted by courtesy of the Royal Meteorological Society.)

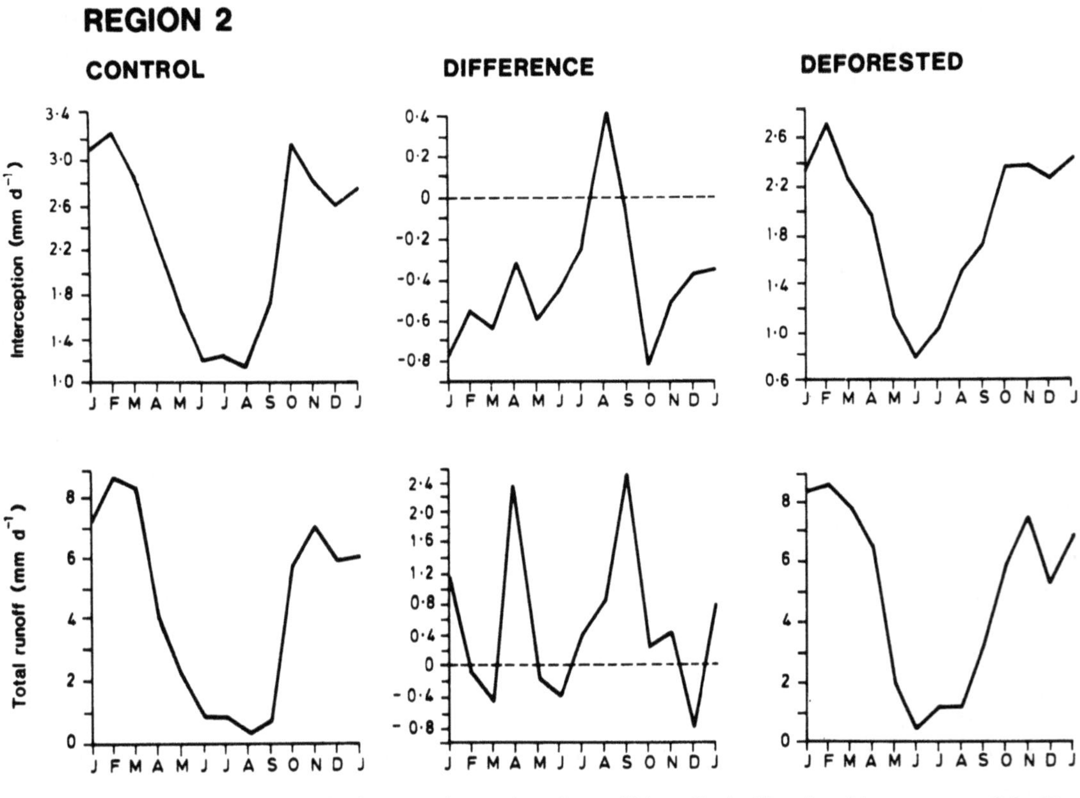

FIGURE 2.9. As Figure 2.8, except for interception and total runoff (mm/day). (Reprinted by courtesy of the Royal Meteorological Society.)

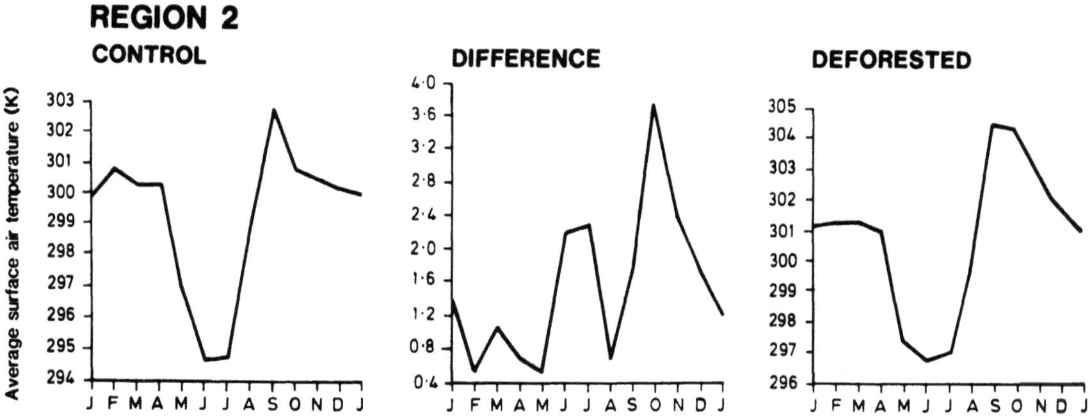

FIGURE 2.10. As Figure 2.8, except for near-surface air temperature (K). (Reprinted by courtesy of the Royal Meteorological Society.)

will increase but much less agreement on the regional distribution of the changes, especially for precipitation. One suggested regional effect has been that soil moisture will decrease in middle latitudes in summer, due to a longer evaporation season following earlier spring snowmelt. Mitchell and Warrilow (1987) demonstrated that this result is sensitive to the treatment of snowmelt runoff. In their standard version of the MO model, snowmelt was added directly to soil moisture, and so helped to saturate the soil and eliminate the model's memory of the increased autumn precipitation obtained with increased CO_2. In a perturbation experiment, snowmelt ran off if either of the two middle layers of soil (out of four) centered at 10 and 41 cm was frozen. The model's memory of the increased autumn rains was no longer lost and increased soil moistures persisted until August (Fig. 2.13). The horizontal structure of the differences in model response is shown in Figure 2.14.

In another experiment with the MO model, feedbacks arising from interactions between soil moisture, temperature, and cloud cover were removed by constraining the soil moisture to follow the control experiment's seasonal variation. The hypothesized feedback mechanism is that decreased soil

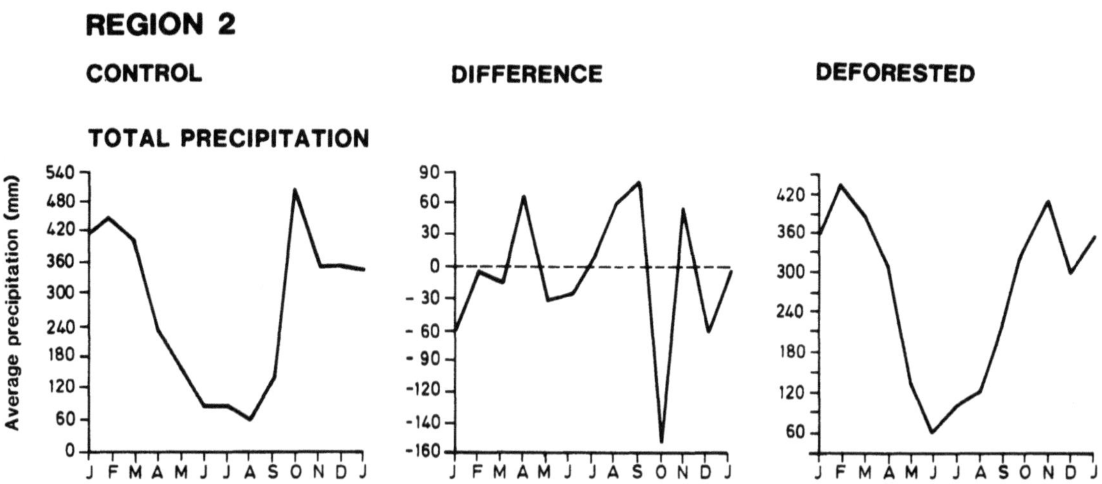

FIGURE 2.11. As Figure 2.8, except for total precipitation (mm/month). (Reprinted by courtesy of the Royal Meteorological Society.)

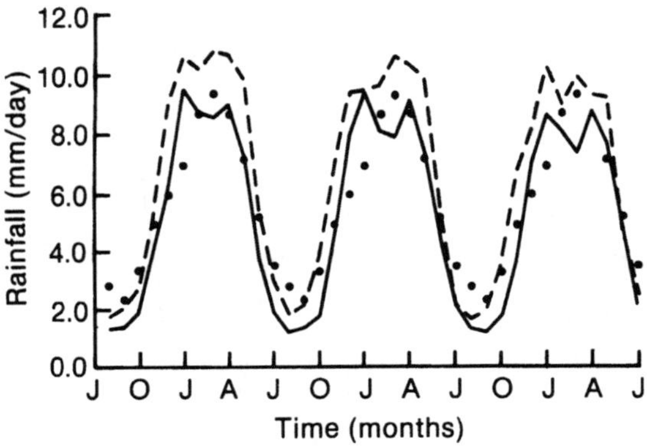

FIGURE 2.12. Simulated monthly mean rainfall (mm/day) from the deforestation experiment by Lean and Warrilow (1989) for the control experiment (dashed), deforested (solid), and observed climatology (dots) averaged over the area 17.5°S–2.5°N, 78.75°–48.75°W.

moisture reduces atmospheric relative humidities, decreases cloud amounts, and so enhances absorption of solar radiation. Over the 5 years of the experiment there was a consistent decrease of 0.5 from 6.7 K, in the warming of June–August global mean land surface temperatures due to doubling CO_2.

6. Uncertainties

There remain many uncertainties in the parameterization of land surface processes for climate models. The specification of land surface characteristics is a major problem and requires considerable work on the interpretation of satellite data. This task is in progress as part of the World Climate Research Programme under the aegis of the ISLSCP (International Satellite Land Surface Climatology Project), but there are many difficulties, not the least of which is the specification of soil characteristics.

Problems with the parameterizations themselves are also numerous. At present, all schemes effectively assume a single value for each parameter for each model gridbox, even though this value may be calculated as a mean based on the distribution of vegetation types in the box. This basically linear process makes little allowance for the nonlinear effects of inhomogeneities such as the spatial variations of moisture availability on the scale of kilometers or less between swampy valleys and well-drained slopes or between forest and ploughed field in a drought (have HAPEX and FIFE told us more about this?); or the early melting of snow on the slopes that receive the most solar radiation; or the increased cloudiness resulting in reduced evaporation in areas which have most precipitation—whether through chance on a diurnal scale over uniform terrain or systematically on long time scales over uplands. Some models (SiB, MO) do allow for the spatial variability of convective precipitation and the consequent enhancement of surface runoff, but this is perhaps the only example of parameterizing to allow for the enormous contrast between the

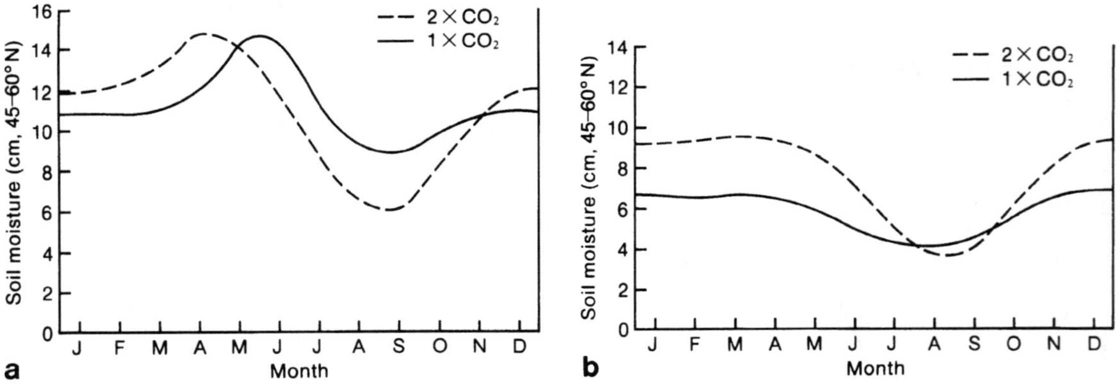

FIGURE 2.13. (a) Simulated monthly mean soil moisture (cm) for a medium soil, averaged over land between 45 and 60°N. Solid line, present CO_2 concentrations; dashed line, doubled CO_2 concentrations. (From Mitchell and Warrilow, 1987.) (b) As (a) but with 100% runoff when the second and/or third of four soil layers are frozen.

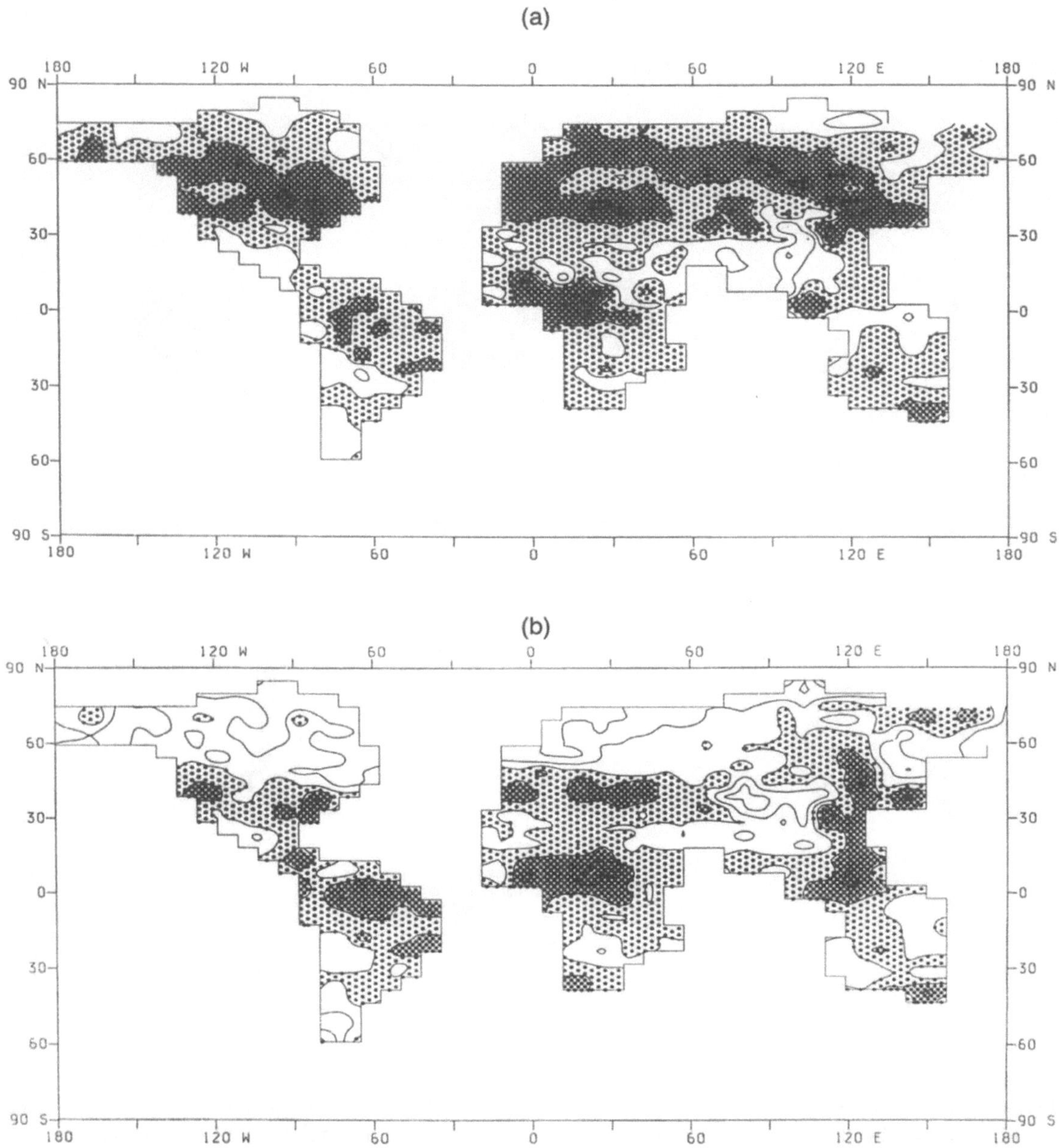

FIGURE 2.14. (a,b) For the same experiments as Figure 2.13, but soil moisture change due to doubling CO_2 (June–August). Contour interval 2 cm. Decreases shaded. (Mitchell and Warrilow, personal communication.)

observational scale of a few meters and the modeling scale of 100 km upward.

Other uncertainties include the relation between z_0 and z_M discussed in Sections 2 and 4. The simulation of snowcover in wooded terrain may be adequately parameterizable using models that explicitly represent the canopy, but the effects of drifting in allowing an early return to snowfree evaporation conditions are nowhere represented.

Lest this discussion generate excessive pessimism, I will close with two depictions of runoff, the annual mean runoff map for a simulation with the GFDL model, with not too different observed estimates (Fig. 2.15, from Manabe and Holloway, 1975), and the modeled and observed seasonal variations of runoff for two subarctic regions (Fig. 2.16), in one of which the model (MO) provides a good simulation—why not in the other?

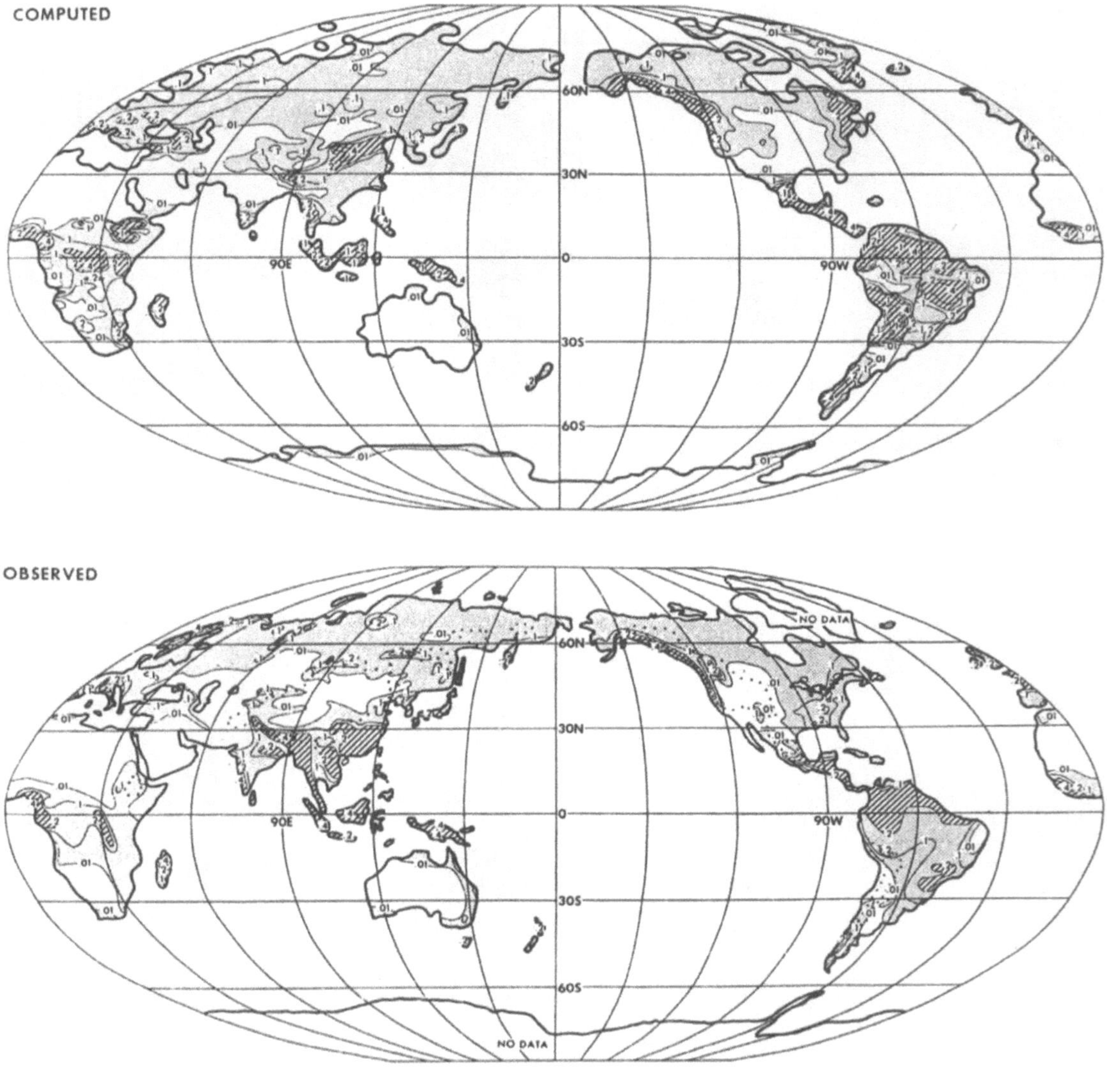

FIGURE 2.15. Global distribution of annual mean rate of runoff simulated by the model (top) compared with an observed distribution (bottom) based on data derived from Lvovitch and Ovtchinnikov (1964), using some smoothing. (From Manabe and Holloway, 1975.)

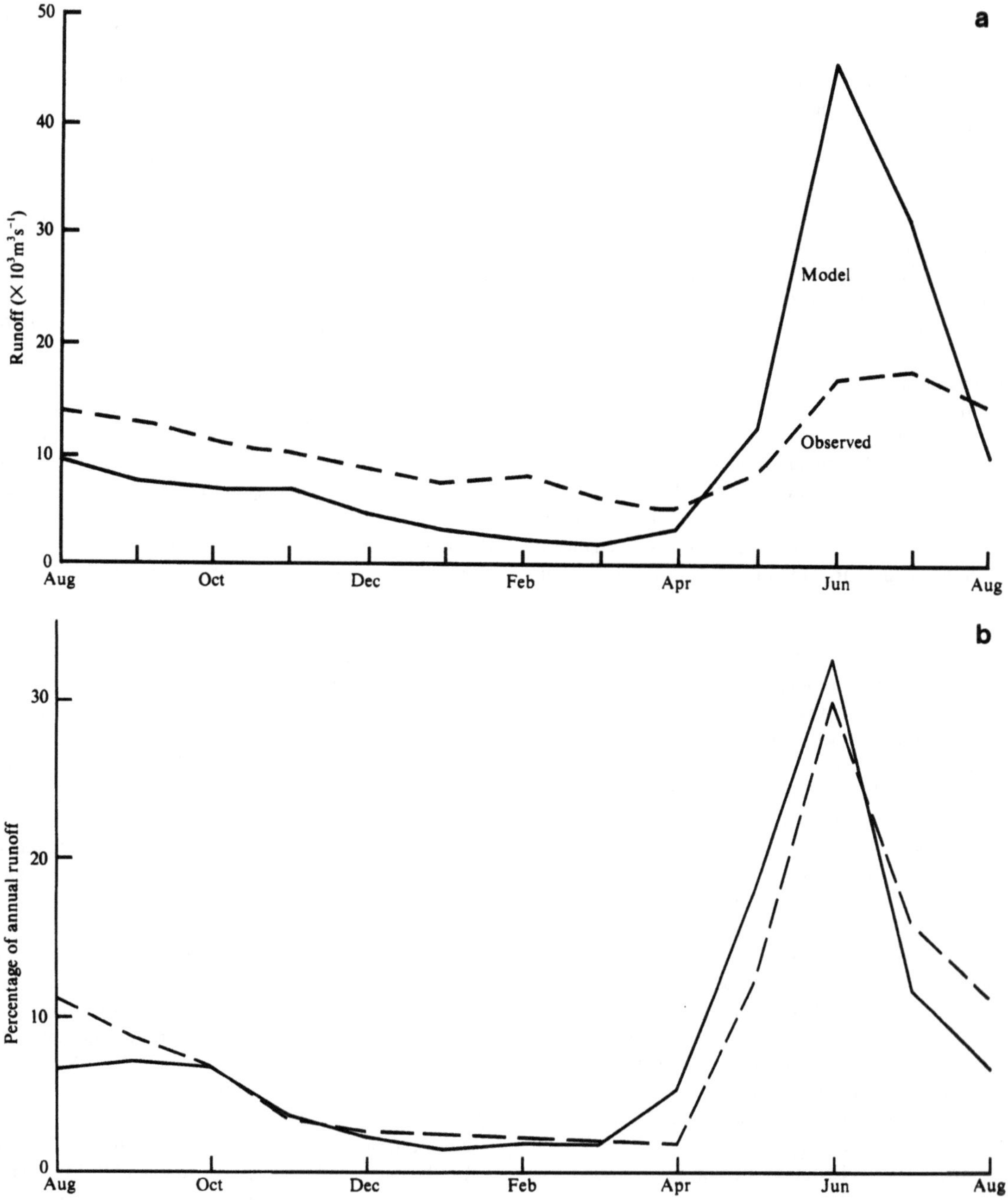

FIGURE 2.16. (a) Modeled (solid) and observed (Mackay and Loken, 1974) (dashed) runoff for the Mackenzie basin (10^3 m^3 s^{-1}). (b) As (a) but as percentage of annual total and for Ob'-Irtysh and Yenisei basins (observed) and 60–105°E north of 50°N (modeled).

References

Arya SPS (1975) Geostrophic drag and heat transfer relations for the atmospheric boundary layer. Quart J R Meteorol Soc 101:147–161.

Carson DJ (1982) Current parameterizations of land-surface processes in atmospheric general circulation models. In "Land Surface Processes in Atmospheric General Circulation Models" (PS Eagleson, ed.), pp 67–108. Cambridge University Press, Cambridge.

Carson DJ, Sangster AB (1981) The influence of land-surface albedo and soil moisture on general circulation model simulations. Research activities in atmospheric and oceanic modelling (I.D. Rutherford, ed.). Numerical Experimentation Programme Report No. 2:5.14–5.21.

Charney JG (1975) Dynamics of deserts and droughts in the Sahel. Quart J R Meteorol Soc 101:193–202.

Charney JG, Quirk WJ, Chow SH, Kornfield J (1977) A comparative study of the effects of albedo change on drought in semi-arid regions. J Atmos Sci 34:1366–1385.

Chervin RM (1979) Response of the NCAR general circulation model to changed land surface albedo. Report of the JOC Study Conference on Climate Models: Performance, Intercomparison and Sensitivity Studies, ICSU/WMO GARP Publications Series No. 22, Volume 1:563–581.

Clapp RB, Hornberger GM (1978) Empirical equations for some soil hydraulic properties. Water Res Res 14:601–604.

Clarke RH, Dyer AJ, Brook RR, Reid DG, Troup AJ (1971) The Wangara experiment—Boundary layer data. CSIRO Div Met Phys Tech Paper No. 19.

Cunnington WM, Rowntree PR (1986) Simulations of the Saharan atmosphere—dependence on moisture and albedo. Quart J R Meteorol Soc 112:971–999.

Deacon EL (1953) Vertical profiles of mean wind in the surface layers of the atmosphere. Geophys Mem No. 91, Meterological Office, 64 pp.

Deacon EL (1973) Geostrophic drag coefficients. Boundary Layer Met 5:321–340.

Deardorff J (1978) Efficient prediction of ground temperature and moisture with inclusion of a layer of vegetation. J Geoph Res 83:1889–1903.

Dickinson RE (1984) Modelling evapotranspiration for three-dimensional global climate models. Geophys Monograph 29, AGU.

Dickinson RE, Henderson-Sellers A (1988) Modelling tropical deforestation: A study of GCM land-surface parameterizations. Quart J R Meteorol Soc 114:439–462.

Dickinson RE, Henderson-Sellers A, Kennedy PJ, Wilson MF (1986) Biosphere-Atmosphere Transfer Scheme (BATS) for the NCAR Community Climate Model. NCAR Technical Note NCAR/TN-275+STR.

Garratt JR (1977) Aerodynamic roughness and mean monthly surface stress over Australia. CSIRO Aust Div Atmos Phys Tech Paper No. 29:1–19.

Hansen J, Russell G, Rind D, Stone P, Lacis A, Lebedeff S, Ruedy R, Travis L (1983) Efficient three-dimensional global models for climate studies: Models I and II. Mon Weather Rev 111:609–662.

Henderson-Sellers A, Gornitz V (1984) Possible climatic impacts of land cover transformations, with particular emphasis on tropical deforestation. Clim Change 6: 231–258.

Kondratyev K Ya, Korzov VI, Mukhenberg VV, Dyachenko LN (1982) The shortwave albedo and the surface emissivity. In "Land Surface Processes in General Circulation Models" (PS Eagleson, ed.), pp 463–514. Cambridge University Press, Cambridge.

Kurbatkin GP, Manabe S, Hahn DG (1979) The moisture content of the continents and the intensity of the summer monsoon circulation. Soviet Meteorol and Hydrol, No. 11:1–6.

Laval K, Ottle C, Perrier A, Serafini Y (1984) Effect of parameterization of evapotranspiration on climate simulated by a GCM. New Perspect Clim Modelling 4:223–247.

Lean J, Warrilow DA (1989) Simulation of the regional climatic impact of Amazon deforestation. Nature (London) 342:411–413.

Lettau H (1969) Note on aerodynamic roughness-parameter estimation on the basis of roughness-element description. J Appl Meteorol 8:828–832.

Lvovitch MI, Ovtchinnikov SP (1964) "Physical-Geographical Atlas of the World" (in Russian). Academy of Sciences, USSR and Department of Geodesy and Cartography, State Geodetic Commission, Moscow.

Mackay DK, Loken OH (1974) Arctic hydrology. In "Arctic and Alpine Environments" (JD Ives, RG Barry, eds.), pp 111–132. Methuen, London.

Manabe S (1969) Climate and ocean circulation: 1. The atmospheric circulation and the hydrology of the Earth's surface. Mon Weather Rev 97(11):739–774.

Manabe S, Holloway JL Jr (1975) The seasonal variation of the hydrological cycle as simulated by a global model of the atmosphere. J Geophys Res 80(12):1617–1649.

Matthews E (1983) Global vegetation and land use: New high resolution data bases for climate studies. J Clim Appl Meteorol 22:474–487.

Mintz Y (1984) The sensitivity of numerically simulated climates to land surface boundary conditions. In "The Global Climate" (JT Houghton, ed.), pp 79–105. Cambridge University Press, Cambridge.

Mintz Y, Serafini Y (1981) Monthly normal global

fields of soil moisture and land-surface evapotranspiration. Symposium on Variations in the Global Water Budget, Oxford.

Mitchell JFB, Warrilow DA (1987) Summer dryness in northern mid-latitudes due to increased CO_2. Nature (London) 330:238–240.

Monteith JL (1973) "Principles of Environmental Physics." Arnold, London.

Oguntoyinbo JS (1970) Reflection coefficient of natural vegetation, crops and urban surfaces in Nigeria. Quart J R Meteorol Soc 96:430–441.

Perrier A (1982) Land surface processes: Vegetation. In "Land Surface Processes in General Circulation Models" (PS Eagleson, ed.), pp 395–448. Cambridge University Press, Cambridge.

Picon L (1983) Étude de l'influence de l'albédo du sol dans les phénomenes de désertification au Sahel a l'aide du modele de circulation générale du L.M.D. Thesis L'université de Paris.

Prabhakara C, Dalu G (1976) Remote sensing of the surface emissivity at 9 μm over the globe. J Geophys Res 81:3719–3724.

Priestley CHB, Taylor RJ (1972) On the assessment of surface heat flux and evaporation using large scale parameters. Mon Weather Rev 100:81–92.

Rockwood AA, Cox SK (1978) Satellite-inferred surface albedo over northwestern Africa. J Atmos Sci 35: 513–522.

Rowntree PR (1988) Review of general circulation models as a basis for predicting the effects of vegetation change on climate. United Nations University Workshop on 'Forests, climate and hydrology—regional impacts,' pp 162–193. Oxford, March 1984.

Rowntree PR, Bolton JA (1983) Simulation of the atmospheric response to soil moisture anomalies over Europe. Quart J Roy Meteorol Soc 109:501–526.

Rowntree PR, Sangster AB (1986) Remote sensing needs identified in climate model experiments with hydrological and albedo changes in the Sahel. Proc ISLSCP Conference, Rome, European Space Agency ESA SP-248, pp 175–183.

Sellers PJ, Mintz Y, Sud YC, Dalcher A (1986) A simple biosphere model (SiB) for use within general circulation models. J Atmos Sci 43:505–531.

Shukla J, Mintz Y (1982) Influence of land surface evapotranspiration on the earth's climate. Science 215:1498–1501.

Shuttleworth WJ et al (1984) Eddy correlation measurements of energy partition for Amazonian forests. Quart J R Meteorol Soc 110:1143–1162.

Shuttleworth WJ et al (1985) Daily variations of temperature and humidity within and above Amazonian forest. Weather 40:102–108.

Stewart JB (1971) The albedo of a pine forest. Quart J R Meteorol Soc 97:561–564.

Stewart JB (1988) Modelling surface conductance of pine forest. Agric Forest Meteorol 43:19–35.

Sud YC, Fennessy M (1982) A study of the influence of surface albedo on July circulation in semi-arid regions using the GLAS GCM. J Climatol 2:105–125.

Sud YC, Smith WE (1985) The influence of surface roughness of deserts on the July circulation: A numerical study. Boundary Layer Meteor 4:383–398.

Tennekes H, Lumley JL (1972) "First Course in Turbulence," p 170. MIT Press, Cambridge, MA.

Thompson N, Barrie IA, Ayles M (1981) The Meteorological Office rainfall and evaporation calculation system: MORECS (July 1981). Hydrological Memorandum 45, Meteorological Office, Bracknell.

Van Zyl WH, De Jager JM (1987) Accuracy of the Penman-Monteith equation adjusted for atmospheric stability. Agric Forest Meteorol 41:57–64.

Walker J, Rowntree PR (1977) The effect of soil moisture on circulation and rainfall in a tropical model. Quart J R Meteorol Soc 103:29–46.

Warrilow DA, Sangster AB, Slingo A (1986) Meteorological Office Met O 20 Tech Note DCTN 38.

Wilson MF (1984) The construction and use of land surface information in a general circulation model. Ph.D. thesis, University of Liverpool.

Wilson MF, Henderson-Sellers A (1985) A global archive of land cover and soils data for use in general circulation climate models. J Climatol 5:119–143.

3
Parameterization of Land-Surface Processes in Numerical Weather Prediction

Christian Blondin

1. Introduction

This chapter introduces the problem of the parameterization of land-surface processes, which is a component of the "physical package" of most of operational Numerical Weather Prediction Models (NWPMs). It is assumed that the reader is familiar with the basic problem of parameterizing turbulent surface fluxes and with at least the background of the most widely used formulations (see, e.g., Rowntree, this volume). So this chapter concentrates on the practical problems of accounting for land-surface processes in NWPMs.

In Sections 2 and 3, general comments are presented, which concern the need for parameterizing land-surface processes in NWPMs, and the various aspects of the implementation of these schemes in operational forecasting systems. The type of schemes used, the specification of surface parameters, and the initialization methods are reviewed (incompletely) in Sections 4, 5, and 6. Section 7 addresses the problem of the operational validation of the parameterizations. Some insight in the foreseeable evolution of the implementation of land-surface processes in NWPMs is given in Section 8. Finally, Section 9 summarizes the main conclusions of this study.

2. The Need for Land-Surface Parameterization in NWPMs

The exchanges of momentum, heat, and moisture at the surface between the atmosphere and the underlying surface is a fundamental physical phenomenon that strongly influences in many ways the dynamics and thermodynamics of the atmospheric system.

The oceans and the continents store the energy available as net radiation at the surface, and deliver it back to the atmosphere after a certain delay. Over land, time delay is very short because most of the net radiation goes back to the atmosphere through turbulent mixing, which has a rather small characteristic time scale and is very efficient. Part of this energy is immediately usable by the atmosphere, since it comes as a flux of sensible heat. The other large part is related to the surface moisture flux, which represents a potential source of energy. Because of the latent heat content of the water vapor, the atmosphere gains energy through condensation processes. This energy, transferred from the surface to the atmosphere by turbulent exchange, compensates for the loss by radiative transfer. Thus, on the global scale, the atmosphere keeps the same level of energy (internal + potential available + kinetic). However, this kind of global energy balance is not achieved on smaller space and time scales, so that atmospheric motions have to transport energy. Clearly the general circulation of the atmosphere depends on the location of the energy sources. The earth's climate is therefore closely connected to the exchange with the surface and with the net radiation distribution. This chapter concentrates on the importance of these processes for numerical weather prediction.

Figure 3.1 shows a typical atmospheric energy budget, revealing that the heat flux due to latent heat release (convection + large scale condensa-

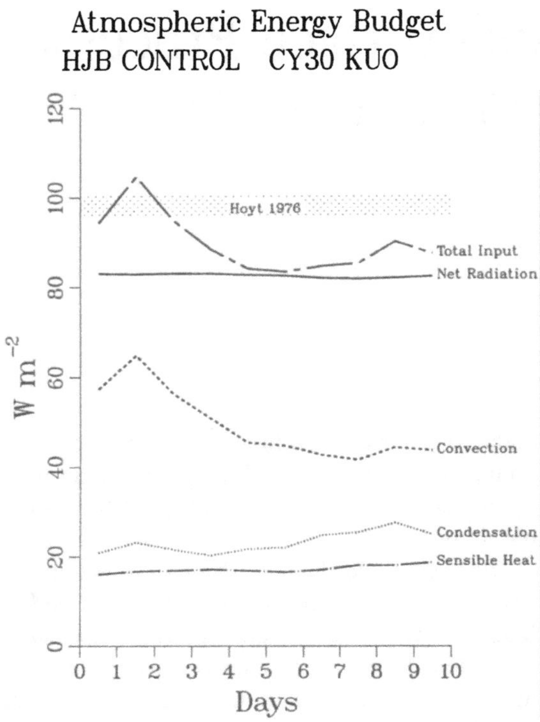

FIGURE 3.1. Global atmospheric energy budget in an example of a 10-day forecast.

tion, around 65 W m^{-2}) is dominant compared to the sensible heat flux (around 15 W m^{-2}). If one compares these sources with the mean global energy (around 2.6×10^9 J m^{-2}), this leads to a recycling time scale of the order of 300 days by the surface heat flux, or an average heating rate of the order of 1 K/day for the whole atmosphere. A systematic error in the estimate of those fluxes may affect the mean atmospheric temperature by a fraction of a degree per day (for a 10-day forecast, this may not be catastrophic though undesirable; for climate simulations this has a much more dramatic effect). The figure is quite different for atmospheric water and for kinetic energy that have characteristic recycling time scales by surface fluxes of the order, respectively, of 10 days and 7 days (Sommeria, 1985). It is therefore essential that surface fluxes be represented with some realism in Numerical Weather Prediction Models (NWPMs).

The continents occupy about 30% of the surface of the globe, but their contribution to the total surface moisture flux is a relatively small fraction compared to that of the oceans, because the partitioning between sensible and latent heat flux is quite different for oceans and continents. Oceans have a Bowen ratio (sensible heat flux divided by latent heat flux) of around 0.1, while the value for continents is between 0.8 and 0.9 (Verstraete and Dickinson, 1986). This indicates that the physical processes controlling the moisture fluxes over land are different from those prevailing over an open free surface of water. They have to be identified and understood to form the basis for a surface scheme in numerical models. Moisture fluxes over land represent a modest contribution to the total surface flux of moisture in the atmosphere, but they can control many aspects of the atmospheric state, due to their short time response to the net radiation input (essentially because of the roughness of land surfaces). Indeed, surface fluxes over land primarily influence the part of the atmosphere in contact with the ground in the so-called Constant Flux Layer (CFL) and in the entire Planetary Boundary Layer (PBL). Through the action of eddy fluxes of moisture the free atmosphere aloft is also responding to shallow and deep convection, which in turn affect the surface conditions by providing precipitation at the surface and by modifying the net radiation.

Despite the increasing use of numerical weather forecasts, and despite the clear illustration of the sensitivity of climate models to the specification of surface parameters and land-surface parameterization (Mintz, 1984), very few studies have been published about the sensitivity of weather forecasts to the parameterization of surface processes.

Rowntree and Bolton (1983) demonstrated that local weather elements like temperature, cloudiness, and precipitation may be affected by local surface features, and that local anomalies can also influence the weather downstream within a few days. Though these arguments support the idea that land-surface processes should be included in the physical package of a NWPM, their parameterization is still very crude compared to the actual knowledge we have about them.

Next we summarize how land-surface processes are implemented in operational NWPMs, and give general reasons that explain the current situation.

3. Land-Surface Parameterization and Their Implementation in Operational NWPMs

3.1 A Short Introduction to the Parameterization of Land-Surface Processes

The aim of this section is mostly to define the vocabulary and the various symbols to be used in the rest of the text.

It is convenient to introduce the parameterization of land-surface processes for homogeneous bare-soil surfaces (Carson, 1986a; parameterizations of land-surface processes in Meterological Office numerical weather prediction and climate models. Unpublished Met Office Report, DCTN 37, Met 0 20). The NWPMs are assumed to have their first level in the CFL and to simulate the surface diurnal cycle. In that case, the Monin–Obukhov similarity theory can be used in a straightforward way (Mahrt et al., 1986).

We follow a classical micrometeorological approach, based on the simple assumption that, on a certain (time and space) scale, all the various parameters and variables referred to in the NWPM, have an unambiguous physical meaning (that is, they can be either estimated directly or derived by appropriate measurements). Properties of statistical homogeneity and stationarity required by the Monin–Obukhov theory are supposed to be fulfilled. The constraints induced by all these assumptions define a certain type of surface for which the following results can be directly applied.

In the following formulas, all physical quantities are expressed in SI units.

Momentum

Because the surface of the earth is rough (in the aerodynamic sense), and because the air is a viscous fluid, a no-slip condition is applied at the surface, $u = 0$. To compensate for removal of momentum by surface friction, there is a downward transfer of momentum in the lowest part of the atmosphere. The measure of this turbulent momentum flux at the surface is the so-called frictional drag

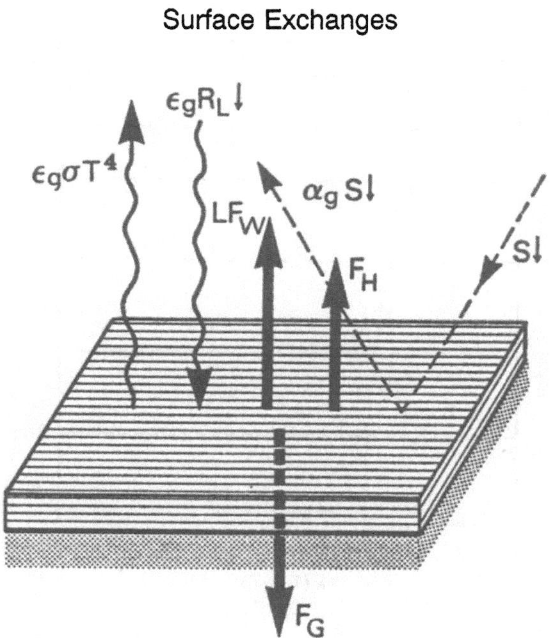

FIGURE 3.2. Schematic illustration of the surface energy budget.

$$\vec{\tau} = -\rho C_D |u^2 + v^2|^{1/2} [u,v]; |\vec{\tau}| = \rho u_*^2 \tag{3.1}$$

where ρ is the air density at the surface, u_* the friction velocity, u and v the two horizontal components (zonal and meridional) of the wind at the first model level, and c_D the frictional drag coefficient (which depends on the vertical stability of the air just above the surface and on an integration constant z_{0m}, the roughness length for momentum).

Heat

Figure 3.2 gives a schematic illustration of the various fluxes controlling the evolution of the surface temperature T_s. Assuming this parameter represents the temperature of a very thin surface layer (air + soil) which cannot store energy, the following balance equation applies:

$$R_N + F_H + LF_W + F_{g0} = 0 \tag{3.2}$$

where R_N is the net radiation

$$R_N = (1 - \alpha) S + \varepsilon (R_L - \sigma T_S^4) \tag{3.3}$$

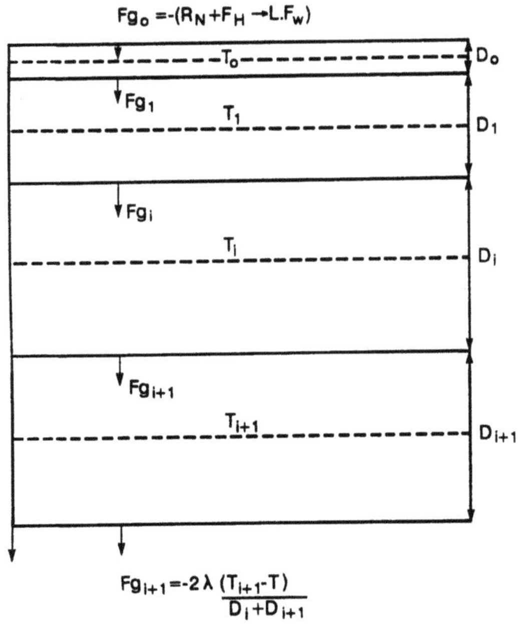

FIGURE 3.3. Schematic representation of a multilayer soil-temperature finite-difference model.

R_L being the downward long-wave radiative flux, S the downward short-wave radiative flux, F_H the turbulent sensible heat flux, F_W the turbulent moisture flux, and F_{g0} the conduction heat flux into the ground (downward fluxes are given positive values, as they tend to increase T_s). σ is the Stephan–Boltzmann constant, ε the long-wave emissivity, α the short-wave albedo, and L the latent heat.

The turbulent heat fluxes, like the turbulent momentum flux, are generally parameterized using Monin–Obukhov similarity theory:

$$F_H = -\rho C_p u_* \Theta_* = -\rho C_p C_h |u^2 + v^2|^{1/2} (\Theta - \Theta_S) \tag{3.4}$$

$$F_W = -\rho u_* q_* = -\rho C_h |u^2 + v^2|^{1/2} (q - q_S) \tag{3.5}$$

where Θ_* and q_* are scaling parameters, Θ_s and q_s the surface values of potential temperature and of specific humidity, Θ and q their values at the first model level, C_p the specific heat at constant pressure, and C_h the bulk transfer coefficient (here assumed to be the same for heat and moisture as it is the case in most NWPMs), which depends on the stability of the air at the surface and on the roughness lengths z_{0m} and z_{0h}, the latter characterizing the exchanges of heat and moisture.

The solution of Equation (3.2), an implicit equation for the skin surface temperature T_s, requires the knowledge of F_{g0} and can be obtained only through iterations. A standard way to by-pass these difficulties is to compute another parameter T_0 that obeys the following prognostic equation:

$$\frac{\partial T_0}{\partial t} = \frac{F_{g0} - F_{g1}}{CD} \tag{3.6}$$

where C is the volumetric heat capacity of the soil.

T_0 is now the mean temperature of a finite depth D of soil, F_{g0} being the heat flux at the surface, and F_{g1} the one at the lower boundary of this soil slab. Most NWPMs have several layers in the ground, and the fluxes F_g at the interfaces (Fig. 3.3) are computed using a finite-difference scheme for the heat conduction equation in the soil

$$F_{gi+1} = -2\lambda \frac{(T_{i+1} - T_i)}{D_i + D_{i+1}} \tag{3.7}$$

where λ is the thermal conductivity of the soil. F_{g0} is then obtained from Equation (3.2) by replacing T_s by T_0 in Equations (3.3), (3.4), and (3.5) (problems related to this are dealt with in Section 4.1).

For the solution of Equations (3.6) and (3.7) we need assumptions for either T or F_g at the lowest soil layer.

Moisture

Figure 3.4 gives a schematic illustration of the fluxes controlling the surface moisture budget. P_R and P_S are the precipitation flux as rain and snow, respectively, M_{sn} is the water flux due to the melting of the surface snow pack S_n, Y_0 is the surface runoff, and M_{g0} is the infiltration flux. From now on, we only consider a snow-free surface.

As before, the balance equation for the moisture flux

$$P_R + F_W + Y_0 + M_{g0} = 0 \tag{3.8}$$

is replaced by a prognostic equation for W_0, the soil wetness of a finite depth D of soil

$$\frac{\partial W_0}{\partial t} = \frac{M_{g0} - M_{g1}}{\rho_{H_2O}} \tag{3.9}$$

in which M_{g0} and M_{gl} are the moisture fluxes at the surface and at the lower boundary of this layer.

Since the specific humidity at the surface is not computed explicitly, or even diagnosed, in NWPMs, the difference $(q - q_s)$ in Equation (3.5) has to be evaluated from known quantities like q and W_0.

If several layers of soil are represented, the moisture fluxes M_g at the interfaces are parameterized using a finite difference approximation of the moisture transfer equation in the soil

$$M_{gi} = \left(-2k\frac{(W_{i+1} - W_i)}{D_i + D_{i+1}} - \gamma\right)\rho_{H_2O} \tag{3.10}$$

in which k is the hydraulic conductivity and γ the hydraulic diffusivity. For the solution of Equations (3.9) and (3.10), we need assumptions for either W or M_g at the lowest soil layer.

For the parameterization of land-surface processes, we need to adapt the above formulas to the general case of inhomogeneous surfaces, to take into account the impact of vegetation. This is done in different ways in NWPMs, which are discussed in the following sections.

NWPMs have to represent primarily synoptic features of characteristic time–space scales of 1 day–1000 km. The current resolution of NWPMs is about a few hundred kilometers, which allows the use of time steps of the order of 15 min.

Though some of the processes parameterized via the previous formulas can have smaller time scales, time truncation errors are not considered to be the major source of error in this kind of scheme. The use of such "local" scheme for horizontal scales of few hundred kilometers is however much more debatable. The problem arises from the nonlinearity of the processes that govern the determination of turbulent fluxes. The mean flux is not related to the "mean" variables in the same way as the "local" flux is related to "local" variables. Surface inhomogeneities also modify the structure of the overlying atmosphere. However, the local effect of such spatial inhomogeneities is generally smoothed out in the PBL by the action of convection, and the fluxes become more and more horizontally homogeneous when approaching the PBL top.

Clearly, the size of NWPMs mesh is one order of magnitude larger than the scale on which the fluxes

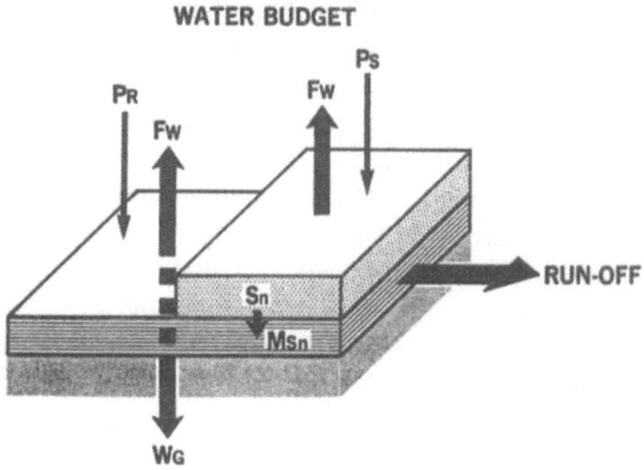

FIGURE 3.4. Schematic illustration of the surface moisture budget.

at the top of the PBL can be assumed to be homogeneous. This means that the effect of a local breeze circulation, as induced by surface inhomogeneities, or of orographic circulations, as induced by subgrid scale orography, should be accounted for when evaluating the fluxes on a scale of a few hundred kilometers.

All current NWPMs compute only grid-averaged parameters: wind components, temperature, specific humidity, and surface pressure, without any information about the distribution of these quantities inside the grid box. It implies that the subgrid-scale effects of local surface inhomogeneities have to be indirectly inferred from various surface characteristics, like the roughness length or the albedo, both of them defined at the model gridscale.

3.2 General Comments About the Implementation of Land-Surface Processes in Operational NWPMs

The strategy for implementing land-surface processes in NWPMs can be divided into four different steps:

1. The choice of a parameterization scheme: This choice is very much dependent on practical constraints connected to the computer environment of each NWPM. Time consumption and memory requirements are generally the strongest limiting factors in the definition of the selected land-surface parameterization scheme. How-

ever, the lack of data at the required horizontal scale, either for specifying the various physical constants or for initializing the soil variables, makes the use of complex land-surface schemes quite questionable.

2. The specification of the noninteractive parameters: Current NWPMs use physical constants or variables to describe the surface that are fixed in time during the forecast (and even in space for some of them, like emissivity), and that represent areal averaged values. For some parameters, only local estimates are known at a rather limited number of locations, and the specification of areal averaged values on a large scale is uncertain. For other parameters, this problem does not exist since data sets with a smaller spatial resolution than the NWPM grid size are available (this is the case for orography, vegetation types, and soil types) or because measurements can help to derive a value for the parameter at the model resolution (e.g., albedo).
3. The initialization of the time-dependent parameters: Both the lack of surface data for surface model variables and, if they exist, their lack of representativeness at the model's resolution, make initialization very difficult. Most of the observed parameters (e.g., temperature at 2 m) are not model variables, but quantities related to model variables (e.g., surface temperature). Therefore, these observed parameters contain relevant information about model variables that can then be used to initialize the latter. It implies that analogues of the observed parameters have to be derived from model variables. This computation is called postprocessing (and generally uses a complex operator). In practice, initialization of surface variables is done in a very crude way.
4. The validation or the evaluation of the quality of the parameterization scheme: It is important to design methods that can help verify model outputs. The ultimate goal has to be a comparison with observations. Therefore, the validation and the initialization are two closely connected aspects (both have to go through postprocessing).

A short review of these different aspects is presented in the next section, in which the differences and the resemblances between NWPMs in various meterological centers are presented. An overview of the current practice in Numerical Weather Prediction is also given.

4. Review of Parameterization Schemes Used in NWPMs and Their Recent Evolution

All the NWPMs mentioned below have many important common features as far as the parameterization of land-surface processes is concerned. They all simulate the diurnal cycle, and all of them have their first level inside the CFL. They all use the Monin–Obukhov similarity theory. They all use Louis's (1979) formulation of the surface exchange coefficients, except the British Meteorological Office (Carson, 1986b; An introduction to the parameterization of land-surface processes, Unpublished Met Office Report, DCTN 45, Met 0 20). They all use only one roughness length for momentum, heat, and moisture.

The heat balance, Equation (3.2), is replaced in all cases by a prognostic equation for T_0, but the number of levels in the soil and the conditions for the lowest model level are generally different from one model to the other. They also have only one parameter to describe the surface thermal state (no supplementary energy budget for canopy layer for example). But almost all have different parameterizations of the surface moisture flux.

4.1 Determination of the Surface Temperature T_0

The surface temperature T_0 is a crucial parameter because it controls not only the energy fluxes but also the moisture fluxes, through the term q_{sat} (T_0, p_s), which is used in all the parameterizations of these fluxes. It enters the radiation computations and the calculation of the heat flux within the ground. The temperatures involved in those different processes are not the same: skin temperature for the energy flux and the net radiation, temperature of the interior of the stomata cells for the plant transpiration, and temperature at a certain depth within the soil, for the heat flux into the ground. This ambiguity of the exact physical meaning of NWPM surface temperature is clearly a problem, especially when dealing with initialization or verification. This is further complicated by the fact that, at the model grid scale, T_0 represents an "areal" mean between vegetated areas and bare soils, mountainous, hilly, or flat terrains.

The number of levels within the soil to compute the surface temperature is 1 in the global NWPM of the UK Met-Office, 2 in the Nested grid model at the National Meterological Centre (simply referred as NMC later) and in both the global and regional French models, 3 in the ECMWF model, and 4 in the Japanese and in the regional NWPM of the UK Met-Office. A typical value for the volumetric heat capacity C of the soil is 2.4×10^6 J m^{-3} K^{-1} and 7.5×10^{-7} m^2 s^{-1} for the thermal diffusivity. The NWPMs consider a total soil depth which varies from 8 cm (5 cm of water to be more correct) in the global UK model to 2.335 m in the regional UK model. The depths of the various slabs are also very different (Fig. 3.5). Among the NWPMs with more than 1 layer in the soil, some assume a fixed temperature of the deepest layer (ECMWF, NMC, Japan, UK) while others involve a rate equation for predicting it (France). Temperature in the first layer is assumed to account for the daily variations, and the intermediate layers for temperature variations on a time scale of 1 week or so.

All the NWPMs have constant soil physical constants (i.e., independent of the soil wetness) and do not distinguish between various surface temperatures (skin, bare ground, canopy, etc.). This has been the case for many years and has not changed in individual NWPMs recently, except in the NMC model, which initially had almost no soil physics. This is partly linked to the difficulties of initializing and verifying this parameter. The introduction of the diurnal cycle in NWPMs is also a recent feature (1984 in the ECMWF model for instance).

4.2 Parameterization of the Evapotranspiration

In NWPMs, the evapotranspiration flux FW reads

$$F_W = \beta E_{pot} \qquad (3.11)$$

where β can be called the evapotranspiration efficiency. It varies between 0 and 1, and accounts for the fact that the soil may not be able to evaporate at the potential rate because of various constraints. It is worth mentioning that the definition of E_{pot}, an equation similar to Equation (3.5), namely

$$E_{pot} = -\rho C_h |u^2 + v^2|^{1/2} [q - q_{sat}(T_0, p_s)] \qquad (3.12)$$

is questionable. This is because the temperature in q_{sat} should be the one of a surface evaporating at the potential rate all the time, which can differ substantially from T_0 in dry conditions. But it is a fairly common practice to formulate the problem that way.

The first attempt to represent the surface moisture flux and its interaction with the soil hydrology has been based on the so-called bucket model (Budyko, 1956), in which the current value of W_0 is limited by a value W_{max} generally called the field capacity, and in which β depends only on W_0. This approach is still in use in most NWPMs. Even the NMC and Japan model keep this parameter constant in time during each forecast (but it has different values depending on the location). The global UK and the two French NWPMs have an efficiency coefficient that reads

$$\beta = \begin{cases} 1, & \text{if } W_0 > W_{cr} \\ W_0/W_{cr}, & \text{if } W_0 < W_{cr} \end{cases} \qquad (3.13)$$

where W_{cr} is a critical value above which the evaporation is at the potential rate and independent of W_0. Current values for hydraulic properties are

- Saturation soil moisture concentration, 0.44 m^3 m^{-3},
- Soil moisture concentration at field capacity, 0.25 m^3 m^{-3},
- Hydraulic conductivity, 10^{-7} m s^{-1}.

This type of parameterization is clearly very crude. It represents only partly the control the vegetation imposes on the surface moisture flux (the moisture stress), and it does not represent the correct evaporation regime over bare soils.

A direct way of partitioning the evaporation can be achieved by assigning a certain distribution of vegetated and bare ground areas inside each grid box of the model over land. This is done by specifying a vegetation cover C_v so that the total moisture flux is

$$F_W = C_v E_{vegetation} + (1 - C_v) E_{bare\ ground} \qquad (3.14)$$

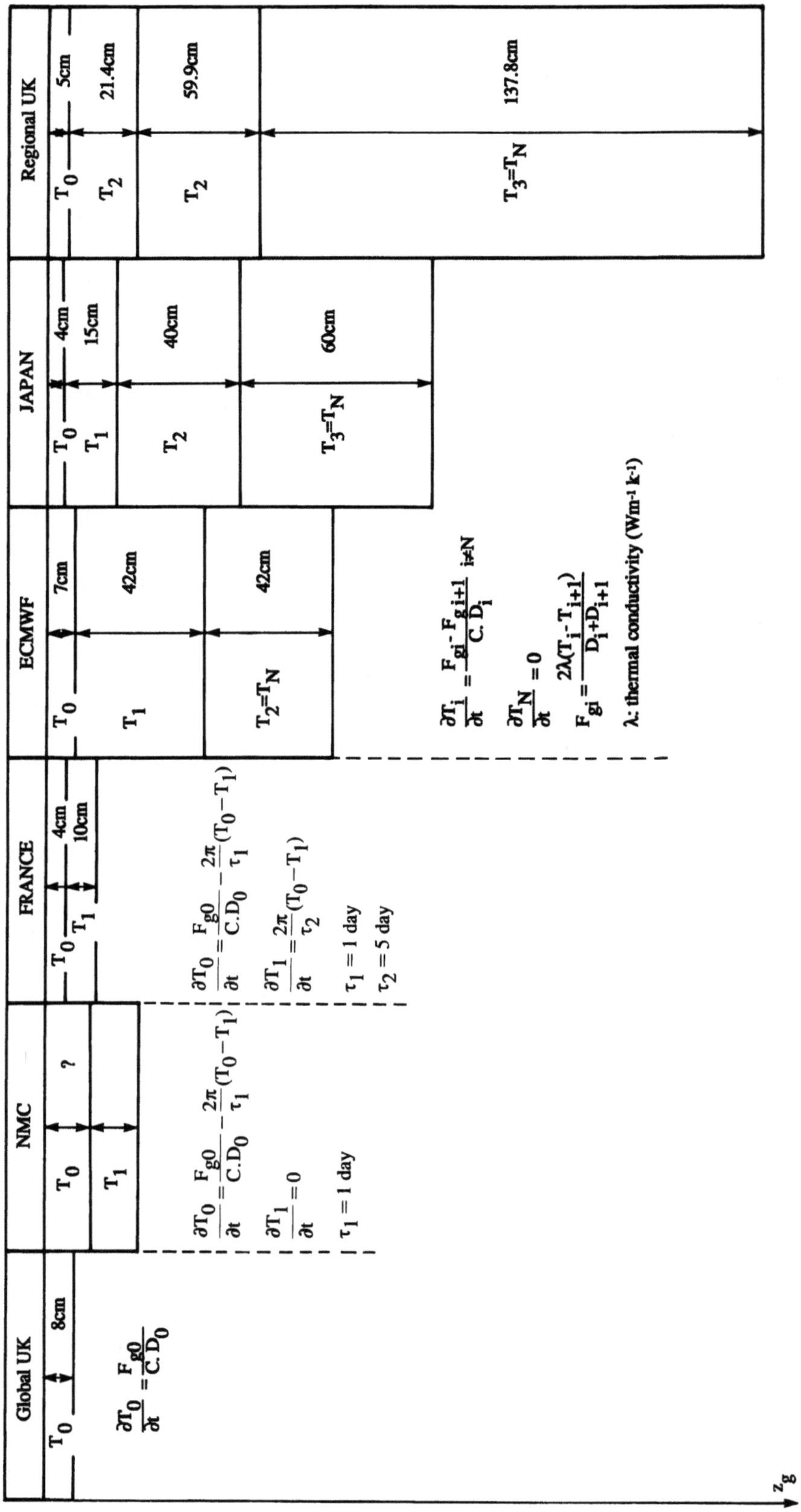

FIGURE 3.5. Comparison of multi-layer soil models used in various NWPMs.

with:

$$E_{\text{bare ground}} = -\rho C_h |u^2 + v^2|^{1/2} [q - hq_{sat}(T_0, p_s)] \quad (3.15)$$

where h is the relative humidity of the ground. No moisture flux is allowed from the atmosphere to the ground unless dew conditions $[q_{sat}(T_0,p_s) < q]$ are fulfilled.

This scheme has been introduced in the French and the ECMWF model. However, the French models still use Equation (3.11) to compute the evapotranspiration of vegetated parts. But the major change (and improvement) has come from incorporating a formulation of surface resistance into evapotranspiration (Monteith, 1965):

$$E_{\text{vegetation}} = \rho \frac{[q_s - q]}{R_a} = \rho \frac{[q_{sat}(T_0,p_s) - q_s]}{R_c} = \rho \frac{[q_{sat}(T_0,p_s) - q]}{R_a + R_c} \quad (3.16)$$

where R_a is the aerodynamic resistance, equal to the inverse of the drag, and R_c is the canopy resistance. R_c characterizes the physiological control of water loss by the vegetation.

This canopy resistance depends on many features and can be written as

$$R_c = R_{c0} \cdot g_C \cdot g_{SM} \cdot g_T \cdot g_R \cdot g_D \quad (3.17)$$

where R_{c0} is the minimum canopy resistance, g_c a canopy (seasonal) factor, which varies with the phenological state of the plant, g_{SM} a soil moisture stress factor depending on the water availability in the root zone, g_T a temperature stress factor (corresponding to the fact that there exists an optimum temperature of life for the plant), g_R a radiation stress factor (the photosynthetic activity is responsible for the opening of the stomata through which the plant exchanges moisture with the surrounding air), and g_D a saturation deficit stress controlling the rate of transfer between the interior of the stomata and the air surrounding the leaves (all stress factors g are larger than, or equal to, 1).

However, the regional UK NWPM uses the following formula:

$$F_W = \rho\beta \frac{[q_{sat}(T_0,p_s) - q]}{R_a + R_c} \quad (3.18)$$

where R_C is fixed (60 s m^{-1}) and where β is the inverse of g_{SM} and has a similar form as in Equation (3.13).

The ECMWF scheme has recently evolved from a scheme using Equation (3.11) to a slightly more complex model in which g_{SM} and g_R are explicitly taken into account (Blondin and Böttger, 1987), but where the remaining stress terms are ignored. g_R is a function of the PAR (photosynthesis active radiation) taken from Sellers et al. (1986). g_{SM}^{-1} is taken as

$$g_{SM}^{-1} = \begin{cases} 1, & \text{if } W > W_{cr} \\ \dfrac{W_{cr} + W}{W_{cr} - W_{pwp}}, & \text{if } W_{pwp} < W < W_{cr} \\ 0, & \text{if } W < W_{pwp} \end{cases} \quad (3.19)$$

where W is now a mean wetness of the soil in the root zone (which can extend through all the soil layers), W_{cr} is a critical value above which the evapotranspiration is not affected by the moisture stress, and W_{pwp}, the so-called permanent wilting point, below which it is assumed that the plant is not able to pump the water from the root zone to the stomata cells, and that transpiration ceases and the plant ultimately dies. Furthermore the interception of precipitation is taken into account, a canopy reservoir being filled by precipitation and dew deposition and evaporating at the potential rate. The content W_l of this canopy reservoir defines the wet fraction of the leaves, $C_{v,wet}$ by

$$C_{v,wet} = W_l / W_{l,max} \quad (3.20)$$

where $W_{l,max}$, the maximum canopy reservoir content, is given by

$$W_{l,max} = LAI\ W_{l,m} \quad (3.21)$$

LAI is the leaf area index (4 in the ECMWF model) and $W_{l,m}$ is taken as 0.2 mm. If this canopy reservoir is full, the excess of precipitation or dew is added to the nonintercepted precipitation or dew (simulates dripping), which forms the water flux at the surface.

Additional processes are also accounted for at the surface, such as a surface runoff due either to a precipitation rate exceeding a maximum infiltration rate depending on the soil hydraulic properties or to the slope of the terrain. It is empirically related to the model subgrid scale variance of the orography.

The impact of the vegetation on the surface and subsurface temperature diurnal cycles through the shading effect is parameterized by reducing the soil thermal diffusivity during daytime.

When looking at all those schemes for the surface moisture flux, the first striking point is their elementary simplicity compared to the current knowledge of vegetation control on water loss and surface hydrology. They are also very simplified compared to the BATS model (Biosphere–Atmosphere Transfer Scheme, Dickinson et al., 1986) developed for the NCAR Community Climate Model, or to the SiB model (Simple Biosphere Model, Sellers et al., 1986). There are many reasons for this. Meterologists who have developed NWPMs have initially underestimated the role of soil physics and plant physiology. It is also largely believed that present model errors are mostly due to other deficiencies of the model (e.g., in the parameterization of the atmospheric part of the hydrological cycle). It is therefore sensible to keep the current surface schemes at this degree of simplicity, since the main forcings (radiation and precipitation) of the surface still exhibit large defects. However, all centers running NWPMs and that have generally, at one stage or another, used Equation (3.11) and (3.13) to compute the surface evaporation have come to the conclusion that this formulation gives too large surface moisture fluxes over land. Most of them have moved or are likely to move to the use of Equation (3.14); which gives more realistic surface fluxes at least on the global scale.

A better representation of physical processes in NWPMs depends on the capability of current and planned earth observing systems to provide means to specify the various physical parameters of the surface schemes. This has been taken into consideration when designing the ECMWF scheme: the vegetation is described only by the vegetation cover C_V, the leaf area index LAI, the minimum resistance R_{c0}, and the root profile. The first three parameters can, in principle, be inferred from satellite measurements with a global coverage, while the last one could be monitored using vegetation data sets.

The other limiting factor in using too complex schemes is clearly the horizontal scale problem. Apart from the difficulty of defining soil and vegetation characteristics at an NWPM resolution, it is worth stressing that key parameters in the formulation of surface exchanges are u, v, T, q, C_D, or C_h and T_0, which undergo large spatial variations inside each model grid box. As soon as subgrid scale orography features are present, they are very difficult to analyze. The analysis of the atmospheric specific humidity suffers from large errors, with data-type-dependent biases (for example, satellite retrievals are drier than radiosondes measurements). Information about the initial water content of the atmosphere and clouds is not efficiently used in current data assimilation schemes. Consequently, the atmospheric forcing inside the CFL, and the overall structure of the PBL, may not be properly represented in the early stage of the forecast.

It is still an open question as to what extent the current formulation using Monin–Obukhov similarity theory holds at the NWPM resolution. Furthermore, the actual accuracy of the parameterization of surface fluxes is still not known. In this respect, field experiments like HAPEX-MOBILHY (André et al., 1986) will provide very valuable information to indicate where to look for progress and improvement. Mesoscale models that can realistically simulate the subgrid scale variability of atmospheric parameters at the NWPM's resolution should be used to tackle this problem as well.

5. Surface Parameters Specification: The Current Practice

The only geographically variable surface parameters are the surface geopotential, the subgrid scale variance of orography, the roughness length, the albedo, the mean climatic value of soil temperature and wetness, the evapotranspiration efficiency, when used as in the Japanese model, and the vegetation cover, when incorporated into the surface scheme. All other properties (emissivity, thermal and hydraulic properties, vegetation properties) are taken as constant, and may have been tuned to give good estimations of one or several parameters, e.g., in a specific area of interest, or adjusted to provide a good fit with some global data (energy or moisture budgets at the surface, radiative balance at the top of the atmosphere, etc.). This exercise is model-dependent but, by and large, the values in question are very similar from model to model. In this field, progress may occur because of the increasing capacities of computers and because of improvement in the interpretation of data coming from remote observing systems. Nevertheless, there will always be a conflict between the use of a priori defined value for a given parameter on the one hand and its actual current value on the other hand; for instance, the emissivity of land surface, which is taken as constant in NWPMs (0.996 in the ECMWF one), could be assigned more realistic values depending on soil type (quartz has an emissivity lower than 0.9, so the emissivity in desert areas should correspond to an area of minimum emissivity), or vegetation type.

5.1 Surface Moisture

As far as the surface moisture flux is concerned, the most important parameters in the aforementioned list are the albedo and the roughness length. Of course, when the efficiency factor β (also called the moisture availability factor) is prescribed, the surface flux is controlled in a very efficient way. "Climatological" values are then imposed. But in most NWPMs this parameter is interactive, its value depending on the soil wetness itself. The values of wetness in the deepest reservoir in the ECMWF and the regional UK models do not have a major effect on surface fluxes, unless the rooting zone extends into this layer. However, it is desirable that these values reflect the large interannual variability of the water content at a depth of 50 cm or so, linked to the natural variations of the precipitation regime from year to year, and do not keep a fixed climate value.

5.2 Orography

Orography is not treated explicitly in the parameterization of surface evaporation, but it influences directly the atmospheric parameters of the models. In mountainous, or even simply hilly, terrains the thermodynamic effects of surface inhomogeneities (terrain slope, variation of vegetation type with height, etc.) and atmospheric circulations (valley winds, breezes) have a strong impact on surface fluxes. These effects are very difficult, if not impossible, to reproduce properly in models at the current resolution. The specification of surface characteristics becomes a very delicate problem in such areas. The problem is even complicated for those models that use a so-called envelope orography.

5.3 Albedo

Concerning the albedo, two approaches have been used so far. Albedo maps have been compiled on the basis of *in situ* measurements from individual sites and for different surface types, combined with land-cover and land-use data sets. This can result in either the use of an annual average surface albedo as a background field in NWPMs (which is the current practice), or in the specification of a variable albedo depending on soil wetness (Wilson and Henderson-Sellers, 1985). Maps have also been obtained using satellite-observed surface albedos, which can be derived with a good accuracy on a monthly basis (Dedieu et al., 1987). The annual average albedo used at ECMWF was obtained using the minimum albedo method of Preuss and Geleyn (1980).

The use of albedos with better time sampling (typically 1 month) is envisaged. New requirements about the albedo specification will also come from the development of the parameterization of surface evaporation, in particular, in connection with the possible refinement in the description of vegetation and of its interaction with the underlying soil. For instance, the parameterization proposed by Deardorff (1978), already implemented in some mesoscale models, needs values for soil and leaf albedo. We have already mentioned the contribution of the PAR to the stomatal resistance, which pleads for having a better spectral definition of the optical properties of natural surfaces.

5.4 Roughness Length

The definition of an appropriate "areal" average roughness length Z_{oeff} had received much attention in recent years (André and Blondin, 1986; Taylor, 1987; Mason, 1986). André and Blondin and Taylor have addressed the problem of defining an effective roughness length from the subgrid scale distribution of this parameter. The results of their studies both suggest that at first order a simple area-weighted mean of $\log(z_o)$ gives a reasonable estimate of $\log(z_{oeff})$, but they can only be used as a guidance for flat terrains. Mason's study is more complete and deals with the problem of dense or moderate slopes, isolated bluff elements and reaches some practical recommendations, valid on scales up to 10 km. This means that in practice the specification of roughness lengths at the NWPM resolution is still an open problem. One of the problems is that large variations of local roughness and orography induce mesoscale circulation inside the PBL. Therefore, the flow becomes in equilibrium with terrain features only at the top of the PBL. This means that the use of the classical CFL flux–gradients relationship in the framework of the Monin–Obukhov similarity theory may not be appropriate at the NWPM scale.

At the microscale, it is common practice to introduce a displacement height in the computation of the surface fluxes, which corresponds to a vertical shift of the origin of the vertical axis. This feature is generally not included in NWPMs (in fact it is more appropriate in numerical models to assume that the surface geopotential includes both the displacement height and the roughness length, though this is not done in many NWPMs).

Another aspect concerns the possible difference between the roughness length to be used to compute momentum stress on the one hand, and heat and moisture fluxes on the other hand. At the microscale there is theoretical and experimental evidence that these two quantities are different (Brutsaert, 1982). Very often natural surfaces are partly permeable or offer greater surface area for heat and water exchanges than for momentum, corresponding to smaller roughness for the former. Since orographic effects like the pressure drag induced by the subgrid scale orography are generally accounted for via the vertical diffusion scheme, the roughness length for momentum is generally empirically increased with regard to its microscale value. The use of the same roughness length for heat and moisture transfer is thus even less justified.

The specification of the roughness length is still empirical. Most NWPMs use a constant value over land for instance ($z_0 = 0.1$ m in UK models). In the NMC model z_0 is vegetation dependent and varies from small values over deserts to larger values over forests. The French and the ECMWF models base the specification of the roughness on the 10′ by 10′ information of the U.S. Navy data set, provided by NCAR, merged with vegetation data from Baumgartner et al. (1977).

6. Surface Variables Initialization: Current Methods

The more interactive a surface parameterization scheme is, the more important it is to initialize those variables with sufficient accuracy. The major difficulty concerning the initialization of the soil wetness comes from the lack of direct measurements of this quantity. Only brightness surface temperatures can be obtained from satellite measurements. However, it is obvious that surface observations like temperature (at 2 m) and dew-point departure, precipitation, insolation, etc., contain information about the state of the underlying surface. Despite the good coverage of the continents by measurements of at least the two first parameters, only the French global model has built in an analysis scheme which makes use of them. In most NWPMs the initial values of the soil variables are obtained via very simple methods.

NWPMs have to be given initial values of their atmospheric variables as well. Both continuous (as in the global UK model) or intermittent (as in the ECMWF model) data assimilation schemes are in use. To obtain the best initial state at some time t_0, observations (at t_0, and those made slightly before and after t_0) are used, as well as the most recent model forecast valid for that time (first guess). The question is: How should these values be used to provide the required initial values?

6.1 Japan

In the Japanese model, β is given a climatological value. The initial soil temperatures are obtained as follows. First, monthly mean climatological soil temperatures of the years 1985 and 1986 are used. Then the 1-year harmonic component of these data is used to obtain the initial value of T_i by the following equation ($i = 1,2,3$, respectively, for the three deepest soil layers, from top to bottom).

$$T_i = \hat{T} + A_i \exp(-z_i/d) \cos\left[\frac{2\pi}{365}(D-P_i) - z_i/d\right] \tag{3.22}$$

where $\hat{T}$ is a mean soil temperature, A_i and P_i are the amplitude and phase of the 1-year component, z_i the depth of the layer, $d = 2.65$ m the e-folding depth, and D, the day of the year. For the initial T_0, the predicted value of the previous run is used.

6.2 NMC

In the NMC model, the surface energy budget reads

$$C_{\text{eff}} \frac{\partial T_0}{\partial t} = -(R_N + F_H + LF_W) - C_{\text{eff}}(T_0 - T_1)/\tau \quad (\tau = 1 \text{ day}) \quad (3.23)$$

The subsoil temperature T_1 is fixed at the start of the forecast. To compute this value, two interim values for the surface temperature T_0 are calculated by iteratively solving the surface energy budget under the conditions of zero C_{eff}, once for local midnight and once for local noon values of the zenith angle. The radiation code is used here only to produce the downward long-wave reaching the surface and the short-wave radiation absorbed at the surface, both of which are independent of T_0. These two values for T_0 are averaged to give T_1, which is then held constant during the forecast. The initial field of T_0 is obtained by solving Equation (3.23) for that T_0 which will yield a zero value for the left-hand side. As in the Japanese model, β is derived from climatological data.

6.3 UK

The UK models temperature and moisture initialization is performed in a way very similar to the Japanese one.

6.4 ECMWF

At ECMWF, the initialization of the soil temperatures consists in taking the values provided by the first guess. The deepest layer temperature, which is given a climatological value and held constant during the forecast, is updated every month. The initialization of the soil wetness has changed with the implementation of the current scheme. In the days when the "βscheme" was used, an analysis of surface precipitation was first performed. Then the soil wetness at the previous analysis cycle (i.e., 6-hr before) was updated by using the soil wetness model equations with a 6-hr time step (in particular, the evaporation rate was kept constant). The soil wetness in the deepest reservoir was updated every month, as well as the temperature. This procedure has been abandoned with the implementation of the new surface scheme that uses the resistance approach. First-guess values for both the temperature and the soil wetness are used as initial values. In contrast to the previous method, no surface observations at all are used. Clearly, both methods present weaknesses and new approaches are now being tested.

6.5 France

The French models make an intensive use of surface observations. An objective analysis of the 2 m temperature and relative humidity is performed. The idea is that the variations of these weather elements are strongly correlated with variations in the state of the surface. The first-guess run provides estimates for the surface variables (Geleyn, 1988). The differences between analyzed and first-guess values, increments of 2 m temperature and relative humidity, are then used to correct the first-guess values of the soil parameters (a weak relaxation toward climate values is applied). This method has some defects. However, the temporal variations of the soil parameters are realistic both at the time scale of 1 day and for longer times.

It is clear that the current methods are not optimal. They do not make use of all the observations available and they ignore the physical content of the surface scheme. It is evident that the initialization of surface variables requires more sophisticated approaches to overcome these two major problems.

7. Operational Validation of Surface Schemes

We must evaluate the performance of the surface parameterization to understand its contribution to both the errors in individual forecasts and to the model systematic errors. This can be done with three types of approaches.

First, a continuous and close scrutiny of NWPMs performances in forecasting surface weather elements such as temperature, wind, and precipitation can provide a valuable feedback. At ECMWF all parameters are checked against a fixed set of reliably

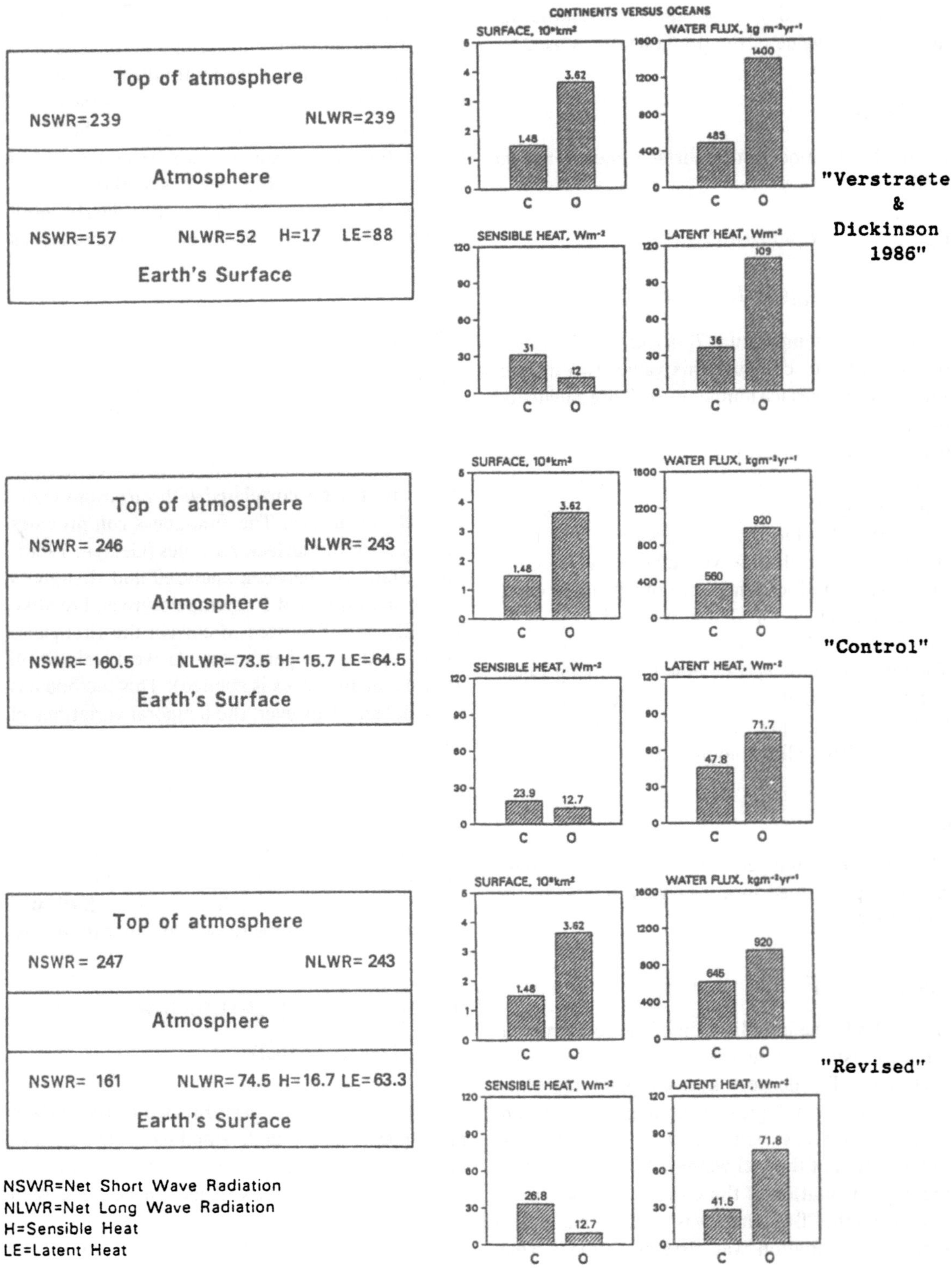

FIGURE 3.6. Globally and annually averaged energy fluxes into and out of the atmosphere. Comparison between oceans and continents is in term of area, evaporation, and sensible and latent heat annual mean fluxes.

reporting synoptic stations in Europe and North America. Verification results are accumulated monthly for each station and are presented in charts or statistical summaries. Though it is not always easy to separate errors coming from postprocessing and data representativeness from model defects, biases in 2 m temperatures or precipitation can point to surface scheme deficiencies.

Second, a posteriori assessment can be done by using more sophisticated algorithms. For example, monthly mean surface temperatures and daily cycles can be derived from satellite observations (Chahine et al., 1986). Data sets from the International Satellite Land-Surface Climatology Project (Bolle and Rasool, 1985) or the International Satellite Cloud Climatology Project (WMO-ICSU, 1985), and also from field experiments such as FIFE (Sellers and Hall, 1987) and HAPEX-MOBILHY should be compared with model analyses and forecasts. In regions where data are of sufficient quality and density to provide an estimate of the surface fluxes using, e.g., methods proposed by Taconet et al. (1986), much could be gained by comparing these "observed" fluxes with their spatial and temporal distribution as predicted by NWPMs. This type of study needs much cooperation between the satellite and the meteorological community and is still in a preliminary stage.

The most straightforward studies to diagnose and possibly understand deficiencies in surface parameterization schemes are sensitivity studies with NWPMs. These are generally done by running the forecasting model at a lower resolution than the operational model, but on longer periods than the standard forecast range. Such experiments resemble climate studies, and diagnostics are carried out in a similar way. Figure 3.6 illustrates results obtained with the ECMWF model run at T42 (truncation of a spectral NWPM whose maximum zonal wave number is 42) resolution and averaged over two periods of 90 days (northern hemispheric winter and summer) for surface fluxes (both over continents and oceans), and compared with climate estimates (Verstraete and Dickinson, 1986). In this figure a comparison is done between runs using the "β-scheme" (called control) and the resistance-based scheme (called revised).

The results indicate the reduction of evaporation over land, when using the resistance approach, leading to a value of the Bowen ratio closer to the climate estimate. Nevertheless, this type of study determines only partly the impact of the surface scheme upon the forecasts, because the full impact can be seen only through the combined effect of initialization and forecasting.

The "β-scheme" and the resistance-based scheme have been used in a complete data assimilation and 10-day forecast cycle for a period of 5 days. From a synoptic point of view, the two runs have produced very similar analyses and forecasts as judged from conventional products like the 500 hPa geopotential map. However, substantial differences in the surface fields have been obtained. Figure 3.7 illustrates the consistent difference between the impacts on temperature, soil wetness, moisture flux, and convective precipitation over tropical Africa, for both short-range (first guess) and medium-range forecasts. The resistance-based scheme (NEW in the figures) generates higher surface temperatures because of reduced surface moisture fluxes.

Convective precipitation which is strongly dependent on the water moisture supply by surface evaporation is decreased. The surface soil wetness increases, however, because the evaporation corresponds to a rather efficient root extraction in the intermediate reservoir. A posteriori assessments over longer periods have demonstrated that the overall impact of the modification of the surface parameterization in the ECMWF forecasting system can be described from the whole experimentation using both long runs, individual forecasts, and a few full data assimilation and forecasts cycles. However, some aspects of the local impacts and the behavior of the scheme in extreme climate conditions are often difficult to anticipate. For instance, a continuous monitoring of the ECMWF surface scheme has indicated that the root uptake in the deep climate reservoir, possible only if the scheme uses a nonzero value of root density in the deepest soil model layer, can be detrimental in simulating persisting drought conditions, as it happened over part of the United States during spring and summer 1988.

Because the root uptake in this reservoir is not causing any soil moisture depletion, the scheme maintains the transpiration at unrealistically high values. A possible remedy is to change the root profile to confine the root extraction to the two upper interactive reservoirs. Figure 3.8 shows the impact of such a change after 6 days of integration in a region (the United States) where the initial

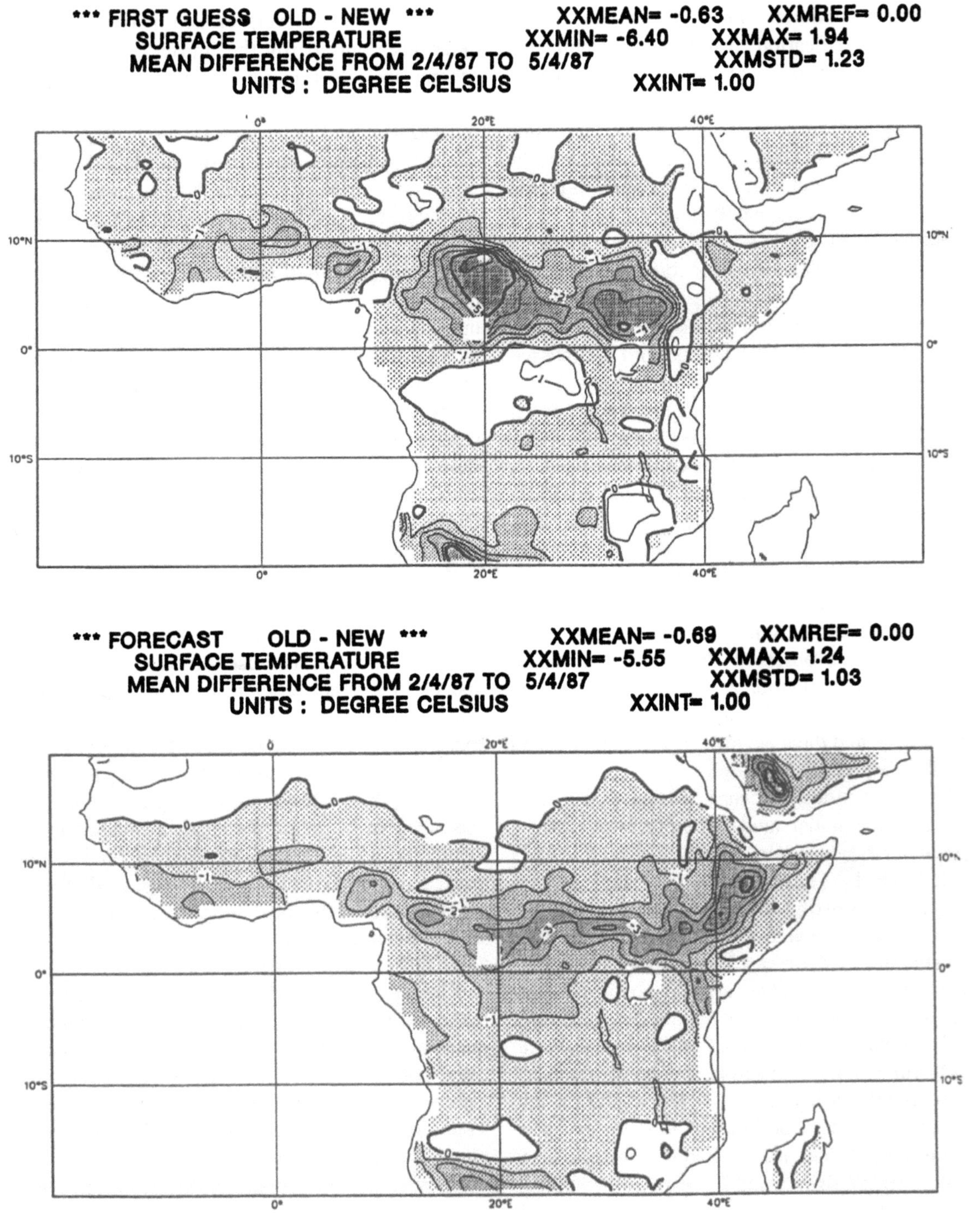

(a)

FIGURE 3.7. Difference in surface fields during first guess runs and 10-day forecasts over tropical Africa produced by the ECMWF model using the "β-scheme" and the "resistance scheme." (a) surface temperature; (b) surface soil wetness; (c) surface evaporation; (d) surface convective precipitation.

*** FIRST GUESS OLD - NEW *** XXMEAN= -1.10 XXMREF= 0.00
SURFACE SOIL WETNESS XXMIN= -11.84 XXMAX= 4.33
MEAN DIFFERENCE FROM 2/4/87 TO 5/4/87 XXMSTD= 2.48
UNITS : MM OF WATER XXINT= 1.00

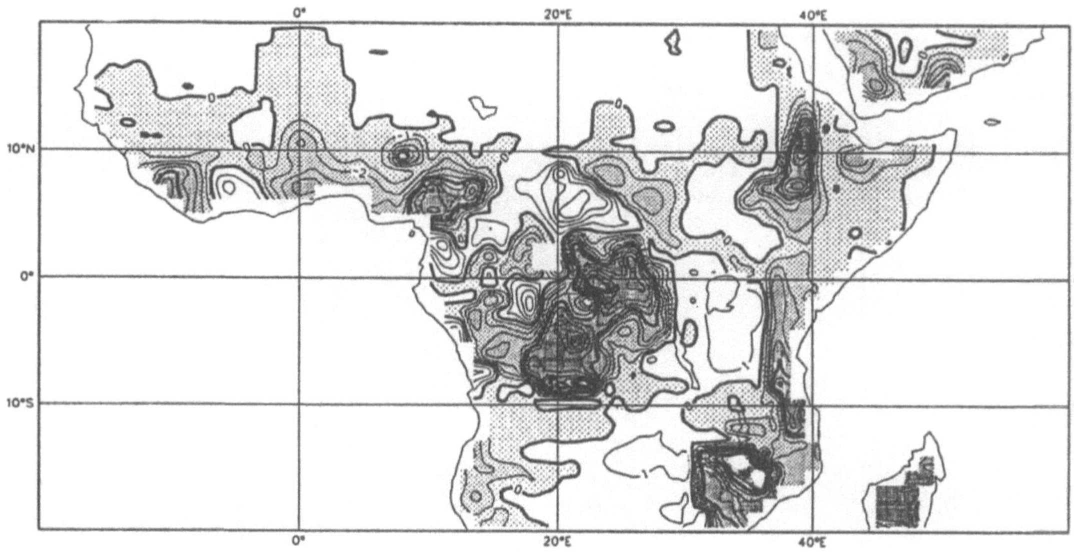

*** FORECASTS OLD - NEW *** XXMEAN= -2.07 XXMREF= 0.00
SURFACE SOIL WETNESS XXMIN= -8.41 XXMAX= 2.44
MEAN DIFFERENCE FROM 2/4/87 TO 5/4/87 XXMSTD= 2.12
UNITS : MM OF WATER XXINT= 1.00

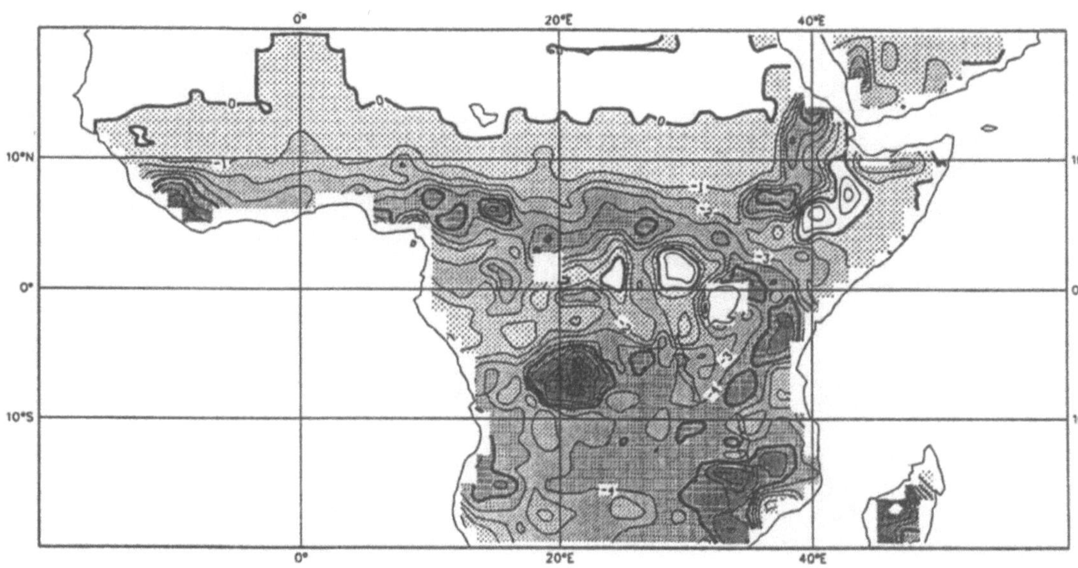

(b)

FIGURE 3.7. (Continued).

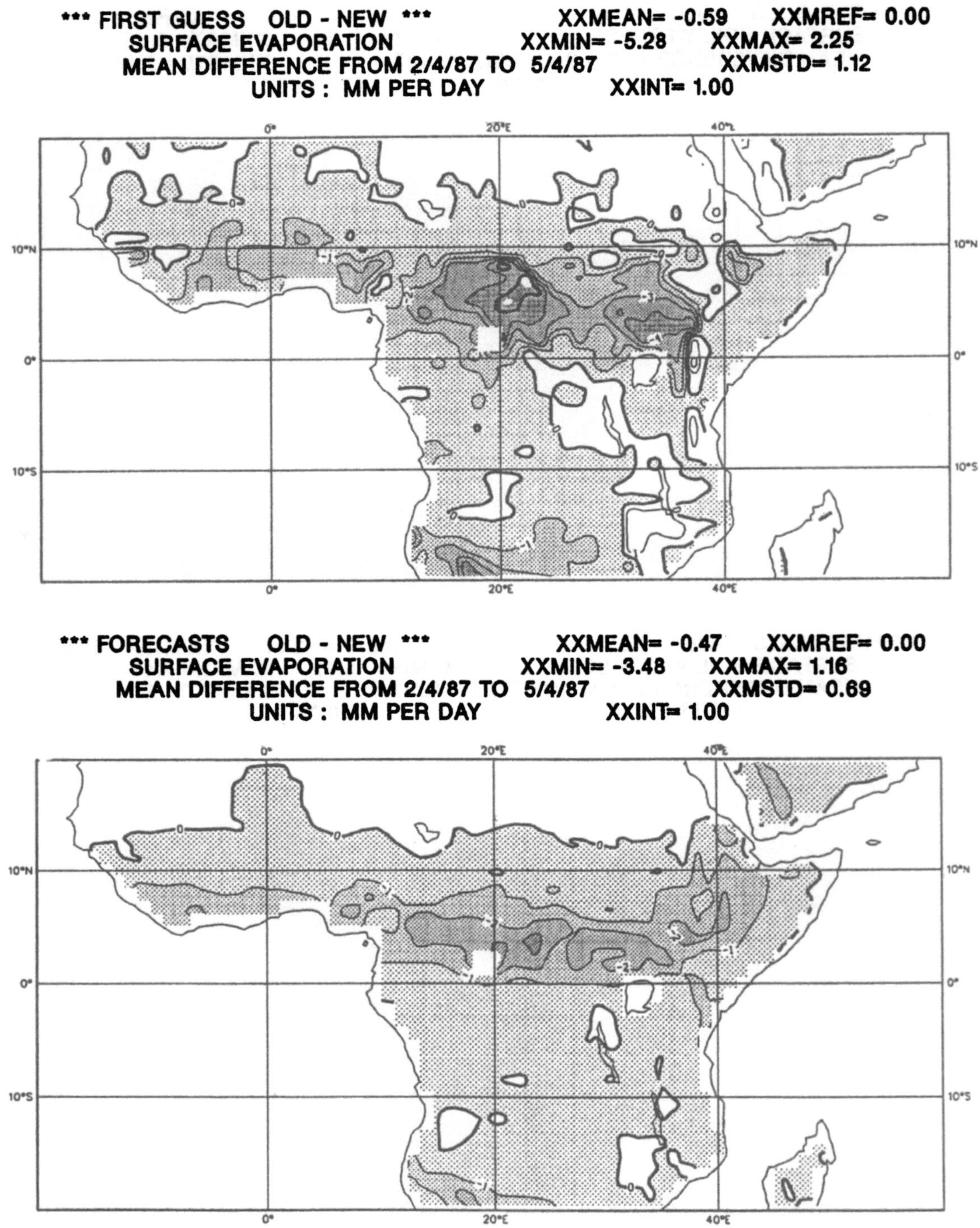

(c)

FIGURE 3.7. (Continued).

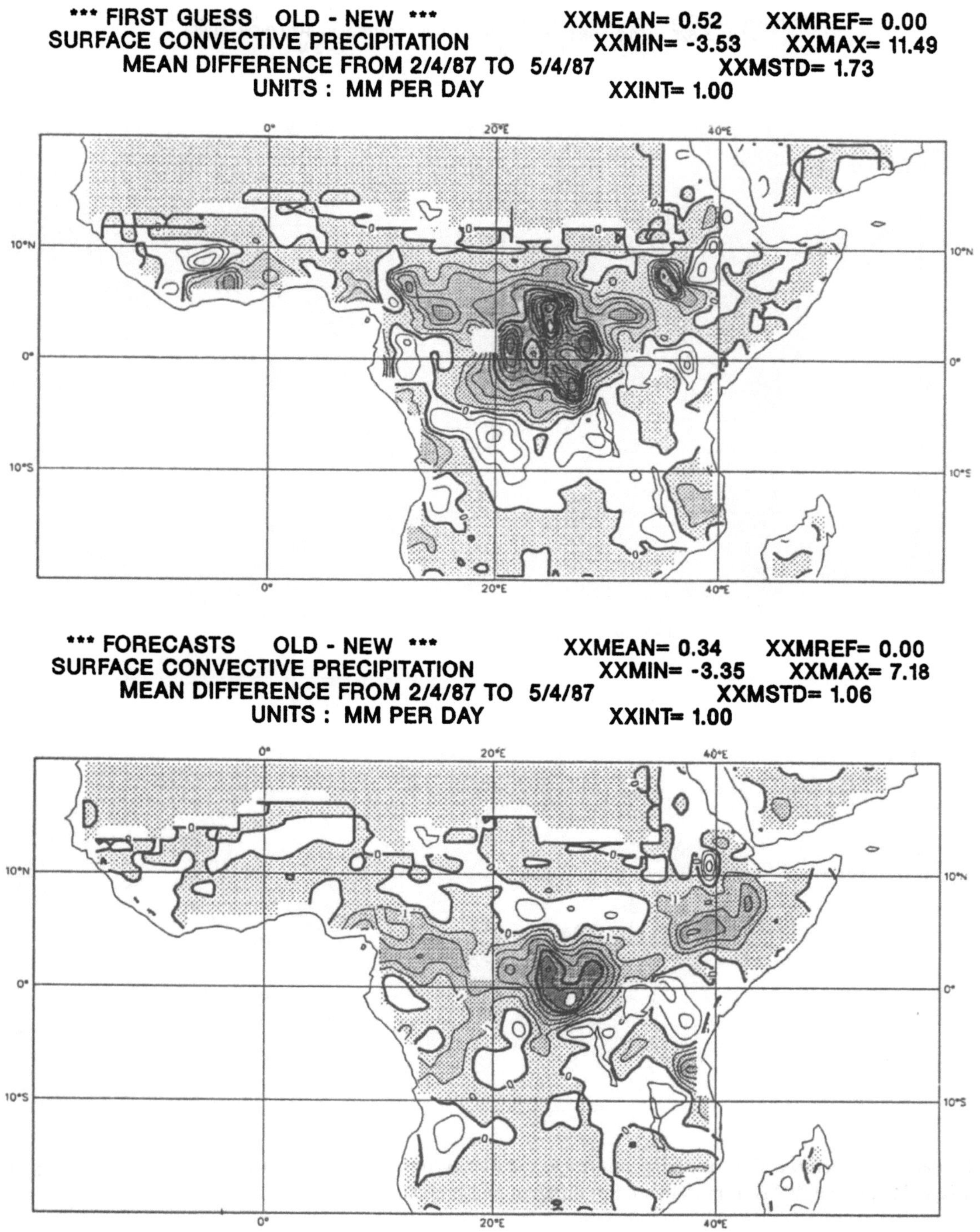

(d)

FIGURE 3.7. (Continued).

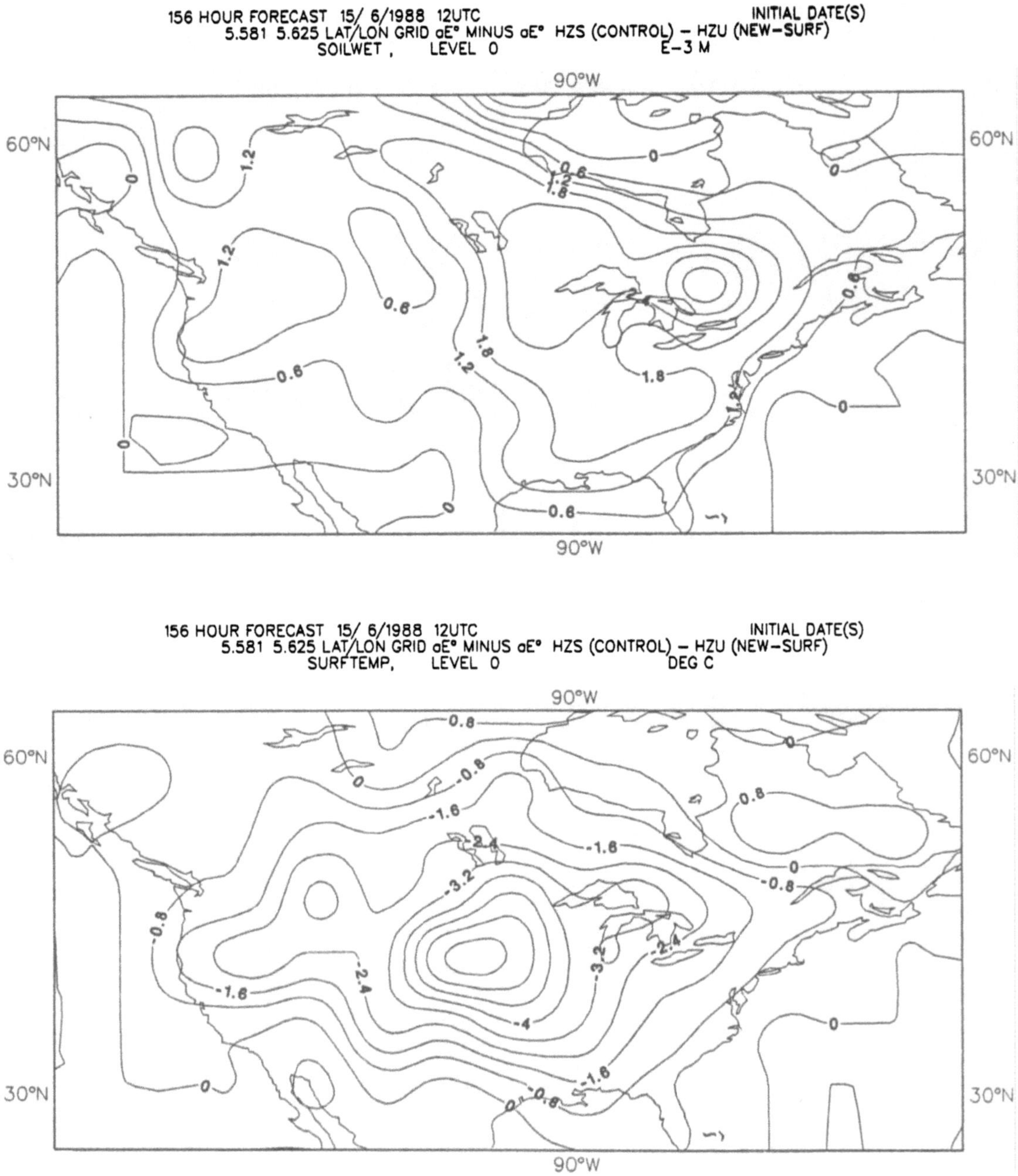

FIGURE 3.8. Difference of surface temperature and soil wetness over the United States forecasted at day 6 using the T42 version of the ECMWF model with two different root profiles.

values of the two top soil layers are close to the permanent wilting point. Differences as large as 6 K in surface temperatures are obtained. A much faster drying of the surface is also clear. Atmospheric temperatures and the large-scale circulation are much less affected. This experiment illustrates the limit of the current possibilities and limitations of forecasting systems.

8. Foreseeable Evolution

The schemes used in NWPMs are generally simple and are dependent on physical parameters, the specification of which is very difficult. The previous sections have mainly illustrated the role of the vegetation and demonstrated that an improvement is obtained when using schemes, even quite crude, including a canopy layer.

Two large avenues can be followed to improve the situation: the use of more physically based schemes, fed with more appropriate physical surface characteristics, and the implementation of powerful analysis schemes that can make use of all observations containing information about the surface conditions. These approaches are still being developed independently.

More advanced surface schemes than the one currently used in NWPMs have already been developed (Deardorff, 1978; Dickinson, 1984; Sellers et al., 1986; Noilhan and Planton, 1989). They have been validated most of the time against micrometeorological data (Sellers and Dorman, 1987) and their sensitivity to various input parameters has also been studied in one-dimensional (vertical) soil-atmosphere column models and in Global Circulation Models (GCM) (Wilson et al., 1987; Dickinson and Henderson-Sellers, 1988). These experiments are useful in defining the parameters to which the model is the most sensitive, and, accordingly, in helping to select the types of observations that are required both for initialization and verification purposes. However, results of such studies have to be adapted to the NWP context, in which the model is run from a data assimilation cycle where the atmospheric state is very close to the actual state and where extreme situations are much more likely to happen than in climate runs. Nevertheless, there is scope for large improvements in the parameterization of surface processes, but this will be converted to improvement in the forecast quality only if the specification of physical input parameters and the initialization of surface variables make similar progress.

As far as data are concerned, much work is currently devoted, especially inside the satellite community, to improve retrieval algorithms and methods to better describe surface characteristics (see Becker et al., 1987). At the same time theoretical and practical work is to be done to improve the current understanding and parameterization of mesoscale circulations in relation to surface inhomogeneities and their influence on the definition of the atmospheric forcing on the surface at the NWPM resolution (Mahrt, 1987; Sud and Smith, 1984).

Numerical weather forecasting is, however, facing other important questions because most of the information about the surface state within operational data assimilation comes from conventional reports (SYNOPs) done at synoptic stations. In almost every case (including satellite information) the raw observations do not concern direct model variables but quantities that have to be derived through a postprocessing. In principle, any observation can be efficiently used in an analysis scheme, providing statistics about first-guess postprocessed and observation errors are known with sufficient accuracy.

For meteorological applications, the most widely used analysis method is the multivariate optimal interpolation (OI). In this approach, model variables are analyzed using a linear combination of observations in such a way that the analyzed state minimizes (in a statistical sense) the deviation from the observations knowing a priori the statistics of the observation errors and of the model errors (Lorenc, 1981; Hollingworth and Lonnberg, 1986). The major drawbacks of the OI come from the inability of the method to deal properly with model variables that are not linear combinations of observations, and from the difficulty of including new data in the assimilation. Furthermore, there is no warranty that the model forecast will be optimal when the short-range model errors are at minimum in some statistical sense.

Variational analysis (Le Dimet and Talagrand, 1986) should overcome these problems. In the context of surface analysis, this approach provides the best method to use all available surface state information and produces an analyzed soil state that is the most consistent with the model parametrization of land-surface processes. In this respect there is a warranty that the best surface scheme will produce the best results at least as far as the succession of analyzed states is concerned because they will produce the best fit to all available information.

One crucial point of this method is the quality of the postprocessing operator. In theory, this operator can be split into two parts: one part gives a "primary" physical quantity at the location of the measurement knowing the model variables (for example, a radiative flux at a certain wave length or in a certain spectral interval at the top of the atmosphere, knowing the vertical profile of temperature and humidity); the second one converts this "primary" physical quantity into a more elaborate one, directly related to the raw answer of the apparatus doing the measurement (e.g., the conversion of radiative flux into electric impulsions into a radiometer). If the second part of the postprocessing operation is linear, the "primary" quantity can be considered a genuine observation. But if it is not, then it is important that the physics of the measurement be known accurately and be modeled numerically so that its adjoint can be computed; the use of the primary quantity as an observation will otherwise not be appropriate within the variational analysis.

This stresses that real improvement in the initialization of soil parameters within NWPMs must come from a coordinated effort of atmospheric modelers and experimentalists. This is a key aspect in improving the use of satellite data especially.

9. Conclusion

In this paper various aspects of the parameterization of land-surface processes in NWPMs have been reviewed. Some aspects have been deliberately ignored, like physical processes associated with snow and ice-covered land surfaces. It has been shown that current surface schemes used in NWPMs are very crude and far from satisfactory. Potential improvement can be expected in different domains:

1. A basic improvement of the physics accounted for in the parameterization scheme itself (for instance, the representation of the role of vegetation);
2. A refinement of land-surface properties both in terms of a geographically varying distribution and a better monitoring of their possible time evolution. The use of already available data sets has to be encouraged along with continuous efforts to update and develop them;
3. An improvement of our understanding of the impact of mesoscale circulations on the atmospheric forcing at the resolution of NWPMs (validity of similarity theory, filtering role of convective PBLs, etc.);
4. A larger use of available information about the surface state, both inside analysis and initialization schemes and within verification and diagnostics studies. This is likely to involve efficient use of new mathematical tools like variational analysis, which requires a coordinated effort of modelers and experimentalists; and
5. An overall improvement of atmospheric models, especially their ability to handle the radiation forcing, at different wavelengths and for cloudy and cloud free atmospheres, and to parameterize condensation processes so that both surface precipitation and atmospheric specific humidities are well predicted within NWPMs.

Acknowledgments. This review would not have been possible without information received from several centers responsible directly for Numerical Weather Predictions or working in close collaboration with the former ones. I would like to express my gratitude to Drs. W.E. Baker and J. Hoke from NMC, P. Rowntree and C. Wilson from UK Met-Office, Dr. A. Segami from Japan Meteorological Agency, R. Dickinson from NCAR, and J.F. Geleyn and G. Rochas from Direction de la Météorologie (France) who sent me documents about the operational implementation of land-surface processes in NWPMs. I also wish to thank J. Pailleux from ECMWF for providing information on satellite and P. Courtier from ECMWF for discussions about variational analysis.

References

André JC, Blondin C (1986). On the effective roughness length for use in numerical three-dimensional models. Bound-Layer Meteorol 35:231–245.

André JC, Goutorbe JP, Perrier A (1986). HAPEX-MOBILHY: A hydrological atmospheric experiment for the study of water budget and evaporation flux at the climate scale. Bull Am Meteorol Soc 67:138–144.

Baumgartner A, Mayer H, Metz W (1977). Weltweite Verteilung des Rauhigkeitparameters ZO mit Anwendung auf die Energiedissipation an der Erdoberflache. Meteorol Rdsch 30:43–48.

Becker F, Bolle HJ, Rowntree PR (1987). The international satellite land-surface climatology project. ISLSCP Report No. 10.

Blondin C, Böttger H (1987). The surface and sub-surface parameterization scheme in the ECMWF forecasting system: Revision and operational assessment of weather elements. ECMWF Res Dept Tech Memo No. 135.

Bolle HJ, Rasool SI (1985). Development of the implementation plan for the International Satellite Land-Surface Climatology Project, Phase I WCP-94 World Meteorological Organization, Geneva.

Brutsaert WH (1982). "Evaporation into the Atmosphere." Reidel, Dordrecht.

Budyko MI (1956). "Heat Balance at the Earth's Surface." Gidrometeoizdat Leningrad (in Russian).

Chahine MT, Haskins R, Suskind J, Reuter D (1986). Remote sensing of land-surface temperature from HIRS/MSU data. Proc ISLSCP Int Conf Parameterization of Land-Surface Characteristics; Use of Satellite Data in Climate Studies; First Results of ISLSCP, Rome, pp 215–221.

Deardorff JW (1978). Efficient prediction of ground surface temperature and moisture with inclusion of a layer of vegetation. J Geophys Res 20:1889–1903.

Dedieu G, Deschamps PY, Kerr IH (1987). Satellite estimation of solar irradiance at the surface of the earth and of surface albedo using a physical model applied to METEOSAT data. J Clim Appl Meteorol 26:79–87.

Dickinson RE (1984). Modeling evapotranspiration for three-dimensional global climate models. "Climate Processes and Climate Sensitivity." Geophysical Monograph, 29, Maurice Ewing Vol. 5, American Geophysical Union, pp 58–72.

Dickinson RE, Henderson-Sellers A (1988). Modelling tropical deforestation: A study of GCM land-surface parameterization. Quart J R Meteorol Soc 114:439–462.

Dickinson RE, Henderson-Sellers A, Kennedy PJ, Wilson MF (1986). Biosphere/Atmosphere Transfer Scheme (BATS) for the NCAR Community Climate Model. NCAR Tech Note, Boulder, CO, 80307.

Geleyn JF (1988). Interpolation of wind, temperature and humidity values from model levels to the height of measurements. Tellus 40A:347–351.

Hollingsworth A, Lonnberg P (1986). The statistical structure of short range forecast errors as determined from radiosonde data. Part I: The wind errors. Tellus 38A:111–136.

Le Dimet F, Talagrand O (1986). Variational algorithms for analysis and assimilation of meteorological observations: Theoretical aspects. Tellus 38A:97–110.

Lorenc AC (1981). A global three-dimensional multivariate statistical interpolation scheme. Mon Weather Rev 109:701–721.

Louis JF (1979). A parametric model of vertical eddy fluxes in the atmosphere. Bound-Layer Meteorol 17: 187–202.

Mahrt L (1987). Grid-averaged surface fluxes. Mon Weather Rev 115:1550–1560.

Mahrt L, Berthou C, Marquet P, André JC (1986). Time averaging the flux-profile relationship. Ann Geophys 4B:411–416.

Mason PJ (1986). On the parameterization of orographic drag. Workshop on "Observation, Theory and Modelling of Orographic Effects," pp 167–194. European Center for Medium-range Weather Forecasting, Reading, UK.

Mintz Y (1984). The sensitivity of numerically simulated climates to land-surface conditions. "The Global Climate" (J. Houghton, ed.), pp 79–105. Cambridge University Press, Cambridge.

Monteith JL (1965). Evaporation and environment. Symp Soc Exp Biol 19, Swansea, 8–12 September 1964, 205–236.

Noilhan J, Planton S (1989). A simple parameterization of land-surface processes for meteorological models. Mon Weather Rev 17:536–549.

Preuss JH, Geleyn JF (1980). Surface albedos derived from satellite data and their impact on forecast models. Arch Meteorol Geoph Biokl A29:345–356.

Rowntree PR, Bolton JA (1983). Simulation of the atmospheric response to soil moisture anomalies over Europe. Quart JR Meteorol Soc 109:501–526.

Sellers PJ, Dorman JL (1987). Testing the Simple Biosphere model (SiB) using point micrometeorological and biophysical data. J Clim Appl Meteorol 26: 622–651.

Sellers PJ, Hall FG (1987). FIFE Experiment plan. ISLSCP Report No. 8.

Sellers PJ, Mintz Y, Sud YC, Dalcher A (1986). A Simple Biosphere model (SiB) for use within general circulation models. J Atmos Sci 43:505–531.

Sommeria G (1985). Parameterization of land-surface processes. Sem Phys Parameterization Numerical Models Atmos, 233–264.

Sud YC, Smith WE (1984). Ensemble formulation of surface fluxes and improvement in evapotranspiration and cloud parameterizations in a GCM. Bound-Layer Meteorol 29:185–210.

Taconet O, Bernard R, Vidal-Madjar D (1986). Evapotranspiration over an agricultural region using a surface flux/temperature model based on NOAA-AVHRR data. J Clim Appl Meteorol 25:284–307.

Taylor P (1987). Comments and further analysis on effective roughness lengths for use in numerical three dimensional models. Bound-Layer Meteorol 39:403–418.

Verstraete M, Dickinson RE (1986). Modelling surface processes in atmospheric general circulation models. Ann Geophys 4B:357–364.

Wilson MF, Henderson-Sellers A (1985). Cover and soil data sets for use in general circulation climate models. J Climatol 5:119–143.

Wilson MF, Henderson-Sellers A, Dickinson RE, Kennedy PJ (1987). Sensitivity of the biosphere-atmosphere transfer scheme (BATS) to the inclusion of variable soil characteristics. J Climatol Appl Meteorol 26:341–362.

WMO-ICSU (1985). International Satellite Cloud Climatology Project (ISCCP)—Description of reduced resolution radiance data. WMO/TD, No. 58, World Meteorological Organization, Geneva.

4
Parameterization Schemes of Land-Surface Processes for Mesoscale Atmospheric Models

P. Bougeault

1. Introduction

For approximately the last 20 years the incentive for developing advanced parameterizations of land-surface processes has been provided by the great issues of climate modeling: Are we able to monitor the short-term global water budget of the planet and its long-term global change? However, mesoscale modelers have long been aware of the importance of correctly specifying the surface energy budget that forms the lower boundary condition of their models, since the thermal circulations that are induced by the spatial variability of the heat and moisture fluxes into the atmosphere are an integral part of mesoscale meteorology.

Within recent years, considerable effort has been made to increase the realism and the physical basis of the parameterizations that are used in mesoscale models, and the natural path to this has been to adapt the work of climate modelers to the specific needs of the mesoscale.

Therefore this chapter addresses the following objectives:

1. To identify the specific needs of mesoscale models;
2. To review the most well-known existing parameterization schemes and their results; and
3. To discuss the possibilities of testing these schemes with existing data, and to identify the most necessary supplementary data.

2. The Specific Needs of Mesoscale Atmospheric Models

2.1 Generalities

Atmospheric mesoscale models use horizontal grid sizes that range from $\Delta x = 10$ km (meso-β-scale) to $\Delta x = 50$ km (meso-α-scale) and time steps that range from less than 1 min to a few minutes. They are integrated over domains ranging from a few hundred to a few thousand kilometers, and for a time span lasting from 12 to 36/48 hr. The atmospheric phenomena described by these models are summarized in Figure 4.1 from Orlanski (1975). These phenomena are essentially determined by the *atmospheric initial conditions* and by the diurnal cycle of surface temperature and heat flux at each grid point of the model. This surface forcing may indeed significantly alter the evolution of atmospheric conditions during the characteristic time scale of the model integration. Spatial contrasts of surface forcing generate thermal circulations of the breeze-type, or more complicated effects. We will now review these processes.

2.2 Circulations Induced by the Surface Forcing

Pielke (1984) has given a detailed review of observational studies of mesoscale circulations generated by spatial contrasts in surface temperature and

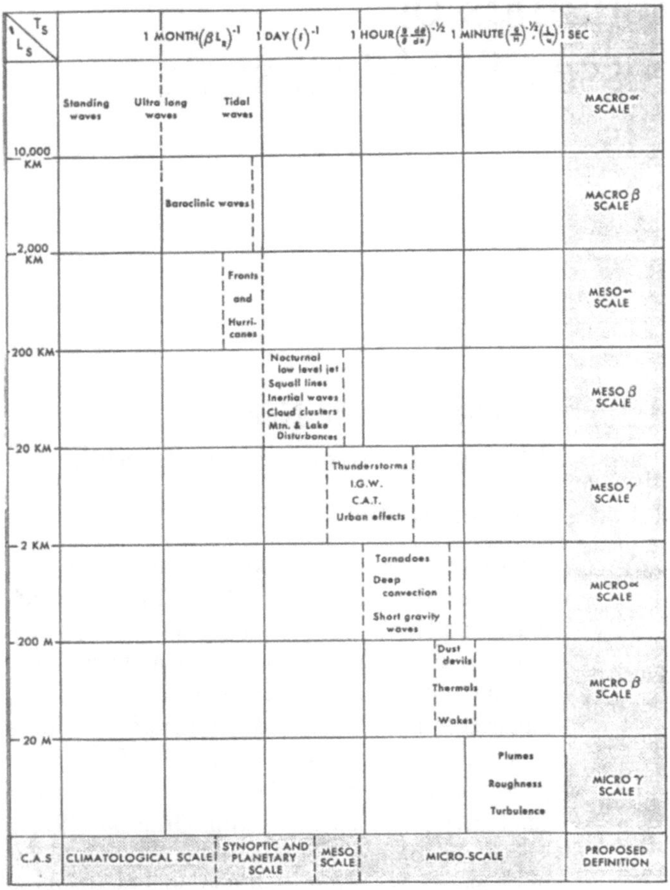

FIGURE 4.1. Space and time scales of interest for atmospheric mesoscale models. (Reproduced from Orlanski, 1975, Bulletin of the AMS, Vol. 56, No. 5, copyright by American Meteorological Society.)

heat flux. According to his conclusions, in the tropics and mid-latitudes, a horizontal gradient of less than about 10 W m^{-2} per 30 km has only minor influence on local wind patterns. With a gradient of 100 W m^{-2} per 30 km, significant effects are discernible from the statistical evaluation of observational data, whereas at 1000 W m^{-2} per 30 km, the influence on local wind patterns is very pronounced in case by case studies.

Ookouchi et al. (1984) investigated the soil moisture effects on the generation and modification of circulations in a mesoscale model. They used a two-dimensional (2D) version of the model of Pielke (1974), with a very simple parameterization of the latent heat flux:

$$E = E^{pot} A \tag{4.1}$$

when E is the evaporation flux, E^{pot} the potential evaporation, and A a specified *moisture availability parameter*. The grid size of the model was 5 km and the time step 90 sec.

The experiments consisted in generating a thermal circulation from a contrast between two regions in the model with different moisture availability, with or without additional effects linked to a moderate mountain. A summary of the different cases treated is given in Table 4.1. Five degrees of soil wetness contrast are examined (F_1 to F_5) plus the sea breeze (SB) case for reference. The differences between F_5 (moisture availability equals 1 on the western section) and SB are that the roughness length is different on the sea and the land, and that the surface temperature increases during the day over a land surface even with $A = 1$. The morphol-

TABLE 4.1. Summary of the cases simulated by Ookouchi et al. (1984) regarding soil moisture availability and domain topography.[a]

	Soil moisture availability			
Case	Western section	Eastern section	Slope angle (tan α)	Topography
SB	(sea)	0.05	0	Flat terrain
F_1	0.1	0.05	0	Flat terrain
F_2	0.2	0.05	0	Flat terrain
F_3	0.3	0.05	0	Flat terrain
F_4	0.5	0.05	0	Flat terrain
F_5	1.0	0.05	0	Flat terrain
M_1	0.05	0.05	2/100	Symmetric mountain[b]
M_2	0.5	0.5	2/100	Symmetric mountain[b]
M_3	0.5	0.05	2/100	Symmetric mountain[b]
S_1	0.05	0.05	2/100	Single slope
S_2	0.5	0.5	2/100	Single slope
S_3	0.5	0.05	2/100	Single slope
S_4	0.05	0.5	2/100	Single slope
S_4	0.05	0.5	1/100	Single slope

[a]Reprinted with permission from Monthly Weather Review, Vol. 112, No. 11, copyright by American Meteorological Society.
[b]Top of mountain is 800 m.

ogy of the predicted flow for case F4 is given in Figure 4.2. At 1600 LST a pronounced circulation is predicted with a noticeable return flow. The veering of the induced flow due to the Coriolis effect is reflected through the V component. Following sunset (2000 LST) an intensification of the circulation and its eastward progress is evident; this is explained by the collapse of low-level turbulence because of stabilization of the PBL. At midnight, the model predicts an additional eastward penetration of the circulation and a strong rotation of the wind. When this conversion to a predominantly V component flow has occurred, the eastward propagation of the circulation is terminated.

With a milder contrast of soil moisture availability, similar patterns to those described above are obtained. However, as expected, the intensities of the circulation are reduced. In Figure 4.3, the maximum wind speed at 5 m height for the various cases is shown, to provide some idea of the effect of soil moisture contrast on the mesoscale flow. This figure shows that a boundary between saturated and dry soils can induce a circulation whose intensity is close to that of a sea breeze. The results also suggest that even a mild contrast in soil moisture availability (0.1/0.05) may exert a measurable effect in the air flow. The soil surface temperature difference between the wet and the dry domains, are shown in Figure 4.4. They provide an indirect indication of the expected circulation intensities. Note that the maximum temperature contrast $\Delta T_{\max}$ generally occurs about 4 hr before the maximum wind velocity $U_{\max}$. Furthermore, the model results support the relation proposed by Walsh (1974) (in the absence of synoptic flow):

$$U_{\max}{}^2 \simeq \beta \, \Delta T_{\max} \qquad (4.2)$$

with $\beta = 1.5$ m²/KS² (see Fig. 4.5).

Ookouchi et al. (1984) also investigated the circulations induced by a symmetric mountain in their model (Fig. 4.6). The reduction of the sensible heat flux along the wet slopes (case M2) results in considerably weaker upslope circulation than obtained in case M1. As mountains sometimes have a relatively wet slope on one side contrasted by a relatively dry slope on the opposite side, the development of the daytime circulation in this case has been studied in case M3. In addition to the development of upslope winds on each side of the mountain, an overflow from the wet side toward the dry slope occurs. The overflow is induced by a net pressure gradient between both sides, resulting from the overall difference of heating along the slopes. Finally, we examine the upward penetration of moisture in these three cases. Figure 4.7 suggests that for M2 considerable moisture is avail-

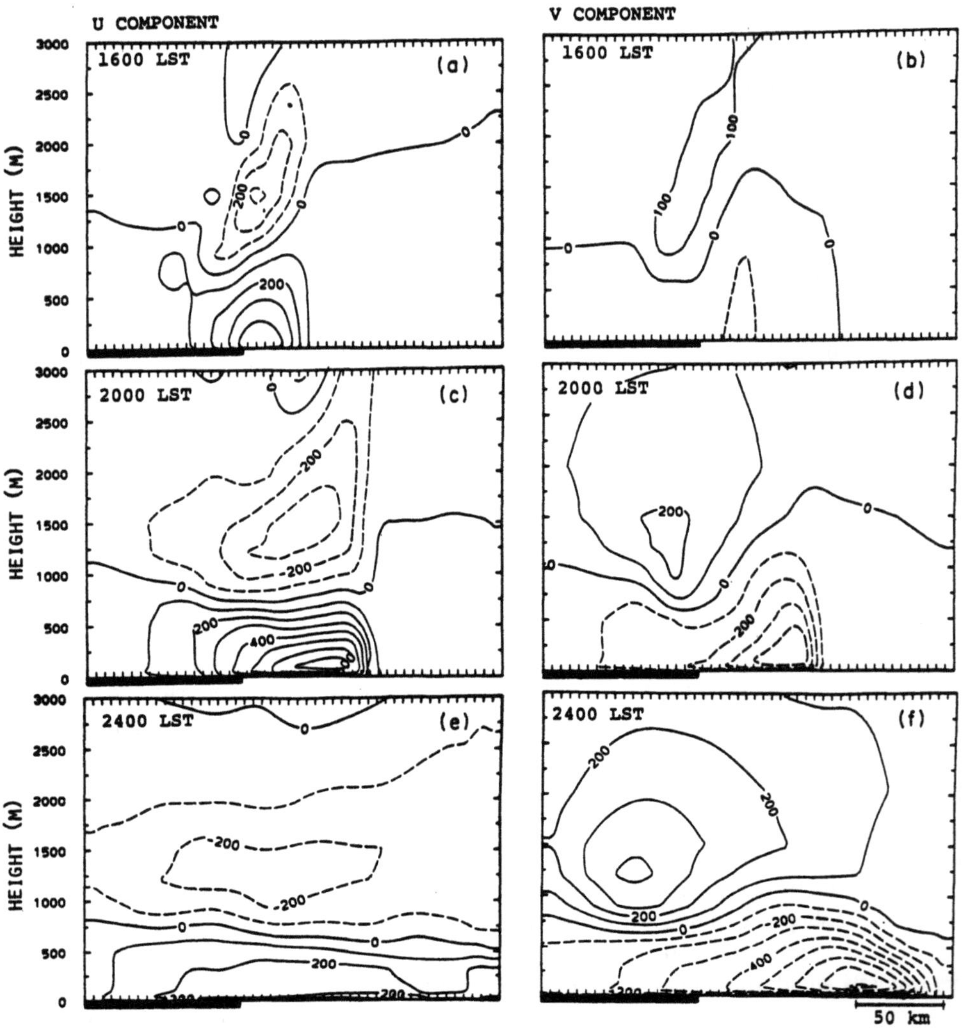

FIGURE 4.2. West–east vertical cross section of the simulated domain presenting the U (west–east) and V (south–north) components (cm s^{-1}) for case F_4 for several selected hours (dashed contours indicate negative component—northerly or easterly). The heavy line indicates the wet portion of the domain ($A = 0.5$) which is adjacent to a dry soil ($A = 0.05$). (Reproduced from Ookouchi et al., 1984, Monthly Weather Review, Vol. 112, No. 11, copyright by American Meteorological Society.)

able, although its upward advection is weak. The sharp vertical gradients of humidity indicate the top of the PBL, which is relatively low, because of the evaporative cooling of the surface. In the dry case, the upward penetration of moisture is deeper, but there is less moisture available. Case M3 provides a larger opportunity for the occurrence of convective clouds, since moisture appears to be both available from the wet side and transported upward by the upslope circulation on the dry side.

One of the limitations of the study by Ookouchi et al. is that there is no way of knowing if the imposed contrasts in moisture availability are realistic. A more recent study by Mahfouf et al. (1987) provides some answers to this question, since it uses a more realistic parameterization set, based on McCumber and Pielke (1981) and Deardorff (1978). The methodology is the same as before, the model is 2D, with $\Delta x = 10$ km, and $\Delta t = 10$ sec. A summary of the experiments of Mahfouf et al. (1987) is shown in

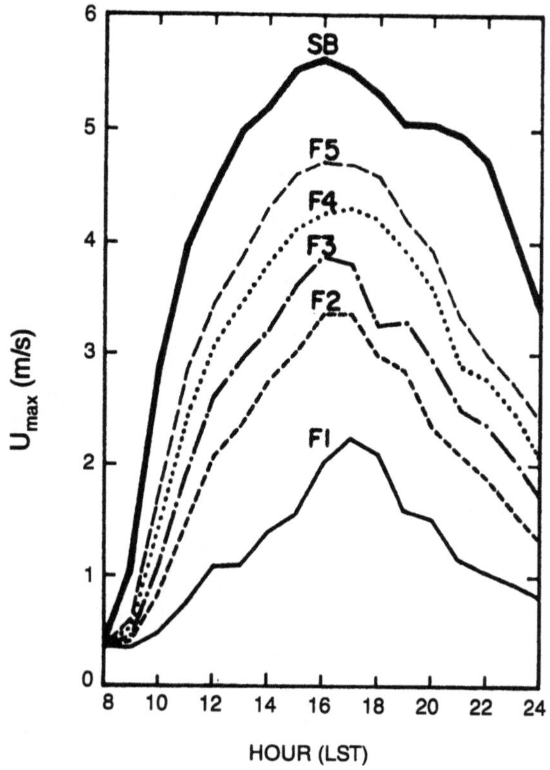

FIGURE 4.3. The maximum wind speed (m s^{-1}) at 5 m height along the domain cross section for various soil availability contrasts as a function of time. Flat terrain cases. (Reproduced from Ookouchi et al., 1984, Monthly Weather Review, Vol. 112, No. 11, copyright by American Meteorological Society.)

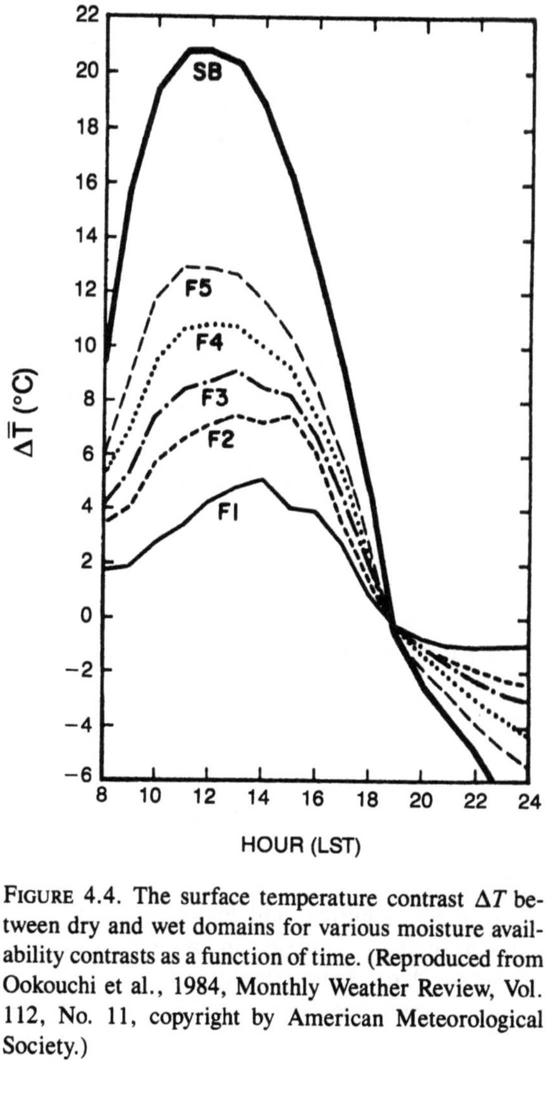

FIGURE 4.4. The surface temperature contrast ΔT between dry and wet domains for various moisture availability contrasts as a function of time. (Reproduced from Ookouchi et al., 1984, Monthly Weather Review, Vol. 112, No. 11, copyright by American Meteorological Society.)

Table 4.2. Instead of imposing the moisture availability, they impose the soil type, the vegetation type, and the initial soil moisture content and let the parameterization compute the moisture availability. The horizontal profile of wind speed at a height of 150 m predicted by the model is shown in Figure 4.8 for different times of the day, in a reference sea-breeze case. The maximum velocity reaches 8 m/sec, in agreement with expectations. The morphology of the flow is also comparable to that obtained by Ookouchi et al., (1984) as well as other modeling studies.

The influence of soil texture is next investigated by using a contrast consisting of sand and clay, no vegetation and an identical ratio of soil moisture content (W) to its saturation value (W_{sat}). Results are shown in Figure 4.9. The intensity of the circulation that develops is quite close to that of a seabreeze flow. This is explained by the different behavior of water in each of these two soils. In sand, the water can be removed easily by capillary rises and can be evaporated at the surface, leading to enhanced evaporation. In clay soil, on the other hand, the water is more tightly bound to the soil particles and the capillary rises are much weaker. The small amount of water available for surface evaporation ensures that almost all the incoming radiative energy is used to warm the soil. Thus, this experience confirms the findings of Ookouchi et al. (1984), but demonstrates that moisture availability does not depend solely on the relative soil

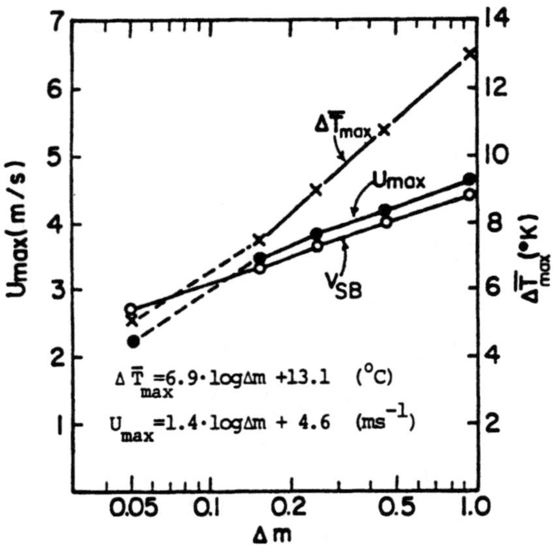

FIGURE 4.5. Log linear dependence of U_{max} and ΔT_{max} as a function of the difference in the soil moisture availability Δm. The open circles indicate wind speed values obtained from Equation (2). (Reproduced from Ookouchi et al., 1984, Monthly Weather Review, Vol. 112, No. 11, copyright by American Meteorological Society.)

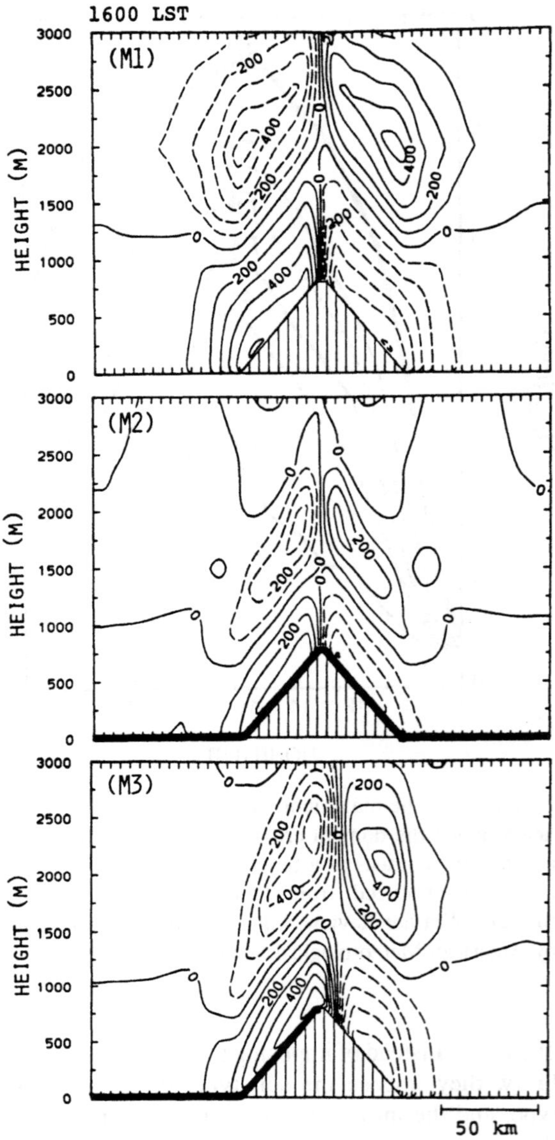

FIGURE 4.6. Vertical cross section of the simulated domain presenting the U (west–east) component (cm s^{-1}) for cases M_1, M_2, and M_3 at 1600 LST (dashed contours indicate a negative component—easterly). The heavy line indicates the wet portion of the domain. (Reproduced from Ookouchi et al., 1984, Monthly Weather Review, Vol. 112, No. 11, copyright by American Meteorological Society.)

moisture content, such as hypothesized in simple parameterizations.

The influence of vegetation is next investigated by using two identical soil textures (loam) with identical moisture content, and a contrast of bare soil versus dense grass (shielding factor of 0.82). Again, the model predicts a significant contrast in the Bowen ratio between the two zones. The induced circulation (Fig. 4.10) reaches 3 m/sec. At the beginning of the simulation, up to 1000 LST, the bare soil warms up very slowly because there is moisture available in the surface soil layer. The low level circulation is from the bare soil to the vegetation-covered area. After 1000 LST, the soil surface becomes very dry, and then warms up more than the grass area, leading to a reverse circulation. Finally, the combined effect of soil and vegetation is investigated in three experiments with different shielding factors for the vegetation. Figure 4.11 compares the wind profiles obtained at 1800 LST. When the two soils are covered with a fully dense foliage, a very weak current circulates from the wet zone to the dry one. The difference of the temperature of the foliage in this case between the wet and the dry zone is 1 K. The authors explain this result by the fact that the vegetation resistance to evapotranspiration is nearly independent of the soil moisture, as long as this one is

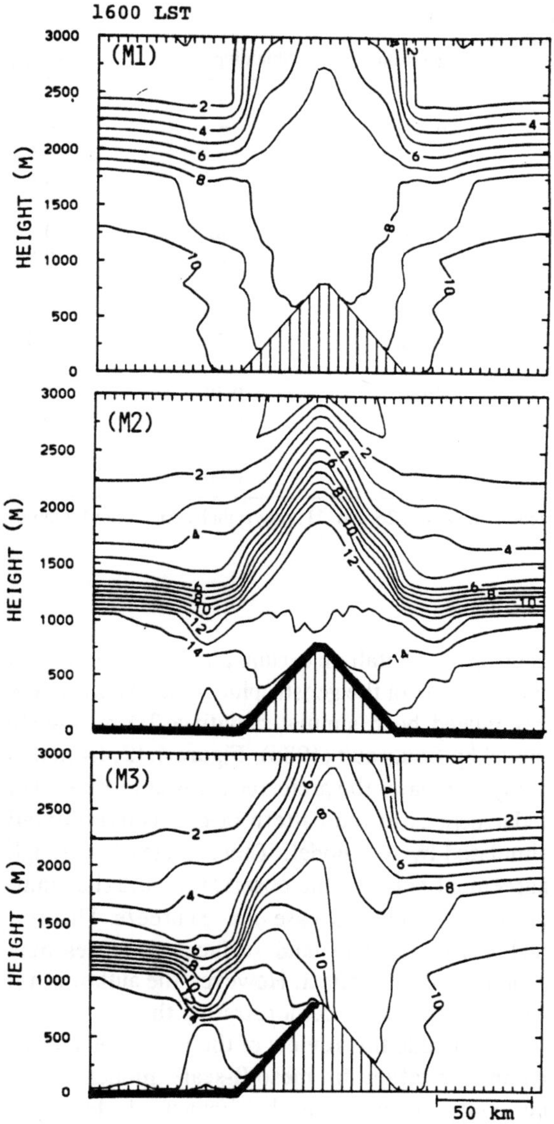

FIGURE 4.7. Vertical cross section of the simulated domain presenting the specific humidity (g kg^{-1}: cases M_1, M_2, and M_3. (Reproduced from Ookouchi et al., 1984, Monthly Weather Review, Vol. 112, No. 11, copyright by American Meteorological Society.)

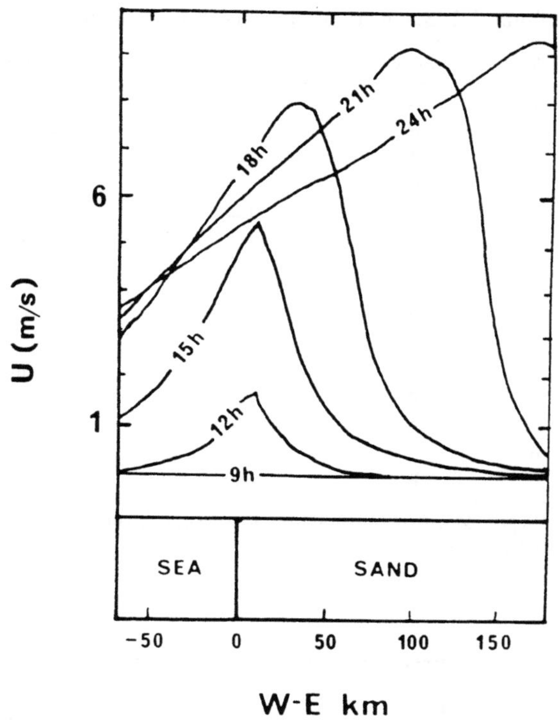

FIGURE 4.8. Horizontal profiles of horizontal wind speed (m s^{-1}) at a height of 150 m at 0900, 1200, 1500, 1800, 2100, and 2400 LST for experiment 1. (Reproduced from Mahfouf et al., 1987, Journal of Climate and Applied Meteorology, Vol. 26, No. 11, copyright by American Meteorological Society.)

larger than the wilting point. Thus a dense vegetation will screen soil moisture contrasts in so far as it is not exposed to a water stress.

We conclude that we should probably not rely too much on results that have been obtained with parameterizations describing bare soil only. Most of the time, the vegetation will damp out the spatial contrasts of soil moisture or soil type. The most easily observable actual effect will be the contrast between bare soil and vegetated areas, or between different types of vegetation. An interesting consequence of these observations has been drawn by Anthes (1984), who hypothesized that planting bands of vegetation in semiarid regions could result in increasing convective precipitation through the generation of mesoscale circulations. Using a linear model of thermal flow development, he concluded that the optimal width of these bands would be of order 50–100 km.

The idea that circulations generated by thermal contrast at the surface may induce convective rainfall has gained ample support by the numerical simulations of Pielke (1974) over Florida. Pielke showed that the 3D configuration of the coastline acted to increase the convergence in the sea breeze mechanism, and thereby focused the formation of

TABLE 4.2. Summary of the numerical simulations of Mahfouf et al. (1987).[a]

No.	Zone	Extention (km)	Surface type	Vegetation	Shielding factor	W/W_{sat}
1	A	70	Sea	–	–	–
	B	180	Sand	–	–	0.3
2	A	125	Sand	–	–	0.6
	B	125	Clay	–	–	0.6
3	A	125	Loam	–	–	0.4
	B	125	Loam	Grass	0.82	0.4
	A	70	Peat	Forest	1.00	0.8
4			Loamy			
	B	180	Sand	Forest	1.00	0.3
	A	70	Peat	Forest	0.50	0.8
5			Loamy			
	B	180	Sand	Forest	0.50	0.3
	A	70	Peat	–	0.00	0.8
6			Loamy			
	B	180	Sand	–	0.00	0.3

[a] Reprinted with permission from Journal of Climate and Applied Meteorology, Vol. 26, No. 11, copyright by American Meteorological Society.

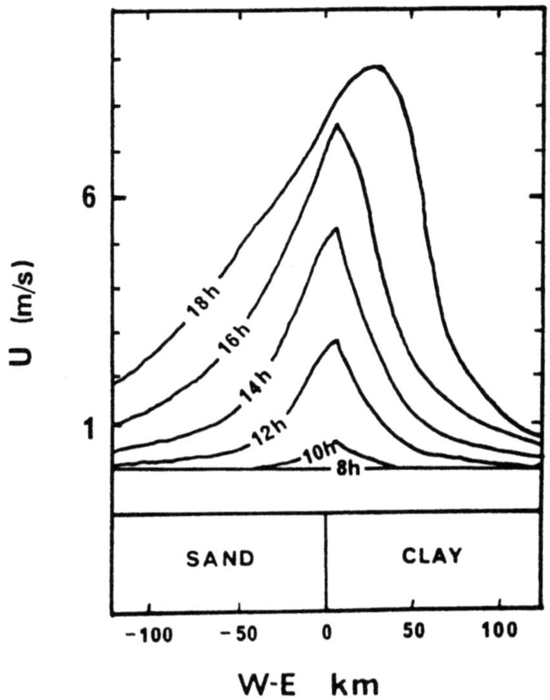

FIGURE 4.9. Horizontal profiles of horizontal wind speed (m s^{-1}) at a height of 150 m at 0800, 1000, 1200, 1400, 1600, and 1800 LST for experiment 2. (Reproduced from Mahfouf et al., 1987, Journal of Climate and Applied Meteorology, Vol. 26, No. 11, copyright by American Meteorological Society.)

convective rainfall on certain places, depending on the direction of the geostrophic wind. This process has indeed been simulated with a 3D mesoscale model by Bougeault (1986). Figure 4.12, from this study, compares the preferred areas of formation of radar echoes to the pattern of convective rainfall simulated by the model. There is clearly a rough agreement between the observation and the simulation. This is for the case of a sea breeze. Theoretically, a contrast in land surface properties may produce a similar effect. However, the author is not aware of any study demonstrating this.

To conclude this review of the impact of surface parameterization in meso-β-scale models, it is necessary to mention the impact of the rapid evaporation of rainwater intercepted by the vegetation. Indeed, in advanced schemes, allowance is made for the interception of a few millimeters of rainwater at the surface of the foliage, which is therefore directly available for evaporation, even if the surface soil layer is relatively dry. This is illustrated in Figure 4.13, from Mahfouf (1988). This study uses the Peridot model of the French Weather Service. The mesh size is 10 km and the region of interest is the one of the HAPEX-MOBILHY experiment. In this situation (June 5, 1986), a cold front went across the domain during the afternoon, leading to 5–10 mm integrated rainfall on a 12-hr period. This rainfall, intercepted by

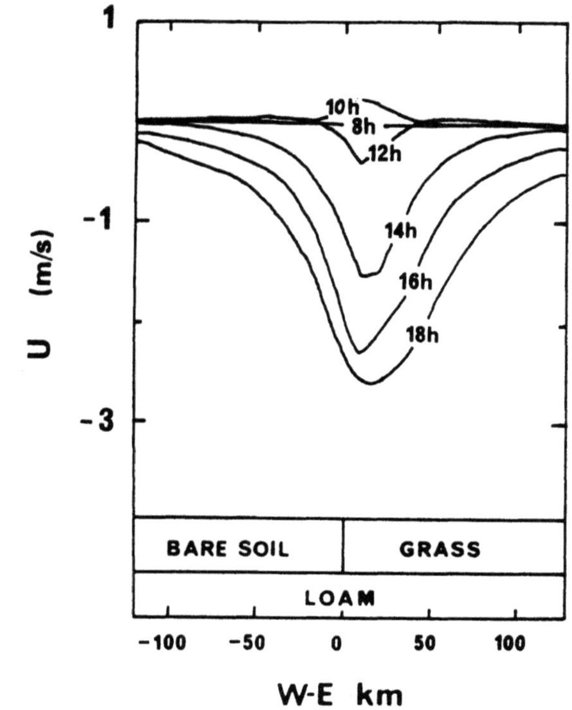

FIGURE 4.10. Horizontal profiles of horizontal wind speed (m s^{-1}) at a height of 150 m at 0800, 1000, 1200, 1400, 1600, and 1800 LST for experiment 3. (Reproduced from Mahfouf et al., 1987, Journal of Climate and Applied Meteorology, Vol. 26, No. 11, copyright by American Meteorological Society.)

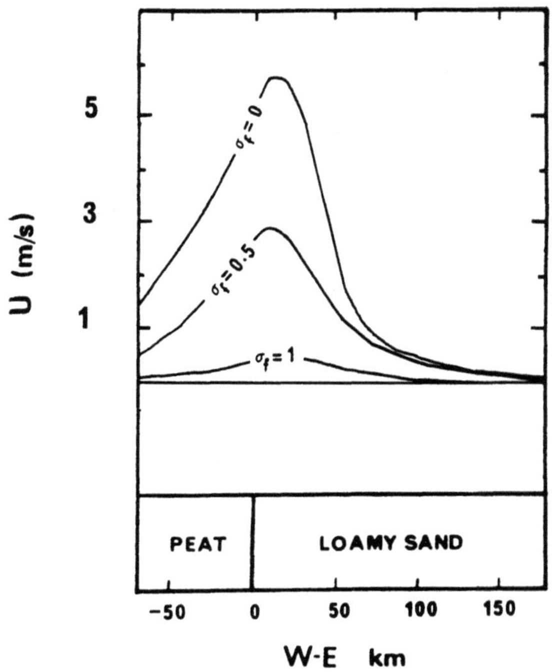

FIGURE 4.11. Horizontal profiles of horizontal wind speed (m s^{-1}) at a height of 150 m at 1800 LST for experiments 4, 5, and 6. σ_f is the shielding factor. (Reproduced from Mahfouf et al., 1987, Journal of Climate and Applied Meteorology, Vol. 26, No. 11, copyright by American Meteorological Society.)

the tall vegetation of the forested areas of Landes and Massif Central, led to an increase of the latent heat flux during the late afternoon of June 5. Two different numerical experiments are shown on Figure 4.13, one with the former surface scheme, which does not allow for rainfall interception, and the other with the new scheme (Noilhan and Planton, 1989), which does. It can be seen that the new scheme predicts a significant increase of both evaporation flux over the forested areas and rainfall rates at 1600 LST, the second being a logical consequence of the first.

All the results presented so far came from meso-β-scale models. At the meso-α-scale, the short-term influence of the surface energy budget is thought to be less important, as the induced thermal circulations are much weaker. However, Benjamin and Carlson (1986) demonstrated the importance of surface parameterization in a meso-α-scale model in an original way. They noted, on several severe weather cases over the South Central United States, that a prominent capping inversion or "lid," which focused the potential for severe thunderstorm activity in a particular area, was formed as an elevated mixed layer moved over potentially cooler air. This effect provided an important alternative to subsidence as the cause of the restraining inversion. This is illustrated in Figure 4.14, by isentropic cross sections across the Mexican Plateau and Texas. A layer of near neutral stability ($-\partial\theta/\partial p < 1$ K/50 mb), surmounting a strong inversion at 800 hPa, is predicted to lie over eastern Texas. This layer was shown to originate over the Mexican Plateau, where it was strongly heated. This stabilizing effect was shown to control the onset of deep convection over central Texas. The experiment also suggested that the lid is not uniform in space and may contain mesoscale

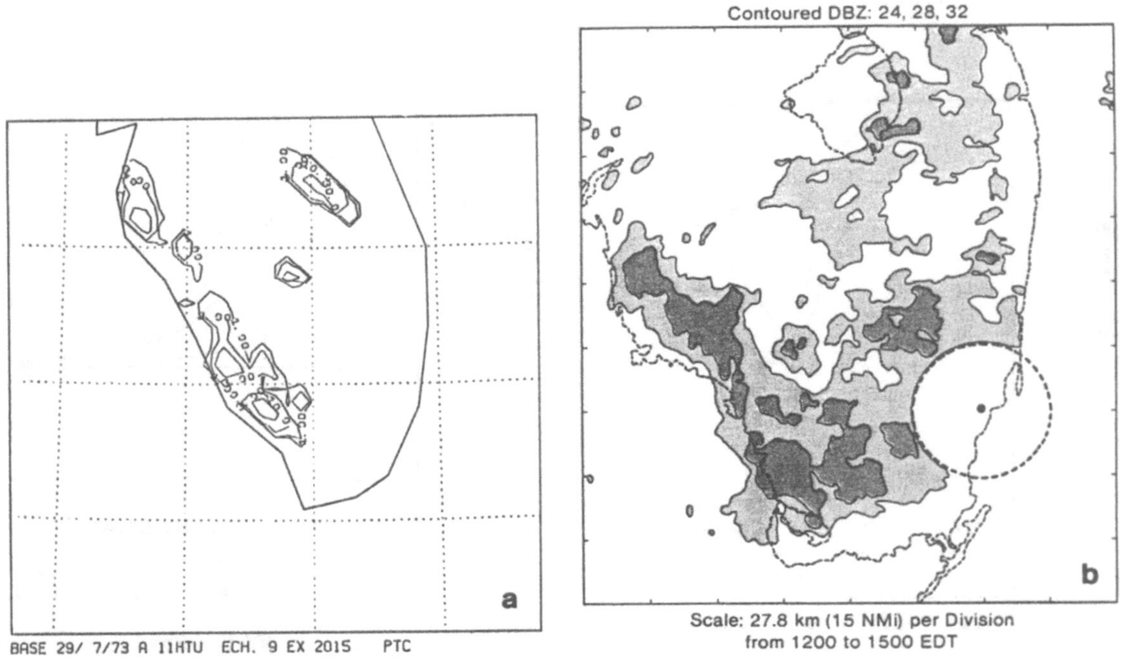

FIGURE 4.12. (a) Rainfall (mm) cumulated from 1200 to 1500 LST, as predicted by an integration of a 10 km grid size mesoscale atmospheric model. (Reproduced from Bougeault, 1986.) (b) For comparison, the convective patterns over South Florida, from composite maps of radar reflectivities restricted to synoptic situations similar to the one simulated by the model. (Reproduced from Blanchard and Lopez, 1985, Monthly Weather Review, Vol. 113, No. 8, copyright by American Meteorological Society.)

gradients of temperature and regions of lower stability, the results of the periodic nature of surface heating. This effect was further analyzed by Benjamin (1986) in the framework of a 2D model. In this study, the model behavior when an elevated mixed layer is formed over a large-scale heated plateau and advected over potentially cooler air is examined. The case for flat heated land is also treated and shows some analogy with the heated plateau. The differential heating between the dry plateau and moist lowlands is shown by Benjamin to produce a daytime intensification of the leeside surface pressure trough in the presence of flow across the plateau. When differential heating is allowed, the magnitude of the surface lee trough increases by 2–4 mb. This variation, in turn, forces a diurnal modulation of the low-level flow ahead of the lee trough, which is often a major source of moisture in outbreaks of convective storms. This effect is related by the author to a differential mixing of momentum under different stability.

To conclude this review of the significance of surface heating in meso-α-scale models, Garrett (1982) reported significant improvements in forecasting thunderstorms in a meso-α-scale model when appropriate surface parameters were used.

2.3 The Concept of Moisture Availability and Its Limitations

The material reviewed in the preceding section clearly demonstrates the necessity of having a correct partition of available energy between sensible and latent heat flux at each horizontal grid point of a model. The simplest way to do it, as proposed by Halstead et al. (1957) or Manabe (1969), relies on the concept of *moisture availability*. Two variants of this idea have been explored by Nappo (1975). Let C_H be a reasonable value of the heat exchange coefficient in the surface layer, U_a the wind speed, T_a the air temperature, q_a the air humidity, and q_s

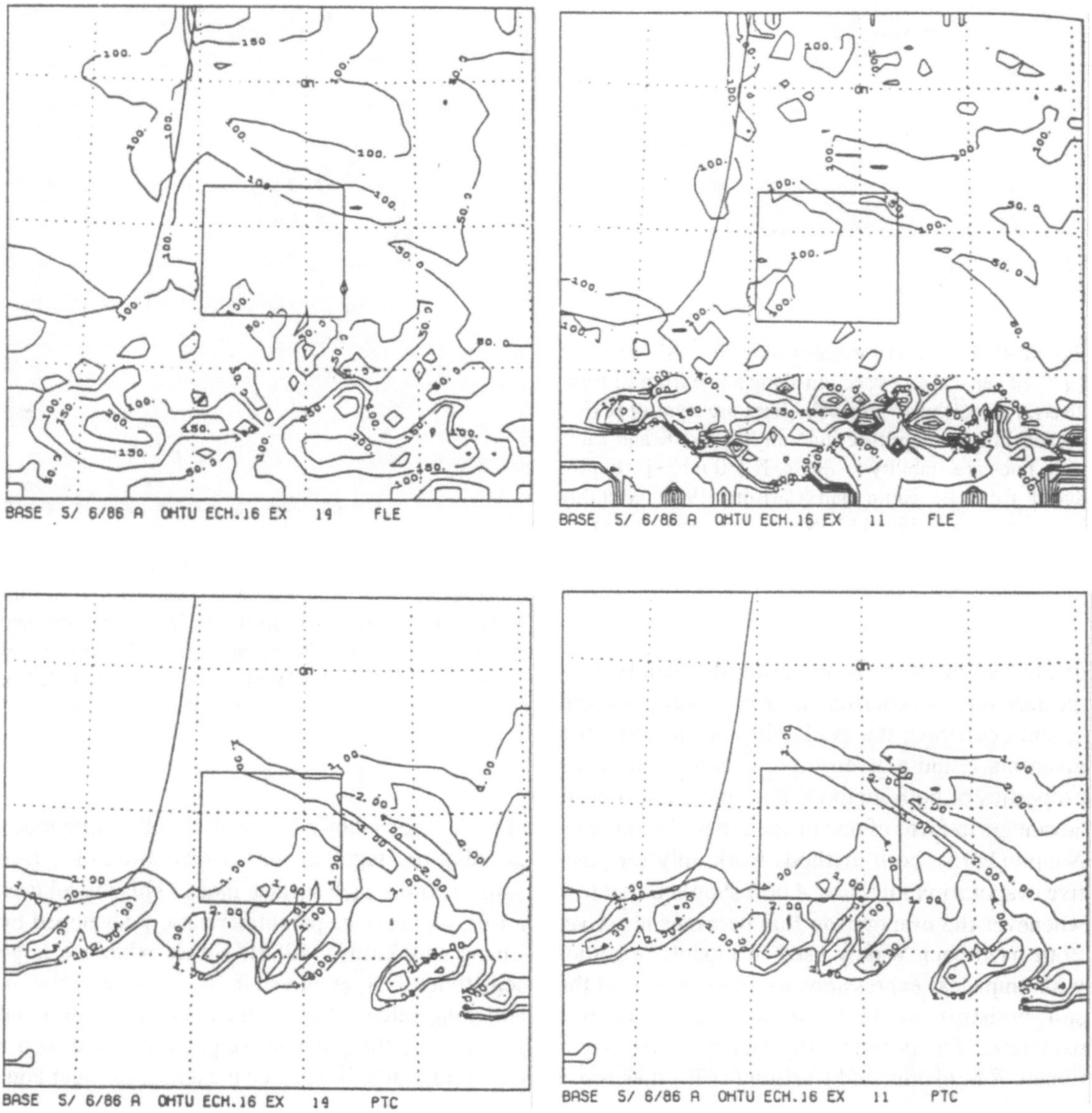

FIGURE 4.13. Simulation of the 5 June 1986 HAPEX-MOBILHY day by the French Weather Service meso-β-scale model. Comparison between the former surface scheme (left) and the Noilhan and Planton (1989) scheme (right), which includes a treatment of interception. Top: Latent heat flux (W/m²) at 1600 LST, i.e., just after a cold front crossed the domain. Bottom: Rainfall (mm) cumulated between 1500 and 1600 LST. (Reproduced from Mahfouf, 1988.)

(T_s) the humidity saturation value at the surface temperature T_s; the evaporation flux E is expressed as

$$E = A\, \rho C_H U_a\, [q_s(T_s) - q_a] \qquad (4.3a)$$

when referring implicitly to the exchange over a vegetation, or as

$$E = \rho C_H U_a\, [h q_s(T_s) - q_a] \qquad (4.3b)$$

when referring to the exchange over bare soil.

A is called the moisture availability and h the equivalent relative humidity of the surface. They are really two manifestations of the same idea that all the complicated processes that enter into the

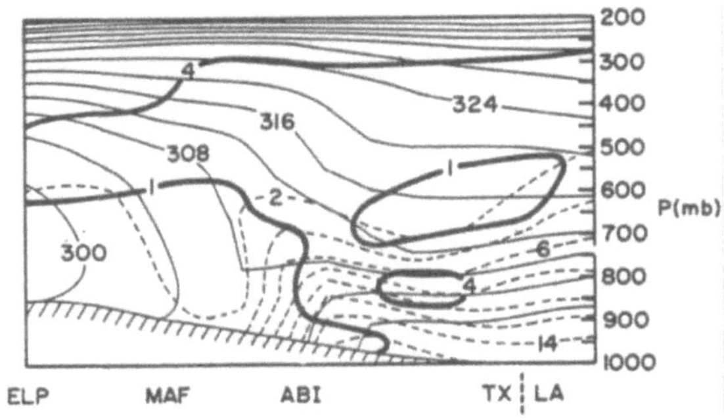

FIGURE 4.14. A 12-hr forecast valid at 00 GMT 11 April 1979, of vertical cross section along a line from El Paso, Texas, to Louisiana. Thin solid lines are isentropes (K), dashed lines are specific humidity (g kg^{-1}), and thick solid lines are stability [$-\partial\theta/\partial p$, K (50 mb)$^{-1}$]. (Reproduced from Benjamin and Carlson, 1986, Monthly Weather Review, Vol. 114, No. 2, copyright by American Meteorological Society.)

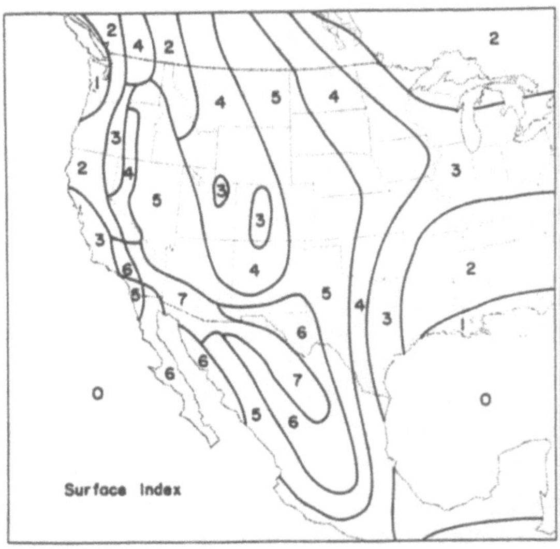

FIGURE 4.15. Surface index distribution for 3D numerical simulations. A table (Table 4.3) was used to specify surface characteristics from the surface index assigned to each grid point. (Reproduced from Benjamin and Carlson, 1986, Monthly Weather Review, Vol. 114, No. 2, copyright by American Meteorological Society.)

determination of the evaporation flux can be synthesized into one coefficient varying between 0 and 1, and describing the availability of moisture at a given place and a given time. Comparison of the properties of both methods does not show a clear advantage in favor of one or the other. As noted by Nappo (1975), both methods work only for positive evaporation rate, and A or h should be set to 1 whenever the evaporation flux becomes negative. Both have been widely used in models, together with empirical expressions to relate A or h to the soil moisture content, or with maps of this parameter. For instance, the model already mentioned of Benjamin and Carlson (1986) makes use of such a map (shown in Fig. 4.15) of surface index distribution, together with a tabulation of A (see Table 4.3) for each surface index. This model uses a PBL and surface parameterization developed by Zhang and Anthes (1982), which exhibits a large sensitivity to A, as shown in Figure 4.16. Thus, in assigning values of the surface index to individual grid points, the greatest weight was given to the moisture availability, since it exerts a stronger con-

TABLE 4.3. Surface characteristics used for each surface index type.[a]

Surface index	Type of surface	Surface albedo (%)	Soil moisture	Thermal inertia (cal cm^{-2} $s^{-0.5}$ K^{-1})	Surface emissivity at 9 μm ε_f	Surface roughness z_0 (cm)
0	Ocean	6	1.0	0.1	1.00	0.01
1	Rain forest, swamp	8	0.83	0.08	0.97	100.0
2	Woodland	10	0.75	0.07	0.95	50.0
3	Mixed grass and trees	12	0.60	0.06	0.92	30.0
4	Crops and grass (pasture land)	15	0.30	0.05	0.90	20.0
5	Semiarid (shrubs and trees)	20	0.10	0.04	0.85	10.0
6	Arid (shrubs)	25	0.05	0.03	0.80	5.0
7	Desert (sparse grass)	30	0.02	0.02	0.75	1.0

[a] Reproduced from Benjamin and Carlson (1986), Monthly Weather Review, Vol. 114, No. 2, copyright by American Meteorological Society.

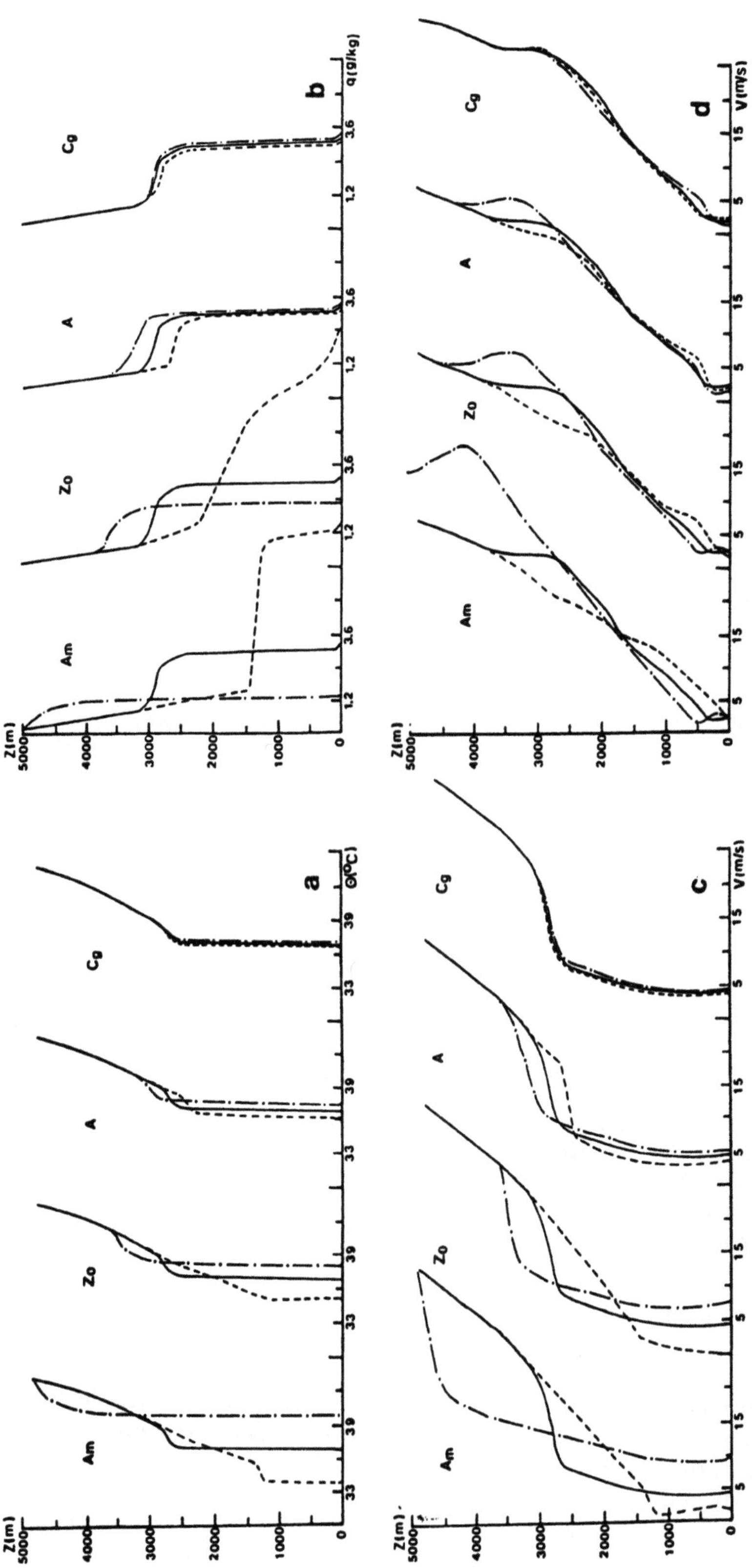

FIGURE 4.16. Vertical profiles of (a) potential temperature, (b) specific humidity, (c) wind speed at 12 hr, and (d) wind speed at 24 hr for various values of moisture availability A_m, roughness parameter Z_0, albedo A, and thermal capacity C_g. The reference values of each parameter are denoted by solid lines, high values by dashed lines, and low values by dash-dot lines. (Reproduced from Zhang and Anthes, 1982, Journal of Applied Meteorology, Vol. 21, copyright by American Meteorological Society.)

(a)

FIGURE 4.17. An example of combination of remote-sensing and inverse modeling. The maps represent the Los Angeles area on 31 May 1978. (a) Surface temperature (°C) at 1330 LST obtained directly from satellite; (b) moisture availability retrieved from inverse modeling (in %); (c), sensible heat flux (w/m² × 0.1), and (d) latent heat flux (w/m² × 0.1) at approximately 1330 LST. (Reproduced from Carlson et al., 1981, Journal of Applied Meteorology, Vol. 20, No. 1, copyright by American Meteorological Society.)

trol on PBL development than other surface characteristics.

One advantage of the concept of moisture availability is that it is easy to obtain by a combination of remote sensing and inverse modeling. Carlson et al. (1981) realized one of the first determinations of surface moisture availability by using infrared satellite temperature measurements in conjunction with a one-dimensional boundary-layer model. They used the Carlson and Boland (1978) model, which has essentially two adjustable parameters, the moisture availability A and the thermal inertia of the ground P. They noted that in applying the model on a single diurnal cycle, it was possible to simulate any desired surface temperature response by appropriately setting these parameters. Thus solutions of the model were generated by cycling through a 24-hr period with equally incremented

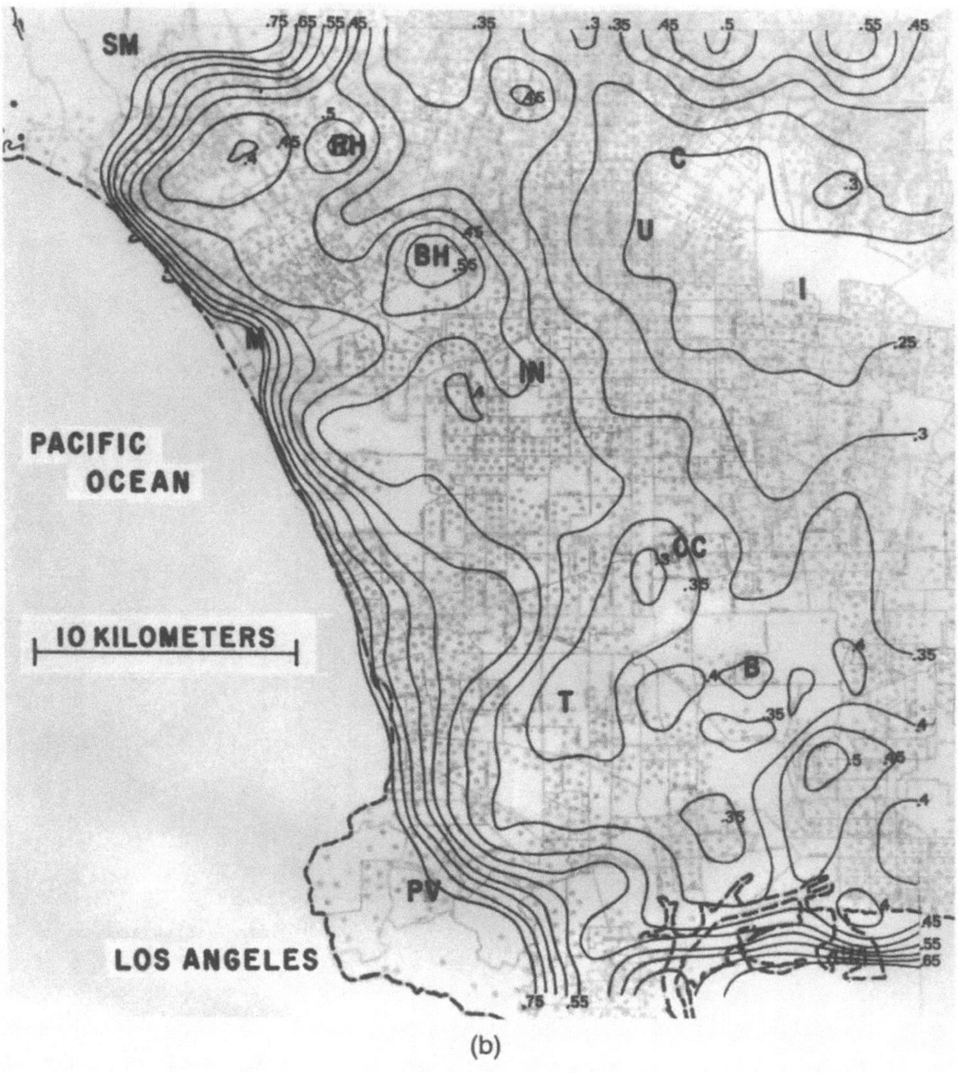

FIGURE 4.17. (Continued).

values of A and P, until 16 daily temperature cycles were completed over the maximum likely ranges for A and P. The model results for surface temperature were archived at two specific times during each of these cycles, corresponding to times of day and night satellite passes. These results were subsequently used to determine a set of regression equations with surface temperatures as predictors and the surface fluxes and parameters A and P as the predicted quantities. The last step was to transform temperature images to images of A, P, and E by substituting the measured temperatures into the regression equations for all pixels. The results are illustrated in Figure 4.17. A strong correlation can be seen between the daytime observed surface temperature and either A or E. There is little doubt that if used in a mesoscale model with the same parameterization that was used for their determination, such maps would provide a correct pattern of spatial variations of the PBL development and of the induced circulations. This approach has been refined by Wetzel et al. (1984).

The simplicity of the approach and the possibility to obtain maps of A from remote sensing are two major advantages of the moisture availability concept for the mesoscale. However, although this

(c)

FIGURE 4.17. (Continued).

concept may, in the future, have some utility for very short-term forecast or theoretical studies, it is more or less condemned by having no physical basis. It is very difficult, for instance, to relate the moisture availability to the quantity of water present in the soil without using physical models described in the next section. This renders the initialization of moisture availability nearly impossible when remotely sensed maps are not available. Also, the rapid modification of A during the course of a model integration, which may be caused by localized rainfall or modification of the surface soil moisture by evaporation, is in general impossible to capture in an interactive way with this approach. Finally, even if the time scale of each forecast is about 1 day, mesoscale models may also be used to monitor the long-term evolution of the soil water content with a spatial resolution commensurate with their grid. This is clearly possible only if the budget of water of the soil has a physical meaning in the models, and pleads in favor of parameterization schemes for evaporation that supply the correct partition of energy when the right soil moisture content is used (and in this case only) and give an accurate description of the short-term response of this partition when rainfall occurs. Such a gener-

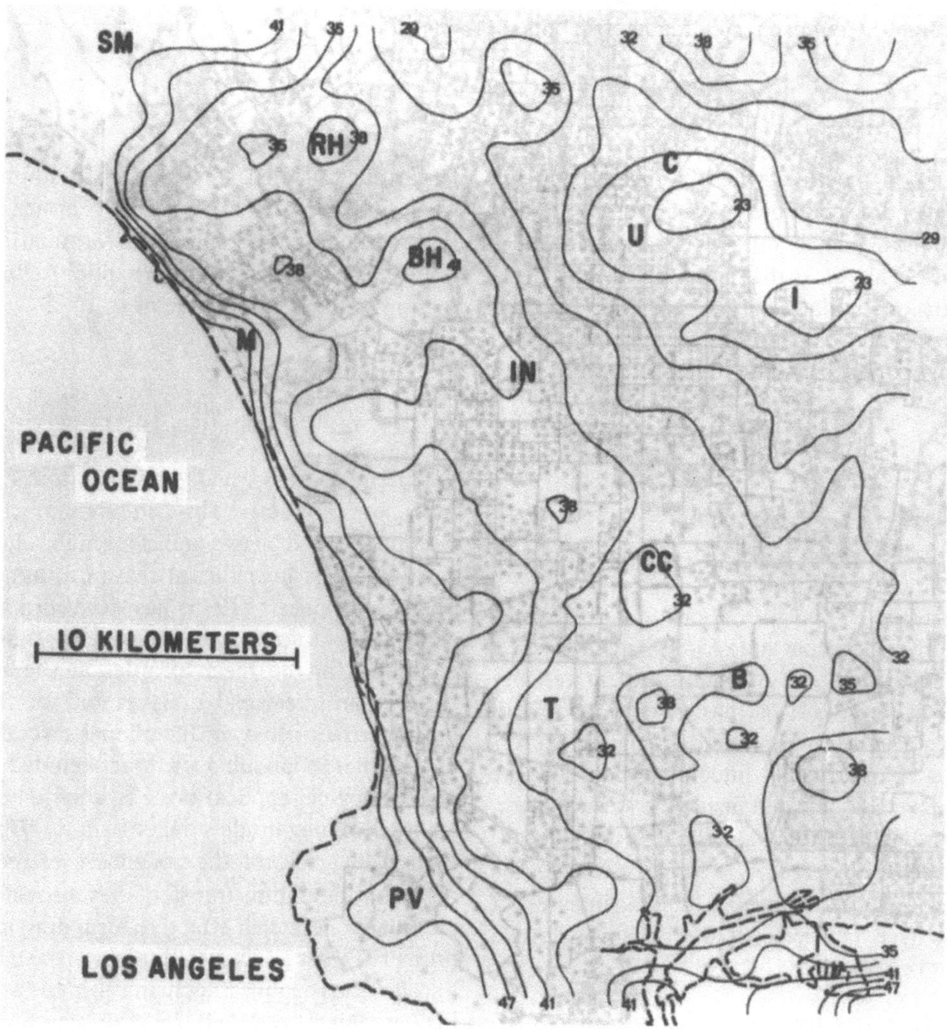

(d)

FIGURE 4.17. (Continued).

ation of schemes is now emerging and will be reviewed in the next section.

3. Advanced Land-Surface Processes Parameterizations Currently Used in Mesoscale Models

We will from now on restrict the discussion to the case of snow and ice-free ground surface. Also, it is not necessary to treat extreme climatic conditions as far as mesoscale models are concerned since those are developed mostly for mid-latitude, densely populated areas. This assumption, however, may have to be relaxed in the future.

Any parameterization used to describe the land surface processes more or less is divided into three subschemes describing, respectively, the surface boundary layer, the soil, and the vegetation layer. Roughly speaking, the role of the surface boundary layer subscheme is to predict the fluxes for given equivalent values of moisture and temperature at the roughness height, whereas the role of the soil and vegetation subschemes is to jointly provide these equivalent values. The task of the surface boundary layer subscheme is therefore essentially to determine the atmospheric drag coefficients.

3.1 The Surface Boundary Layer

It is common practice, when parameterizing the surface layer drag coefficients for mesoscale models to rely on the similarity theory of Monin and Obukhov (1954). This states that in the surface layer, the nondimensional values of the gradients of wind, temperature, and moisture are universal functions of the height divided by the Monin- Obukhov length

$$\frac{\partial u}{\partial z} = \frac{u_*}{kz}\phi_M\left(\frac{z}{L}\right), \; \frac{\partial \theta}{\partial z} = \frac{\theta_*}{kz}\phi_H\left(\frac{z}{L}\right),$$
$$\frac{\partial q}{\partial z} = \frac{q_*}{kz}\phi_v\left(\frac{z}{L}\right) \tag{4.4}$$

with $u_* = (\overline{u'w'})^{1/2}$, $\theta_* = -H/\rho u_*$, $q_* = -E/\rho u_*$, k the von Karman constant, $L = -u_*^3/[k(g/T)(H + 0608T_vE)]$, and u',w' are the turbulent velocity fluctuations. Dyer (1974) has given a review of useful expressions for ϕ_M and $\phi_{H,V}$. These expressions can be vertically integrated between the reference level Z_a, located inside the surface layer, corresponding to the height of the first model level, and *the roughness height* Z_0 to supply a value of the drag coefficient. Defining Φ_M, $\Phi_{H,V}$ as integrals of ϕ_M, $\phi_{H,V}$ (see Paulson, 1970, for an analytical expression), this reads

$$U_a = \frac{u_*}{k}\left\{\Phi_M\left(\frac{z_a}{L}\right) - \ell n\left(\frac{z_0}{L}\right)\right\}$$
$$\theta_a - \theta(z_{OH}) = \frac{\theta_*}{k}\left\{\Phi_H\left(\frac{z_a}{L}\right) - \alpha_\theta^{-1}\,\ell n\left(\frac{z_{OH}}{L}\right)\right\}$$
$$q_a - q(z_{Oq}) = \frac{q_*}{k}\left\{\Phi_V\left(\frac{z_a}{L}\right) - \alpha_q^{-1}\,\ell n\left(\frac{z_{Oq}}{L}\right)\right\} \tag{4.5}$$

where $\alpha_{\theta,q}$ are the turbulent Prandtl numbers and are experimentally determined. Thus

$$H = -\rho u_*\theta_* = \rho C_H\, U_a[\theta(z_{OH}) - \theta_a]$$
$$E = -\rho u_* q_* = \rho C_V\, U_a[q(z_{Oq}) - q_a] \tag{4.6}$$

with

$$C_{H,V} = \frac{k^2}{\left\{\Phi_M\left(\frac{z_a}{L}\right) - \ell n\left(\frac{z_0}{L}\right)\right\}\left\{\Phi_{H,V}\left(\frac{z_a}{L}\right) - \alpha_{\theta,q}^{-1}\,\ell n\left(\frac{z_{OH,q}}{L}\right)\right\}} \tag{4.7}$$

Unfortunately, this expression is still implicit in L. Therefore, it is solved for H and E either by iterative methods or by simplified formulation (e.g., Louis, 1979), allowing for an explicit stability correction to the neutral value of $C_{H,V}$

$$(C_{H,V})_n = \frac{k^2}{\ell n\left(\frac{z_a}{z_0}\right)\ell n\left(\frac{z_a}{z_{OH,q}}\right)} \tag{4.8}$$

Furthermore, it is usually assumed that $Z_0 = Z_{0H} = Z_{0q}$ for simplicity. This "common practice" has been questioned on two points for high values of Z_0. First, there is no physical reason to assume this identity. Brutsaert (1975) has derived a theoretical expression for $Z_{0,H}$ based on the assumption of continuity in the mean profiles at the interface between an interfacial sublayer and the fully turbulent surface sublayer. Examining several models of the interfacial sublayer, he concluded that for meteorological applications $Z_{0,H}$ should be at least an order of magnitude smaller than Z_0. This is due to the dual role of the roughness elements. As regards momentum transfer, they provide obstacles that generate an effective form drag involving pressure forces. However, as regards the transfer of a scalar, the roughness elements provide a "sheltering" or "wind break" effect, inhibiting transfers at the surface.

In the case of the exchange between the surface layer and a surface covered by vegetation, the situation is still more complicated. Brutsaert (1979) developed an approach to formulate bulk mass or heat transfer coefficients for the case of vegetation with a constant value of the scalar of interest (heat or water) at the foliage surface. The result can be expressed in terms of an equivalent roughness height $Z_{0,c}$ for the scalar quantity C. Examination of several experimental cases shows that $\ln(Z_0/Z_{0,C})$ is of order 2 to 3.

The second weakness of the standard approach to computing the drag coefficient is the failure of the Monin–Obukhov similarity theory at heights corresponding to $Z_a/Z_0 < 50$ to 100, as observed by Thom et al. (1975), Garratt (1978), Raupach (1979), and Cellier (1986). This traditional limitation to the theory takes an important practical significance for Z_0 of order 1 m, as can be observed over a forest or a region of mixed trees and shrubs, since the level of reference Z_a used for the computation of the fluxes is therefore below the zone of validity of the theory.

Garratt (1980) has proposed a modification of the theory to accommodate this effect. Defining a new parameter Z_*, as the height of the transition layer below which the usual similitude fails, his proposal reads

$$\phi^*_{H,M,V}(z) = \phi_{M,H,V}\left(\frac{z}{L}\right) f\left(\frac{z}{z_*}\right),$$

$$\text{with } f\left(\frac{z}{z_*}\right) = \exp\left[-\alpha_1\left(1-\frac{z}{z_*}\right)\right], \quad \alpha_1 \sim 0.5 \tag{4.9}$$

According to the author's knowledge, no formulation has yet been proposed to take these findings into account in the parameterization of the drag coefficients for mesoscale models. One may hope that the two errors partially compensate for each other. More presumably, the precise adjustment of the formulation is precluded by the quite obscure nature of the definition of Z_0 in a mesoscale model, since this quantity usually accounts, in addition to the micrometeorological roughness, for the subgrid scale variability of the terrain height and surface nature.

3.2 The Soil

As the second component of a land-surface parameterization, the soil subscheme has the task of predicting the values of temperature and humidity at the surface of hypothetical, vegetation-free layer of soil, as well as the evolution of its moisture content over a depth of about 1 m.

The basic equations that govern the evolution of temperature and moisture inside a homogeneous soil can be found, for instance, in McCumber and Pielke (1981). The dependent variables are the temperature T and the volumetric moisture content W. For the temperature, a diffusion equation can be written:

$$C\frac{\partial T}{\partial t} = \frac{\partial}{\partial z}\lambda\frac{\partial T}{\partial z} \tag{4.10}$$

where the volumetric heat capacity C is defined as

$$C = (1 - W_{sat})C_1 + WC_{H_2O} \tag{4.11}$$

The saturation value of W, W_{sat} and the dry volumetric heat capacity of the soil C_1 both depend mainly on the texture of the soil (size of the particles). The thermal conductivity λ can vary over at least one order of magnitude depending on the soil moisture and the soil texture. However, Al Nakshabandi and Kohnke (1965) demonstrated that the use of the moisture potential ψ allows a definition of λ that is nearly independent of the soil texture:

$$\lambda = \max[\exp(-\log_{10}|\psi| + 2.7)419, 0.172] \tag{4.12}$$

The moisture potential is a measure of the work that is necessary to extract water from the soil, against gravity and capillary forces. It is usually expressed as a negative number of meters, and is equal to the depth from which water could be elevated to the surface for an equivalent work against the gravity alone. Therefore, the higher the magnitude of ψ, the more difficult it is to extract water from the soil. ψ is a unifying concept that allows the available moisture from different types of soils to be compared. Values of ψ vary from a few centimeters (very wet soils) to a few hundred meters (very dry soils).

Measurements indicate that there exists a simple relation between ψ and W for each type of soil, which reads (Clapp and Hornberger, 1978)

$$\psi = \psi_{sat}(W_{sat}/W)^b \tag{4.13}$$

ψ_{sat}, W_{sat} are the saturation values of ψ and W, and b is a constant that varies between 4 and 11. C_1, ψ_{sat}, W_{sat}, and b have been tabulated, for several textural classes, by Clapp and Hornberger (1978) and later generalized by Cosby et al. (1984).

The conservation equation for the soil moisture content can be written equivalently in terms of ψ or W. The formulation with W reads

$$\frac{\partial W}{\partial t} = \frac{\partial}{\partial z}K_\eta\frac{\partial}{\partial z}(\psi + z) \tag{4.14}$$

where K_η is the hydraulic conductivity. By defining the diffusivity

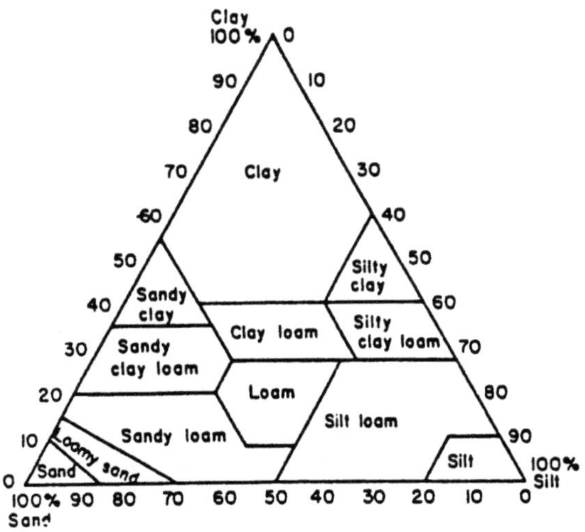

FIGURE 4.18. The USDA textural classification of soil. (Reproduced from Cosby et al., 1984, Water Resources Research, Vol. 20, 1984, copyright by the American Geophysical Union.)

$$D_\eta = K_\eta \frac{\partial \psi}{\partial W} \tag{4.15}$$

Equation (14) reverts to

$$\frac{\partial W}{\partial t} = \frac{\partial}{\partial z} D_\eta \frac{\partial W}{\partial z} + \frac{\partial K_\eta}{\partial z} \tag{4.16}$$

K_η and D_η are also related to W through simple expressions:

$$K_\eta = K_{\eta s}(W/W_{sat})^{2b+3}$$

$$D_\eta = -(b\, K_{\eta s}\, \psi_{sat}/W)\,(W/W_{sat})^{b+3} \tag{4.17}$$

with $K_{\eta s}$ tabulated as a function of the type of soil by Clapp and Hornberger (1978). This tabulation is based on the USDA textural classification and Cosby et al. (1984) demonstrated that 11 classes of this classification (silt is considered separately) allow for a satisfactory statistical description of the thermal and hydraulic properties of a large number of soil samples. Note, however, that for all these parameters, the tabulated value is the mean for a large number of samples of each type of soil and that the variability inside each soil textural class may be as large as the mean difference from one class to another. A schematic diagram of this textural classification is shown in Figure 4.18.

With the equations given above, it is easy to construct a detailed 1D model of a soil layer and to study the evolution of soil temperature and moisture under various atmospheric forcings. This has been done, e.g., by McCumber (1980), Mahfouf (1986), and Noilhan (1987). To give one example, Figure 4.19 considers the diurnal evolution of the temperature of a sandy clay soil with uniform initial moisture $W = 0.27$ (Mahfouf, 1986). The model computes a realistic diurnal temperature wave, that reaches a depth of approximately 40 cm. The sensitivity of this kind of result to the input parameters has been widely investigated. For a diurnal time scale, the most sensitive

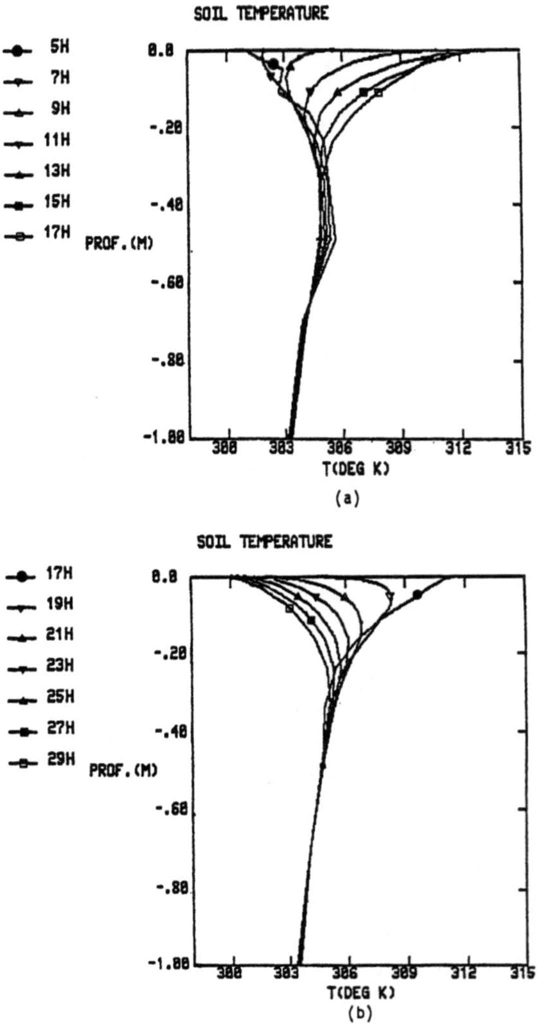

FIGURE 4.19. The diurnal variation of the soil temperature profile. Case of a sandy clay soil with constant initial moisture. (Reproduced from Mahfouf, 1986.)

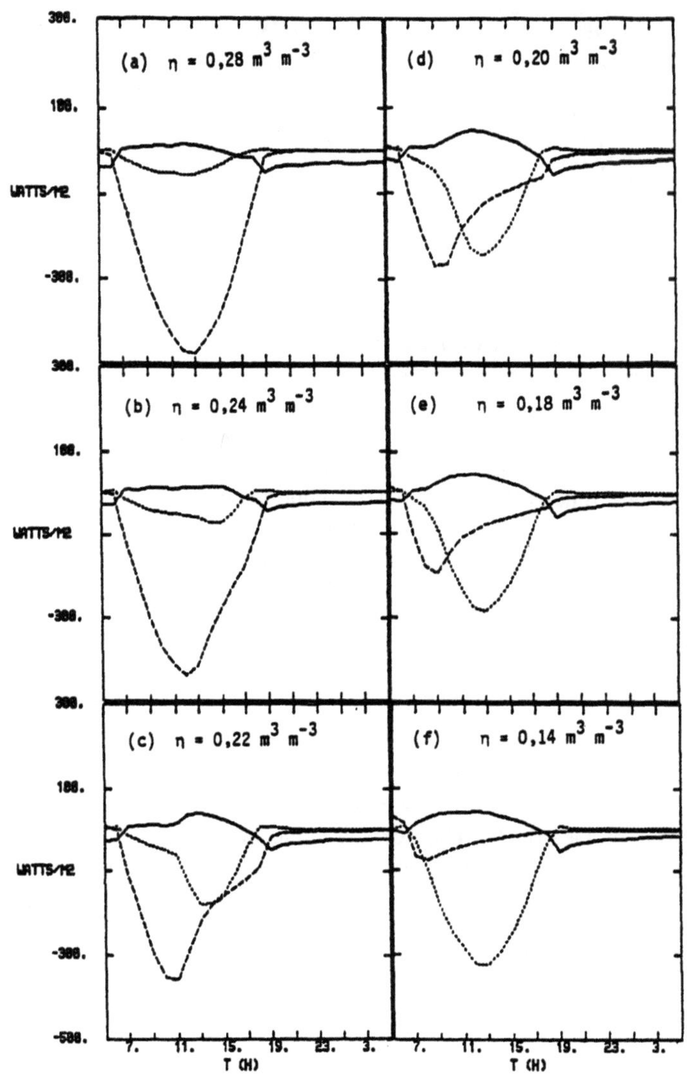

FIGURE 4.20. The influence of the initial soil moisture on the diurnal evolution of the surface energy budget. Solid line: ground heat flux; dotted line: sensible heat flux; dashed line: latent heat flux. (Reproduced from Mahfouf, 1986.)

parameter is found to be the initial soil moisture *W*. As an illustration, the different components of the heat budget at the surface are shown for several values of *W* in Figure 4.20 (from Mahfouf, 1986). A drastic change can be seen in going from nearly saturated to very dry soil. This has been confirmed by McCumber and Pielke (1981) and Zhang and Anthes (1982). Note, however, that this is true only for the case of bare soil.

The system of Equations (4.10 and 4.17) has been incorporated with success in 3D mesoscale models by McCumber (1980) and Mahfouf (1986). However, a significant amount of work has been devoted to the elaboration of a simpler approach based on two-layers models. This work is motivated by the need to increase the computational efficiency of soil parameterizations in 3D models and by the belief that simpler bulk models may be easier to calibrate for use at the scale of a grid-square that encompasses a large number of different types of soil.

Following these thoughts, Bhumralkar (1975) and Blackadar (1976) showed that it is possible to achieve qualitatively good behavior if the soil is divided into only two layers: one surface layer of very small depth (typically 10 cm) that follows the diurnal cycle, and one deep layer (typically 1 m) that does not. This so called force-restore method

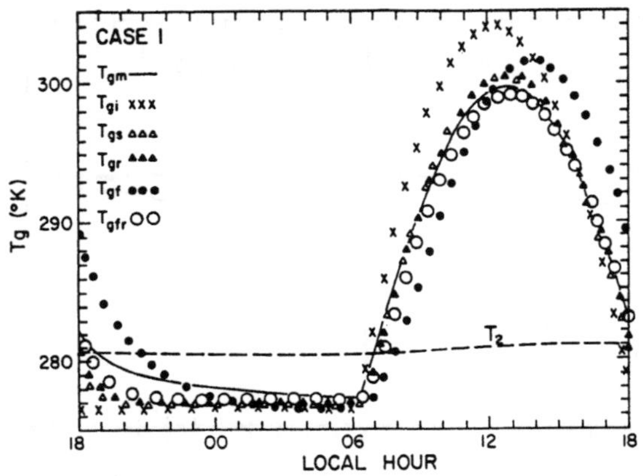

FIGURE 4.21. Ground surface temperatures calculated for the second 24 hr of a 2-day period, as a function of the local time of the day. Different simplified methods are tested. T_{gm} is the reference computation. T_{gfr} is the "Force-restore" method explained in the text. (Reproduced from Deardorff, 1978, Journal of Geophysical Research, Vol. 83, 1978, copyright by the American Geophysical Union.)

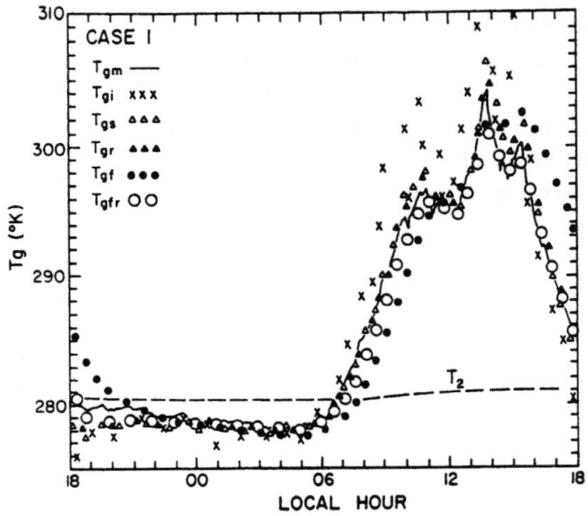

FIGURE 4.22. As in Figure 4.21, but for partially random atmospheric forcing.

has gained some popularity since Deardorff (1977) demonstrated its superiority over several other simplified formulations. We adopt here the notations of Deardorff. The surface value of the soil temperature T_g is predicted by the simple equation

$$\frac{\partial T_g}{\partial t} = -\, 2\pi^{1/2}\, [\rho_G C_G(\lambda\tau)^{1/2}]^{-1}\, H_A - 2\pi\tau^{-1}\,(T_g - T_2) \qquad (4.18)$$

with notations as before, the diurnal time constant, $\tau = 86{,}400$ sec and H_A is the total flux toward the atmosphere. The deep layer temperature T_2 can be predicted by

$$\frac{\partial T_2}{\partial t} = -\, H_A/\rho_G C_G(\lambda\tau_2)^{1/2}, \quad \tau_2 = 365\tau_1 \qquad (4.19)$$

The first term in Equation (4.18) expresses the atmospheric forcing, whereas the last one is seen to restore T_g exponentially toward the deep soil temperature, that changes with an annual time scale only. Constants in Equation (4.18) have been chosen to yield the exact solution for a sinusoidally varying forcing of a 1-day period.

Deardorff has tested Equations (4.18 and 4.19) on several types of soil and several types of atmospheric forcings. Thermal properties were assumed independent of soil moisture content. A sample of his results is shown in Figure 4.21. This corresponds to a constant atmospheric forcing, except for solar radiation. It is apparent in Figure 4.21 that most of the simplified methods are not able to reproduce accurately the "true" solution [detailed vertical model based on Equation (10)]. The force-restore method succeeds in reproducing the phase and amplitude of the diurnal forcing. In another test, both a random atmospheric forcing and random solar radiation perturbed by clouds were imposed. This is shown in Figure 4.22. Again, the result of the force-restore method is seen to be acceptable.

The success of the simplified formulation based on the force-restore method for the prediction of ground surface temperature has inspired a similar effort for the prediction of the ground-surface moisture content. Experimental studies on bare soil and numerical simulations based on detailed models show indeed a large diurnal variation of the ground-surface moisture content, which is rapidly attenuated with depth. Moreover, these variations determine a similar variation of the soil albedo. Deardorff (1977) proposed a parameterization of this process based on an analogy with Equation (4.18):

$$\frac{\partial W_g}{\partial t} = -\, C_1(E_g - P)/\rho_W d_1' - C_2(W_g - W_2)/\tau, \qquad 0 \le W_g \le W_{sat} \qquad (4.20)$$

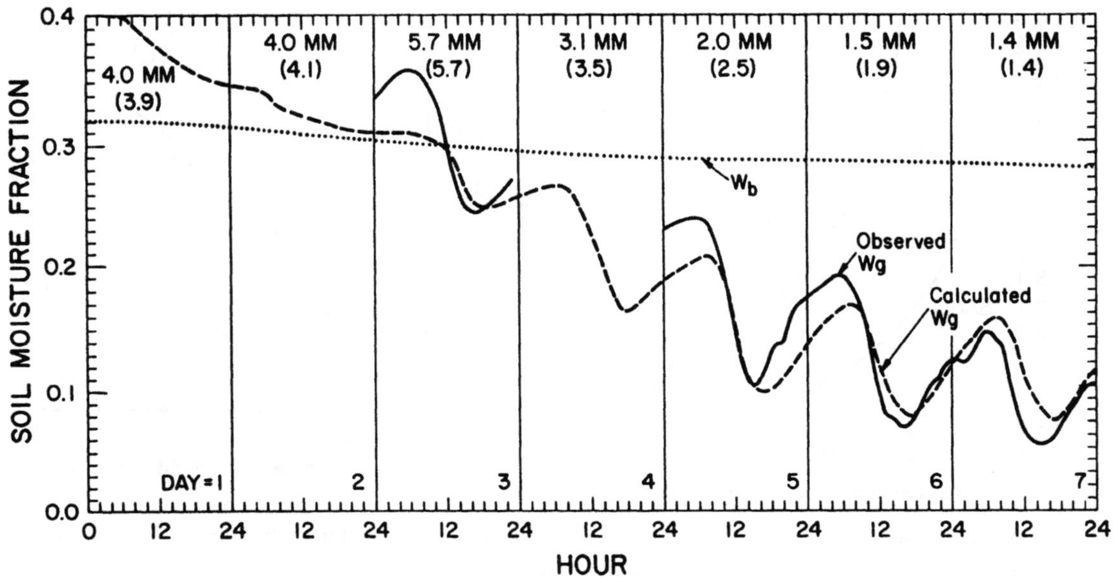

FIGURE 4.23. Predicted values of W_g (ground-surface moisture fraction, dashed curve) and of W_2 (bulk-soil moisture fraction, dotted curve), and smoothed observed values of W_g (solid curve) for 7 days following irrigation. Calculated daily values of evaporation are given in millimeters; observed values are listed in parentheses. (Reproduced from Deardorff, 1977, Journal of Applied Meteorology, Vol. 16, No. 11, copyright by American Meteorological Society.)

W_g is a representative value of W for a thin slab of soil, C_1, C_2 are dimensionless constants, E_g the evaporation rate from the soil surface, P the precipitation rate, W_2 a representative value of W for the upper half-meter of soil, and d_1' a typical depth to which the diurnal cycle extends. Deardorff suggested $d_1' = 10$ cm. W_{sat} is the saturation moisture fraction. C_1 will depend, in general, on the soil properties and on the choice of depth d_1'. The first term on the right of Equation (4.20) provides drying of the soil surface when significant evaporation occurs or wetting of the soil surface when precipitation occurs. The last term tends to restore W_g exponentially toward the deep soil value in the absence of precipitation or evaporation. As a tentative estimate of C_1 and C_2, derived from a single data set, Deardorff proposed

$$C_2 = 0.9$$

$$C_1 = \begin{cases} 0.5 & \text{if } W_g/W_{sat} > 0.75 \\ 14 - 22.5\,(W_g/W_{sat} - 0.15) & \text{if } 0.15 < W_g/W_{sat} < 0.75 \\ 14 & \text{if } W_g/W_{sat} < 0.15 \end{cases} \qquad (4.21)$$

The evolution of the deep soil layer moisture W_2 can be computed through a simple conservation equation

$$\frac{\partial W_2}{\partial t} = -\,(E_g - P)\,/\,\rho_W\, d_2' \qquad (4.22)$$

with $d_2' = 50$ cm. The scheme has to be completed by an assumption to compute the evaporation. The simplest possible form is

$$E_g = \begin{cases} E^{pot}\,(W_g/W_{max}) & \text{if } W_g \le W_{max} \\ E^{pot} & \text{if } W_g > W_{max} \end{cases} \qquad (4.23)$$

with

$$W_{max} = 0.75\, W_{sat}$$

The scheme was tested by Deardorff on a single data set with the satisfactory result shown in Figure 4.23. The parameterization is able to simulate both the diurnal cycle of W_g and the continued drying of soil surface for several days. The partial recovery of soil-surface moisture content at night is due to the moisture transfer from the lower layers, as parameterized by the "restore" term in

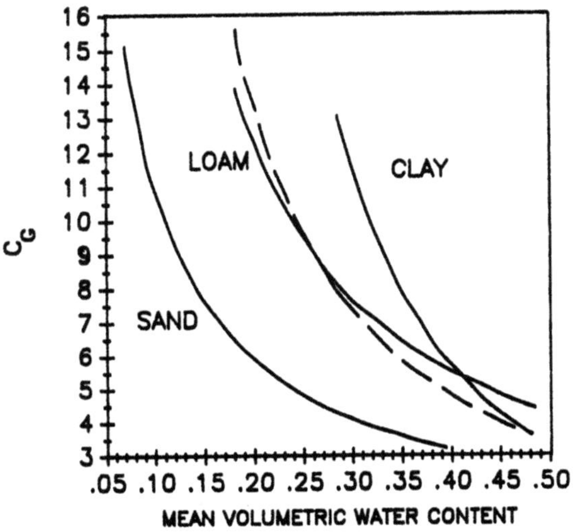

FIGURE 4.24. Variations of C_g (in 10^6 K m^2 J^{-1}) versus W_2 given by Equation (26) for sand, silt loam, and clay. The dashed line corresponds to a calculation from thermal properties of McCumber and Pielke (1981) and hydraulic properties of Clapp and Hornberger (1978), in the case of silt loam. (Reproduced from Noilhan and Planton, 1989, Monthly Weather Review, Vol. 117, No. 3, copyright by American Meteorological Society.)

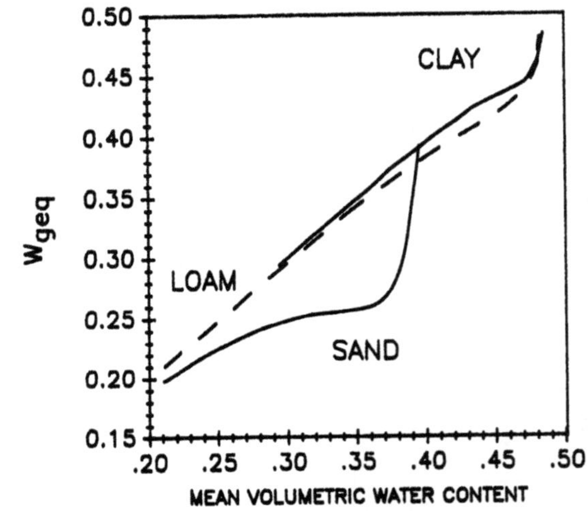

FIGURE 4.25. Variation of W_{geq} (in m^3 m^{-3}) versus W_2 given by Equation (29) for sand, silt loam, and clay; the dashed line corresponds to the exact balance between capillarity and gravity, using the hydraulic properties given by Clapp and Hornberger (1978), in the case of silt loam. (Reproduced from Noilhan and Planton, 1989, Monthly Weather Review, Vol. 117, No. 3, copyright by American Meteorological Society.)

Equation (4.20). Hence, the parameterization appears to describe the processes on bare soil quite realistically.

The force-restore method appears therefore as a viable and economic substitute for the detailed approach. The major weakness in the formulation previously discussed is that no account is taken of the different types of soil and of the influence of the soil water content on the heat capacity, diffusivity, and conductivity properties. These parameters can, however, vary by several orders of magnitude for different types of soil and for different soil moisture contents as shown, for example, by Mahrt and Pan (1984). It has been recognized that this variability in soil properties should be taken into account in the simplified formulations. Significant work in this direction has been accomplished by Mahrt and Pan (1984) and Noilhan et al. (1987). We will describe the second approach in some detail.

In the bare soil version of Noilhan and Planton's scheme the equation for the ground surface temperature T_g retains the simple form of the force-restore method

$$\frac{\partial T_g}{\partial t} = -C_G H_A - 2\pi(T_g - T_2)/\tau \quad (4.24)$$

with

$$C_g = 2\pi^{1/2}/(\lambda C_g \tau)^{1/2} \quad (4.25)$$

where λ is the thermal diffusivity and C_g the soil heat capacity. Both λ and C_g depend on the soil type and the volumetric water content. To simplify this dependence, a parametric formula, depending only on W_2 and W_{sat}, has been worked out:

$$C_G = C_{Gsat}\left(\frac{W_{sat}}{W_2}\right)^{b/2\log(10)} \quad (4.26)$$

where W_{sat} and b are the same as in the classification of Clapp and Hornberger (1978) and C_{Gsat} has been evaluated as a function of the soil type only. Three examples of this adjustment are given in Figure 4.24 for sand, loam, and clay. For a given value of W_2, the finest texture exhibits higher values of C_G because of the reduction of its thermal capacity; on the other hand, for a given texture, C_G decreases when W_2 increases because of the increase in thermal capacity, which will limit the diurnal variations of T_g.

The evolution of the ground surface water content is treated in Noilhan and Planton's scheme in a similar way, with some differences with respect to the Deardorff (1977) formulation:

$$\frac{\partial W_g}{\partial t} = C_1(P - E_g)/\rho_W d_1 - C_2(W_g - W_{geq})\tau, \quad 0 \le W_g \le W_{sat} \qquad (4.27)$$

$$\frac{\partial W_2}{\partial t} = (P - E_g)/\rho_W d_2, \quad 0 \le W_2 \le W_{sat} \qquad (4.28)$$

It can be seen that W_g is now restored toward W_{geq}, the equilibrium value of W_g, by the second term of Equation (4.27). This equilibrium value is not necessarily equal to W_2 (the bulk value of W) because the equilibrium between gravity and capillary forces does not lead to a constant profile of W. The two coefficients C_1 and C_2 and W_{geq} have been calibrated for different soil texture and moisture content as discussed below.

The values of W_{geq} versus W_2 have been calculated using Clapp and Hornberger's (1978) specification for hydraulic properties and the condition of balance between capillary and gravity forces in the unsaturated case. When saturation occurs, ψ was limited to its maximum value ψ_{sat}. Then a polynomial has been adjusted:

$$W_{geq}/W_{sat} = x - ax^p(1 - x^{8p}), \; x = W_2/W_{sat} \qquad (4.29)$$

The parameters a and p have been calculated and tabulated for the different types of soil. Figure 4.25 shows the result of this adjustment for sand, silt loam, and clay, along with the exact solution in the case of silt loam. For the calibration of C_1 and C_2, the detailed model of Noilhan (1987) was used as described by Noilhan and Planton (1989). The results of this model on a large number of cases suggest the following expression:

$$C_1 = C_{1sat}\,(W_{sat}/W_g)^{b/2+1} \qquad (4.30)$$

$$C_2 = C_{2ref}[W_2/(W_{sat} - W_2 + W_1)] \qquad (4.31)$$

with W_1 a small, numerical value that limits C_2 at saturation. C_{1sat} and C_{2ref} have been tabulated for the different soils. Some values of C_1 determined in this way are shown in Figure 4.26. C_1 is generally increasing when the soil is drying since the hydraulic diffusivity is then reduced, and the response of W_g to the atmospheric forcing is amplified. Similarly, the values of C_2 as a function of W_2 are displayed in Figure 4.27. C_2 is decreasing when the soil is drying, as the diffusivity is reduced.

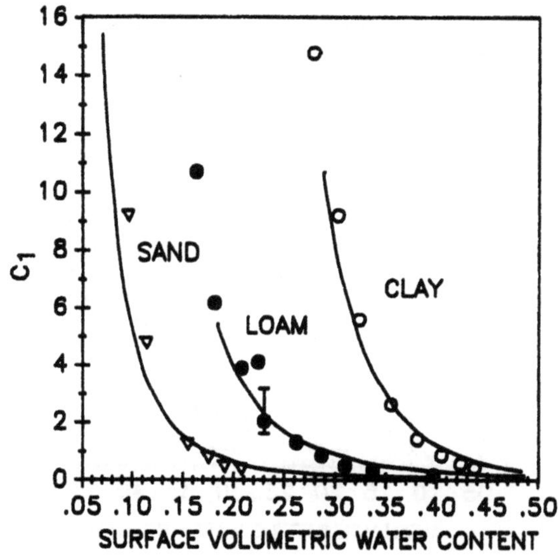

FIGURE 4.26. Variations of the dimensionless coefficient C_1 versus W_g given by Equation 4.30) for three different soil textures. The symbols correspond to results given by a reference multilayer soil model. The error bar corresponds to different initializations of the multilayer model and is only indicative. (Reproduced from Noilhan and Planton, 1989, Monthly Weather Review, Vol. 117, No. 3, copyright by American Meteorological Society.)

3.3 The Vegetation

As the third and final component of land-surface parameterizations, the vegetation subscheme has the task of predicting the foliage temperature, and the specific moisture in the immediate vicinity of the leaves. This requires consideration of very simplified biological mechanisms.

The first parameterization of the vegetation based on physical concepts was proposed by Deardorff (1978). This parameterization, although intended for climate models, was later used in mesoscale models (e.g., McCumber, 1980; Garrett, 1982; Mahfouf, 1986) and it is therefore useful to discuss

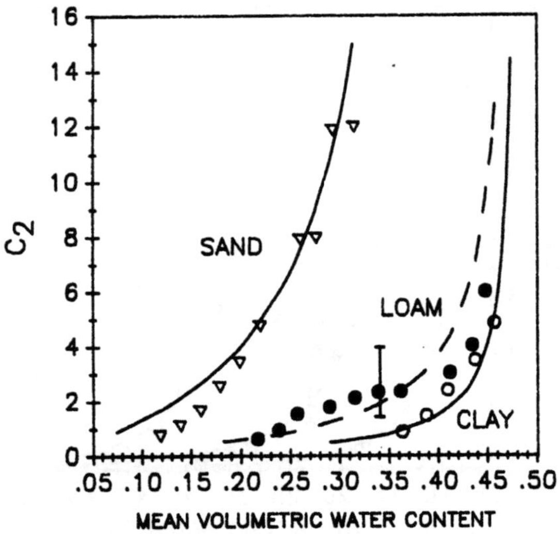

FIGURE 4.27. As in Figure 4.26, but for C_2 versus W_2 given by Equation (4.31).

it in some detail. Several of the concepts used by Deardorff have been widely borrowed by more recent works, and need to be defined.

A single layer of vegetation that has negligible heat capacity is assumed to be present. Its horizontal density is characterized by the single quantity σ_f, which is an area-averaged shielding factor associated with the degree to which the foliage prevents short wave radiation from reaching the ground and can vary between 0 and 1. $\sigma_f = 0$ corresponds to the bare soil case and $\sigma_f = 1$ corresponds to complete radiative blocking.

Three different determinations of the temperature and the humidity are considered. T_a and q_a represent the temperature and humidity of the air above the canopy (at level Z_a) and are assumed to be supplied by the atmospheric model. T_g and q_g are ground surface properties that are determined, respectively, by the energy budget at ground level (T_g) and the soil parameterization (q_g). Finally T_f and q_f are the representative temperature and moisture of the canopy itself. T_f is determined from an equilibrium energy budget of the canopy, and q_f is the equivalent moisture in the immediate vicinity of the leaves, as defined below. We note therefore that two different energy budgets are solved, which makes it necessary to specify separate values for the ground and foliage emissivities and albedo (ε_g, ε_f, α_g, α_f). Additionally, the properties of the air inside the canopy T_{af}, q_{af} are defined as an empirical linear combination of the three basic determinations. Similarly, the air velocity inside the canopy, U_{af} is taken as different from, but a function of, the air velocity outside the canopy U_a.

The exchange of heat and moisture between the canopy and the atmosphere is assumed to be governed by U_{af}, $T_f - T_{af}$, $q_{sat}(T_f) - q_{af}$. For a single leaf, one assumes

$$H_{leaf} = \rho_a C_p C_f U_{af}(T_f - T_{af}) \qquad (4.32)$$

where C_f is a dimensionless heat transfer coefficient, which is given the expression

$$C_f = 0.01[1 + 0.3 \text{ ms}^{-1}/U_{af}]$$

The second term is present to allow for a certain amount of heat transfer even if the air speed is zero (free convection).

For the entire canopy, this expression must be weighted by the vertical density of the foliage. This leads to a definition of the leaf area index *LAI* as the total one-sided leaf area over a unit ground area. Then the sensible heat flux from the canopy to the air is approximately

$$H_f = 1.1\, LAI\, \rho_a\, C_p\, C_f\, U_{af}\, (T_f - T_{af}) \qquad (4.33)$$

(the factor 1.1 accounts for that part of the biomass that exchanges heat without transpiring).

The evaporation flux, again for a single leaf, is given by

$$E_{leaf} = \rho_a\, C_f\, U_{af}\, (q_f - q_{af}) \qquad (4.34)$$

and similarly $E_f = LAI \times E_{leaf}$.

The necessity to determine q_f leads to some central assumptions. q_f, the moisture in the immediate vicinity of the leaves, will in general depend only on the amount of water retained on the foliage (from dew or rain) and on the physiology of the transpiration of the plant, which is synthesized by a single parameter known as the "generalized stomatal resistance" r_s. E_f is split into two parts: E_{tr} corresponding to the transpiration of the water by that part of the leaves that is not covered by water and $E_f - E_{tr}$, corresponding to that part of the leaves that is covered by water.

It is assumed that the quantity of water retained on the foliage, W_{dew} can vary between 0 and a maximum value W_{dmax} depending on the *LAI* and on σ_f

$$W_{max} = 0.2\ (\text{mm})\ \sigma_r\ LAI \qquad (4.35)$$

The evolution of W_{dew} is given by a simple conservation equation

$$\frac{\partial W_{dew}}{\partial t} = \sigma_f P - (E_f - E_{tr}) \qquad (4.36)$$

Knowing W_{dew} and W_{dmax}, the fraction of the foliage δ that is covered by the water can be parameterized simply as a function of W_{dew}/W_{dmax}. Deardorff formulated this function as

$$\delta = (W_{dew}/W_{dmax})^{2/3} \qquad (4.37)$$

as a simple way to allow for the total evaporation of W_{dew} in a finite time in Equation (4.36).

Then it is assumed that on the fraction δ of the leaf area evaporation occurs at the potential rate

$$E_f - E_{tr} = \rho_a\ \text{LAI}\ \delta\ C_f\ U_{af}\ [q_{sat}(T_f) - q_{af}] \qquad (4.38)$$

On the remaining fraction $(1 - \delta)$ the moisture flux originates from the transpiration of the plant through the stomata. To evaluate this flux, an electrical "resistance" analogy is used. The value of the moisture is $q_{sat}\ (T_f)$ inside the stomata, q_f in the immediate vicinity of the leaf, and q_{af} somewhere in the canopy air. The conservation of moisture implies that the flux from the interior of the stomata is equal to the flux from the vicinity of the leaf to the air aloft. Hence

$$C_f\ U_{af}\ (q_f - q_{af}) = \frac{q_{sat}\ (T_f) - q_f}{r_S} \qquad (4.39)$$

where r_S is the stomatal "resistance." We define the atmospheric resistance

$$r_a = (C_f\ U_{af})^{-1} \qquad (4.40)$$

and rewrite Equation (4.39) as

$$\frac{q_f - q_{af}}{r_a} = \frac{q_{sat}(T_f) - q_f}{r_s} = \frac{q_{sat}(T_f) - q_{af}}{r_a + r_s} \qquad (4.41)$$

The flux of transpiration is therefore

$$E_{tr} = \rho_a\ LAI\ (1 - \delta)\ \frac{(q_{sat}(T_f) - q_{af})}{r_a + r_s} \qquad (4.42)$$

Combining Equations (4.38) and (4.42), one finds

$$E_f = \rho_a\ LAI\ C_f\ U_{af}[q_{sat}(T_f) - q_{af})r'' \qquad (4.43)$$

with

$$r'' = 1 - (1 - \delta)\frac{r_s}{r_a + r_s} \qquad (4.44)$$

Deardorff further noted that this computation is only valid if $q_{sat}(T_f) > q_{af}$, i.e., if the moisture flux is directed from the leaves toward the air. If the contrary applies, dew is forming, and the stomatal resistance r_s is no longer a significant concept. This can be incorporated simply in Equation (44) by using

$$r'' = 1 - \delta_c(1 - \delta)\frac{r_s}{r_a + r_s} \qquad (4.45)$$

with $\delta_c = 1$ if $q_{sat}(T_f) > q_{af}$, and 0 otherwise.

Finally, we note that Equation (4.43) can be rewritten

$$E_f = \rho_a\ LAI\ C_f\ U_{af}(q_f - q_{af}) \qquad (4.46)$$

if we adopt the definition

$$q_f = r''\ q_{sat}\ (T_f) + (1 - r'')q_{af},\ q_r < q_{sat}\ (T_r) \qquad (4.47)$$

which is a convenient definition of q_f.

To complete the expression of the latent and sensible heat fluxes for use in the atmospheric model, the remaining fluxes, coming from the ground, must be considered. They are further divided into fluxes coming from the ground through the vegetation (to be weighted by σ_f), and from the soil directly $(1 - \sigma_f)$. This is done in a logical and simple way.

In summary, to implement this approach, one has to know three basic parameters: the shielding factor σ_f, the leaf area index *LAI*, and the generalized stomatal resistance r_s (one assumes implicitly that all the other empirical coefficients will have only a minor effect on the scheme's results).

Although Deardorff suggested that *LAI* and σ_f may be related by at least a rough relation, this is probably true only for the case of homogeneous canopy. When using this scheme to describe a patchwork of different vegetation types and bare soil inside a grid-box of a mesoscale model, as is

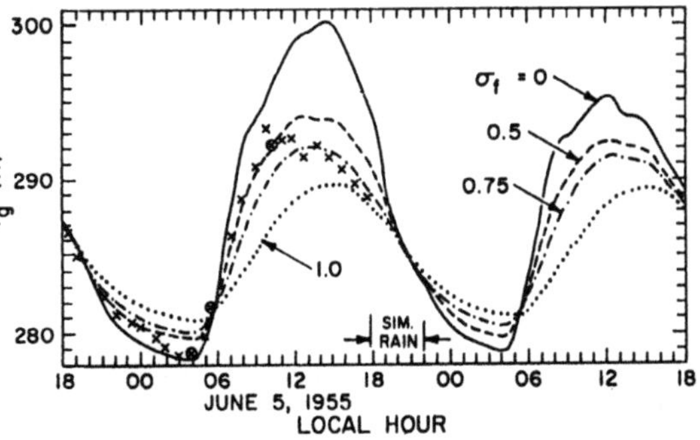

FIGURE 4.28. Variation of the ground surface temperature calculated by the force-restore method and vegetation parameterization over a 2-day period. Results for four different shielding factors, σ_f, are shown, along with extrapolated observed values, denoted by crosses. (Reproduced from Deardorff, Journal of Geophysical Research, Vol. 83, 1978, copyright by the American Geophysical Union.)

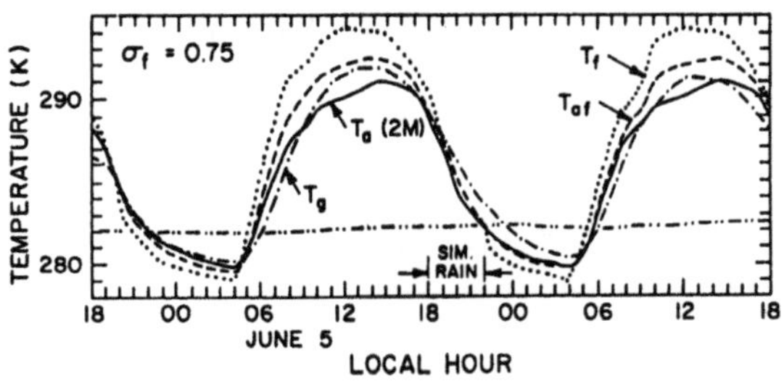

FIGURE 4.29. Variation of the predicted temperatures T_f, T_{af}, T_g, and T_a over the 2-day period for a shielding factor of 0.75. (Reproduced from Deardorff, 1978, Journal of Geophysical Research, Vol. 83, 1978, copyright by the American Geophysical Union.)

nearly always the case, there is little hope that such a relation may be valid. In fact, the leaf area index may be more usefully related to the type of vegetation or to the age of the crop, whereas σ_f will describe the spatial heterogeneity of land use.

Last is the problem of parameterizing r_s, as the single information on the plant physiology. There seems to be a very large agreement on the fact that r_s depends mainly on the following factors (e.g., Jarvis, 1976):

- A minimum resistance r_{smin}, which depends on the type of vegetation and of its age (physiological development).
- The available solar energy, as the stomata may close if not enough energy is available, leading to an increase of r_s.
- The availability of water in the root zone. In some situations not enough water is available. Then the plant will be stressed implying an increase of r_s.
- The deficit of water vapor pressure in the air (this seems to be mainly true for coniferous forests).
- The temperature, as plants are living organisms, that reduce their activity for very hot or very low temperatures.

As a first approximation, Deardorff proposed the following simple form:

$$r_s = r_{smin}\left\{\frac{s_{max}^{\downarrow}}{s_{\downarrow} + 0.03\, s_{max}^{\downarrow}} + \varphi + \left(\frac{W_{wilt}}{0.9\, W_2 + 0.1\, W_g}\right)^2\right\} \tag{4.48}$$

where $S_{max}^{\downarrow}$ is the maximum incoming solar radiation that can be achieved, φ is a seasonal dependence, incorporating the aging of crops, and W_{wilt} is a "wilting point" value of the soil moisture. Thus, r_s can increase to very large values when either the available solar radiation $S_{\downarrow}$ is very small, or W_g is much smaller than the wilting point. The smallest possible value of r_s in this approach is r_{smin}. It will be reached, generally speaking, near noon on clear days, for growing crops on soils with an ample moisture supply. It is still dependent on the type of plant. Note, however, that if $r_{smin} \ll r_a = (c_f U_{af})^{-1}$, the scheme will automatically supply the potential evaporation. More generally speaking, the ratio E^{pot} will mainly depend on r_a/r_s.

Some of the results of a test of the scheme published by Deardorff (1978) are now discussed. The test extends over a two-day period, with general conditions of no cloudiness, but a simulated rainfall event of 2 cm during the night. Observed air temperature, specific humidity, and wind speed are imposed at the Z_a level. The simulated ground surface temperature T_g is shown at Figure 4.28 for several values of σ_f. The parameterization predicts a 2.5-fold reduction in the diurnal range of T_g from $\sigma_f = 0$ to $\sigma_f = 1$, and a corresponding lag in the time of the occurrence of peak ground temperature

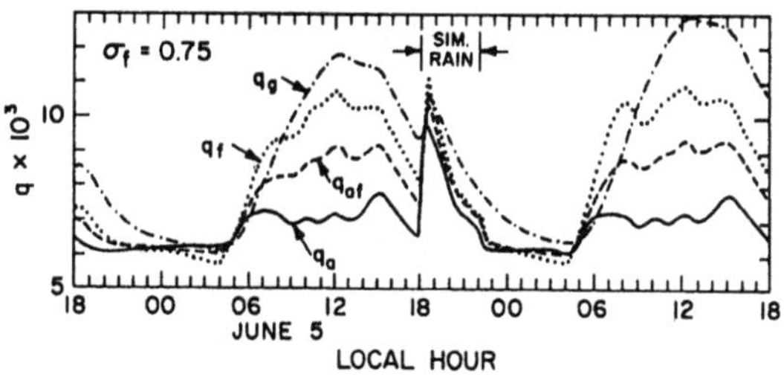

FIGURE 4.30. As in Figure 4.29, but for q_f, q_{af}, q_g, and q_a.

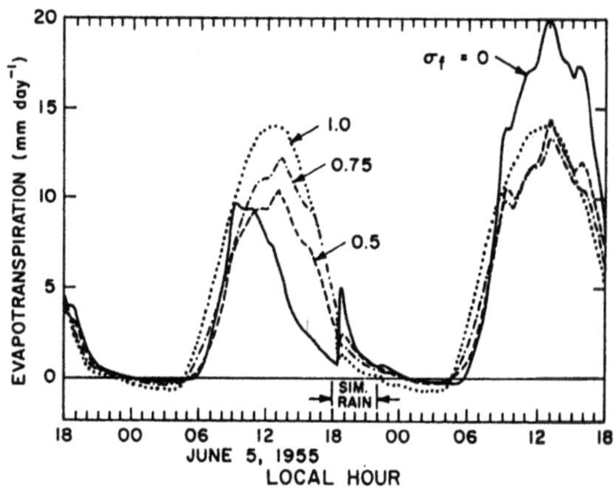

FIGURE 4.31. Predicted variation of the evapotranspiration rate over the 2-day period for four values of the shielding factor. (Reproduced from Deardorff, 1978, Journal of Geophysical Research, Vol. 83, 1978, copyright by the American Geophysical Union.)

of about 1 hr. The maximum bare ground temperature is 5 K less on the second day because of the effect of simulated rainfall. Available observations seem to support the value of $\sigma_f = 0.75$ as adequate for this case. For this value, the diurnal variation calculated for T_f, T_{af}, T_g, and the prescribed variation of T_a are shown in Figure 4.29 and those for q_f, q_{af}, q_g, and q_a in Figure 4.30. The representative foliage temperature T_f is several degrees larger than T_g at midday, however, smaller than the T_g value for bare soil. Deardorff noted that these values are dependent on the value of r_a, and with a value of r_a increased by a factor of 100, T_f reaches values 5.5 K warmer. During the night, q_f becomes less than q_g and q_{af}, then promoting condensation onto the foliage. Finally, the evapotranspiration rate is shown in Figure 4.31 for several values of σ_f. The net 24-hr evapotranspiration is predicted to be 1.9 times larger for $\sigma_f = 1$ than for $\sigma_f = 0$. This result follows from the enhanced dryness of the soil surface, which develops during the day in the absence of vegetation. To capture this effect, a model needs to include both a foliage parameterization and a prediction of the soil moisture surface content. For $\sigma_f = 0$, a pronounced peak in E occurs during the simulated rainy period, because the ground becomes wet while it is still relatively warm. The bare soil evaporation rate is much increased on the day following the simulated rain, whereas the transpiration rate is little affected.

All these results are considered to be very realistic, and it is especially rewarding to see that the model is able to predict the difference of temperature between the foliage, the ground, and the air with such accuracy. This is probably the reason why most recent attempts to parameterize evapotranspiration fluxes over a vegetation have been based on similar concepts, with the notable exception of Sellers et al. (1986). In fact, most recent works have been directed toward improving the weakest points of Deardorff's parameterization, mainly the dependence of stomatal resistance r_s on the soil moisture, and the parameterization of the atmospheric resistance r_a.

It has been recognized that the parameterization of the atmospheric resistance $r_a = (C_f U_{af})^{-1}$ has values that are too large in Deardorff's formulation, leading to excessive leaf temperatures. Alternative formulations have been proposed by Dickinson (1984), Taconet et al. (1986), Sellers et al. (1986), Wetzel and Chang (1987), and Pinty and Mascart (1988). It seems to be a somewhat controversial issue!

Concerning the stomatal resistance, Dickinson (1984) proposes the following formulation:

$$r_s = r_{smin} R_1 S_1 M_1 \tag{4.49}$$

R_1 gives the dependence on solar radiation. It varies between about 1 for overhead sun and r_{smax}/r_{smin} at night (r_{smax} is then the cuticular resistance of the leaves $\sim$ 5000 s/m):

$$R_1 = \frac{1 + f}{f + r_{smin}/r_{smax}}, \varphi \quad f = R_s^{\downarrow}/R_{sref}^{\downarrow} \tag{4.50}$$

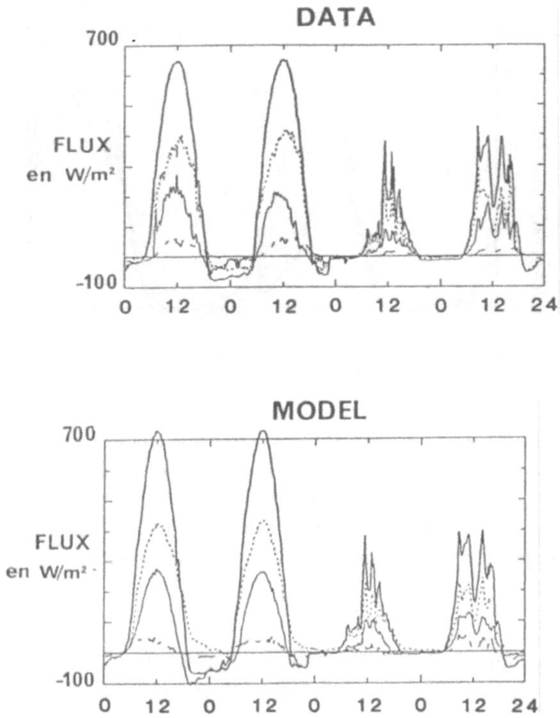

FIGURE 4.32. Comparison of observed and surface energy budget from 15 to 18 June 1986 over a field of maize. Thick line: net radiation; thin line: sensible heat flux; short-dashed: latent heat flux; long-dashed: ground heat flux. (Reproduced from Pinty and Mascart, 1988.)

$R^{\downarrow}_{sref}$ is a reference value of the photosynthetically active part of the downward solar flux which is set to 30 W/m² for trees and 100 W/m² for grassland and crops.

S_1 is a seasonal dependence factor that includes such effects as the increase of r_s when $T<273$ K.

M_1 depends on the soil moisture and the ability of plant roots to extract water from the soil for a given level of root moisture. M_1 varies from 1 when there is ample water supply to large values when the plant transpiration exceeds a maximum value depending on the soil moisture and the root resistance.

Most recent attempts to determine the dependence of r_s on the soil moisture are based on the concept of critical value of the liquid water potential inside the soil and the root–leaf system (see, e.g., Wetzel and Chang, 1987; Pinty and Mascart, 1988). The idea is that the circulation of water inside the plant is governed by the equation

$$E_{tr} = \rho_W (\psi_f - \psi_{sr} + h_f)/(R_f + R_r) \qquad (4.51)$$

where ψ_{sr} is the matric potential of water in the root zone, h_f the mean height of the vegetation, R_f the internal resistance to water flow inside the plant, R_r the resistance to water flow through the soil to the root surface, and E_{tr} the transpiration flux. The potential of water inside the leaf, ψ_f, must adjust to satisfy Equation (4.51) and supply a transpiration value E_{tr}. It must therefore take values below ψ_{sr}. However, it is observed that for some critical value ψ_{fc}, around -200 m, this is no longer possible, and the leaf stomata begin to close, thereby increasing R_s, which leads to a smaller value of E_{tr}. This has been exploited by Choudhury and Idso (1985), who defined a stress factor

$$M_1 = 1 + (\psi_f/\psi_{fc})^{5.5} \qquad (4.52)$$

Thus, if ψ_{sr}, h_f, R_f, R_r, and ψ_{fc} are known, Equations (4.51) and (4.52) can be used to determine acceptable values of ψ_f and E_{tr} at each time step. This scheme has been used by Pinty and Mascart (1988). The advantage of this method is that it relies on a concept of very general significance, the water potential ψ, which is quite well known inside the soil (through the relations discussed in the previous section), and can be measured inside the leaf (although this is a difficult measurement!). Furthermore, according to Wetzel and Chang (1987), it allows for a rational approach to the problem of spatial averaging of plant properties and soil heterogeneities in water content. The drawbacks are that new quantities R_f and R_r are introduced, which need further parameterization. One may, however, hope that the order of magnitude of R_f and R_r does not depend too much on the type of soil or of plant. A more difficult problem is that Equation (4.52) will introduce a very strong dependence of r_s on ψ_{fc}, which is a quite poorly known parameter.

An example of the results obtained by Pinty and Mascart (1988) with this approach on HAPEX-MOBILHY data is now discussed. The first case concerns a field of corn in the growing phase on a sandy soil with ample moisture supply (data from Lubbon 2 station). The parameters used for the model are $h_f = 0.68$ m, $\alpha_f = 0.17$, $\varepsilon_f = 0.98$, $LAI = 1.5$, $\sigma_f = 0.8$, $r_{smin} = 50$ s/m, $\psi_{fc} = -100$ m. The observed and modeled energy budget is reproduced in Figure 4.32. The Bowen ratio is about 0.5, which is reasonable for a crop in the growing stage. The evolution of r_s, ψ_f, and ψ_{sr} is next shown in Figure 4.33. In response to the strong atmospheric demand, the leaf potential is rapidly decreasing during the daytime toward values ~ -150 m, well below its initial value of -100 m. Thus, through Equation (4.51), the value of E_{tr} becomes controlled by the ability of the root system to extract water. This can be seen by the increase of r_s during afternoon hours, well above its minimum value of $r_{smin} = 50$ s/m. Note that this is taking place even with ample soil moisture supply ($\psi_{sr} \sim -5$ m). Another example comes from measurements at Courrensan for wheat on a relatively dry soil. The parameters used are $h_f = 0.83$, $\alpha_f = 0.1$, $\varepsilon_f = 0.98$, $LAI = 1.5$, $\sigma_f = 0.8$, $r_{smin} = 50$ s/m, and $\psi_{fc} = -100$ m. The budget of energy shown in Figure 4.34 exhibits a Bowen ratio of nearly 2. The value of the water potential in Figure 4.35 shows a moderate hydric stress. The soil water potential ψ_{sr} is around -60 m. It is possible in this case to make a difference between the potential inside the roots (ψ_r) and ψ_{sr}, since for

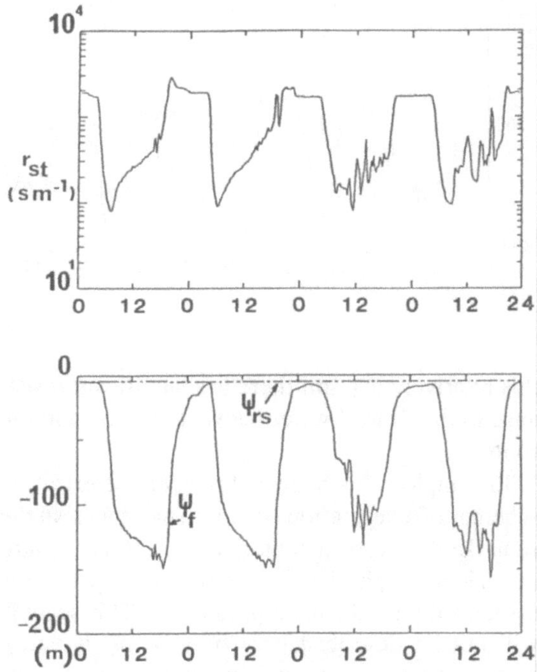

FIGURE 4.33. The evolution of the stomatal resistance (top) and of the water potential in the leaves (ψ_f) and in the soil (ψ_{rs}) (bottom) for the case shown in Figure 4.32. (Reproduced from Pinty and Mascart, 1988.)

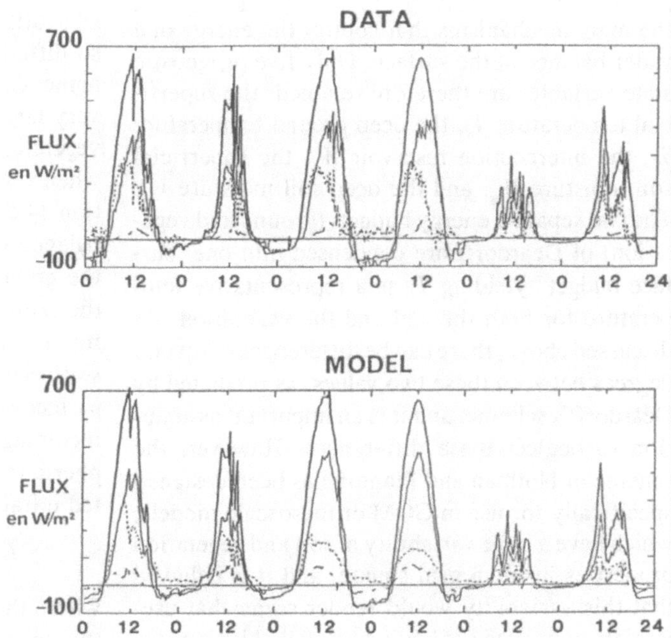

FIGURE 4.34. As in Figure 4.32, but for period from 13 to 18 June 1986, over a field of corn. (Reproduced from Pinty and Mascart, 1986.)

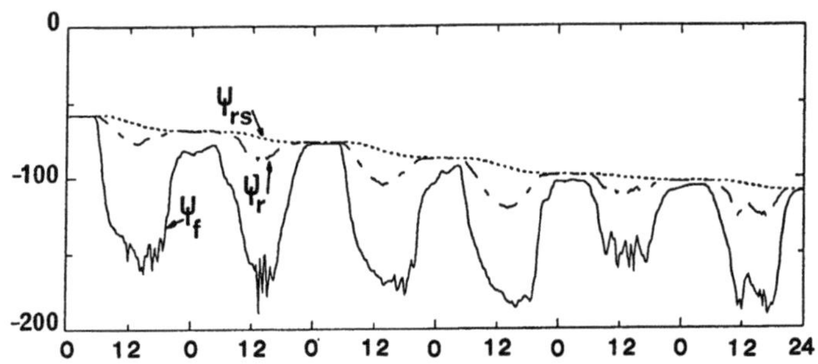

FIGURE 4.35. The evolution of the water potential for the case shown in Figure 4.34. Legend as in Figure 4.33. (Reproduced from Pinty and Mascart, 1988.)

this relatively dry soil there is a significant resistance to the flow of water between the soil and the roots.

To complete this review of the parameterization schemes of a vegetation layer we shall mention the remarkable effort of Sellers et al. (1986) toward increasing the physical description of the processes associated with the plant physiology. This scheme will not be described here because of its vastly increased complexity and because it has not yet been used in mesoscale models. However, there is no reason why it should not perform correctly in such a model. At the opposite extremity of the complexity spectrum, Noilhan and Planton (1989) proposed a simplified version of Deardorff's parameterization. This effort was guided by the concern to keep as low as possible the number of parameters describing the physics while preserving the main mechanisms that control the energy and water balance at the surface. Only five prognostic state variables are therefore retained: the superficial temperature T_s, the deep ground temperature T_2, the interception reservoir W_r, the superficial soil moisture W_g, and the deep soil moisture W_2. The two separate energy budgets (ground and vegetation) of Deardorff are condensed into one "surface budget" yielding T_s as a representative temperature for both the soil and the vegetation. As discussed above, there can be differences of several degrees between these two values, as predicted by Deardorff's scheme, and it is an important assumption to neglect these differences. However, the scheme of Noilhan and Planton has been designed specifically for use in GCM or mesoscale models, which have a large variability of soil and vegetation properties inside a grid square, and it is believed that this variability would render somewhat useless, the advantage of having a detailed knowledge of T_g and T_f as in Deardorff's scheme. Also, infrared satellite pictures can only provide characteristic temperature of the earth's surface, and a scheme with only one temperature appears well suited for comparison with infrared pictures, or for the inverse modeling approach. The equation for the surface temperature in Noilhan and Planton's scheme is [generalized form of Equation (4.24)]

$$\frac{\partial T_s}{\partial t} = - C_T H_A - 2\pi (T_s - T_2)/\tau \quad (4.53)$$

with, in the general case of soil with vegetation

$$C_T = [(1 - \text{veg}/C_G + \text{veg})/C_V]^{-1} \quad (4.54)$$

where veg is the horizontal density of the vegetation, varying from 0 to 1. This is formally the same parameter as σ_f of Deardorff (1978), but with possible different interpretation since the schemes are so different; thus it is safer to give it a different name. C_G has been described above and C_v has the very large value 10^{-3} K m^2 J^{-1} to indicate that the vegetation has negligible heat capacity. Thus, when veg becomes close to 1, $C_T = C_V$ and Equation (4.53) reduces to a classical surface energy balance with no heat flux into the ground. When the ground is partially covered by the vegetation, the expression of C_T combines C_V and C_G allowing for linearization of the heat flux within the soil/vegetation medium. The net radiation at the surface is the sum of the absorbed fractions of the incoming solar radiation R_G and of the atmospheric infrared radiation R_A, reduced by the emitted infrared radiation

$$R_N = R_G (1 - \alpha) + (R_A - \varepsilon\sigma T_s^4) \quad (4.55)$$

where the albedo α and the emissivity ε combine the soil and the vegetation information and σ is the

Stefan–Boltzman constant. Again, this makes the scheme able to use direct information from teledetection studies for α and ε, without having to solve the problem of which part comes from the soil and which from the vegetation (see, e.g., Dedieu et al., 1987).

The turbulent fluxes are calculated by means of classical aerodynamic formulas. For the sensible heat flux

$$H = \rho_a C_H U_a (T_s - T_a) \tag{4.56}$$

where C_H is the surface layer drag coefficient, depending on the thermal stability of the atmosphere as defined in Section 3.1. The water vapor flux E divides into two parts: E_g from the soil surface and E_v from the vegetation.

$$E_g = (1 - \text{veg})\, \rho_a C_H U_a \,[h_u q_{sat}(T_s) - q_a] \tag{4.57}$$

$$E_v = \text{veg}\, \rho_a C_H U_a [q_{sat}(T_s) - q_a] h_v \tag{4.58}$$

where h_u is the relative humidity of the ground surface and is parameterized by

$$h_u = 0.5\,(1 - \cos 4W_g/3W_{sat}\pi) \quad \text{if } W_g < 3/4\, W_{sat}$$
$$= 1 \quad \text{if } W_g > 3/4\, W_{sat} \tag{4.59}$$

The "halstead coefficient" h_v takes into account the direct evaporation from the fraction $\delta = (W_r/W_{rmax})^{2/3}$ of the foliage covered by intercepted water, as well as the transpiration from the fraction $1 - \delta$, in a way which is reminiscent of Deardorff's formula [Equation (4.44)]:

$$h_v = 1 - (1 - \delta)\, R_s/(R_a + R_s) \tag{4.60}$$

where $R_a = (C_H\, U_a)^{-1}$ is the atmospheric resistance and R_s the generalized stomatal resistance. This quantity is, again, the single parameter taking into account the physiology of the vegetation. It is parameterized according to

$$R_s = R_{smin}\, (LAI)^{-1}\, F_1 F_2 F_3^{-1} F_4^{-4} \tag{4.61}$$

The factor F_1 measures the influence of the solar radiation following Dickinson (1984):

$$F_1 = \frac{1 + f}{1 + R_{smin}/R_{smax}}, \quad f = 0.55(R_G/R_{GL})(2/LAI) \tag{4.62}$$

The factor F_2 takes into account the effect of the water stress on the surface resistance. It varies between 0 and 1 when W varies between W_{wilt} (wilting point value) and 0.75 W_{sat}.

$$F_2 = \begin{cases} 1, & \text{if } W_2 \geq 0.75 W_{sat} \\ \dfrac{W_2 - W_{wilt}}{0.75 W_{sat} - W_{wilt}} & \text{if } W_{wilt} < W_2 \leq 0.75 W_{sat} \\ 0, & \text{if } W_2 \leq W_{wilt} \end{cases} \tag{4.63}$$

The factor F_3 represents the effect of the vapor pressure deficit of the atmosphere, following Jarvis (1976) (for coniferous only):

$$F_3 = 1 - g\,[q_{sat}(T_a) - q_a] \tag{4.64}$$

Finally, the factor F_4 introduces a dependence on the air temperature following Dickinson (1984):

$$F_4 = 1 - 0.0016(298.0 - T_a)^2 \tag{4.65}$$

The Noilhan and Planton (1989) scheme has been extensively tested against data collected during the HAPEX-MOBILHY field experiment. At the present time, the most extensive ensemble of tests may be found in Jacquemin (1988), some of which will be described in the next section.

4. Methods and Necessary Data for Testing the Land-Surface Representations

4.1 Testing a One-Dimensional Version Against *In-Situ* Data

It should be first mentioned that the problem of testing the schemes that have been described above is an ill-posed one. Although the ultimate goal is to obtain a formulation that will represent the average behavior of a grid-square of, say, 10 × 10 km², containing a large variety of soil and vegetation types; the only available reference data at present are local data that are at most representative of a single field. Furthermore, the schemes themselves are based on concepts that have local meanings, some of which are very difficult to measure (e.g., the shielding factor σ_f or the generalized stomatal resistance r_s). There is some hope that the use of remotely sensed data may provide regional data sets useful for this problem sometime in the future. However, the currently developed approaches to make use of these data are based on "inverse modeling" concepts, which rely themselves on parameterization schemes that are assumed correct. We are therefore faced with a difficult problem! In the author's opinion, the only tractable way is to test the schemes at the local scale against *in situ* data, and resort to more or less empirical procedures to derive grid-averaged behavior that would automatically revert to the local value if the grid-square were homogeneous. This point will be further discussed in a companion chapter (Bougeault et al.) in this volume and we will concentrate here on the testing against *local* data in the context of a one-dimensional version of the schemes.

In any parameterization scheme, it is possible to make a partition of the quantities in four categories: Those quantities that describe a more or less permanent characteristic of the place of study, with regard to the time scale of the mesoscale flow of interest (1 day or so), may be called *parameters*. Those quantities that will eventually become the prognostic variables of the mesoscale model, and usually have a significant variation on the same reference time scale, may be called the *state variables*. Those quantities that are only very indirectly influenced by the surface scheme, like radiation or rainfall, may be called the *atmospheric forcings*. Finally, those quantities that are computed as diagnostics and will eventually be fed back to the mesoscale model are clearly *the results*.

Among the parameters, for most schemes described above, we need values for soil type and depth, the vegetation type, σ_f and *LAI*, and resistance, the roughness height, the albedo and emissivity, etc. These quantities clearly need to be specified from observations. However, some of them like σ_f, *LAI*, and r_s might be so difficult to measure that they will eventually be used as additional degrees of freedom to tune the results. One must, in this case, check that the optimal value found by the trial-and-error procedure falls within the range of possible values for this parameter. The state variables, as defined above, will contain the soil moisture content, temperature, and the atmospheric values U_a, T_a, and q_a at the lowest level of the model. These quantities are initialized with the observed values, and should be verified against the observed values during the time of the test forecast. The atmospheric forcings clearly must be prescribed during the entire time of the forecast, as a 1D model cannot be expected to forecast rainfall or clouds correctly. Finally, the results will generally comprise the fluxes of latent and sensible heat toward the atmosphere, the flux of heat into the ground, the surface temperature of the soil and of the foliage, and the atmospheric value of T, u, and q at the screen level. These quantities must be verified against observations during the forecast.

4.2 Interactive Versus Noninteractive Models Tests

Two variants of this procedure have been used to test the schemes in the way described above. In the so-called interactive method, the surface parameterization is coupled to a PBL parameterization. Thus, U_a, T_a, and q_a are actually predicted by the 1D model. The interaction of the surface scheme with the atmosphere is fully represented, and works the same way as in a 3D mesoscale model. To recover a realistic behavior of the state variables at the lowest level despite the simplified 1D framework, it is then necessary to impose a "dynamic forcing" to the model, which takes the form of horizontal and vertical advection terms. Those may be deduced from a mesoscale analysis. This method has been advocated by Mahfouf (1986), André and Bougeault (1988), Jacquemin (1987), and Noilhan and Planton (1989). Its major advantage is to provide a test of the surface scheme in exactly the same conditions as will be present in the 3D model. Thus if some tuning of poorly known parameters is needed, as is always the case, it ensures that the optimal value will also be optimal for the 3D model. Another advantage is that using this method, the values of the parameters at screen level are definitely *a result* of the model, and provide an *independent check* against observational data. Indeed, it will, in general, be possible to obtain a correct partition of the energy between latent and sensible heat flux with any scheme, given an adequate choice of some of the parameters. However, if the scheme is able to predict simultaneously a correct value of the fluxes and of the screen level variables, one may gain some confidence about its physical basis. On the other hand, a disadvantage of the interactive method is that it will in general be difficult to obtain atmospheric values close to those observed, for reasons that have nothing to do with the surface scheme. One of these reasons is the weak current quality of mesoscale analysis systems. Another one is more fundamental. The observed values of T_a, q_a, and U_a are influenced not only by the parcel over which the measurements are taken, but by the mesoscale environment, which frequently includes very different types of surface. Thus by making the idealization of a 1D model over an infinite homogeneous surface that has the same properties as those of the catchment under study, one definitely makes an error over T_a, q_a, and U_a. The only way to remedy this problem is to force the surface scheme with observed values of T_a, q_a, and U_a during the entire test integration. This so-called noninteractive method has been more widely used in the past because of its simplicity. Its shortcoming is that because the parameters at the screen level are so close from the imposed atmospheric values, it does not provide an independent check of the scheme quality. Also, because of the stabilizing effects of imposing the observed values, schemes tested in this manner might be more prone to numerical or physical instabilities, when included in a 3D model. Finally, the optimal values of the parameters deduced this way might no longer be optimal when the state variables are allowed to vary. Given the ill-posed nature of the problem, there is probably no definite choice between these two methods and both of them should be used alternatively, depending on the problem and of the specific case study.

The *length* of the test should also be given some discussion here. Most authors have shown tests runs lasting 1 or 2 days, with the notable exception of Pinty and Mascart (1988) who performed tests lasting 6 days. Although the time scale useful for mesoscale models is clearly 1 to 2 days, it is of interest to make long test runs even under repetitive conditions to investigate the sensitivity of the schemes results to small variations in the initial values. This problem is quite similar to the so-called "spin-up" problem for atmospheric models. Also, it is absolutely necessary to test the long-term evolution of the soil moisture content. This will provide an excellent check of the systematic errors of the schemes. This is easy to do with the noninteractive method since the atmospheric parameters

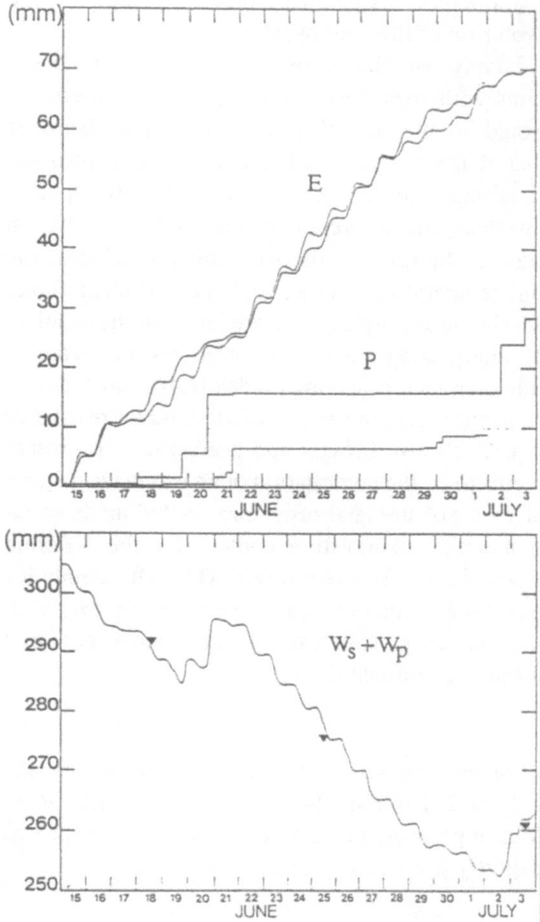

FIGURE 4.36. The long-term evolution of the forecasted and observed cumulated evaporation (top, in mm), precipitation, and soil moisture (bottom). Symbols are observations by neutron sounding, taken once a week during the HAPEX-MOBILHY field phase. (Reproduced from Jacquemin, 1988.)

are imposed, and longer term bias might not be introduced. In the interactive method, however, it will in general be impossible to obtain a significant long-term evolution of the atmospheric variables. One way to proceed is to cycle on a large number of 1 day predictions, by initializing the soil moisture content of forecast n by the model prediction of forecast $n - 1$. An example of such a check from Jacquemin (1988) is shown in Figure 4.36. Care should be taken, however, while making long tests runs, to account for any slow evolution of the "parameters," as could be produced, e.g., by the evolution of the canopy physiology.

Finally, we should present some recommendations with regard to which kind of measurements would be most useful in the future. If we consider that it has become standard practice to measure simultaneously at the local scale the atmospheric variables, the different components of the radiation budget, the rainfall, the soil moisture content, the soil temperature profile, and the turbulent fluxes within the atmosphere, we are left with the problem of interpreting these measurements in terms of shielding factor, leaf area index (*LAI*), atmospheric resistance (r_a), and generalized stomatal resistance (r_s). The most difficult and needed task at present seems to be the compilation of an extensive data set of values of stomatal resistance, including its variation with soil moisture content for the different types of vegetation of interest. This calls mainly for direct measurements of this quantity, together with soil moisture, soil textures, and water potential inside the soil and the leaves.

Acknowledgments. I gratefully acknowledge the help of J. Noilhan, B. Jacquemin, and J.F. Mahfouf in preparing this review. The manuscript was expertly prepared by D. Lacarrere.

References

Al Nakshabandi G, Kohnke H (1965) Thermal conductivity and diffusivity of soils as related to moisture tension and other physical properties. Agric Meteorol 2:271–279.

André JC, Bougeault P (1988) On the use of Hapex data for the validation and development of parameterization schemes of land-surface processes. WCP publication series No. 126, Appendix B.

Anthes RA (1984) Enhancement of convective precipitation by mesoscale variations in vegetative covering in semiarid regions. J Clim Appl Meteorol 23:541–554.

Benjamin SG (1986) Some effects of surface heating and topography on the regional severe storm environment. Part II: Two dimensional idealized experiments. Mon Weather Rev 114:330–343.

Benjamin SG, Carlson TN (1986) Some effects of surface heating and topography on the regional severe storm environment. Part I: Three dimensional simulations. Mon Weather Rev 114:307–329.

Bhumralkar CM (1975) Numerical experiments on the computation of ground surface temperature in an atmospheric general circulation model. J Appl Meteorol 14:1246–1258.

Blackadar AK (1976) Modeling the nocturnal boundary layer. Preprints, Third Symposium Atmospheric Turbulence, Diffusion and Air quality, Raleigh, Amer. Meteor. Soc., pp 46–49.

Blanchard DO, Lopez RE (1985) Spatial patterns of convection in South Florida. Mon Weather Rev 113: 1282–1299.

Bougeault P (1986) Le modèle Péridot: une étude de qualification à méso-échelle. *Note EERM*, No. 168.

Brutsaert W. (1975) The roughness length for water vapor sensible heat, and other scalars. J Atmos Sci 32:2028–2031.

Brutsaert W (1979) Heat and mass transfer to and from surfaces with dense vegetation or similar permeable roughness. Bound Layer Meteorol 16:365–388.

Carlson TN, Boland FE (1978) Analysis of urban rural canopy using a surface heat flux temperature model. J Appl Meteorol 17:998–1013.

Carlson TN, Dodd JD, Benjamin SG, Cooper JN (1981) Satellite estimation of surface energy balance, moisture availability and thermal inertia. J Appl Meteorol 20:67–87.

Cellier P (1986) On the validity of flux gradient relationship above very rough surfaces. Bound Layer Meteorol 36:417–419.

Choudhury BJ, Idso SB (1985) An empirical model for stomatal resistance of field-grown wheat. Agric Meteorol 36:65–82.

Clapp R, Hornberger G (1978) Empirical equations for some soil hydraulic properties. Water Resour Res 14: 601–604.

Cosby BJ, Hornberger GM, Clapp RB, Ginn TR (1984) A statistical exploration of the relationships of soil moisture characteristics to the physical properties of soils. Water Resour Res 201:682–690.

Deardorff JW (1977) A parameterization of ground-surface moisture content for use in atmospheric prediction models. J Appl Meteorol 16:1182–1185.

Deardorff JW (1978) Efficient prediction of ground-surface temperature and moisture, with inclusions of a layer of vegetation. J Geophys Res 83 C4:1889–1903.

Dedieu G, Deschamp PY, Kerr YH (1987) Satellite estimation of solar irradiance at the surface of the earth and of surface albedo using a physical model applied to meteosat data. J Clim Appl Meteorol 26:79–87.

Dickinson RE (1984) Modeling evapotranspiration for the three-dimensional global climate models. In "Climate Processes and Climate Sensitivity." Geophysical Monograph, 29, Vol. 5.

Dyer AJ (1974) A review of flux profile relationships. Bound Layer Meteorol 7:363–372.

Garratt JR (1978) Flux profile relations above tall vegetation. Quart J R Meteorol Soc 104:199–211.

Garratt JR (1980) Surface influence upon vertical profiles in the atmospheric near surface layer. Quart J R Meteorol Soc 106:803–819.

Garrett AJ (1982) A parameter study of interactions between convective clouds, the convective boundary layer, and a forested surface. Mon Weather Rev 110: 1041–1059.

Halstead MM, Richman R, Covey W, Merrman J (1957) A preliminary report on the design of a computer for micrometeorology. J Meteorol 14:308–325.

Jacquemin B (1987) Test des paramétrisations des processus de surface dans le modèle Péridot de mésoéchelle. Note ENM.

Jacquemin B (1988) Validation d'une paramétrisation des processus de surface dans le modèle Péridot mésoéchele. Rapport de DEA, Université Paul Sabatier, Toulouse.

Jarvis PG (1976) The interpretation of the variations in leaf water potential and stomatal conductance found in canopies in the field. Philos Trans R Soc London Ser B 273:593–610.

Jarvis PG, James GB, Landsberg JJ (1976) Coniferous forest. In "Vegetation and the Atmosphere" (J.L. Monteith, ed.), Vol. 2. Academic Press, New York.

Louis JF (1979) A parametric model of vertical eddy fluxes in the atmosphere. Bound Lay Meteorol 17: 187–202.

Mahfouf JF (1986) Contribution à la définition d'une paramétrisation des transferts entre le sol, la végétation et l'atmosphère: analyse de sensibilité et insertion dans un modèle mésoéchelle. Thèse de doctorat, Université de Clermont II.

Mahfouf JF (1988) Etude de situations perturbées de Hapex-Mobilhy. Note ENM.

Mahfouf JF, Richard E, Mascart P (1987) The influence of soil and vegetation on the development of mesoscale circulations. J Clim Appl Meteorol 26:1483–1495.

Mahrt L, Pan H (1984) A two-layer model of soil hydrology. Bound Lay Meteorol 29:1–20.

Manabe S (1969) Climate and the ocean circulation. 1-The atmospheric circulation and the hydrology of the earth's surface. Mon Weather Rev 97:739–774.

McCumber MC (1980) A numerical simulation of the influence of heat and moisture fluxes upon mesoscale circulations. Ph.D. dissertation, University of Virginia, Charlottesville.

McCumber MC, Pielke RA (1981) Simulation of the effects of surface fluxes of heat and moisture in a mesoscale numerical model. Part I: Soil layer. J Geophys Res 86:9929–9938.

Monin AS, Obukhov AM (1954) Basic laws of turbulent mixing in the ground layer of the atmosphere. Trudy Inst Geophys Acad Sci USSR 24:163–187.

Nappo CJ (1975) Parameterization of surface moisture and evaporation rate in a planetary boundary layer model. J Appl Meteorol 14:289–296.

Noilhan J (1987) Une modélisation unidimensionnelle des échanges énergétiques dans le sol, la végétation et la couche limite atmosphérique. Note EERM, No. 171.

Noilhan J, Planton S (1989) A simple parameterization of land surface processes for meteorological models. Mon Weather Rev 17(3):536–549.

Noilhan J, Planton S, Jacquemin B (1987) Une paramétrisation des processus desurface pour les modèles atmosphériques de moyenne et grande échelle. Note EERM. No. 200.

Ookouchi Y, Segal M, Kessler RC, Pielke RA, (1984) Evaluation of soil moisture effects on the generation and modification of mesoscale circulations. Mon Weather Rev 112:2281–2292.

Orlanski I (1975) A rational subdivision of scales for atmospheric processes. Bull Am Meteorol Soc 56: 527–530.

Paulson CA (1970) The mathematical representation of wind speed and temperature profiles in the unstable atmospheric surface layer. J Appl Meteorol 9:857–861.

Pielke RA (1974) A three-dimensional numerical model of the sea breeze over South Florida. Mon Weather Rev 102:115–139.

Pielke RA (1984) Mesoscale Meteorological Modeling. Academic Press, New York.

Pinty JP, Mascart P (1988) Evaluation d'un modèle d'interface sol/végétation/atmosphère à l'aide de données SAMER et HYDRA de la campagne Hapex-Mobilhy 1986. Note OPGC, No. 96.

Raupach MR (1979) Anomalies in flux-gradient relationships over forest. Bound Lay Meteorol 16:467–486.

Sellers PJ, Mintz Y, Sud YC, Dalcher A (1986) A simple biosphere model (SIB) for use within general circulation models. J Atmos Sci 43:505–531.

Taconet O, Bernard R, Vidal-Madjar D (1986) Evapotranspiration over an agricultural region using a surface flux/temperature model based on NOAA-AVHRR data. J Clim Appl Meteorol 25:284–307.

Thom AS, Stewart JB, Oliver HR, Gash JHC (1975) Comparison of aerodynamic and energy budget estimates of fluxes over a pine forest. Quart J R Meteorol Soc 101:93–105.

Walsh JE (1974) Sea breeze theory and application. J Atmos Sci 31:2012–2026.

Wetzel JP, Chang JT (1987) Concerning the relationship between evapotranspiration and soil moisture. J Clim Appl Meteorol 26:18–27.

Wetzel JP, Atlas D, Woodward PH (1984) Determining soil moisture from geosynchronous satellite infrared data: A feasibility study. J Clim Appl Meteorol 23:375–391.

Zhang D, Anthes RA (1982) A high-resolution model of the planetary boundary layer—sensitivity tests and comparisons with SESAME 79 data. J Appl Meteorol 21:1594–1609.

5 Evaporation Models in Hydrology

W. James Shuttleworth

1. Introduction

Almost as soon as mankind began to wear, or at least wash, clothes, he realized that evaporation was strongly related to weather; that they dried best on sunny, windy days, particularly if the air was "fresh" and dry, with low humidity. When hydrologists first began to model natural evaporation, now more than 40 years ago, this is where they started. Weather variables measured near the ground often exhibit some statistical intercorrelation; sunny days tend to be warmer, for instance. As soon as evaporation rate could be measured, it was therefore easy to observe and formulate empirical relationships with meteorological variables, either individually or in groups. Hydrologists created a whole range of evaporation "models" as a result.

Over the years we have reached a more profound understanding of the physical and physiological aspects that determine the natural evaporation process, and overviewing this is a major objective of this chapter: it provides the content of Section 2. From the perspective of this better understanding it is, moreover, also possible to impose structure in the otherwise bewildering range of hydrological models, and this forms the subject of Section 3. The concluding section again draws on this same perspective to define those (few) hydrological models likely to be of use to meteorologists in models of global climate.

2. The Evaporation Process

Natural evaporation occurs when liquid water at or just below the earth's surface is converted into water vapor and transferred in this form into the atmosphere. The process can occur only if there is an input of energy either from the sun or from the atmosphere itself, and is controlled by the rate at which the energy, in the form of water vapor, can diffuse away from the earth's surface. Diffusion occurs by the two rather different physical processes of molecular diffusion and turbulent diffusion at subsequent stages in the transfer path, but in hydrological models both of these processes are most commonly and conveniently described by analogy with electrical resistance. The process of evaporation from a natural surface is then expressed, on a physical basis, by models that describe the effect of molecular and turbulent diffusion resistances on the partition of energy from the sun or the atmosphere.

This representation is only approximate; in particular the mechanisms of turbulent diffusion within and just above stands of vegetation are often poorly described by the simple diffusion equation (see, for example, Raupach, 1989). At the same time the numerical consequences of assuming this equation in the canopy when calculating evaporation are usually small (see Shuttleworth, 1989), and the assumption is in any case often superseded by more lax assumptions, such as neglecting canopy depth altogether.

In this section we outline the basic physical concepts, and how they are combined to form an adequately realistic description of the evaporation process. Section 3 explores the progressive simplification of this to give simpler, but less precise, models of evaporation that also have an important, practical role in hydrological science.

2.1 Energy

2.1.1 *Latent Heat*

The molecules making up liquid water are close, with a separation of just over one molecular diameter. At such distance there is a short range force between them, which falls off rapidly as the separation increases.

In water vapor, the molecules are very much further apart, typically 10 or more molecular diameters, depending on the vapor pressure. At this separation the intermolecular force is very small indeed. To create water vapor from liquid water it is necessary that the separation between all the molecules increase. To do this it is necessary to do work against the force holding them together; that is to supply energy. The amount of energy required is of course directly related to the number of molecules, which is in turn proportional to the mass of water involved. The amount of energy per unit mass of liquid water is called the latent heat of vaporization, λ, and is 2.47×10^6 J kg^{-1} at 10°C. It changes slightly with temperature, by about 0.1% per °C, because the initial separation of the molecules making up the liquid varies with temperature.

The flow of water vapor molecules away from an evaporating water surface implies a transfer of energy away from the surface in the form of latent heat. The energy transferred is numerically equal to the product of the mass flow, i.e., the evaporation, E, in kg m^{-2} s^{-1}, and the latent heat of vaporization.

2.1.2 *Transfer at a Water Surface: Saturation*

Water vapor arises by the exchange of molecules with a free water surface somewhere in the interface between the earth and the atmosphere. This free water surface can either be inside the vegetation or soil, in "dry" conditions, or on the surface of the vegetation or soil, during and just after rainfall.

The energy of water molecules at the surface of a liquid follows a statistical distribution that is dictated by the surface temperature, T_s. To leave the surface a molecule must have an energy greater than (λ/n), where n is the number of molecules per unit volume of liquid water. The rate, B, at which molecules "boil off" from the surface is given by

$$B = k_1 \exp\left(-\frac{\lambda/n}{kT_s}\right) \qquad (5.1)$$

where k_1 is a constant, k is Boltzmann's constant, and T_s is in degrees Kelvin.

At the same time water vapor molecules approach the liquid at a rate that is directly proportional to the vapor pressure, e, of the water vapor in contact with the surface. A certain fraction, r, of these is reflected on collision, while the remaining fraction $(1 - r)$, is absorbed. The net evaporation rate, E, is the difference between B and this rate of vapor absorption (Shuttleworth, 1975), and is given by

$$E = k_1 \exp\left(-\frac{\lambda/n}{kT_s}\right) - k_2(1 - r)e \quad (5.2)$$

where k_2 is a second constant. If molecules are able to diffuse freely away from the surface then the vapor pressure adjacent to the surface can remain low, and the difference between these terms finite. In this situation it is possible to maintain a persistent flow of molecules from the liquid to the vapor, with an equivalent flow of latent heat.

If, on the other hand, the volume of air above the liquid is sealed off, it is no longer possible for the water molecules to diffuse away from the surface. As more molecules leave the surface, the concentration of the water vapor and its equivalent vapor pressure increases until the two rates in Equation (5.2) are equal and there is no longer any evaporation. The air is then said to be "saturated" and cannot absorb any more water molecules. At a given temperature this situation occurs at a particular vapor pressure, which is called the "saturated vapor pressure," e_s. The continuation $E = 0$ in Equation (5.2) gives

$$e_s = \frac{k_1}{k_2(1 - r)} \exp\left(\frac{\lambda/n}{kT_s}\right) \qquad (5.2a)$$

so that this vapor pressure is a defined function of T_s, which can be measured experimentally.

Figure 5.1 shows the variation of saturated vapor pressure with temperature. It is important in physical models of evaporation that not only is the variation of e_s a known function of temperature, but so is the gradient $(\partial e_s/\partial T)$. This gradient, or more particularly Δ, which is the mean gradient between two temperatures T_1 and T_2,

$$\Delta = \frac{e_s(T_2) - e_s(T_1)}{T_2 - T_1} \qquad (5.3)$$

recurs frequently in equations describing evaporation, as we shall see later.

FIGURE 5.1. Saturated vapor pressure of water as a function of the temperature of the water surface. (The fact that this function, and the average gradient between two temperatures, T_1 and T_2, are well known, is much used in models of evaporation.)

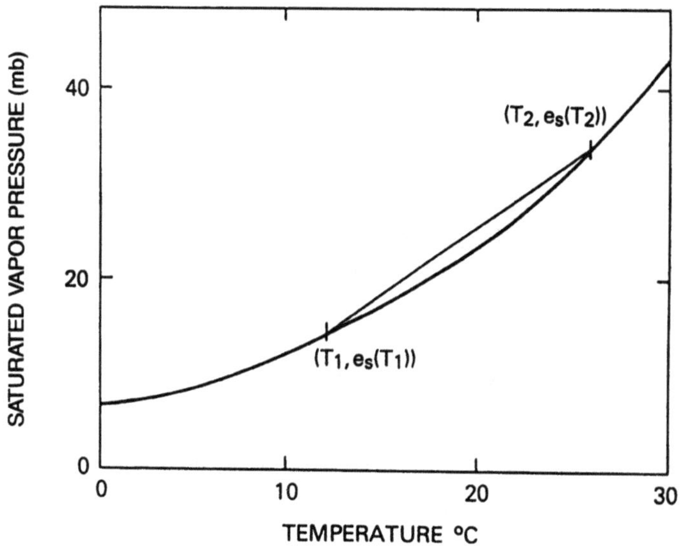

2.1.3 Sensible Heat

There are other ways in which the radiant energy input to the earth's surface can be used apart from supporting evaporation. In practice large quantities of energy are exchanged between the ground and the atmosphere by thermal conduction and thermal convection. Thermal conduction occurs only close to the surface; further away, the primary exchange mechanism is thermal convection. Convection occurs when mass motions of the air result in the transport and mixing of the heat content of the air between different levels in the atmosphere. We speak of the flow of energy between the surface and the atmosphere by the combined processes of thermal conduction and thermal convection as "sensible" heat flux, because it is this transfer that determines air temperature, a property of the air that we can sense.

The flux of sensible heat is very commonly away from the surface in daytime conditions, when the temperature of the ground is usually greater than that of the atmosphere, but it is usually toward the surface at night, when it supports the outward flux of radiation. If the surface of the vegetation is wet, evaporation rate is often high, particularly for tall vegetation, and even during the day can exceed the energy reaching the surface as radiation. In such situations there will be an inward flux of sensible heat to provide the additional energy.

2.1.4 Radiation

The sun is the primary source of radiant energy at the earth's surface. The radiation it provides is equivalent to that of a full radiator of about 6000 K. The spectrum is modified through absorption by atmospheric gases, particularly water vapor; and by scattering from the individual molecules of the atmosphere in clear sky conditions, and additionally from clouds when these are present. Most of the radiation is confined to short wavelengths, in the band 0.3–3 μm, the particular spectrum depending on the fraction of the total shortwave energy input, S_T, reaching the ground in the direct solar beam. A significant part of the shortwave energy, S_D, reaches the ground in a diffuse (or nondirectional) form after scattering. The fraction varies but is typically 15–25% in clear sky conditions, approaching 100% in overcast conditions.

On reaching the ground some of the shortwave radiation is reflected by the surface. The reflection coefficient, or albedo, depends on the angle of incidence of the solar beam and the type of vegetation. The daily mean reflection coefficient is commonly about 0.25 for short, agricultural crops, but is more typically about 0.15 for taller crops such as tobacco, sugar cane, and natural or man-made forest. This difference arises because tall vegetation is more able to absorb the solar beam by multiple reflections within the canopy.

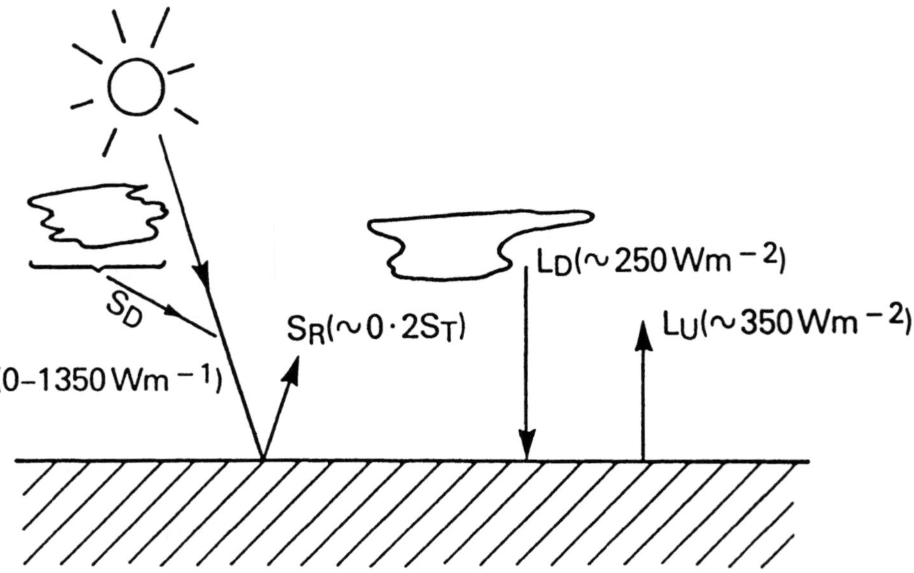

FIGURE 5.2. Components of the radiation balance at the earth's surface. (The total incoming solar radiation, S_T, comprises a diffuse component, S_R, and is partly reflected as the component, S_R. The downward longwave radiation, L_D, is less than the upward longwave radiation, L_U, emitted from the surface.)

There is a significant exchange of radiant energy between the earth's surface and the atmosphere in the form of radiation at wavelengths in the range 3–100 μm. Both emit "black body" radiation with a spectrum characteristic of their respective temperatures. The spectrum of incoming longwave radiation near the surface is greatly modified by absorption by atmospheric gases, particularly water vapor and carbon dioxide. Partly because of this, but mainly because the temperature of the earth's surface is usually greater than that of the atmosphere, there is a net loss of energy from the ground as longwave radiation.

Figure 5.2 illustrates schematically the radiation balance at the earth's surface. The amount of incoming solar radiation, S_T, in the range 0–1350 W m^{-2}, varies with the altitude of the sun, and therefore varies with time, as does the fraction of shortwave radiation reflected. The upward longwave radiation varies with surface temperature, but is typically in the order of 350 W m^{-2}, while the downward longwave is typically in the order of 250 W m^{-2}. The difference, $L_U - L_D$, is usually in the range 87–107 W m^{-2} in clear sky conditions. The situation in cloudy conditions is more complex: for a more extensive description of radiation exchange, see Monteith (1973). This radiation balance gives rise to a net radiation exchange, R_N.

2.2 Molecular Diffusion

Diffusion is the process by which those properties that characterize a particular volume of fluid (e.g., its heat content, momentum content, or the concentration of its constituents) are moved from one position in the fluid to another. Such movement takes place when there are variations in the concentration of these characteristics, and occurs because, at normal temperatures, the individual molecules making up the fluid are in permanent and random motion. There is random exchange of molecules between adjacent positions, and the characteristic entity is transferred along with this mass exchange. If the concentration of the entity at these two positions was originally different, there will be a net transfer, or flow, in response to the different concentration.

Molecular motion in air is extremely rapid; molecules have a velocity of about 500 m s^{-1} but have path length between collisions of around 5×10^{-8} m. Diffusion occurs as a direct consequence of this molecular agitation; molecular movement is responsible for the transfer, while the high collision rate ensures rapid equilibration at each location. The rate of flow, or "current," of an atmospheric entity is directly related to the gradient in the equivalent concentration. If we consider the movement of water vapor in one dimension, then the transfer expressed as a mass flux per unit area, E, is proportional to, χ, the concentration gradient (in kg m^{-3}) according to the equation

$$E = - D_V \frac{\partial \chi}{\partial z} \tag{5.4}$$

in which D_V is a molecular diffusion coefficient with dimensions of m^2 s^{-1}.

It is very common to treat the evaporation as the flow of latent heat. Equation (5.4) is then rewritten in terms of the latent heat flux, λE, which is related to the gradient of atmospheric vapor pressure by the equation

$$\lambda E = - \frac{\rho c_p}{\gamma} D_V \frac{\partial e}{\partial z} \tag{5.5}$$

In this equation γ is the so called "psychrometric constant," being the combination of terms ($c_p\rho/\lambda 0.622$) and having a value of 0.66 mb/°C at 20°C and 1000 mb.

In an analogous way the concentration of heat in air is related to air temperature, T, through the volumetric heat capacity (ρc_p), where ρ is the density of air and c_p its specific heat, so that the heat transfer H per unit area is given by

$$H = - D_H \frac{\partial(\rho c_p T)}{\partial z} = - \rho c_p D_H \frac{\partial T}{\partial z} \tag{5.6}$$

Similarly, if u is the velocity of the air perpendicular to the z axis, then the air possesses momentum in this plane with a concentration (ρu), and there can be momentum transfer, giving rise to a viscous force or shearing stress per unit area, τ, with

$$\tau = D_M \frac{\partial(\rho u)}{\partial z} = \rho D_M \frac{\partial u}{\partial z} \tag{5.7}$$

Because the same process is responsible for all of these different transfers, the diffusion coefficients D_V, D_H, and D_M are similar in size, around 0.18 m^2 s^{-1}, and all increase by about 0.7%/°C at normal temperatures.

2.3 Turbulent Diffusion

The wind blowing over a natural surface is retarded by an interaction with that surface. In principle, very smooth surfaces do exist in nature, but in practice virtually all natural surfaces are aerodynamically "rough," even at low windspeeds. The interaction of moving air with a rough surface gives rise to an apparently random and haphazard movement in which packets of air, of varying size, are continuously being created and destroyed, but move in an ill-defined yet coherent way during their transient existence. This phenomenon, known as turbulence, is initiated at the surface, but propagates upward into the atmosphere in a way determined by the height above the ground, and, in the case of a crop, also downward into the vegetation. The mixing it generates is a very efficient mechanism for transferring entities through the atmosphere, much more efficient than molecular diffusion. It is the primary process responsible for the exchange of mass, momentum, and heat between air close to the ground and that at higher levels in the atmosphere.

Much turbulence is produced by the frictional retardation of the wind, but the transfer properties of this frictional turbulence are enhanced, or diminished, if there is a gradient in mean air temperature along a direction away from the surface. In the absence of any mean wind at all, turbulent transfer can still occur by free convection if there are temperature differences in the atmosphere. In general, frictional turbulence and free convection exist together in a hybrid form known as "mixed convection." It is usual to treat this situation theoretically by treating frictional turbulence as the primary transfer mechanism, with the effect of free convection as a semiempirical correction. In recognition of this, it is convenient to consider first a "neutral" atmosphere, that is, one in which there is no mean temperature gradient.

2.3.1 Forced Convection

The windspeed $u(z)$ at a height z above an extensive horizontal surface uniformly populated with roughness elements (usually vegetation) is observed to increase through a boundary layer in which shearing stress is conserved following a logarithmic relation which has the general form

$$u(z) = \frac{k}{u_*} \ell n[(z-d)/z_0] \tag{5.8}$$

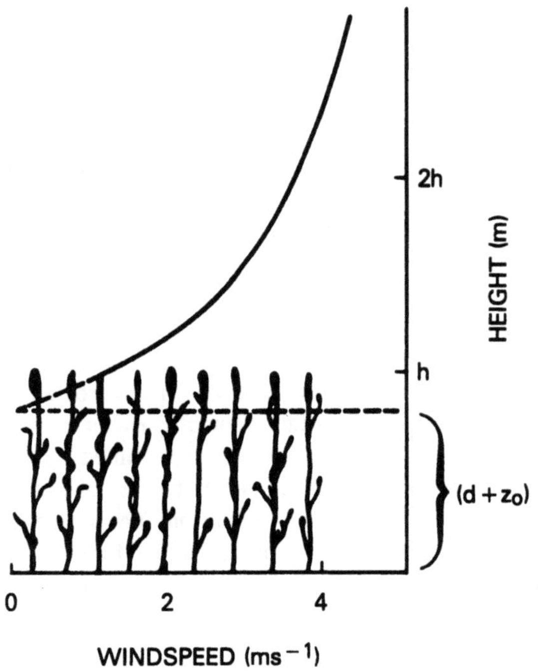

FIGURE 5.3. Variation in windspeed observed, in neutral conditions, above a crop with zero plane displacement, d, and roughness length, z_0.

The parameter k in this equation is a scaling constant, known as *von Karman's constant*, a property of the turbulent wind field, which is assumed independent of the structure of the surface, and is usually assigned the value 0.4. The parameters d and z_0, on the other hand, can be regarded as characteristic of the surface cover, to be determined by experiment, and are known as the *zero plane displacement* and *roughness length*, respectively. Their values vary with the height of the roughness elements, h, and their spacing, but d is often in the range $0.6h$ to $0.8h$, and z_0 typically $0.1h$. u_* has dimensions of velocity and is known as the *friction velocity*; it is defined in terms of the shearing stress by

$$\tau = \rho u_*^2 \tag{5.9}$$

The form of Equation (5.8) is illustrated in Figure 5.3.

Just as the transfer of momentum by molecular diffusion is represented by Equation (5.7), it is usual to describe the transfer by turbulent diffusion in the same way, with an equation of the form

$$\tau = \rho K_M \frac{\partial u}{\partial z} \tag{5.10}$$

in which K_M is the eddy diffusivity for momentum. The expression for K_M, which is consistent with Equations (5.8), (5.9), and (5.10), is

$$K_M = ku_*(z-d) \tag{5.11}$$

In the same way it is possible to define similar entities for the transfer of sensible and latent heat from the equations

$$\lambda E = -\frac{\rho c_p}{\gamma} K_V \frac{\partial e}{\partial z} \tag{5.12}$$

and

$$H = -\rho c_p K_H \frac{\partial T}{\partial z} \tag{5.13}$$

which have obvious analogy with Equations (5.5) and (5.6). The turbulent transfer mechanism responsible for the exchange of all of these entities is considered the same. It is therefore usual to assume $K_M = K_H = K_V$ over the height range of the near-surface turbulent boundary layer, providing there is no temperature gradient away from the aerodynamically rough surface.

2.3.2 Mixed Convection

If there is a gradient of temperature away from the surface, the transfer by forced convection is modified. When air temperature decreases with height, any parcel of air created and moved upward by forced convection will tend to continue its ascent, because it will be warmer, lighter, and therefore more buoyant, than the air into which it is moved. In this situation the atmosphere is described as "unstable." When the temperature of the atmosphere increases with height, the opposite occurs. In this case buoyancy forces act against the further upward motion, and the atmosphere is said to be "stable."

Turbulent transfer in stable and unstable conditions differs from forced convection in neutral conditions, but it might be expected that this difference can be described empirically in terms of parameters that reflect the relative efficiency of the free and forced convection mechanisms. For this reason, enhancement (or otherwise) of forced convection, in other than neutral conditions, has very often been parameterized in terms of a dimensionless parameter R_i, known as the Richardson number, which is related to the ratio of the production of free and forced convection, and is defined by

$$R_i = \frac{g}{T}\frac{\partial T}{\partial z} \bigg/ \left(\frac{\partial u}{\partial z}\right)^2 \qquad (5.14)$$

where g is the acceleration due to gravity and T is the temperature in degrees Kelvin.

The transfer coefficients for momentum, and sensible and latent heat in a general mixed convection situation are redefined from Equation (5.11) as

$$K_{M,V,H} = ku_*(z\text{-}d)\phi_{M,V,H} \qquad (5.15)$$

The functions ϕ_M, ϕ_V, and ϕ_H express the stability-dependent description of mixed convection as a modification of forced convection, and are empirical. In stable conditions, available data, e.g., Webb (1970) indicate an interrelationship of the form

$$\phi_V = \phi_H = \phi_M = (1 - 5R_i)^{-1} \ldots R_i \text{ +ve} \qquad (5.16)$$

while in unstable conditions, Dyer and Hicks (1970) concluded that

$$\phi_V = \phi_H = \phi_M{}^2 = (1 - 16R_i)^{-1/2} \ldots R_i \text{ −ve} \qquad (5.17)$$

2.4 Resistance Networks

It is a common useful simplification to express differential diffusion equations for energy and momentum movement through the atmosphere in integrated form. The integration is very straightforward in cases where the (one-dimensional) flux can be treated as a constant in the direction specified by the coordinate z. If Equation (5.4) is used to describe the molecular diffusion of latent heat between two points at which the vapor pressure is e_1 and e_2, respectively, then this description can be integrated to the form

$$\frac{\rho c_p}{\gamma}[e_2 - e_1] = -\lambda E\left[\int_{z_1}^{z_2} \frac{dz}{D_V}\right] \qquad (5.18)$$

Clearly the integration is independent of the detailed mechanism responsible for the diffusion, and Equation (5.12) can be treated in exactly the same way. Equation (5.18) has an obvious similarity to Ohm's law for electrical current, viz.,

(Potential difference) = (Current) × (Resistance)

and on the basis of this analogy it is usual to use the name "resistance" for the two entities

$$\left[\int_{z_1}^{z_2} \frac{dz}{D_{M,V,H}}\right] \text{ and } \left[\int_{z_1}^{z_2} \frac{dz}{K_{M,V,H}}\right] \qquad (5.19)$$

The introduction of the concept of resistance into models of the exchange between vegetation and the atmosphere allows a description that is mathematically similar at each stage in the transfer path, even though the physical mechanism changes. Turbulent diffusion in the vertical direction is represented by "eddy diffusion" resistances. There are a great many of these in the case of a multilayer simulation of the interaction, but only a few in simpler models. Molecular diffusion is the transfer mechanism that is significant elsewhere in the interaction. In particular, diffusion through the laminar boundary layer around vegetation elements is represented by "boundary layer" resistances, diffusion of water vapor through the stomatal openings in leaves (transpiration) is represented by "stomatal" resistances, while diffusion through a drying layer of soil at the ground may be represented by a "substrate" resistance.

The network of resistances involved in the partition of available energy into sensible heat and evaporation is complex for real stands of vegetation, but computer models have been built that describe the process in one dimension (e.g., Waggoner and Reifsnyder, 1968; Sinclair et al., 1971; Halldin and Lindroth, 1986). Such models involve the simultaneous, interactive solution of energy balance equations at several heights in the canopy, and require estimates of the resistances operating at each level. Figure 5.4 illustrates the network of resistances commonly used.

Typically the model starts by calculating the energy available for evaporation at each height, and then uses the assumed resistance chain to calculate canopy profiles of temperature and humidity. The procedure is described in some detail by Waggoner (1975). Given correct input data and a good description of through-canopy turbulent transfer, such a model would provide a realistic description of detailed behavior by simulating measured behavior in real canopies. Use of such models is unlikely to advance the basic physical understanding used in their synthesis, but they do

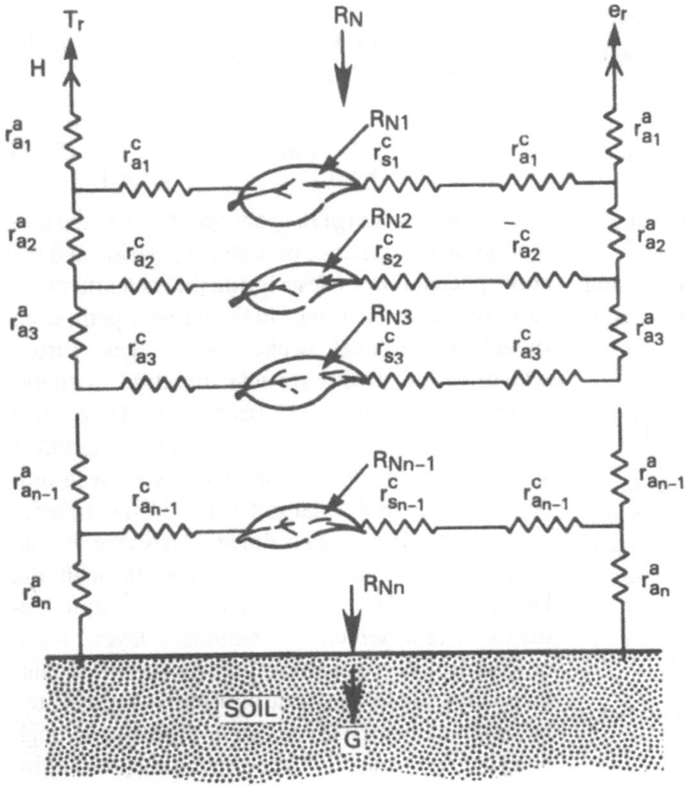

FIGURE 5.4. Schematic diagram of the resistance network that simulates the partition of intercepted net radiant energy, R_N, into latent and sensible heat in numerical simulation models.

provide a check on whether this understanding is sufficiently advanced. The input data required restrict their predictive use; however, once tested, such models can provide a useful yardstick against which to test simpler models, more suitable to predictive application (e.g., Sinclair et al., 1976). In practice it is generally possible to provide a useful and realistic physical description of the evaporation process without needing to resort to the use of a multilevel description. It is often found possible to combine the in-canopy resistances, the stomatal and boundary layer resistances, assuming they acted in parallel at a single level in the canopy, at an "effective source/sink height." This is plausibly considered to occur at a level somewhere close to the effective sink for momentum, i.e., at $(z_0 + d)$, these values being those appropriate when canopy cover is complete.

Figure 5.5 illustrates the single source model for the more general case in which soil evaporation is included. Analysis of this resistance network (Shuttleworth and Wallace, 1985) yields an equation that expresses the net evaporation from the soil–plant–atmosphere system in terms of the saturation deficit, D, and other meteorological variables (which are assumed to be measured at some height above the canopy), A and A_s, the total energy available for evaporation in the canopy and at the soil, and the component resistances illustrated in Figure 5.5. This equation takes the form

$$\lambda E = C_c PM_c + C_s PM_s \tag{5.20}$$

where PM_c and PM_s have the form

$$PM_c = \frac{\Delta A + \{\rho c_p D - \Delta r_a^c A_s\}/(r_a^a + r_a^c)}{\Delta + \gamma\{1 + r_s^c/(r_a^a + r_a^c)\}} \tag{5.21}$$

$$PM_s = \frac{\Delta A + \{\rho c_p D - \Delta r_a^s (A - A_s)\}/(r_a^a + r_a^s)}{\Delta + \gamma\{1 + r_s^s/(r_a^a + r_a^s)\}} \tag{5.22}$$

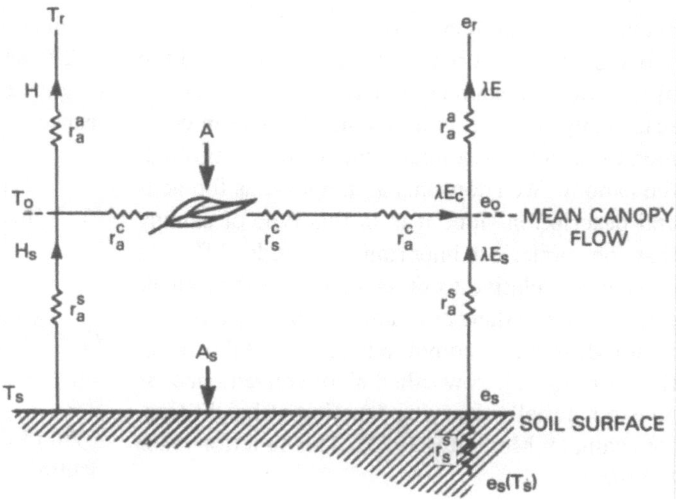

FIGURE 5.5. The resistance network used in the general, single-source model that includes soil interaction (Shuttleworth and Wallace, 1985).

and the coefficients C_c and C_s are given by the expressions

$$C_c = \{1 + R_cR_a/R_s(R_c + R_a)\}^{-1} \quad (5.23)$$

and

$$C_s = \{1 + R_sR_a/R_c(R_s + R_a)\}^{-1} \quad (5.24)$$

where

$$R_a = (\Delta + \gamma)r_a^a \quad (5.25)$$

$$R_s = (\Delta + \gamma)r_a^s + \gamma r_s^s \quad (5.26)$$

$$R_c = (\Delta + \gamma)r_a^c + \gamma r_s^c \quad (5.27)$$

In the particular situation that most of the available energy is absorbed by the canopy (and the soil is assumed to play little role in the interaction), Equation (5.20) simplifies to the more familiar Penman–Monteith equation. This is strictly only relevant to dense, closed canopies, and takes the form

$$\lambda E = \frac{\Delta A + \rho c_p D/(r_a^a + r_a^c)}{\Delta + \gamma\{1 + r_s^c/(r_a^a + r_a^c)\}} \quad (5.28)$$

In this particular situation, Figure 5.5 could be drawn more simply by omitting all the resistances and fluxes below the effective source height, and the combination $(r_a^a + r_a^c)$ considered a single "aerodynamic" resistance, r_A^c.

2.5 Component Resistances

For the remainder of this section, the general, single-source model that yields Equation (5.20) (illustrated in Fig. 5.5) is adopted as the most realistic description of evaporation suitable for practical application. The objective is now to provide formulas for the component resistance that adequately, but simply, represent the transfer processes commonly observed in the field.

2.5.1 *Eddy Diffusion Resistance*

Hydrological and meteorological problems usually require some working compromise between practicality and precision, and the representation of turbulent transfer through and immediately above vegetation is no exception. The assumption that "K-theory" holds here, i.e., that Equations (5.10), (5.12), and (5.13) apply, has always been dubious, and is now known to be particularly so for tall, sparse vegetation with a concentrated and elevated canopy source/sink distribution (Raupach, 1989). It is surely just good luck that the numerical significance of K-theory failure on the calculation of integrated evaporation is limited in this particular situation (Shuttleworth, 1989). The present analysis proceeds in cognizance of the weakness of K-theory through and just above vegetation, but in recognition that our objective is to provide a working description of eddy diffusion resistances for a model that has more profound simplicity, in that it assumes the canopy has no depth at all.

If the vegetation is uniform and dense, and the distribution of sources and sinks in the canopy reasonably constant with height, it has been common to assume one-dimensional diffusion through the canopy. We adopt this assumption as the general description since it is in this type of canopy that the numerical importance of eddy diffusion resistance (relative to other resistances) is greatest, and where their best description is, therefore, required. It is also common to represent the variation of $K_{M,V,H}(z)$, now called K for convenience, as an exponentially decaying function of height (see, for example, Monteith, 1973). This function takes the form

$$K = K_h \exp[-\, n(1 - z/h)] \qquad (5.29)$$

where n is an empirical constant, with a typical value of 2–3. The value of K_h in this equation is that at the top of the canopy, given by Equation (5.11) when $z{=}h$, where h is the vegetation height.

Shuttleworth and Wallace (1985) set the effective source/sink for energy flux at a height Z, fixed at the effective sink of momentum in a neutral atmosphere for a "closed canopy." The (arbitrary) criterion for this last condition is that the leaf area index, $L{=}4$. Shuttleworth and Gurney (1989) adopt this definition of Z, but supersede the simple leaf area dependence of r_a^a and r_a^s given by Shuttleworth and Wallace (1985). The leaf area index dependence of z_0 and d derived from a second order closure model by Shaw and Pereira (1982) are simulated empirically using the equations

$$d = 1.1h \ln[1 + (0.07L)^{1/4}] \qquad (5.30)$$

and

$$z_0 = z_0' + 0.3h(0.07L)^{1/2} \quad L < 2.85 \qquad (5.31)$$

or

$$= 0.3h(1 - d/h) \qquad L > 2.85 \qquad (5.32)$$

where z_0' is the roughness length of the (soil) substrate.

These last equations are assumed to provide general, though approximate, estimates of z_0 and d for varying crop height and leaf area, and, in particular, of Z_0 and D, the preferred values when $L{=}4$, which determine the fixed source/sink position, $Z = Z_0 + D$. Formulas for r_a^a and r_a^s are obtained by integrating Equations (5.10) and/or (5.29) as appropriate, from Z up to the screen height, z_R, and from Z down to the ground, respectively. The resulting equations are

$$r_a^a = \frac{1}{ku_*}\left[\ln\frac{(z_R - d)}{(h-d)}\right] + \frac{h}{nK_h}\left[\exp\{n(1-Z/h)\}-1\right] \qquad (5.33)$$

$$r_a^s = \frac{h\exp(n)}{nK_h}\left[\exp\{-nz_0'/h\} - \exp\{-nZ/h\}\right] \qquad (5.34)$$

All of the above analysis applies in conditions of neutral atmospheric stability. First-order corrections can be made for the effect of stability using interactive techniques, with Equations (5.33) and (5.34) providing initial estimates of the temperature of the canopy and soil: about two iterations are usually required (Choudhury and Monteith, 1988).

2.5.2 *Boundary Layer Resistance*

Moving air within vegetation interacts with the individual elements making up the canopy, and the flux is exchanged through a boundary-layer resistance, which, in the case of sensible and latent heat, might be envisaged as the result of molecular diffusion through a laminar boundary layer close to the surface of the element. The situation for momentum is different since in this case transfer is enhanced by the additional interaction of pressure forces on the element. In consequence, the so-called boundary layer resistance for momentum, $r_{a,M}^c$, is usually significantly less than that for the heat fluxes, r_a^c.

The value of r_a^c changes with windspeed, and is therefore a function of height in the canopy. Jones (1983) gives the dependence as

$$r_a^c(z) = 100[w/u(z)]^{1/2} \qquad (5.35)$$

where w is a representative leaf width. Integrating this function through the canopy (see, for example, Choudhury and Montieth, 1988), the mean boundary layer resistance for momentum for a canopy of leaf area L area is given by

$$r_a^c = \frac{A}{L} u_h \qquad (5.36)$$

where u_h is the windspeed at the top of the canopy calculated from Equation (5.8). The value of A is roughly proportional to $w^{1/2}$ [see Equation (5.35)], but a typical value is 25 s m^{-1}.

The additional aerodynamic resistance seen by energy fluxes that cannot be transferred by pressure interactions, the so called "excess resistance," r_B, is conventionally expressed in terms of the nondimensional parameter B^{-1} (Chamberlain, 1966) such that

$$r_B = \frac{B^{-1}}{u_*} \quad (5.37)$$

The size and variation of the excess resistance have been studied by Chamberlain (1966) and Thom (1972). These studies suggest that the value of B^{-1} (typically in the order of 4) is not strongly related to surface roughness, but depends on u_*. Thom (1972) further suggests that an empirical relationship of the form

$$B^{-1} = Cu_*^{1/3} \quad (5.38)$$

"may provide a first approximation," with C in the order of 6, when u_* is in units of m s^{-1}.

Assuming the excess aerodynamic resistance occurs mainly in the boundary layer, then the boundary layer resistance for momentum heat may be estimated from that computed for energy fluxes using the relationship

$$r^c_{a,M} = r^c_a - r_B = r^c_{a,M} - 6u_*^{-2/3} \quad (5.39)$$

2.5.3 Substrate (Soil) Surface Resistance

It is only recently that representing evaporation from the surface of soil by a combination equation of the Penman–Monteith type, which requires a surface resistance, has begun to gain acceptance (Monteith, 1981; Shuttleworth and Wallace, 1985). The optimum parameterization of r^s_s is not yet defined, but two approaches have been attempted. The first is to define a simple empirical function of near surface soil moisture: two examples are the equation of Shu Fen Sun (1982), thus

$$r^s_s = 3.5(\Theta_{sat}/\Theta) + 30 \quad (5.40)$$

and that of Camillo and Gurney (1986), thus

$$r^s_s = 8.4 - 41.4\Theta \quad (5.41)$$

In these equations the constants are those optimized using data for particular soils; Θ is the volumetric soil moisture of a shallow ($\simeq$ 0.5 cm) surface layer, and Θ_{sat} that at saturation.

The second approach is to attempt some simple representation of the physics at the soil surface. Choudhury and Monteith (1988), for example, assume evaporation occurs from a wet soil below a dry layer (assumed isothermal), whose depth increases as drying proceeds. The resulting form of r^s_s is

$$r^s_s = (\tau \ell)/(pD_V) \quad (5.42)$$

where τ and p are the tortuosity and porosity of the soil, D_V is the molecular diffusion coefficient for water vapor, and ℓ is the depth of the dry layer, which increases between the times t and $(t + \delta t)$ by an amount

$$\ell(t + \delta t) - \ell(t) = E_s \delta t/\Theta_{wet} \quad (5.43)$$

where E_s is the calculated substrate (soil) evaporation in the model, and Θ_{wet} is the volumetric soil water content of the (wet) soil beneath this dry layer.

2.5.4 Canopy Surface Resistance

It is convenient to draw a distinction between r^c_s, the *surface* resistance of vegetation, and the stomatal resistance r_{ST}. Surface resistance is the more general concept depending on the presence, or otherwise, of surface water on the vegetation. It is equal to r_{ST} when the canopy is dry, and effectively zero (Shuttleworth, 1975) when it is completely wet, with a smooth transition between the two. Shuttleworth (1978) explores this transition in detail and writes

$$r^c_s = \left[\frac{W}{r^c_a(\Delta/\gamma + 1)} + \frac{1 + W}{r_{ST} + r^c_a(\Delta/\gamma + 1)}\right]^{-1} - r^c_a(\Delta/\gamma + 1) \quad (5.44)$$

In practice W, the fractional wetted *surface area*, is difficult to use in models of evaporation; it is more convenient to represent the evaporation rate through conditions of intermediate canopy wetness in terms of (C/S) the ratio of the (*volume* of) water present on the canopy at any particular time (C) to the maximum water storage, or canopy capacity

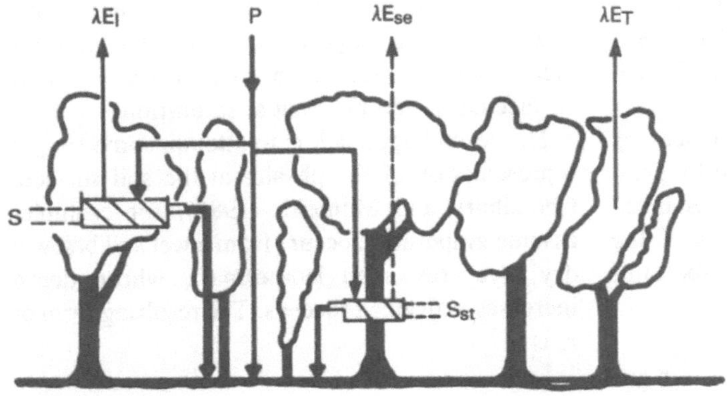

FIGURE 5.6. Schematic diagram of the Rutter model of canopy water balance used to compute the interception and transpiration components of evaporation through rainstorms.

(S). Evaporation is then assumed to occur from wet and dry portions of the canopy at rates λE_I and λE_T, respectively, these being given by

$$\lambda E_I = (C/S)\lambda E_W \tag{5.45}$$

$$\lambda E_T = (1 - C/S)\lambda E_D \tag{5.46}$$

where λE_W and γE_D are calculated rates assuming the canopy is totally wet ($r_s = 0$), and totally dry ($r_s^c = r_{ST}$), respectively. The total evaporation is the sum of these two. This scheme has (some) validation for tall vegetation (Shuttleworth, 1978); therein is implied an implicit relationship between (C/S) and W which, when introduced into Equation (5.44), gives values for r_s^c that are consistent with micrometeorological observations made over a drying forest.

In building a realistic model of evaporation that explicitly acknowledges the interception of rainfall by the canopy, it is therefore necessary to know the amount of water (C) held on the vegetation. The Rutter model (Rutter et al., 1971, 1975), which is illustrated in Figure 5.6, has been successfully used to provide this for forest vegetation, and probably has wider relevance.

The value of C is initially set to zero, corresponding to a dry canopy, and is then obtained by maintaining a running water balance, following the equation:

$$dC/dt = P(1 - p - p_t) - \lambda E_I - D_r \tag{5.47}$$

where P is the incoming precipitation rate, p is the proportion of rain falling directly to the ground through gaps in the canopy, p_t that diverted to the stems, and D_r is the rate of canopy drainage (millimeters per minute), given by Rutter et al. (1971) as

$$D_r = a \exp\{b(C - S)\} \tag{5.48}$$

where a is typically 0.002 mm/min and b typically 4 mm^{-1} for forest vegetation. A similar water balance can be made for the water stored on (or in transit down) the trunks or stems if this is considered necessary (see Fig. 5.6).

2.5.5 *Bulk Stomatal Resistance*

To complete the suite of resistances in the general single source model it is necessary to provide a submodel of r_{ST}, the bulk stomatal resistance for the dry canopy, in order to compute λE_D in Equation (5.46).

The sensible heat and momentum fluxes arising from the plant canopy in a soil–crop–atmosphere system can be considered to originate on the surface of vegetation elements making up the canopy. However, if the canopy is dry, the latent heat flux largely arises as a result of evaporation from cell walls inside stomatal cavities within the vegetation. It escapes to the leaf surface by molecular diffusion through the stomatal pores. Diffusion through the stomatal aperture represents an additional resistance, applicable only to latent heat, which is called stomatal resistance. Figure 5.7 illustrates the diffusion process responsible.

A typical leaf is usually considered small enough for temperature gradients across its surface to be ignored. The atmosphere in the stomatal cavities within the leaf can be assumed to be at saturation vapor pressure corresponding to leaf temperature.

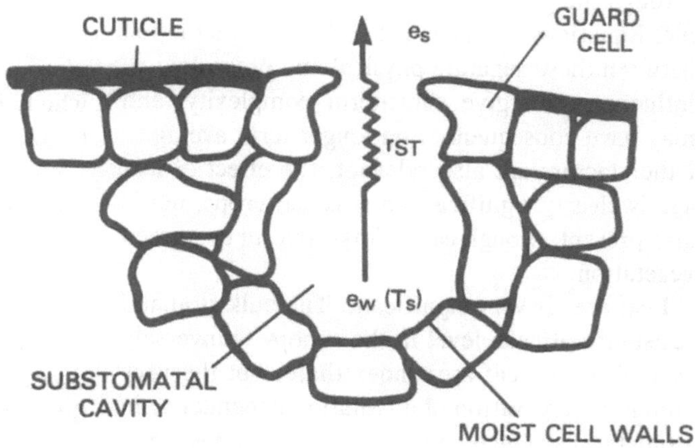

FIGURE 5.7. Schematic diagram of the molecular diffusion process responsible for stomatal resistance in transpiring, dry vegetation.

Water potentials in leaves are rarely less than −40 bars, and at this potential the relative humidity is still 97% of saturation. With these simplifications it follows that the flow of water vapor to the leaf surface, i.e., the transpiration, can be described by the equation

$$\lambda E = \frac{\rho c_p}{\gamma} \frac{e_s(T_s) - e_s}{r_{ST}} \qquad (5.49)$$

where r_{ST} is the stomatal resistance for unit surface area of leaf, and T_s and e_s are here the temperature and vapor pressure immediately adjacent to the leaf surface.

There is no consensus on a general mathematical formula to describe the stomatal resistance of a single leaf, or indeed of the whole canopy. This mainly reflects the species diversity of this essentially biological entity. It is also, in part, because some hydrological and meteorological problems can be modeled by quite a simple assumption, even an assumed constant value; while others, particularly those in which conditions of restricted soil moisture occur, or where description at the hourly time scale is required, necessarily require a more complex form. Even at the simplest level some broad differences between classes of vegetation should probably be acknowledged. Closed-canopy, forest vegetation, for instance, very commonly has a bulk stomatal resistance (say) 50% greater than that of many fully grown, agricultural crops.

In general, a dependence on soil water availability is clearly appropriate since (in reality and in models) it is through a relationship between surface resistance and soil moisture that aqueous mass conservation is maintained in the transpiration process. Some semiempirical descriptions of stomatal resistance assume a multiplicative factor that includes soil moisture explicitly (e.g., Stewart, 1988; Gash et al., 1989); others incorporate an indirect dependence inside a multiplicative factor that is explicitly framed in terms of leaf water potential (e.g., Sellers et al., 1986); still others ignore soil water restrictions until a critical evaporation is reached, and then set a value for stomatal resistance appropriate to an evaporation rate controlled by root resistance (e.g., Dickinson et al., 1986). In practice the net behavior of these schemes is fairly similar and reflects typical field response, namely an initially weak relationship to decreasing soil moisture, followed by a rapid increase in stomatal resistance as the available water approaches a limiting value that changes according to the soil type. Bearing in mind the form of this interrelationship, the fact that soil can be as important as vegetation, and that the detailed functional form is in any case prone to additional complexity (see Rutter et al., 1975), argument as to the relative merits of these schemes is hardly justified. No doubt all can be used to provide an adequate description of particular calibration data within experimental errors.

When modeling evaporation at the hourly time scale, the response to rapidly changing meteorological variables is significant. Incident (solar) radiation, atmospheric saturation deficit, and, to a less extent, temperature are often found to change stomatal resistance, but also have a direct effect through the physics of the process [see, for example, Equations (5.21) and (5.22)]. The interaction between these separate physical and physiological influences can give short-term complexity, and may have consequences on longer term averages. Other factors are also relevant; the effect of leaf age is clearly significant in seasonal crops, but is also present, though easily forgotten, in evergreen vegetation.

Leaf area is very significant. The bulk stomatal resistance at each level in the canopy is inversely related to the leaf area index there; but the very common observation of a radiation dependence in stomatal resistance complicates the synthesis of a whole canopy average value, since local radiation is related to the integrated leaf area above. In many types of canopy the irradiance below a leaf area index of L' can be written as

$$S'(L') = S \exp(-\alpha' L') \qquad (5.50)$$

where S is the shortwave flux at the top of the canopy and α' is an attenuation constant, typically 0.6 ± 0.3. At any (leaf area) depth in the canopy, the mean irradiance of a leaf, S_1, is equivalent to the attenuation of the vertical flux and is

$$S_1 = -\partial S'(L')/\partial L' = \alpha' S(L') \qquad (5.51)$$

In many problems it is more convenient to work with the reciprocal of stomatal resistance, the stomatal conductance, here represented by c_{ST} for a single leaf and C_{ST} for the whole canopy. *In the absence of any other environmental stress* [see Equation (5.56)], the stomatal conductance has a response to solar irradiance which often takes the form

$$c_{ST} = c_0 + \frac{c_1 S_1}{1 + c_2 S_1} \qquad (5.52)$$

though the constants c_0, c_1, and c_2 may change significantly between canopies and species. In this equation, c_0 is the cuticular resistance, assumed light independent, and is typically at least an order of magnitude less than the second term for daytime radiation levels in the field.

In the simplest case, when c_0, c_1, and c_2 are assumed independent of leaf position, and the bulk stomatal conductance of the complete canopy, c_{ST}, set equal to that of all the leaves acting in parallel; then, with a total leaf area of L,

$$C_{ST}(S,L) = \int_0^L c_{ST}(L')\, dL' \qquad (5.53)$$

which, by combining Equations (5.50), (5.51), and (5.52) and integrating, means

$$C_{ST}(S,L) = c_0 L + \left(\frac{c_1}{c_2 \alpha'}\right) \ln\left[\frac{1 + c_2 \alpha' S}{1 + c_2 \alpha' S \exp(-\alpha' L)}\right] \qquad (5.54)$$

Figure 5.8 illustrates the typical form of the radiation dependent functions c_{ST} and C_{ST} with $\alpha' = 0.6$. The light response of a single leaf (of a C3 plant) usually involves an initial, rapid rise from the cuticular conductance, followed by leveling off at solar irradiances greater than (say) 400 W m^{-2}, see Figure 5.8a. Equation (5.54) predicts that a canopy made up of similar leaves has a less moderated response at high radiation, as more leaves deeper in the canopy achieve light saturation. Figure 5.8b illustrates the response of C_{ST} to increasing leaf area. Since a higher proportion of the leaves have light limited stomatal conductance at higher leaf area, the average stomatal conductance of the leaves falls, and the relationship is not linear.

In practice it is usually possible to represent the functional behavior of Equation (5.54) more simply (to within a few percent), and define a normalized function relative to the value when $S = 1000$ W m^{-2} using the expression

$$C_{ST} = c_0 L + \frac{S(1000 + X_R)}{1000(S + X_R)} \qquad (5.55)$$

where X_R is an empirical constant whose value might be derived by optimization against field data. Its value can also be estimated from the canopy parameters α', L, and c_2 by equating the ratio of the radiation dependent terms in Equations (5.54) and (5.55) at $S = 1000$ W m^{-2} and (say) $S = 200$ W m^{-2}.

The notion of representing bulk canopy stomatal resistance as a product of normalized "stress" functions is popular. In a recent review of forest micrometeorology, Shuttleworth (1989) writes the whole-canopy stomatal resistance for forest canopies in a form that includes its most important and commonly observed environmental dependence, thus

$$C_{ST} = c_0 c_V c_R c_D c_T c_{SM} \qquad (5.56)$$

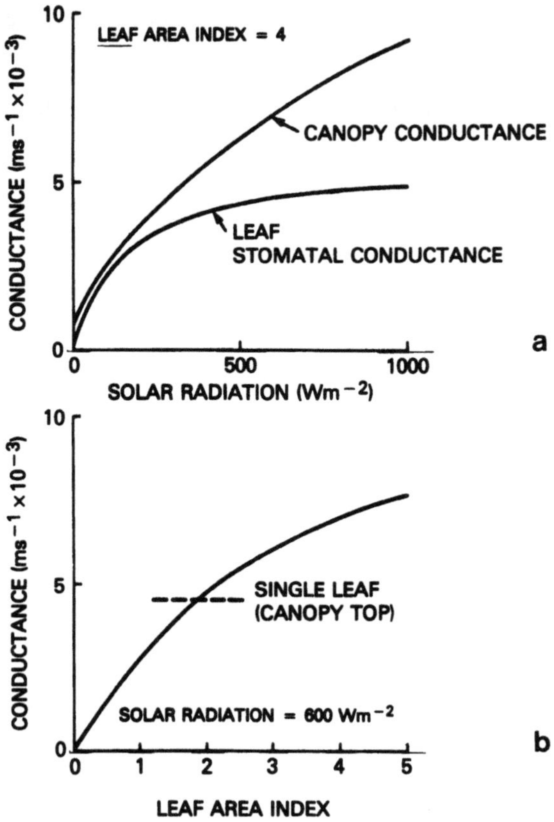

FIGURE 5.8. (a) Typical examples of the change in stomatal conductance with increasing solar radiation for a single leaf, and for a complete canopy made up of these leaves with a leaf area index of four; (b) change in the whole canopy stomatal conductance as a function of total leaf area for a canopy of the same leaves, with an incident, above-canopy solar radiation of 600 W m^{-2}.

where c_0 is a value characteristic of the forest, corresponding to the unstressed value at a specified leaf area index, green leaf fraction and leaf age, and an incident solar radiation of 1000 W m^{-2};

c_V is a vegetation cover function, representing the combined effect of seasonal changes in leaf area index, green leaf fraction, and foliage aging;

c_R is a radiation stress function, which is set to unity for an incident solar radiation of 1000 W m^{-2} [see Equation (5.55)];

c_D is an atmospheric saturation deficit stress function, which is unity for a saturated atmosphere;

c_T is a temperature stress function, which is unity at a (specified) optimum needle temperature, and falls on either side of this; and

c_{SM} is a soil moisture stress function, which is unity when the soil is at field capacity through the rooting zone.

The parametric content of the functions in Equation (5.56) changes in detail from one forest calibration to the next, but the size and form remains broadly similar. Some of these are illustrated for Shuttleworth's four "calibration forests" in Figure 5.9. Equation (5.56) might prove to have wider relevance, and the general behavior of component functions, illustrated in Figure 5.9, might not be untypical of other classes of vegetation.

3. The Spectrum of Evaporation Models

The multiplicity of techniques purporting to model evaporation is daunting, and the variation in their type and form disconcerting to those unfamiliar with the problem. The situation is further complicated by the fact that hydrologists have found it useful to create hypothetical evaporation rates, namely *potential evaporation*, *potential evapotranspiration*, and *reference crop evaporation*, which attempt to provide some measure of the atmosphere's ability to support the evaporation process. Some models attempt to estimate one or more of these conceptual entities, rather than actual evaporation. We make some attempt at clarification by expanding briefly on these.

3.1 Standard Evaporation Rates

It is clear from the physical description of evaporation given in the previous section that the rate of natural evaporation is in reality a complex function of atmospheric, soil, and vegetation factors. In an attempt to effect some simplification, it has been considered convenient to define certain standard rates of evaporation, which are designed to give a measure of the meteorological or climatological control on the evaporation process at a particular location. These standard rates are conceptual, in the sense that they represent idealized situations; they

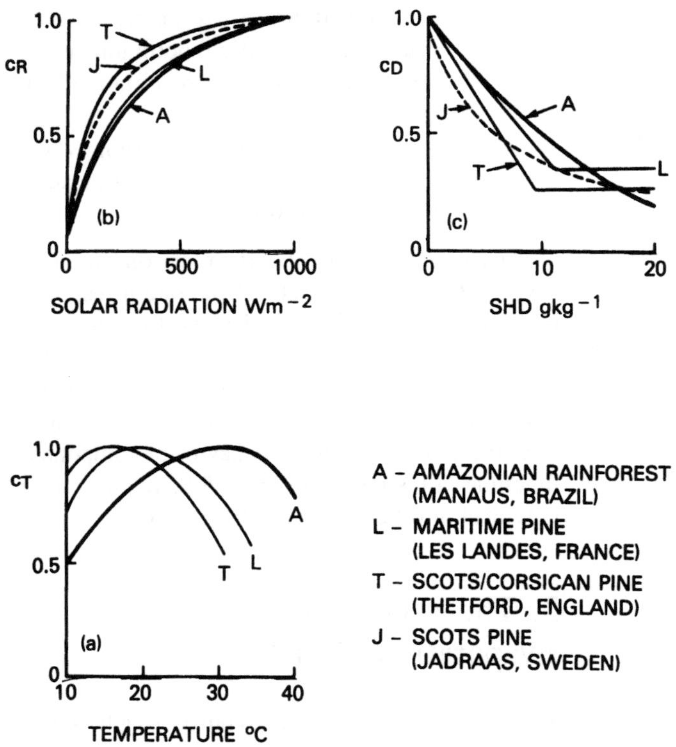

FIGURE 5.9. Typical behavior of normalized "stress factors" for temperature, c_T, solar radiation, c_R, and atmospheric saturation deficit, c_D, for forest vegetation (redrawn from Shuttleworth, 1989).

also maintain a level of imprecision in the definition of certain aspects, particularly those related to the surface, where this is considered advantageous.

Penman (1948) created the concept of potential evaporation, which might be defined as "the quantity of water evaporated from an idealized, extensive free water surface, per unit area, per unit time, under existing atmospheric conditions." This definition is imprecise, in that it does not specify certain criteria that would influence such evaporation in a real situation (e.g., the extent of energy transfer within the water, or the level of turbulent transfer above the water surface). Potential evaporation is nonetheless a conceptual entity that can be easily visualized, which will bear some reasonably close relationship to actual evaporation from particular free water sources and which will give some measure of the meteorological control on evaporation.

On the basis of the experimental evidence available, for many years it was believed that the type and form of vegetation cover on the earth's surface (even whether it was there or not) had little effect on the rate of natural evaporation, providing this was limited by the energy supplied to the surface, and not by the availability of surface water. In the presence of this belief, it is reasonable to conceive an entity, potential evapotranspiration, which might be conservative at a particular location and determined mainly by meteorological conditions. It can be defined (Gangopadhyaya et al., 1966) as "the maximum quantity of water capable of being lost, as water vapor, in a given climate, by a continuous, extensive stretch of vegetation covering the whole ground, when the soil is kept saturated." It includes both evaporation from the soil and from the vegetation for a specified region over a given time interval. The concept was (and is) used as a "scale" on which the influence of surface control can be superimposed, usually as a multiplication factor, often parameterized in terms of soil moisture deficit. The term remains in common use today, and still has some validity as a conservative entity, serving as a climatological index against which medium term measurements of actual evaporation can be interrelated at different geographic locations. It is increasingly recognized, however, that it is necessary to acknowledge an effect of vegetation type, and even local position, to preserve the entity's essential conservative nature as a local meteorological scale; an empirical estimation formula used to describe it requires local calibration where possible (Tanner, 1967).

Gradually it has become clear that the initial observation of a lack of dependence on vegetation cover when water is not limiting, and the idea that potential evapotranspiration represents a maximum rate (which implies that energy advection is a scarce and transient phenomenon) might be related to the fact that many of the original studies took place over short crops. With such crops the control exerted by the atmosphere is at a maximum, since it not only dictates the driving potential in the diffusion process, but also generates the dominant, possibly controlling, resistance to vapor transfer. This realization, coupled with the continuing desire to create a conservative (but surface independent) entity, has given rise to the creation of a better defined standard evaporation rate, reference crop evapotranspiration (Doorenbos and Pruitt, 1977). (Note: it should be remembered that in Penman's original work, λE_T referred to the evaporation from *turf*.) This entity is defined as "the rate of evapotranspiration from an extensive surface of 8 to 15 cm tall, green grass cover of uniform height, actively growing, completely shading the ground, and not short of water."

There is a considerable overlap between the concept of potential evapotranspiration and reference crop evapotranspiration, particularly in regard to the empirical formulas used to estimate them, but the better definition of reference crop evaporation minimizes the problems of vegetation control and advection, and therefore increases the probable universality of locally derived empirical equations.

All these "standard" evaporation rates are meant to be a measure of meteorological control at a particular location, with as little dependence on surface effects as possible; they choose to ignore the fact that meteorological parameters near the surface are themselves influenced by upwind surface exchange rates. In this respect, it should be remembered that if they are to fulfill their intended role and, in the form of a medium term average, remain conservative entities at a given location, there is an implicit assumption that there will be no drastic change in average surface cover for some distance upwind.

3.1.1 Crop Coefficients

In most practical applications it is actual evaporation which is required, but many evaporation models provide estimates of one of the standard rates, generally λE_{RC}. The conventional practice is then to multiply this standard rate by an additional factor, i.e.,

$$\lambda E = K_c \lambda E_{RC} \tag{5.57}$$

The factor K_c is called the crop coefficient. For the particular case of complete canopy cover, it can be seen [from Equation (5.28)] to be a time average measure of

$$K_c = \frac{\Delta + \gamma}{\Delta + \gamma(1 + r_s^s/r_A^c)} \tag{5.58}$$

where r_A^c and r_s^s are here the "effective" aerodynamic resistance and surface resistance of the crop. Such a term is, of necessity, extremely complex. It will contain dependence on the vegetation structure, through r_A^c, and its stomatal behavior, since this is the primary source of variation in r_s^s in dry conditions. It will also contain some explicit dependence on meteorological parameters, i.e., temperature in Δ, windspeed in r_A^c, and rainfall in r_s^s (indirectly in the amount of time the canopy is wet or partially wet). The stomatal control is itself also related to current or past meteorological parameters. Clearly the use of such a complex parameter should only be made with a proper realization of the risks involved. Although the basis and definition of the crop coefficient are essentially scientific, its complexity is such that its application is, perhaps, an art.

In practice, a great many hydrological applications of evaporation models concern the implementation of irrigation for agricultural crops. The objective is usually to supply water that is adequate, in that it does not limit growth, but not excessive, so that the soil surface is not always wet. Such conditions are essentially those used to specify potential evapotranspiration (except there is some implicit, poorly defined suggestion of a restriction to low rainfall conditions). In this case an entity, which we will call a "potential crop coefficient," K_{co}, might be relevant, and is used in the equation

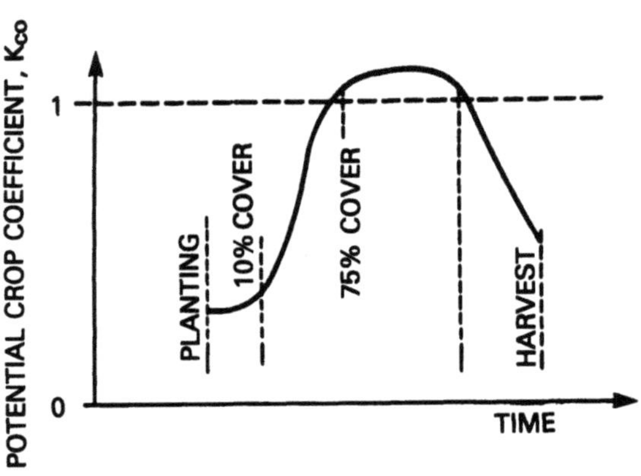

FIGURE 5.10. Typical behavior of the potential crop factor, K_{co}, as a function of time for a (seasonal) agricultural crop.

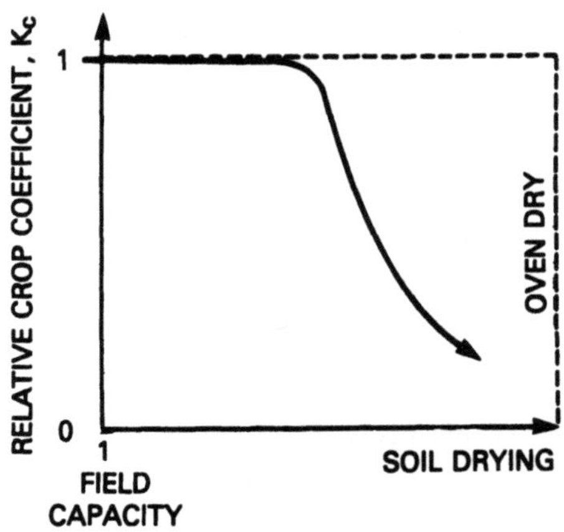

FIGURE 5.11. Typical behavior of the crop factor, K_c, in response to increasing soil dryness over the crop rooting zone.

$$\lambda E_T = K_{co}\, \lambda E_{RC} \tag{5.59}$$

It is quite possible that K_{co} could be less variable than K_c from one (meteorological) situation to the next, since r_s^s is likely to be a purer measure of stomatal resistance; this, in turn, is probably less variable, since the soil moisture deficit remains small. It will of course still have some local meteorological dependence through Δ and r_A^c.

Considerable work has been done on measuring the potential crop coefficient, K_{co}. As might be expected for agricultural crops (with which it is often used), there is a pronounced seasonal variation of the type illustrated schematically in Figure 5.10. Tables of values of K_{co} exist in the literature for certain parts of the world (e.g., United States and Israel), which may be used to provide an estimate of λE_T, using Equation (5.57), for different stages in the development of irrigated annual crops (see in particular pp. 37–44 of Doorenbos and Pruitt, 1977). Irrigated perennial and fruit crops do not have such a marked variation in K_{co}, so that tables summarizing empirical estimates of the potential crop coefficient are more usually presented as a function of the time of year (e.g., pp. 45–52, of Doorenbos and Pruitt, 1977).

If estimates of evaporation are required for nonirrigated crops, the water status of the soil can become important through surface resistance; K_c becomes more relevant in such situations. The problem is often compounded by the fact that such estimates are very commonly required for water resource estimation. This implies the presence of some significant rainfall, and a consequently confused definition of the effective surface resistance, involving both plant and meteorological parameters. In moist climates, where rainfall occurs fairly frequently throughout most of the year, an average (time-dependent) empirical description of K_c (and by implication of r_s) is probably the best that can be attempted. Here the situation is mitigated by the fact that for a great many short crops the actual evaporation rate is commonly quite close to reference crop evaporation most of the time ($K_c \simeq 1 \pm 0.1$). Tall crops (forests) remain the primary exception because of the possibility of advective enhancement of local evaporation rate (Shuttleworth and Calder, 1979).

In climates with large seasonal variations in rainfall, evaporation rates in the rainy season will probably again be close to λE_{RC} for short crops, but might exceed λE_{RC} for tall crops, depending on details of the interception mechanism. In the dry season K_c will usually fall significantly as the surface resistance responds to the increasing soil moisture deficit. Studies have been made of the variation in K_c in response to soil water content. The results can, of course, differ considerably, as might be expected for an empirical parameter of this type, but many workers are in agreement that the overall behavior in drying cycle follows the general pattern illustrated in Figure 5.11.

TABLE 5.1. A classification of evaporation models used in hydrology.

Model class	Evaporation estimated	Data requirement	Current usage
Simulation (e.g. 'SPAM', Waggoner-Reifsnyder)	λE *(actual)*	*Detailed models of physiological response, Detailed information on canopy exchange processes, Detailed information on canopy structure. Short term measurements of meteorological data.*	*Small, stable*
Single source (e.g. Penman-Monteith: Crop only Shuttleworth-Wallace: Crop/soil)	λE (actual)	Submodel of surface resistance, Coarse measurements of canopy structure. Short term measurements of meteorological data.	Small, increasing
Intermediate (e.g. Thom-Oliver-Gash)	λE_T	*Daily meteorological data. Coarse measurements of canopy structure. Information on rainfall pattern.*	*Minimal, stable*
Energy balance (e.g. Penman)	λE_{RC},λE_0	Daily meteorological data. (T, R_N, u, e)	Large, stable
Radiation (e.g. Priestley-Taylor)	λE_{RC},λE_0	Daily meteorological data. (T, R_N)	Medium, increasing
Humidity (e.g. Dalton)	λE_{RC},λE_0	Daily meteorological data. (T, u, e)	Small, decreasing
Temperature e.g. Blaney-Criddle)	λE_{RC},λE_0	Daily meteorological data (T)	Medium, stable

The qualitative behavior seems to be similar for both crops and soil. During the first stage K_c remains quite close to its initial value until a "wilting point" is reached; then it begins to decrease in response to decreasing soil water content. If a simple representation of the separate effects of both crop type and soil water depletion is required, it is expedient to write

$$\lambda E_T = K_c K_{co} \lambda E_{RC} \tag{5.60}$$

In conditions of prolonged drought, the crop (if present) begins to die and the evaporation rate is no longer controlled by meteorological conditions; it depends on soil characteristics, especially hydraulic conductivity.

3.2 Classification of Evaporation Models

At first it is difficult to discern any order or hierarchy in the range of evaporation models (usually just equations) available in hydrology. Here we classify them, mainly through the meteorological input they require. Part of the purpose is to attempt a partial explanation of why some of the formulas, with little apparent physical basis, might give rise to a valid statistical correlation with empirical constants of at least local significance. The classification, and the hierarchy it imposes, is also useful as a basis for understanding why particular models can, in the first instance, only realistically provide direct estimates of one of the standard evaporation rates rather than actual evaporation.

The classification scheme, outlined in Table 5.1, itemizes the various evaporation model classes, together with their data requirement, the type of evaporation they might legitimately provide (e.g., actual evaporation, potential evaporation), and a subjective assessment of current usage, and whether this is growing or declining. The scheme is meant to be general in terms of the broad classes described, but no attempt is made to include every example of the (sometimes numerous) alternative forms within each class. This is partly because doing so would tend to confuse rather than clarify, and partly because it is the author's opinion that argument regarding the relative merits of individual examples in each class is very often semantic in light of their empirical basis.

The reader familiar with evaporation modeling will recognize a deliberate bias toward newer techniques, some the subject of active development. In general, the trend in current research is toward the development of models capable of estimating actual evaporation directly. This usually means that the method has to attempt a closer description of reality by explicitly acknowledging the surface processes in the model itself. Such methods generally require a greater data input. The extent of present use of the various models is partly conditioned by familiarity, and largely by the current availability of relevant historical data. Popular acceptance of a particular evaporation model has often preceded the widespread availability of the data required for its use.

In the following sections we describe the various models in the descending order of Table 5.1, attempting to describe the additional assumptions involved in progressing from one class to the next as we do so.

3.2.1 Simulation Models

Clearly the technique most likely to estimate actual evaporation is to build a model that tries to simulate the physical and physiological processes that actually occur in the real situation. This can be attempted only with the availability of a large digital computer, and in the presence of a high level of detailed understanding of the basic processes giving rise to the energy fluxes. Such models are perhaps best used as a test of the validity of this understanding, rather than as a practical means of estimating evaporation rates.

Usually these models are built in one dimension, and attempt to simulate evaporation from vegetation by including all the information available for the vegetation stand under study, e.g., its structure and form, and submodels of its stomatal behavior in response to meteorological parameters. The model must also be supplied with short-term measurements of the meteorological conditions above the canopy as input, and then simultaneously solves all the equations describing the canopy using these as a boundary condition. In doing so, it generates simulated profiles of temperature, vapor pressure, and the heat fluxes. These can be compared with measurements as a test of the submodels used.

Several models exist of the general form illustrated in Figure 5.4. The vegetation is divided into a finite number of horizontal layers. About 10 layers are usually used, and for each layer the interception of solar and thermal radiation is calculated, and partitioned into sensible heat, latent heat, and photochemical energy. Iterative procedures are used until an energy balance is achieved for all foliage layers. The SPAM model, described in more detail by Sinclair et al. (1971, 1976), is an early, successful example. Such models must be considered the best available method of predicting actual evaporation, given an extremely high data availability; and providing the required submodels are available. It is not certain at present that universal, physiological submodels can be created, although research effort has been applied to the problem; it could be that fairly coarse submodels might suffice.

Shuttleworth (1976) demonstrated that numerical multilayer models of this type can, in fact, be rewritten in analytical form, by taking the limit corresponding to an infinite number of levels. The objective was to demonstrate that the technique of numerical, multilayer simulation on the one hand, and the description of bulk evaporation by a single equation on the other, are actually, and not just numerically, equivalent. Another advantage was that it is possible to exploit the analytic nature of his model in a simplifying process to demonstrate the assumptions involved in using "single source" models as a more practical alternative (Shuttleworth, 1978).

3.2.2 Single Source Models

Single source or "big leaf" models of plant canopies (e.g., Monteith, 1965; Thom, 1972), consider the overall effect of the whole canopy reasonably approximated by a model that assumes all the component elements of the vegetation are exposed to the same microclimate. In the general model (e.g., Shuttleworth and Wallace, 1985; Shuttleworth and Gurney, 1989), the sensible heat and latent heat generated at the "effective source sink height" in the canopy are merged with those from the soil beneath. They then pass through additional resistances to reach some level above the canopy, "the screen height," at which measurements of temperature and vapor pressure are made (see Fig. 5.5). In general, these resistances can differ from the equivalent resistance to momentum flux (Thom, 1972), and they can differ from each other, but, in practice assuming equality for resistances controlling energy transfer is a common and satisfactory assumption.

In dry conditions the latent heat flux arises inside the stomatal cavities and is subject to the diffusive resistance it encounters in negotiating the stomatal opening. If the canopy is totally wet, it arises at the surface of the vegetation, and the effective resistance to vapor exchange at a free water surface is negligible (Shuttleworth, 1975, 1978). The situation is more complex when the canopy is only partly wet.

The single source representation of transpiring, *closed canopies* is a well-established and well-tested model: it is already extensively described in the literature (see, for example, Thom, 1972). The hypothetical numerical similarity (Monteith, 1965) of bulk stomatal resistance and a vertical integration of the component stomatal resistances making up the canopy has been tested experimentally by Black et al. (1970), Brun et al. (1972), Szeicz et al. (1973), and Tan and Black (1976), numerically, against a multilayer model, by Sinclair et al. (1976), and theoretically by Shuttleworth (1976).

When the source of water vapor is a completely wet canopy of the vegetation, the term r_s^c in Equations (5.21) and (5.28) is zero. The effect of a change in this parameter on the observed evaporation is most marked for tall (forest) vegetation, although the effect is still present to some degree for other vegetation types. Most work in modeling the evaporation of intercepted water has been concerned with forest stands, where merging with a model of transpiration is most simple. The evaporation rate for intercepted water is sufficiently high to ensure that partly wet canopies are of short duration.

Although simulation and single source models are superior to all other techniques, in that they provide a direct estimate of actual evaporation, their use is inhibited by the current lack of short-term meteorological data sets, and the submodels of stomatal resistance required for their implementation. The alternative is to estimate one of the standard evaporation rates, and then to modify these by a crop factor, in an attempt to represent the real effect of vegetation (see Section 3.1.1). The remaining models in Table 5.1 are primarily used to provide the initial estimate of such standard rates.

The Penman equation (Penman, 1948), described later was initially created to provide an estimate of potential evaporation, λE_o, but Rijtema (1965) demonstrated that, providing the measured (or estimated) net radiation used in the equation is that relevant to the vegetation itself, it actually provides an estimate of potential evapotranspiration, λE_T, for short green vegetation. In the framework of the definitions used in this report, the Penman formula therefore provides an estimate of reference crop evapotranspiration, λE_{RC}. The recognition of a distinction between λE_{RC} and λE_T is based on the realization that λE_T can differ from crop to crop in response to different transpiration control, and different levels of advective enhancement. All models previously described as estimating λE_T are more correctly described as providing estimates of λE_{RC}.

3.2.3 Intermediate Models

An estimate of (the more variable) potential evapotranspiration would provide a better basis for the estimation of actual evaporation than λE_{RC}: Thom and Oliver (1977) have attempted to modify the original Penman formula to provide an equation including additional terms (with a physical basis), which could be adjusted empirically to yield an estimate of λE_T for different crops. The equation they propose takes the form

$$\lambda E = \frac{\Delta Q_N + \gamma E_a'}{\Delta + \gamma(1+n)} \tag{5.61}$$

E_a' is a modified version of the equivalent term in the Penman equation, namely

$$E_a' = 13.8(e_s - e)(1 + U/100)/\ell n^2(z/z_0) \text{ (mm day}^{-1}\text{)} \tag{5.62}$$

where e and e_s(mb) are the actual and saturation values of vapor pressure measured at a height z above d, the zero plane displacement of the crop ($d \simeq 0.75\, h$ where h is the vegetation height); U is the corresponding wind run in miles per day. z_0 is an estimate of the aerodynamic roughness parameter for the vegetation (of order $h/10$). Q_N (which appears in this equation and recurs in the next two sections) is the net energy available for evaporation at the evaporating surface. Q_N is most commonly a representation of $(R_N - G)$ in this work, where R_N is the net radiation, and G is variously interpreted as the heat flux conducted downward into the soil or free water surface.

The term n in Equation (5.59) is a measure of the control exerted from the crop by virtue of its stomatal resistance. It is defined as $n = r_s/r_a$ in which r_a is the effective aerodynamic resistance, already implicit in E_a', and given by the expression

$$r_a = 4.72[\ell n^2(z/z_0)]\,(1 + 0.54u)\ (\mathrm{s\ m^{-1}}) \qquad (5.63)$$

and r_s is the effective surface resistance estimated by

$$r_s = (1 - I/E)\,r_{sd} \qquad (5.64)$$

In this expression r_{sd} is an estimate of the average stomatal resistance for the crop, which is assumed constant. I is that part of the total evaporation, E, occurring directly from intercepted precipitation.

It is in this last equation that the Thom and Oliver approach loses some of its predictive potential by virtue of the fact that the parameter r_s, correctly, is not purely related to stomatal resistance. Obviously, Equation (5.59) can be calibrated against existing measurements of actual evaporation, to yield a value of r_{sd} and a time-dependent function (I/E); r_{sd} might be relevant in another uncalibrated application, but it is not clear that (I/E) will necessarily be so.

Gash (1978) pointed out that the Thom and Oliver function can be rewritten into a form where it might be used in conjunction with separate estimates (or measurements) of interception loss, to provide a worthwhile predictive equation. Presently such interception estimates are only really available for tall vegetation. The equation he proposed has the form

$$\lambda E = \frac{\Delta Q_N + \gamma E_a'}{\Delta + \gamma(1 + r_{sd}/r_a)} + I(1 - c) \qquad (5.65)$$

where in the additional term, I is the interception loss, and c (about 0.1 for forests) is a correction term which compensates for calculating the transpiration even under wet conditions, and is given by

$$c = (\Delta + \gamma)/[\Delta + \gamma\,(1 + r_{sd}/r_a)] \qquad (5.66)$$

Clearly Equation (5.65) is relevant only if the estimate of interception does not require short-term meteorological data, otherwise the method is merely another example of the (closed canopy) single source technique, with a constant value for surface resistance. In a later paper Gash (1979) proposed the use of a simple model of interception for tall crops which, neglecting evaporation from the trunks, takes the form

$$I = P_s + nS + (E/R)\,(P - P_s) \qquad (5.67)$$

where P is the precipitation input
- P_s is the fraction of precipitation in rain storms less than $S/(1-p)$
- S is the canopy storage
- n is the number of storms with precipitation greater than $S/(1-p)$
- R is the mean rainfall rate in storm conditions
- E is the mean evaporation rate from a totally wet forest canopy in storm conditions at the location under study

The three terms in this equation correspond to small storms (totally lost because they do not wet the canopy), to evaporation after storms, and to evaporation during storms, respectively. Such a description obviously requires some knowledge of the rainfall pattern for the area, and estimates of the parameters P and S. It implies that R can be made available as a climatological entity, and that E will ultimately prove not to have significant dependence on the type and form of forest structure (although it may vary from one broad climatic region to the next).

The Gash model is still in an early stage of development, but the Thom–Oliver–Gash Equation (5.65) might prove a predictive formula of λE_T for tall vegetation, while Equation (5.61) might be useful in this context for short vegetation, given the relevant empirical value for m. Thom and Oliver suggest $m = 1.4$, partly as a value generally consistent with typical values of r_{sd}, but, more particularly, because this is the value required to ensure that the use of Equations (5.61) and (5.62) yield the same annual estimate of λE_{RC} as the original Penman equation in the climate of southeast England where the formula was calibrated.

3.2.4 Energy Balance Models

The analysis of Thom and Oliver (1977) provides an explanation of the link between single source and simpler energy balance models, of which the Penman equation is the original and typical example. This equation takes the form

$$\lambda E_{RC} = \frac{\Delta Q_N + \gamma f(u)\,(e_s - e)}{\Delta + \gamma} \qquad (5.68)$$

and provides an estimate of potential evaporation, λE_0, if Q_N is a measurement of net energy input relevant to a free water surface, or reference crop evaporation, λE_{RC}, if Q_N is relevant to short, green vegetation (Rijtema, 1965). Thom and Oliver (1977) demonstrate that the wind function

$$f(u) = 0.26(1 + U/100) \qquad (5.69)$$

used in Equation (5.66) is implicitly serving two purposes:

1. It contains a reasonable, average description of the effect of thermal stratification on a more rigorous formula for aerodynamic resistance with an assumed value of sensible heat flux of $H = 50$ W m^{-2}, a value typical of the conditions under which the Penman equation was derived.
2. When compared to a physically rigorous formula for aerodynamic resistance in neutral conditions, the implicit (small) value of z_0 it contains is such as to compensate for the absence of an effective surface resistance in the denominator, for crops in which $r_s/r_a \simeq 1.4$, at least in the meteorological situation used to calibrate the original Penman formula.

The presence of this physical justification brings increased confidence to the use of Equation (5.68) as a basis for estimating λE_{RC}, such use being already widespread on the basis of past practical success. The necessary measurements are those of dry bulb temperature, wet bulb temperature, wind run and net radiation. The first three are fairly common observations but, unfortunately, net radiation is not always available as a measurement, and has to be estimated from solar radiation measurements, or measurements of sunshine hours. Doorenbos and Pruitt (1977) provide an excellent and simple description of how the Penman equation can be applied in practice. This is not repeated here.

Equation (5.68) (or minor modifications of it, not discussed here) is the simplest physically based equation which can be used to estimate λE_{RC}. Other meteorologically based classes of estimation techniques, lower in the hierarchy of Table 5.1, implicitly require additional empirical assumptions about the intercorrelation of those meteorological parameters that appear in the Penman equation. Such correlations do exist, and do have some intuitive physical basis, but there must always remain doubt as to the universality of such empiricism in moving from one climatic region to the next. The models lowest in Table 5.1 should be used with care, unless they can be calibrated against actual measurements of reference crop evaporation in local conditions.

3.2.5 Radiation Models

Of the remaining model classes, that which will most likely reserve some of the assumed universality of the Penman equation relates reference crop evaporation to radiation, or more correctly, to a combination of temperature and radiation. The equation we take as representative is that of Priestley and Taylor (1972), an equation with increasingly popular usage, which has the form

$$\lambda E_{RC} = \alpha \frac{\Delta}{\Delta + \gamma} Q_N \qquad (5.70)$$

where α is an empirical constant in the order of 1.26.

The conceptual basis for equations of this type is the presence of an empirical relationship between the two terms in the numerator of Equation (5.66), viz.

$$(e_s - e) \approx (\text{constant}) \frac{\Delta Q_N}{f(u)} \qquad (5.71)$$

Appropriate relationships of this type are observed and are not unreasonable on physical grounds. Increased radiation will indeed tend to generate increased temperature and increased humidity in the atmosphere, as a result of the in-going surface fluxes of sensible and latent heat. Since the relationship between saturated vapor pressure and temperature is nonlinear, a joint increase in both temperature and humidity will very probably give rise to the increase in vapor pressure deficit ($e_s - e$) required by this equation. Moreover, the rate of increase in the deficit caused in this way is greater at higher temperatures, possibly in rough proportion to W, which appears in the numerator of Equation (5.71). The size of the increase will probably be reduced at higher windspeed because the volume of air affected by the surface fluxes is increased. [Note the presence of $f(u)$ in the denominator.]

It should be emphasized immediately that the above argument is really no more than a rationalization, and should not be taken as evidence for Equation (5.70). At the same time it is worth noting Penman's (1956) observation that, for European conditions at least, the first term in the Penman equation is commonly four to five times as great as the second. This observation is consistent with the value $\alpha = 1.26$, and might be taken as some justification for allowing a higher level of empiricism in the less significant second term. The justification for using Equation (5.70) in preference to Equation (5.68) is that it does not require the measurement of humidity, which can be difficult. Such an equation would also be simpler to use in conjunction with satellite data to calculate reference crop evaporation for large, remote areas where few data are currently available.

3.2.6 Humidity Models

It is of course possible to express an assumed relationship between the two terms in the numerator of Penman's equation in the opposite form, and create empirical relationships between reference crop evaporation and vapor pressure deficit. Such expressions, which are modifications of the Dalton equation, have the general form

$$\lambda E_{RC} = f'(u)\,(e_s - e) \qquad (5.72)$$

in which f' is an empirical expression derived for the location of interest. Clearly the universality of such an equation is likely to be less than those involving radiation, since in this case the smaller of the two terms in Equation (5.68) is being used to estimate the greater. Moreover, humidity data are more difficult to obtain than radiation data. When they are available in conjunction with windspeed, it is often possible to use the Penman equation as a superior estimate, with estimates (or measurements) of net radiation.

3.2.7 Temperature Models

Several empirical formulas exist which relate λE_{RC} to temperature. In keeping with the philosophy of this chapter we select one example, the Blaney-Criddle (1950) equation, on no other grounds than that it is widely used. The equation is designed to provide daily estimates of evaporation averaged over a month and takes the form

$$\lambda E_{RC} = C_u d_l\,(0.46T + 8) \qquad (5.73)$$

where C_u is the so called "consumptive use" factor, reported in engineering literature but best determined locally

d_l is the fraction of daylight hours occurring in the month

T is the temperature in °C (Note that the numerical constants in this equation merely reflect the fact that empirical constants have been derived with temperature in °F).

Clearly, the physical basis for this equation is that both the terms in the numerator of the Penman equation are likely to have some, albeit ill-defined, relationship with temperature. Since the radiation term is the larger of the two, it is the correlation between radiation and temperature, which is more important. There is a lag in the yearly temperature cycle with respect to the yearly radiation cycle, and the monthly "consumptive use" factor (unwittingly) includes some allowance for this thermal lag.

The only real justification for using models of this type is that an estimate of evaporation is required on the basis of existing data, and temperature is the only measurement available. In this situation the work of Doorenbos and Pruitt (1977) provides a useful, practical reference.

4. Concluding Remarks

Section 2 of this chapter provides a review of our present understanding of physical and physiological aspects of the natural evaporation process for land surfaces. Section 3 provides an overview of the broad range of models available to hydrologists, and explores their interrelationships. We conclude by commenting briefly on the relevance of these models to meteorology, and speculate on the development of evaporation models at the relevant (larger) scale, and on progress toward providing these.

4.1 Model Options for Meteorology

The selection of viable candidate models of land surface evaporation for use in meteorology from those available in hydrology is most easily made on the basis of Table 5.1. In hydrology, use of models in the lowest classifications of that hierarchy is justified by the need to estimate evaporation from preexisting measurements of a few meteorological variables: the selected model reflects this availability. This is not the case when formulating meteorological models. Synthesizing the near surface meteorology is then part of the exercise: there is no point using a model drawn from a class with less physical (more empirical) justification, merely because it requires less meteorological variables as input. Temperature models, humidity models, and, with less justification, radiation models can be discarded in preference to energy balance models, as the simplest option for use in meteorology. In practice, a meteorological model will usually seek to describe actual evaporation rather than the reference crop potential rate. It will therefore be necessary to resort to Equation (5.60) to take at least some account of soil moisture status, if not crop type.

Starting at the top of Table 5.1, it is not difficult to reject simulation models as a viable option for use in meteorological modeling. Their use would involve a multilayer computation comparable to that in the rest of the climate model, and this could not be justified. A single source model that allows soil evaporation [see Equation (5.20)] is therefore the most sensible compromise between realism and practicality. Recent progress in formulating simple (though approximate) expressions to describe the response of the component resistances to changing leaf area and crop height, reviewed in Section 2.5 of the present paper, encourages this suggestion.

Having specified the most complex, but realistic, option as the single source model, and the simplest, but least realistic, option as the energy balance model, one class of hydrological model, the intermediate model, remains. There is some possibility that this class could be of use to meteorologists, perhaps as an interim step in progress between the already commonly used energy balance model and the single source model, which is becoming more popular.

4.2 Large-Scale Aggregate Descriptions and Their Calibration

Regardless of the complexity of the model option selected to provide a single point description of evaporation, there remains the problem of providing a suitable model description of this same process at the larger area scale use in climate models. In a recent review, Shuttleworth (1988) speculates on aspects of this problem and suggests that the single point model remains the appropriate format (irrespective of the selected model option). At least this format ensures the model captures the essence of the hydrology, and has the correct asymptotic limit. There is, however, an important corollary; the parameters used in the single point model must lose direct physical and physiological relevance and, in general, become a synthesis between local values and subgrid statistics. The canopy storage for rainfall interception is a good example. This is investigated by Shuttleworth (1988). His analysis suggests the *effective* canopy store for a meteorological scale grid of uniform vegetation might be a factor three less than the local value for convective precipitation.

The data necessary to calibrate these equivalent single point models are simple to define but difficult to provide. Land surface parameterizations (LSPs), as they are called, are required to describe surface fluxes of radiant energy, momentum, evaporation, and sensible heat when provided with model-generated time series of "driving variables," these being the incoming radiation and precipitation, and the temperature, humidity, and windspeed at a level close to the ground. To give adequate climate simulation, the LSP is required to operate realistically at the hourly time scale or less. Ideally, therefore, the appropriate calibration data are measured time series of the driving variables and ensuing fluxes, available as grid mesh scale average values, sampled at (say) hourly intervals for an extended period, preferably for several years.

In practice the direct measurement of calibration data with precisely these characteristics is not simple. In principle, the long-term average calibration

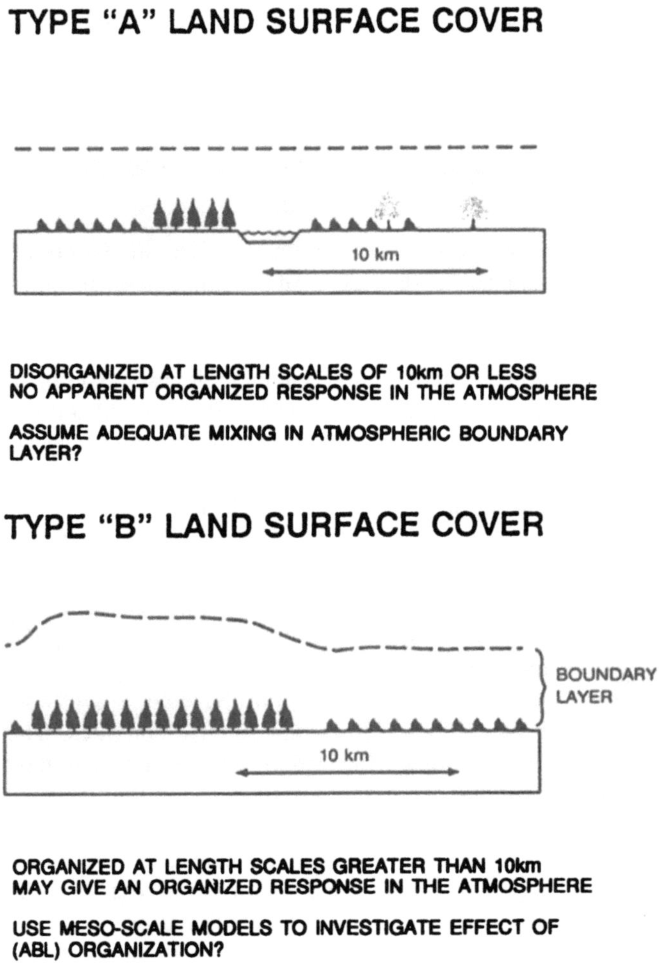

FIGURE 5.12. Types of land surface cover classified according to the scale of organization at the surface and ensuing response in the atmospheric boundary layer.

can be checked against traditional catchment-scale hydrological data (viz. rainfall and runoff) for areas where this is extensively and reliably available. At the other extreme, direct measurements using aircraft ought to be able to provide occasional short-term "snapshots" of the required data at the relevant area scales. Considerable effort is currently being deployed toward building algorithms relating the required data to that available from remote sensing techniques, or from soundings of the atmospheric boundary layer. There has been significant success in remote sensing area-average radiation fluxes, but otherwise such research is still in its infancy. So far, arguably, the best method is through multiple sampling of surface fluxes for characteristic cover types using micrometeorological instrumentation, and then synthesizing area-average values. Usually the manpower and instrumental resources required for this mean such data are collected through field experiments involving international collaboration.

The methodology required to synthesize such short time scale, large area scale data is not yet firmly established but the recent multisite land surface observational studies have proved revealing. An important experiment carried out under the International Satellite Land Surface Climatology Project (ISLSCP), the First ISLSCP Field Experiment (FIFE), studied the interaction of reasonably uniform prairie grass vegetation over a 15 × 15 km area with strong topography. Results suggest that over the 20 sample sites, the proportion of the available radiant energy used for evaporation is the same within an (albeit large) experimental error (Shuttleworth et al., 1989). In this case the appropriate aggregation procedure is thus to form a simple average.

On the other hand, if there is significant variation in surface cover, several different vegetation types, for example, then the appropriate mechanism for synthesizing aggregate fluxes seems to depend on the scale of organization in these surface features (Shuttleworth, 1988). This hypothesis, which is suggested by the results from the Hydrological Atmospheric Pilot Experiment (HAPEX), is illustrated in Figure 5.12. For surface cover disorganized at length scales of about 10 kilometers or less (which is quite common), the turbulent atmospheric boundary layer (ABL) could well provide a mixing zone such that the appropriate aggregation is the average over the sampled vegetation types, weighted by their proportional cover. However, when surface cover is organized at scales bigger than this, there may well be an organized response in the atmosphere such as to modify (usually to moderate) the effect of those land surface properties that help create the response. Numerical mesoscale modeling seems an appropriate mechanism to investigate the significance of this modification over landscapes with organized change at scales greater than 10 km, and to synthesize the required area average values.

References

Black TA, Tanner CB, Gardner WR (1970) Evapotranspiration from a snap bean crop. Agron J 62:66–69.

Blaney HF, Criddle WO (1950) Determining water requirements in irrigated areas from climatological and irrigation data, USDA (SCS) TP-96, p. 48.

Brun JL, Kanemasu ET, Powers WL (1972) Evapotranspiration from soybean and sorgum fields. Agron J 64:145–148.

Camillo PJ, Gurney RJ (1986) A resistance parameter for bare-soil evaporation models. Soil Sci 141(2): 95–105.

Chamberlain AC (1966) Transport of gases to grass and grass like surfaces. Proc R Soc London Ser A 290:236–260.

Choudhury BJ, Monteith JL (1988) A four-layer model for the heat budget of homogeneous land surfaces. Quart J R Meteorol Soc 114:373–398.

Dickinson RE, Henderson-Sellers A, Kennedy PJ, Wilson MF (1986) Biosphere-atmosphere transfer scheme (BATS) for the NCAR Community Climate Model, Tech Note/TN-275 + STR, National Centre for Atmospheric Research, Boulder, CO.

Doorenbos J, Pruitt WO (1977) Crop water requirements, F.A.O. Irrigation and Drainage Paper, 24, Rome, Italy.

Dyer AJ, Hicks BB (1970) Flux-gradient relationships in the constant flux layer. QJR Meteorol Soc 96:715.

Gangopadhyaya M, Uryvaev VA, Omar MH, Nordenson TJ, Harbeck GE (1966) Measurement and estimation of evaporation and evapotranspiration, W.M.O. Technical Note No. 83, Geneva, Switzerland.

Gash JHC, Shuttleworth WJ, Lloyd CR, Andre JC, Goutorbe JP, Gelpe J (1989) Micrometeorological measurements in Les Landes forest during HAPEX-MOBILHY. Agric For Meteorol 46(1–2):131–147.

Gash JHC (1978) Comment on the paper by A.S. Thom and H.R. Oliver "On Penman's equation for estimating regional evaporation." Quart J R Meteorol Soc 104: 532–533.

Gash JHC (1979) An analytical model of rainfall interception by forests. Quart J R Meteorol Soc 105: 43–55.

Halldin S, Lindroth A (1986) Pine forest microclimate simulation using different diffusivities. Boundary Layer Meteorol 35:103–123.

Jones HG (1983) "Plants and Microclimate," pp. 53–54. Cambridge University Press, New York.

Monteith JL (1965) Evaporation and the environment. Symp Soc Exp Biol 19:205–234.

Monteith JL (1973) "Principles of Environmental Physics." Edward Arnold, London.

Monteith JL (1981) Evaporation and surface temperature. Quart J R Meteorol Soc 107:1–27.

Penman HL (1948) Natural evaporation from open water, bare soil and grass. Proc Roy Soc London. Ser A 193, 120–145.

Penman HL (1956) Evaporation. An introductory survey. Natl J Agric Sci 4:9–29.

Priestley CHB, Taylor RJ (1972) On the assessment of surface heat flux and evaporation using large scale parameters. Mon Weather Rev 100:81–92.

Raupach MR (1989) Stand overstorey processes. Phil Trans R Soc London Ser B 324:175–190.

Rijtema PE (1965) An analysis of actual evaporation. Agric Res Reports 659, (Pudoc, Wageningen), p. 107.

Rutter AJ, Kershaw KA, Robins PC, Morton AJ (1971) A predictive model of rainfall interception in forests, I: Derivation of the model from observations in a plantation of Corsican Pine. Agric Meteorol 9:367–380.

Rutter AJ, Morton AJ, Robins PC (1975) A predictive model of rainfall interception in forests. II: Generalization of the model and comparison with observations in some coniferous and hardwood stands. J Appl Ecol 12:367–384.

Sellers PJ, Mintz Y, Sud YC, Dalcher A (1986) A simple biosphere model (SiB) for use within General Circulation Models. J Atmos Sci 43(6):505–531.

Shaw RH, Pereira AR (1982) Aerodynamic roughness of a plant canopy: A numerical experiment. Agric Meteorol 26:1498–1501.

Shu Fen Sun (1982) Moisture and heat transport in a soil layer forced by atmospheric conditions, M.Sc. Thesis, Univ. of Connecticut.

Shuttleworth WJ (1975) The concept of intrinsic surface resistance: Energy budgets at a partially wet surface. Bound Layer Meteorol 8:81–99.

Shuttleworth WJ (1976) A one-dimensional theoretical description of the vegetation-atmosphere interaction. Bound Layer Meteorol 10:273–302.

Shuttleworth WJ (1978) A simplified one-dimensional theoretical description of the vegetation-atmosphere interaction. Bound Layer Meteorol 14:3–27.

Shuttleworth WJ (1988) Macrohydrology: The new challenge for process hydrology. J Hydrol 100:31–56.

Shuttleworth WJ (1989) Micrometeorology of temperate and tropical forest. Phil Trans R Soc London Ser B 324:299–334.

Shuttleworth WJ, Calder IR (1979) Has the Priestley-Taylor equation any relevance to forest evaporation? J Appl Meteorol 18:634–638.

Shuttleworth WJ, Wallace JS (1985) Evaporation from sparse crops—an energy combination theory. Quart J R Meteorol Soc 111:839–855.

Shuttleworth WJ, Gurney RJ (1989) The theoretical relationship between foliage temperature and canopy resistance in sparse crops. Quart J R Meteorol Soc 116:497–519.

Shuttleworth WJ, Gurney RJ, Hsu AY, Ormsby JP (1989) FIFE: The variation in energy partition at surface flux sites. Remote sensing and large-scale global processes (Proc. IAHS 3rd Int. Assembly, Baltimore), IAHS Publ. No. 186:67–74.

Sinclair TR, Allen LH, Jr, Stewart DW (1971) A simulation model for crop-environmental interactions and its use in improving crop productivity, Proc. 1971 Summer Computer Simulation Council, Inc., La Jolla, CA, p. 784.

Sinclair TR, Murphey CE, Knoerr KR (1976) Development and evaluation of simplified models simulating canopy photosynthesis and transpiration. J Appl Ecol 13:813.

Stewart JB (1988) Modelling surface conductance of pine forest. Agric For Meteorol 43:19–37.

Szeicz G, van Bavel CHM, Takami S (1973) Stomatal factor in the water use and dry matter production by sorghum. Agric Meteorol 12:361–389.

Tan CS, Black TA (1976) Factors affecting the canopy resistance of a Douglas fir forest. Bound Layer Meteorol 10:475–488.

Tanner CB (1967) Measurement of evapotranspiration. In "Irrigation of Agricultural Lands, Agronomy II," p. 534–574.

Thom AS (1972) Momentum, mass and heat exchange of vegetation. Quart J R Meteorol Soc 98:124–134.

Thom AS, Oliver HR (1977) On Penman's equation for estimating regional evaporation. Quart J R Meteorol Soc 103:345–357.

Waggoner PE (1975) In "Vegetation and the Atmosphere," (J.L. Monteith, ed.), Vol 1, pp 205–228. Academic Press, London.

Waggoner PE, Reifsnyder WE (1968) Simulation of the temperature humidity and evaporation profiles in a leaf canopy. J Appl Meteorol 7:400–409.

Webb EK (1970) Profile relationship: The log-linear range, and extension to strong stability. Quart J R Meteorol Soc 96:67–90.

6
A Two-Dimensional Model of the Hydrological Response of a Hillslope

Peter J. Camillo, Robert J. Gurney, and Judith E. Devaney

1. Introduction

One important part of the global hydrological system is a catchment, which separates rainfall into evaporation, overland flow, and infiltration. For a heavy rain, infiltration excess reaches the stream first as overland flow. Part of the infiltrated water may then flow rapidly below the surface to reemerge downslope or enter the stream. This is usually referred to as saturated subsurface flow. The rest reaches the unsaturated zone. The flow there is vertical and horizontal, and the latter component may eventually contribute to the stream flow. Another component that can contribute to the stream flow is horizontal flow in a perched water table above the bedrock. These processes are depicted schematically in Figure 6.1.

The primary output of catchment models is the hydrograph, in which the rainfall and fluxes to the stream from each of the above processes are plotted as a function of time. The rainfall rate and the sum of all the output fluxes are the usual data from a catchment, and a primary goal of catchment modeling is to understand the sensitivity of the output to the physical characteristics of the catchment, such as topography, cover type, soil characteristics, and antecedent moisture.

Betson and Ardis (in Kirkby, 1972) define catchment models as being of three basic types, but with overlapping characteristics so they may be considered a continuum. The first is stochastic. These models are statistical, in which time series of measured hydrographs (output) are correlated to rainfall (input) using classical time series analysis techniques. This leads quite naturally to parametric models, their second class, in which the parameters of the stochastic models are related empirically to the physical properties of the catchment. The third class contains deterministic models based on the laws of conservation of energy, mass, and momentum, usually expressed as time- and space-dependent differential equations. Since such models generally contain nonmeasurable parameters that must be calibrated, deterministic models are partly parametric. Other variables are also spatial and/or temporal averages, and are consequently difficult to measure over a complete hillslope. There are many deterministic catchment models, but none of them includes all of the processes in the hydrological cycle. This is due to the extreme complexity and variability of natural catchments (Freeze, 1971; Eagleson, 1970; Freeze, in Kirkby, 1972; Smith and Woolhiser, 1971; Bathurst, 1986).

The concept of partial (or contributing) areas is one basis of the understanding of how catchments distribute rainfall (van de Griend and Engman, 1985). Due to the spatial variability of catchment characteristics (soils, cover, topography), different areas handle the rain in different ways. For example, if the rain rate exceeds the infiltration capacity for a particular area, then the excess rain becomes overland flow. Once the soil is saturated, the water can flow rapidly below the surface and parallel to it. This process is referred to as saturated subsurface flow. The water will reemerge somewhere downslope, adding to overland flow. The areas change over time, so the saturated partial area that contributes to overland flow varies in time as well as in space.

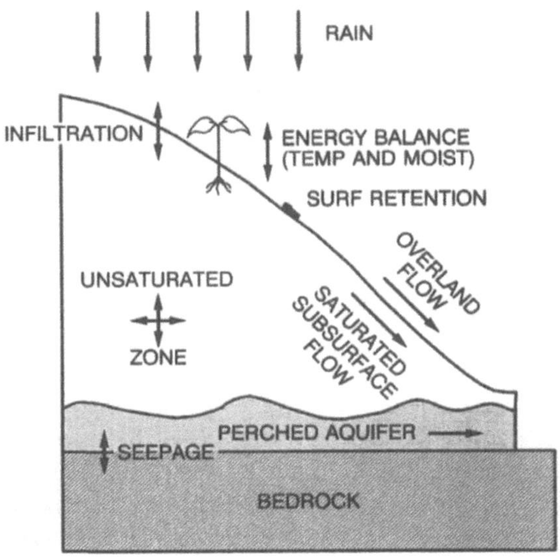

FIGURE 6.1. Typical hydrological processes in a catchment model.

These responses are complex and highly variable from one catchment to another, but after a heavy rain the first peak in the hydrograph is usually due to overland flow, and the second is contributed by saturated subsurface flow. These are the fast responses. Water that reaches the unsaturated zone may flow horizontally to the surface and vertically to the water table. These components may take days or weeks to reach the stream.

This model may be compared to existing catchment models, for instance the Systeme Hydrologique European (SHE) (Bathhurst, 1986) and TOPMODEL (Beven and Kirkby, 1979; Beven et al., 1984).

The SHE model includes the following components:

Surface retention and evapotranspiration (Penman-Monteith equation)
Snowmelt
Overland flow
Vertical flow in the unsaturated zone
Horizontal flow in a single layer unconfined aquifer.

The SHE model is three dimensional. The surface is divided into grid squares (typically 250 × 250 m) with a few soil layers. Each such box is assumed to be homogeneous. Some of the flux components are modeled by the partial differential equations for mass and energy conservation, and the rest by bucket models.

TOPMODEL is also three dimensional, with the surface divided into homogeneous areas. However, there are no soil layers; it is primarily a bucket model. There are no space and time dependent partial differential equations. TOPMODEL includes the following components (Beven et al., 1984):

Surface retention
Evaporation at the potential rate
Near surface retention, fed by infiltration from the surface store
Unsaturated zone retention
Overland flow
Unsaturated zone flow.

The MPP hillslope model includes the following components:

Surface retention
A complete surface energy balance (temperature and moisture) with separate evaporation rates from the soil, plants (with water extraction from the unsaturated zone), and surface retention
Overland flow
Saturated subsurface flow parallel to the surface
Horizontal and vertical flow in the unsaturated zone
Horizontal flow in an unconfined aquifer.

The MPP model is a vertical slice of a hillslope, so it is basically a two-dimensional model. It may be considered three dimensional only if the gradients are all downslope, not across the slope. It is based on a catchment model of Hillel and Hornberger (1979), which is simply a series of uncoupled one-dimensional soil columns placed side by side. We have improved their design by allowing for horizontal flow in the unsaturated zone between the columns, and including the soil and surface temperatures.

It was decided at the beginning of this research effort to create one-, two-, and then three-dimensional models in succession. The one-dimensional model (Devaney et al., 1987) was compared to a similar one which runs on a serial machine (Camillo et al., 1983; Gurney and Camillo, 1984) to make sure the equations are solved correctly on the MPP, and as a timing benchmark. After the two-dimensional model is completely tested, a three-dimensional version will be developed.

There were additional reasons for doing the modeling on a parallel computer such as the MPP. The MPP is a 128 × 128 array of processing units, each with associated memory. It is well known that such computers are theoretically very fast when solving arrays of partial differential equations, particularly on spatial problems. Problems related to global change, such as long-term coupled atmosphere/ocean/land general circulation models, are very computationally intensive and theoretically complex, therefore lend themselves to solution on massively parallel computers. However, memory and I/O problems have kept this class of computers from general adoption yet for this problem. The hillslope problem involves solving many of the same types of equations with the same boundary condition problems, but in a more limited setting. Hence, it was decided to write a coupled heat and moisture flow model on the MPP to explore the problems of programming this class of model in a limited way.

2. The Two-Dimensional Hillslope Model

The specifications for each component of the model given in the first section are described here as flux and continuity partial differential equations. The method of solution is also briefly described.

2.1 Unsaturated Zone

The soil moisture in a cell, θ, is the solution of the continuity equation (Philip and de Vries, 1957)

$$\frac{\delta\theta}{\delta t} = -\nabla\cdot\mathbf{Q} - S = -\frac{\delta(QX)}{\delta x} - \frac{\delta(QX)}{\delta z} - S \tag{6.1}$$

S is water extracted by plant roots. Its calculation is described in Gurney and Camillo (1984). $\mathbf{Q}$ is the unsaturated zone moisture flux. There are two components, one in each direction:

$$QX = -KX(d\Psi/dx) \tag{6.2a}$$

$$QZ = -KZ(d\Psi/dz - 1) \tag{6.2b}$$

KX and KZ are the hydraulic conductivities in the horizontal and vertical directions respectively, and Ψ is the soil matrix potential. These are estimated with the parameterization (Clapp and Hornberger, 1978)

$$K(\theta) = K_s(\theta/\theta_s)^{2b+3} \tag{6.3a}$$

$$\Psi(\theta) = \Psi_s(\theta/\theta_s)^{-b} \tag{6.3b}$$

in which θ_s, K_s, and Ψ_s are moisture content, conductivity, and potential at saturation. The value of b depends on soil texture. K_s may have different values in the x and z directions.

Boundary value fluxes must be specified for moisture at the top and bottom of the hillside (vertical direction) and at the hillslope divide and surfaces (horizontal direction). The top boundary flux is the infiltration or evaporation rate, computed from the surface energy balance. The horizontal flux into the hillslope at the divide is zero. The horizontal flux at the hillslope surface depends on whether that cell is saturated. If it is and if the sum of the vertical fluxes plus the horizontal flux into the cell from the interior of the hillslope would cause soil moisture to exceed saturation, then the flux onto the surface is set to whatever value is needed to keep moisture just at saturation. Otherwise, it is zero. This is the mechanism that allows subsurface return flow.

The temperature profile in the soil is usually modeled with Fourier's equations. However, we have chosen to implement the computationally simpler yet physically adequate force-restore equations (Lin, 1980) because of memory limitations of the MPP. The surface and deep soil temperatures, T_s and $\bar{T}$, are modeled by:

$$\frac{\delta T_s}{\delta t} = \frac{2G}{a} - \frac{2\pi}{\tau}(T_s - T) \tag{6.4a}$$

$$\frac{\delta T}{\delta t} = \frac{G}{aN365\pi} \tag{6.4b}$$

where $a = N\lambda c\tau/\pi$, G is the soil-surface heat flux, λ and c are, respectively, the soil thermal conductivity and heat capacity, and τ is the length of day. T is the temperature at the depth where fluctuations are seasonal rather than diurnal. For most soils this depth is about 2 m. The conductivity and heat capacity are modeled as functions of soil moisture and soil type with the de Vries model (de Vries, 1975).

To solve the continuity equations for θ, T_s, and T, boundary conditions must be supplied for moisture and temperature both at the air/soil interface and in the bottom layer of the profile. In principle, either the fluxes q_θ and G or the variables θ and T could be specified. In the model, soil heat flux and soil evaporation rate are computed by the surface energy balance equation to provide flux boundary conditions.

2.2 Saturated Zone

The water table height in each column is HB. The horizontal flux is QB, and the vertical flux is QZ. The fluxes and vertical boundary conditions are calculated by the one dimensional Boussinesq equation (Pikul et al., 1974):

$$QB = -\frac{1}{2} K_{\text{sat}} \frac{d}{dx}(HB^2) \tag{6.5a}$$

$$\frac{QZ}{\text{top}} = KZ \tag{6.5b}$$

$$\frac{QZ}{\text{bottom}} = QZ_{\text{bot}} \tag{6.5c}$$

where K_{sat} is the horizontal saturated conductivity of the saturated zone, KZ is the flux into the water table from the unsaturated zone and is modeled as the vertical hydraulic conductivity of the layer, and QZ_{bot} is an input parameter representing an impervious layer ($QZ_{\text{bot}} = 0$) or upward or downward seepage ($QZ_{\text{bot}} \neq 0$).

The time derivative is

$$\theta_s \frac{d(HB)}{dt} = -\nabla \cdot Q - \frac{d(QB)}{dx} - QZ_{\text{bottom}} - QZ_{\text{top}} \tag{6.6}$$

where θ_s is the moisture content at saturation in the saturated zone.

QB at the catchment divide is set to zero. At the seepage face the height HB is a fixed input parameter. Therefore the time derivative of HB in Equation (6.6) is zero for the last column, and the discretized form of Equation (6.6) may be solved for the horizontal flux QB at the seepage face. This is the saturated zone flux which contributes to the hydrograph.

2.3 Overland Flow

If the surface water height is larger than a critical value, the overland flow flux is determined by Manning's equation:

$$QT = \frac{N \sin \alpha}{n} HT^{5/3} \tag{6.7}$$

where α is the slope of the surface and n is an input surface roughness coefficient.

The rate of change of the water surface height is

$$\frac{d(HT)}{dt} = -\frac{d(QT)}{dx} + R - I + E_{\text{surf}} \tag{6.8}$$

R, I, and E_{surf} are rain rate, infiltration rate, and the evaporation rate of the ponded water, respectively. Rainfall rates are input and evaporation comes from the surface energy balance. The boundary value for HT is zero at the catchment divide. The overland flow flux which contributes to the hydrograph is calculated directly from Equation (6.7).

The infiltration rate is determined from the movement of water in the soil, described earlier.

2.4 Surface Energy Balance

The energy balance equation provides the surface fluxes:

$$G = R + LE + H \tag{6.9}$$

All fluxes are positive downward. G is the heat absorbed by the soil, R is the net radiation flux, LE is the evapotranspiration energy flux, and H is the sensible heat. After finding the solution, the surface moisture flux q_θ is set equal to the soil evaporation rate, and G is used in the force-restore model [Equation (4)]. The surface temperature needed to evaluate the fluxes is known from the force-restore equation.

The net radiation R is divided into short- and long-wavelength components:

$$R = R_{SHORT} + R_{LONG} \tag{6.10}$$

Either or both components may be estimated or measured. All four options are allowable within the computer program, with standard models such as the Brunt model for longwave radiation:

$$R_{LONG} = \varepsilon_s (\sigma\varepsilon_a T_a^4 - \sigma T_s^4) \tag{6.11a}$$

where σ is the Stefan–Boltzmann constant, ε_s is the surface emissivity, and ε_a is the atmospheric emissivity which may be calculated from

$$\varepsilon_a = 0.74 + 0.005\varepsilon_s \tag{6.11b}$$

where ε_a is the atmospheric vapor pressure in millibars.

If needed, the short wave component may be modeled by (Eagleson, 1970):

$$R_{SHORT} = (1 - ALB) I_0 \mu_0 \exp(-n/\mu_0) \tag{6.11c}$$

where ALB is the solar albedo and is modeled as a polynomial in μ_0, I_0 is the solar radiance at the top of the atmosphere, μ_0 is the cosine of the solar zenith angle, and n accounts for atmospheric absorption. The cosine is calculated from:

$$\mu_0 = \sin\delta \sin\phi + \cos\delta \cos\phi \cos\tau \tag{6.12}$$

where δ is the angle between the plane of the Sun and the celestial equator, ϕ is the latitude, and τ is the hour angle of the Sun, which depends on the time of day.

The latent and sensible heat fluxes are derived from the usual resistance formulations. The soil and vegetation are considered as one surface with temperature T_s. Some surface water storage is allowed. This affects the evaporation rates, because the surface resistance is zero for the fraction of the evaporation that comes from the stored water. With P_{FRAC} the fraction of evaporation from plants and X_s the fraction of the surface covered by water:

$$X_s = HT/HT_{MAX} \tag{6.13}$$

Then

$$LE_{plant} = -P_{FRAC} \frac{\rho c_p}{\gamma} \frac{e_{sat}(T_s) - e_a}{R_2} \tag{6.14}$$

$$LE_{soil} = -(1 - P_{FRAC}) \frac{\rho c_p}{\gamma} \frac{h(\theta, T_s)\, e_{sat}(T_s) - e_a}{R_1} \tag{6.15}$$

where e_{sat} is the saturated vapor pressure at temperature T_s, and $h(\theta, T_s)$ is the relative humidity of the soil. Models for these may be found in Camillo et al. (1983).

The resistances are

$$\frac{1}{R_1} = \begin{cases} \dfrac{X_s}{R_{SA}} + \dfrac{(1 - x_s)}{R_{SA} + R_S} & he_{sat} > e_a \text{ (evaporation)} \\[2ex] \dfrac{1}{R_{SA}} & he_{sat} < e_a \text{ (dew)} \end{cases} \tag{6.16a}$$

$$\frac{1}{R_2} = \begin{cases} \dfrac{X_s}{R_{CA}} + \dfrac{(1 - X_s)}{R_{CA} + R_{CAN}} & he_{sat} > e_a \text{ (evaporation)} \\[2ex] \dfrac{1}{R_{CA}} & he_{sat} < e_a \end{cases} \tag{6.16b}$$

$$R_{SA} = \frac{\ell n^2[(z/z_0(\text{soil})]}{k^2 U_a} \tag{6.16c}$$

$$R_{CA} = \frac{\ell n^2[(z-d)/z_0\ (\text{plant})]}{k^2 U_a} \tag{6.16d}$$

$$R_S = R_{SO} + R_{SI}(\theta_{sat} - \theta) \tag{6.16e}$$

$$R_{CAN} = \left(\frac{R_{CO}}{\text{LAI} \times R_{CI} + R_{C2} \times \mu_0 \times R_{SIN}} \right) / STRESS \tag{6.16f}$$

$$STRESS = 1 - STO[e_{sat}(T_a) - e_a] \tag{6.16g}$$

R_{SIN} is the incoming solar radiation, either measured or modeled. R_{C2} is an input coefficient that converts this to the fraction of R_{SIN} that is photosynthetically active radiation (PAR) in a sense normal to the leaves (Camillo, 1987). *LAI* is the total leaf area index, and R_{CO}, R_{CI}, and STO are empirical coefficients that vary with plant species. The sensible heat flux is measured by

$$H = -P_{FRAC}\, \rho c_p \frac{(T_s - T_s)}{R_{CA}} - (1 - P_{FRAC})\, \rho c_p \frac{(T_s - T_a)}{R_{SA}} \tag{6.17}$$

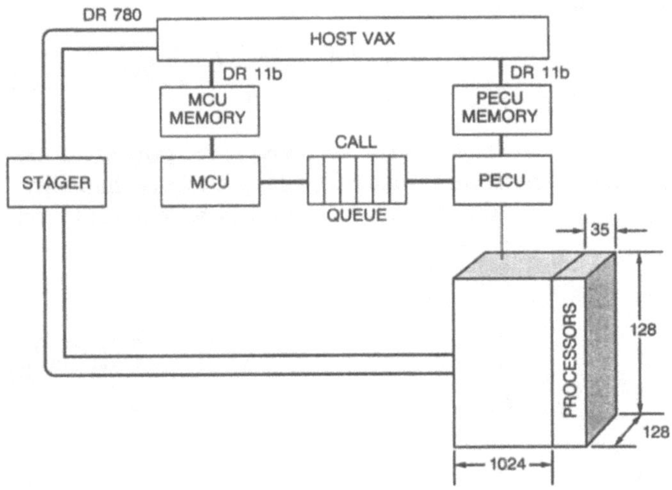

FIGURE 6.2. Block diagram of the massively parallel processor.

2.5 Method of Solution

The soil moisture and temperature continuity equations are solved by calculating the spatial derivatives of the moisture fluxes and then computing the time integral using numerical models.

The soil is divided into N layers and M columns of varying widths, Δz_i and Δx_i, respectively, which are input parameters. At a specified time the fluxes [Equations (6.2), (6.5a), and (6.7)] at the interior boundaries are calculated. The surface energy balance equations are evaluated and all boundary conditions applied. The continuity equations (6.1), (6.4a), and (6.4b) are of the form:

$$\frac{d\mathbf{y}}{dt} = \mathbf{f}(t, \mathbf{y}) \qquad (6.18)$$

The vector **y** represents the state of the system in the unsaturated zone and $f(t,y)$ represents the model equations. This is solved with an Adams–Bashforth predictor–corrector method (Booth, 1957; Teddington, 1958). This solution is described in detail in Camillo et al. (1983). New values of the state vector, $\mathbf{y}(p)$ $(t+\Delta t)$ are predicted in terms of the backward differences of the derivatives; the derivatives are recalculated from the model equations, and then the corrected value of the state vector, $\mathbf{y}(c)$ $(t+\Delta t)$, is obtained.

The difference between $\mathbf{y}(p)$ and $\mathbf{y}(c)$ is a reliable estimate of the discretization error, and the software determines if each element of this difference lies within a user-specified window. If all differences are smaller than this window, the integration step size (Δt) is doubled, leading to increased computational efficiency and reduced roundoff errors. If any difference is too large, the step size is halved. This halving and doubling requires no reevaluation of the model equations. The values of the four backward differences for the new integrator time (whether for halving or doubling of the integration time step) are calculated as linear combinations of the four back values for the old integrator time.

The continuity equations for surface and saturated flow [Equations (6.6) and (6.8)] are solved very simply as

$$H(t+\Delta\tau) = H(t) + \Delta t(dH/dt) \qquad (6.19)$$

where H represents both HB and HT.

3. Mapping the Model on the MPP

As can be seen from Figure 6.2, the MPP consists of several parts that can be roughly divided into a serial processing segment (the Main Control Unit or MCU and its associated memory), a parallel processing segment (the Processing Element Control Unit or PECU and its associated memory, along with the Array Unit, which consists of memory and processors configured into a 128-by-128 array), and a staging memory for parallel

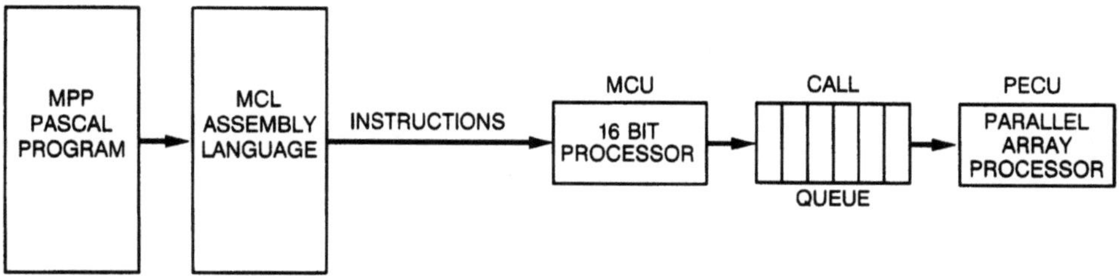

FIGURE 6.3. MPP Pascal system flowchart.

data transfers between the MPP and its host (VAX 8200).

The Main Control Unit (MCU) is a scalar calculator. This is a special-purpose, microcoded 16-bit processor that has a 16-bit hardware multiplier. Parallel calculations are performed with the 16,384 bit-serial Array Unit processors. A total of 1024 bits of memory are available to each processor. The MPP is a single-instruction, multiple data (SIMD) machine. Hence, identical operations are performed on each PE at the same time.

The MCU is programmed via MCL assembly language. The PECU is programmed via PEARL assembly language. A high-level language, MPP Pascal, is compiled into MCL assembly language. Parallel computations are initiated in the PECU by calls to routines in PECU memory from MCL. A call queue exists so that the MCU can stack up to 15 calls deep, which the PECU can process in turn, as shown in Figure 6.3. VAX language front end and subroutines are needed to handle initialization, input and output to disk, as well as to interface

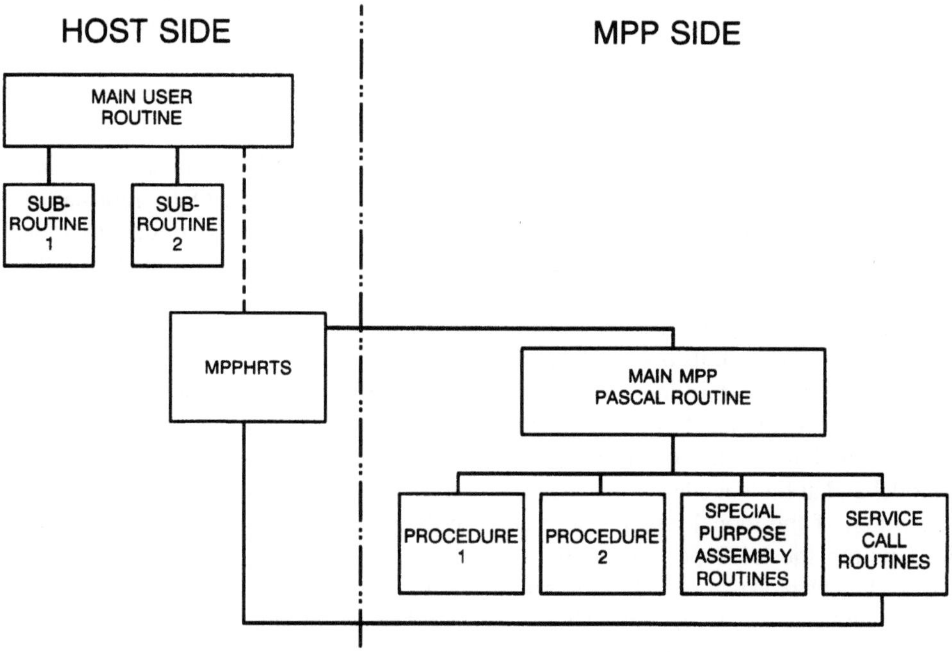

FIGURE 6.4. Structure of the two-dimensional model programming routines.

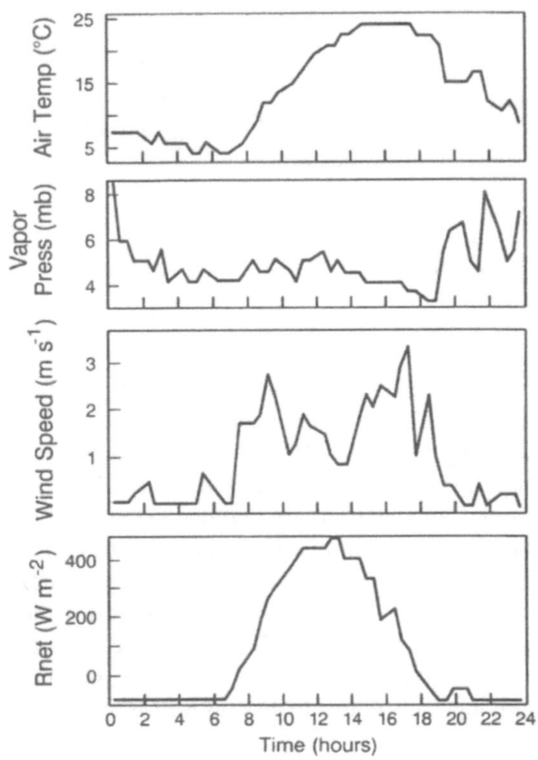

FIGURE 6.5. Meteorological data used to drive the surface energy balance model.

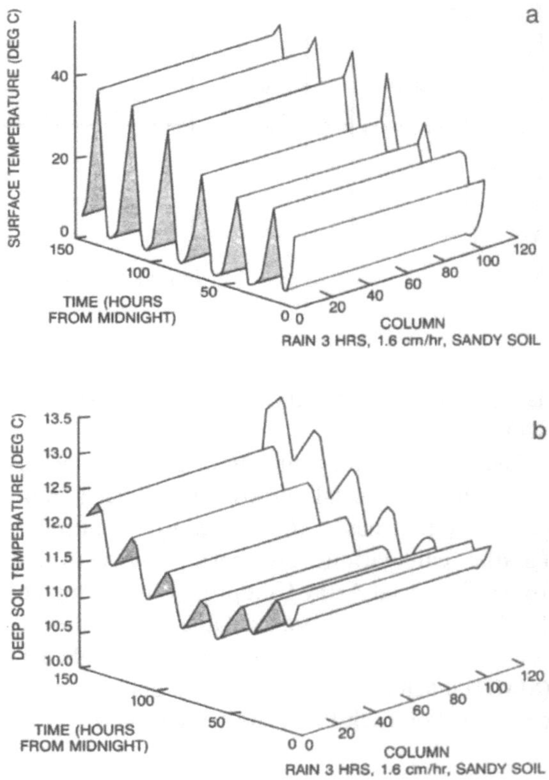

FIGURE 6.6. Force-restore solutions for the temperatures as a function of time and position on the hillslope. (a) Surface. (b) Deep soil.

with the MPP-HOST run time system (MPPHRTS) that accesses the MPP. The Hillslope model was written with a VAX FORTRAN front end and subroutines, an MPP Pascal main routine and procedures, and some special purpose assembly routines, as illustrated in Figure 6.4.

The initial mapping of the hillslope model to the MPP primarily involved determining the scalar, vector, and parallel components of the calculations. Calculations both within the layers of the hillslope model and at the boundaries are required at each time step of the model integrations. The calculations within the hillslope are identical at each layer, so they were arranged to be performed simultaneously in the MPP Array Unit. Thus, the basis for the parallel processing consisted of dedicating single MPP processing elements (PEs) to the processing of individual soil cells. The memory of each PE contained data concerning the state of that soil cell. Only information from its nearest neighbors was needed to complete the processing. This took advantage of the nearest-neighbor connectivity of the MPP architecture.

The boundary conditions involved vector calculations at the top and at the bottom of the hillslope. The MPP has two capabilities that enable this to be done easily. First, an arbitrary part of an array can be accessed via Boolean masks. Thus a mask that selects the processing elements that handle the hillslope surface can be used to do the processing on these elements. Second, when vector values were needed at the surface, the capabilities of the MPP to do row and column broadcasting was used to get the vector values (which were stored in columns) to be at the surface of the hillslope. Both the Boolean selection and the row and column broadcasting are natural procedures for the MPP and go quickly. In addition, because the processors are selected by masks, the possibility of error is reduced. Bottom boundary conditions were done similarly.

Data initialization on the MPP involved initialization of parallel arrays as well as scalar and vector data. These arrays were initialized in FORTRAN arrays and scalars on the VAX and transferred to the MPP scalars and arrays via the DR 780 and DR 11b buffers, respectively (Fig. 6.2). The capability of the stager to permute the data bits between the VAX and Array Unit was used to change the format of the floating point data from the VAX to the MPP format while the data were being transferred between the VAX and the MPP. Because of this, parallel data transmission between the VAX and the MPP appeared translucent. Explicit bit swapping of scalar integer data between the VAX and MCU to accommodate the two separate integer formats was still necessary, however, as these values were transmitted across the DR11b buffer, which does not have data permuting capability.

Special-purpose routines were written to enable the information in the parallel arrays to be output to the VAX. These routines passed parallel arrays into predefined VAX FORTRAN arrays. MPP Pascal callable VAX FORTRAN routines were written that could write out the data in these arrays for user examination of intermediate and final results.

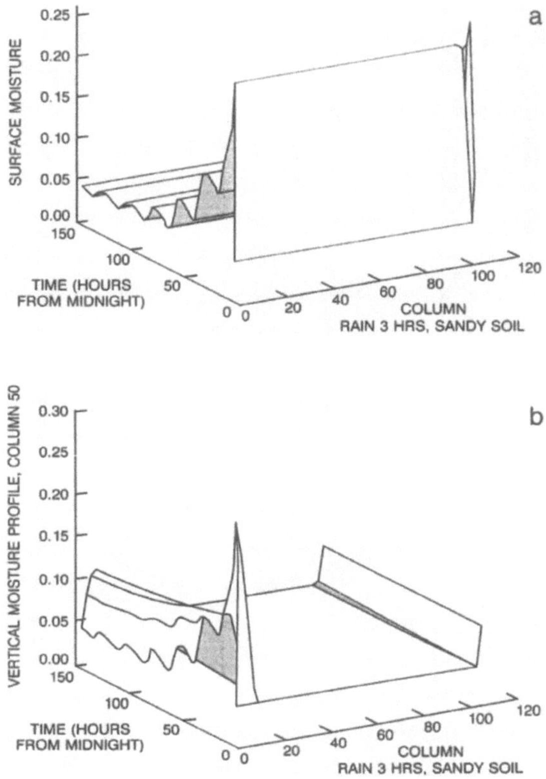

FIGURE 6.7. (a) Surface soil moisture as a function of time and position on the hillslope. (b) Soil moisture profile for column 50 as a function of time.

4. Model Output

The model output is illustrated with one set of test results, which includes the surface energy-balance of and infiltration into an initially very dry sandy loam soil.

The hillslope was divided into 102 columns of width 0.5 m each. The first column had 100 soil layers of thickness 0.1 m with two bottom layers of 0.5 m. The last column had only the two bottom layers. The slope was a line drawn from the top of the first column to the top of the last, so the area modeled was a right triangle with height 11 m and base 60 m. These 5252 soil cells used approximately one-third of the PECU capacity.

The initial volumetric moisture in the unsaturated zone was set to 0.05 $m^3\ m^{-3}$ everywhere. To model a sandy loam the parameters in the hydraulic conductivity and matrix potential models [Equation (6.3)] were set to $\theta_s = 0.375$, $K_s = 2.8 \times 10^{-5}$ m s^{-1}, $\Psi_s = -0.43$ m, and $b = 5$. These values were derived from fitting to the characteristic curves measured during an experiment near Phoenix in 1972 (Jackson, 1973). The meteorological data used to drive the surface energy balance model are shown in Figure 6.5. They were reused for each of the 6 days modeled here. Camillo and Gurney (1986) show how these data were fitted to the surface energy balance model. The rainfall rate was 1.6 cm h^{-1} for the first 3 h.

Perhaps the most important result is that the simulations took approximately 1 min of CPU time per 24-hr period, or 6 min for the entire 6-day run. These days computer runs of an hour or more are

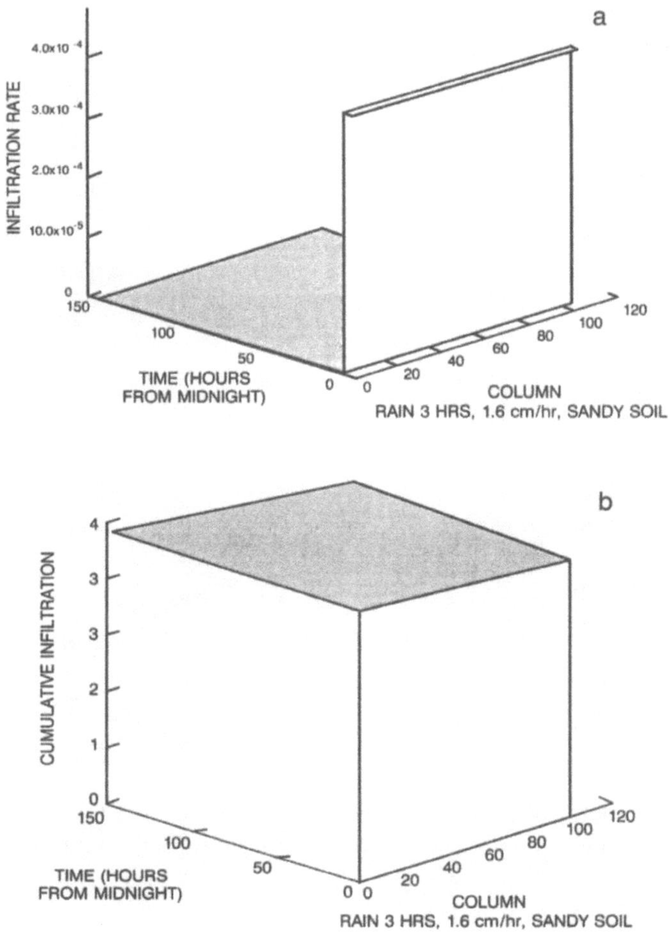

FIGURE 6.8. (a) Infiltration rate. (b) Cumulative infiltration.

not uncommon, so it is feasible to simulate 2 months or more on the MPP. This will allow for simulations of many storms and inter-storm periods.

Figure 6.6a and b shows the force restore solutions to the surface and deep soil temperatures [Equations 6.4a) and 6.4b)] as functions of time and column number. Time zero is the start of the simulation, which here is midnight. Column 1 is at the hillslope divide and column 102 is at the seepage face. It is difficult from these plots to project the daily maximum value onto the time axis, but for each day this occurs at about 2 PM. The temperatures range from 22 to 40°C, increasing as the soil surface dries.

Figure 6.7a shows soil moisture in the top soil layer as a function of time and position. The rapid rise as the initially dry soil absorbs all the rain and the subsequent decline over the next 5 days are physically realistic. Figure 6.7b shows the soil moisture profile in column 50 (halfway down the hillslope) as a function of time. This shows that the moisture never penetrated deeper than about the top five layers, or 0.5 m. It also shows that after 2 days the surface exhibited small oscillations about a value of 0.05 (same as in Fig. 6.7a), increasing to a value of about 0.12 at about 0.3 m then decreasing to an unchanging value of 0.05 below 0.5 m. Thus, the dynamic zone seems to be the top 0.5 m.

Figure 6.8a shows the infiltration rate as a function of time and position. The maximum rate shown here (4.4×10^{-4} cm s^{-1}) equals the rain rate,

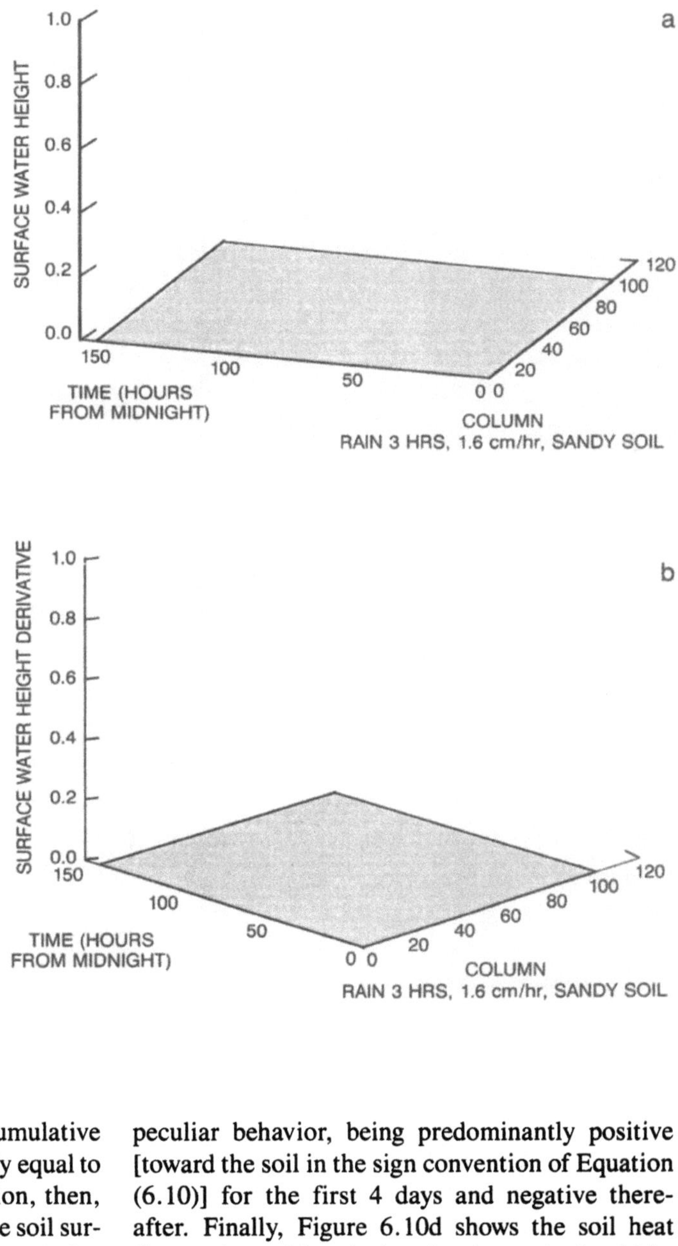

FIGURE 6.9. (a) Depth of surface retention. (b) Time derivative of surface retention.

1.6 cm h^{-1}. Figure 6.8b shows the cumulative evaporation everywhere as 4.8 cm, exactly equal to the cumulative rainfall. For this simulation, then, all the rain immediately infiltrated into the soil surface. Figure 6.8b also shows that the cumulative infiltration calculation is correct. Figure 6.9a and b also supports this; there is no surface retention.

The surface energy balance fluxes are plotted in Figure 6.10a–d. The net radiation (Fig. 6.10a) is the data used to drive the energy balance model. These are the same every day since the 24-h data set was reused each day. The latent heat flux (Fig. 6.10b) decreased each day as the soil dried out. The sensible heat flux (Fig. 6.10c) exhibits peculiar behavior, being predominantly positive [toward the soil in the sign convention of Equation (6.10)] for the first 4 days and negative thereafter. Finally, Figure 6.10d shows the soil heat flux. It is positive during the day as it should be for a soil surface which is getting warmer every day (Fig. 6.6a), but it is also 50% of net radiation, a rather large value. The problem is not in the values for thermal conductivity and heat capacity, as may be seen in Figure 6.11. These vary with soil moisture as they should.

These peculiarities in the surface fluxes are most likely due to the use of the same net radiation every day, which cannot be representative of all

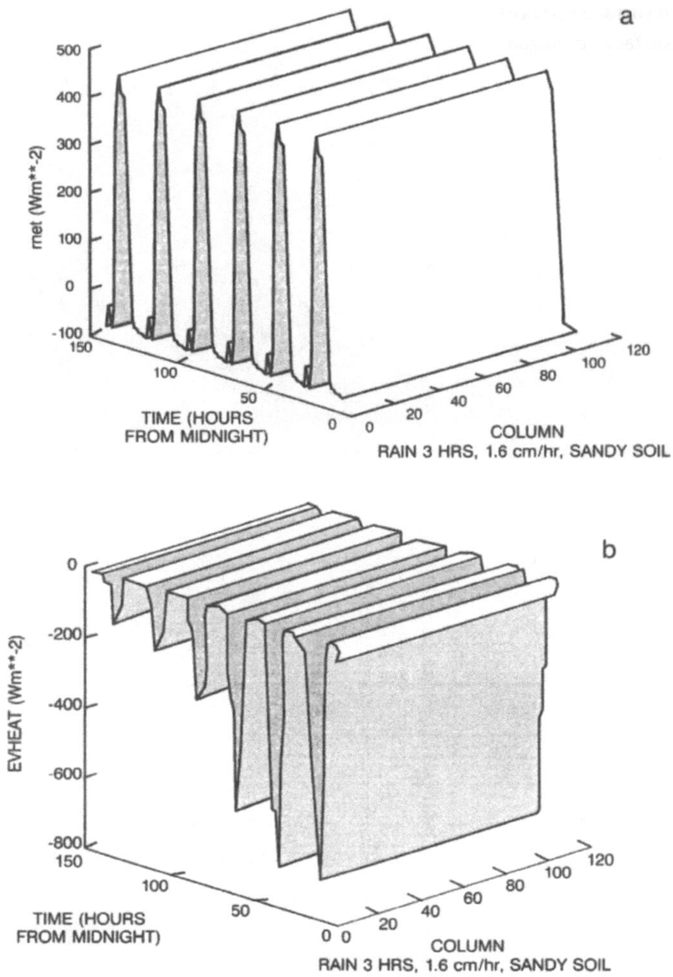

FIGURE 6.10. Surface energy balance fluxes. (a) Net radiation. (b) Latent heat.

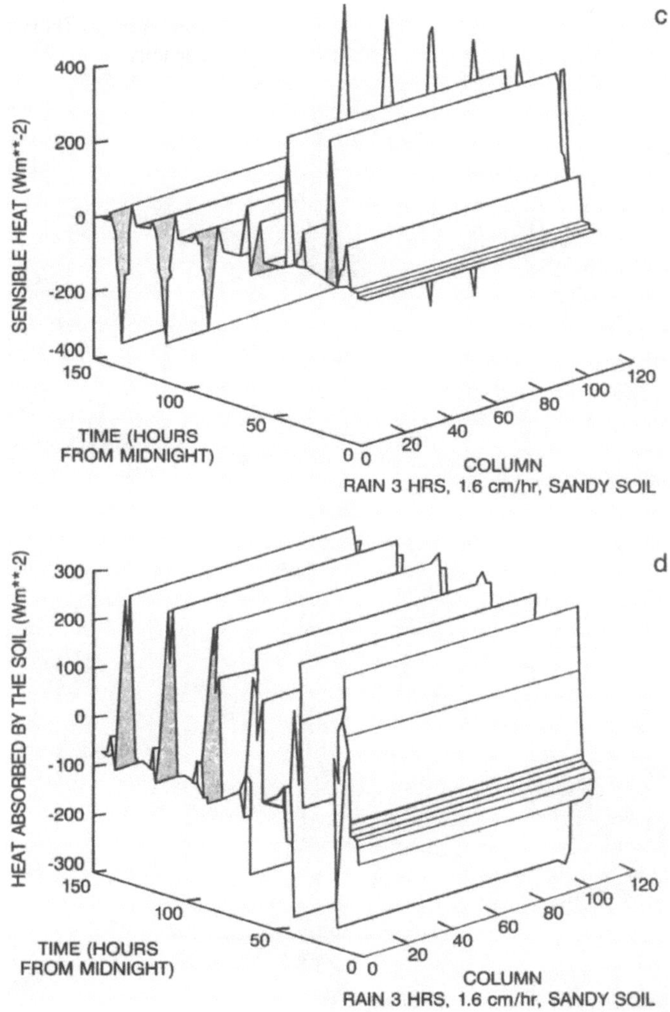

FIGURE 6.10. (c) Sensible heat. (d) Heat absorbed by the soil.

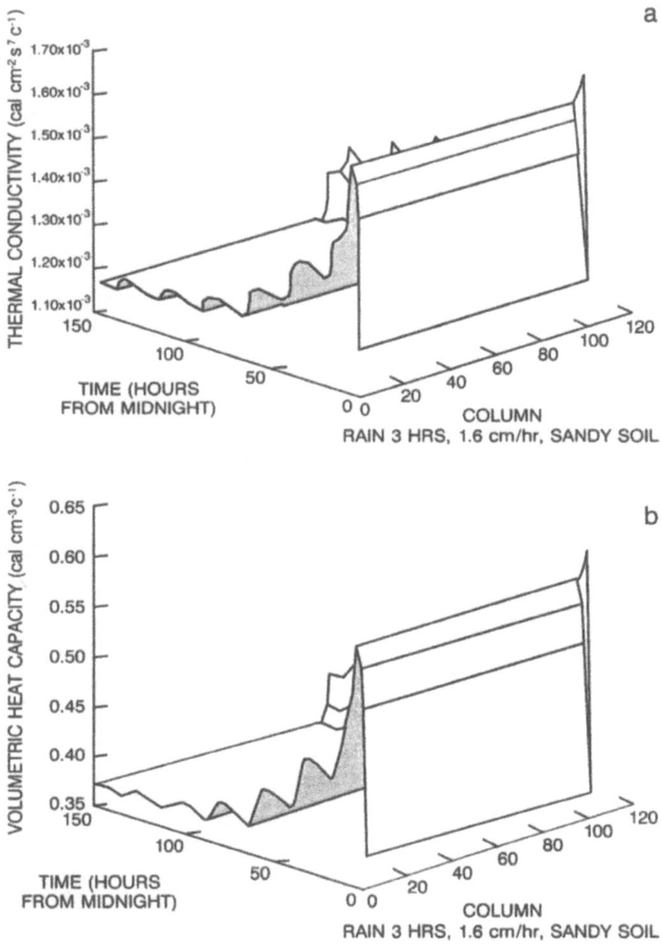

FIGURE 6.11. Thermal properties of the top soil layer. (a) Thermal conductivity. (b) Heat capacity.

the surface conditions modeled here. Further testing against real data is clearly desirable as suitable data become available.

5. Summary

We have presented a new model of the hydrological response of a hillslope to rain. It runs on a SIMD parallel architecture computer, the Massively Parallel Processor, at Goddard Space Flight Center. Its major advantage over other models of its type is its much reduced execution times due to the parallel architecture of the MPP compared to a serial machine. The main conclusion that can therefore be drawn is that useful programs of this type can be written on a highly parallel computer without excessive difficulty. This architecture therefore shows promise for the next generation of models to study wider global change issues.

References

Bathurst JC (1986) Physically-based distributed modeling of an upland catchment using the Systeme Hydrologique Europeen. J Hydrol 87:79–102.

Beven KJ, Kirkby MJ (1979) A physically based contributing area model of basin hydrology, Hydrol Sci Bull 24:43–69.

Beven KJ, Kirkby MJ, Schofield N, Tagg AF (1984) Testing a physically based flood forecasting model (TOPMODEL) for three UK catchments. J Hydrol 69:119–143.

Booth AD (1957) "Numerical Models." Academic Press, New York.

Camillo PJ (1987) A canopy reflectance model based on an analytical solution to the multiple scattering equation. Remote Sens Environ 23:453–477.

Camillo PJ, Gurney RJ (1986) A resistance parameter for bare soil evaporation models. Soil Sci 141:95–105.

Camillo PJ, Gurney RJ, Schmugge TJ (1983) A soil and atmospheric boundary layer model for evaportranspiration and soil moisture studies. Water Resour Res 19:371.

Clapp RB, Hornberger GM (1978) Empirical equations for some soil hydraulic properties. Water Resour Res 14:601.

Devaney JE, Camillo PJ, Gurney RJ (1987) A SIMD implementation of a distributed watershed model. Proceedings of the Second International Conference on Supercomputing, Santa Clara, California, May 3–8.

de Vries DA (1975) Heat transfer in soils. In "Heat and Mass Transfer in the Biosphere." Scripta, Washington, DC.

Eagelson PS (1970) "Dynamic Hydrology." McGraw-Hill, New York.

Freeze RA (1971) Three-dimensional, transient, saturated-unsaturated flow in a groundwater basin. Water Resour Res 7:929–941.

Gurney RJ, Camillo PJ (1984) Modelling daily evaporation using remotely sensed data. J Hydrol 69:305–324.

Hillel D, Hornberger GM (1979) Physical model of the hydrology of sloping heterogeneous fields. Soil Sci Soc Am J 43:434–439.

Jackson RD (1973) Diurnal changes in soil-water content during drying. In "Field Soil Water Regime," R.R. Bruce et al. (eds.). Soil Sci. Soc. Am. Proc., Special Publ. 5.

Kirkby MJ (ed.) (1972) "Hillslope Hydrology." Wiley-Interscience, Chichester, UK.

Lin JD (1980) On the force-restore method for prediction of ground surface temperature. J Geophys Res 85:3251.

Philip JR, de Vries DA (1957) Moisture movement porous materials under temperature gradients. EOS Trans. Am. Geophys. Union, AGU. 38:222–232.

Pikul MF, Street RL, Remson I (1974) A numerical model based on coupled one-dimensional Richard and Boussinesq equations. Water Resour Res 10:295–302.

Smith RE, Woodhiser DA (1971) Overland flow on an infiltrating surface. Water Resour Res 7:899–913.

Teddington A (1958) "Modern Computing Methods." Philosophical Library, New York.

van de Griend AA, Engman ET (1985) Partial area hydrology and remote sensing. J Hydrology 81:211–251.

7
Introduction of a Realistic Soil–Vegetation Component in a Hydrological Model: Application to HAPEX-MOBILHY Experiment

C. Ottlé and D. Vidal-Madjar

1. Introduction

In hydrological modeling, a correct description of the energy and mass exchanges at the interface between land and atmosphere is of primary importance. This is due to the fact that if the watershed boundary is correctly determined, the only water supply is rainfall, and evapotranspiration involves more than the half of the total amount of precipitation in temperate regions.

The level of complexity of the parameterization of soil–vegetation–atmosphere exchanges depends on the final objective of the modeling. If the objective is to simulate the river flows on a decade basis, the most important thing is a good knowledge of the spatial and temporal repartition of the rainfall; the evapotranspiration, in that case, can be described in a very rough manner as well as the surface runoff. On the other hand, if the water budget of the soil unsaturated zone is of interest (application to irrigation, accurate prediction of surface runoff, etc.), correct knowledge of the rainfall patterns is no longer sufficient and one must take into account a more complex scheme for the simulation of the soil–vegetation–atmosphere behavior.

In the framework of HAPEX-MOBILHY (Hydrologic Atmospheric Pilot Experiment and Modélisation du Bilan Hydrique), the main objective of the program was to look at the feasibility of following the atmospheric water budget on a scale compatible with the general circulation model (GCM) and to define suitable experimental methodologies. The role of watershed modeling may be important as far as possible knowing the total regional water discharge and total rainfall input at the proper scale, to estimate the evapotranspiration flux at a climatic temporal scale (from 10 days to several years). However, this cannot be done if the evapotranspiration is taken into account through a rough estimate of the potential evaporation and if the vertical movements of the water within the unsaturated soil is not realistically described as is the case for most of the hydrological models.

In this chapter, we will use the HAPEX-MOBILHY example to show how a distributed model, including a surface scheme that takes into account atmospheric forcing, may be transformed to achieve good monitoring of evapotranspiration.

In Section 2, we will examine the principle of the hydrological model used for the HAPEX campaign; in Section 3 the new surface scheme will be presented; and finally, it will be shown with preliminary examples how the outputs of the model are improved.

2. HAPEX Hydrological Model Principle

This work is done in the frame of the HAPEX-MOBILHY experiment with the hydrological model of the ENSMP (Ecole Nationale Supérieure des Mines de Paris) (Girard, 1974; Ledoux, 1980) implemented over the Adour basin to simulate the streamflows of the region.

2.1 Model Description

The model is a deterministic distributed model. It has two layers: the surface layer and the under-

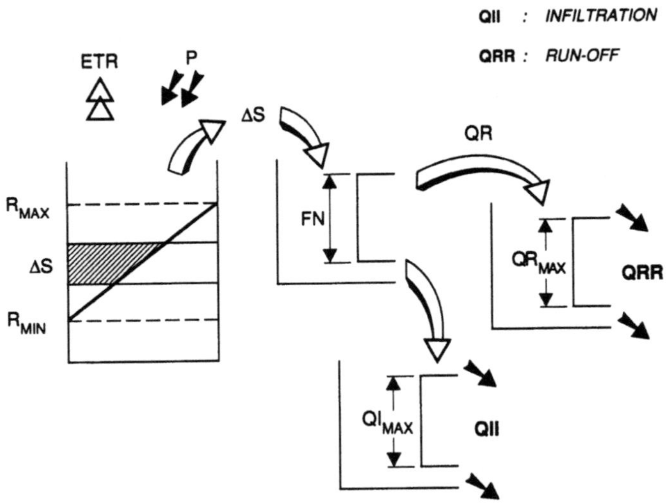

FIGURE 7.1. Surface parameterization of the hydrological model from Girard (1974).

ground. The surface layer is represented in a conceptual way by a reservoir that is filled each day by rainfall and is emptied according to the evapotranspiration, the infiltration rate, and surface runoff. The underground is modeled in a more complex and physical way by the discretization of the hydrodynamic equations and their integration.

The two domains have been divided in meshes of 5 km except for the rivers where the exchanges between the surface and the underground have to be more precisely described and where the mesh is divided by 2.

The hydrologic processes are modeled by three mechanisms. The first scheme called "production" deals with the surface layer and divides the rainfall in storage, infiltration, surface runoff, and evaporation depending on soil coefficients. The second one calculates the surface transport, routing the runoff from each grid point to the closest river mesh according to drainage directions defined from topographic maps. Finally, the last routine calculates the infiltration to the water tables, the transport in the underground, and also the water fluxes from the ground to the surface, which occur only in the river meshes. As the model is designed for decade time periods, the computing time step has been set to 1 day.

2.2 Parameterization of the Surface Layer

The parameterization of Girard (1974) considers the surface layer as a bulk layer 1 m deep. The hydrological budget is calculated at each time step of the model (1 day), according to precipitation, potential evaporation, and the initial state of the reserve. The quantity of water stored in the surface layer depends on two parameters representing the minimum and the maximum levels of the reservoir (R_{min} and R_{max}, see Fig. 7.1). The remaining water (the hatched area) is then shared between infiltration or runoff, depending on an infiltration threshold value set for each type of soil. Two more reservoirs have been introduced to calculate delayed infiltration and runoff.

All these parameters have been adjusted for each type of soil so that the simulated flows reproduce the observations at the outlets. This calibration has been done using 9 years of data.

Obviously this parameterization is not sufficient if one wants to follow the hydraulic state of the surface for irrigation purposes, for instance, or wants to monitor soil moisture or real evapotranspiration to capture its spatial heterogeneity. For the simulation of the energy and water budgets at the surface, a more precise description of the soil layer is needed, which allows the very superficial soil layer (the first centimeters of soil) to be differentiated from those of the bulk layer.

3. The New Surface Scheme

The correct simulation of the soil water content and its variations in space and time are of prior importance for both atmospheric forecast and

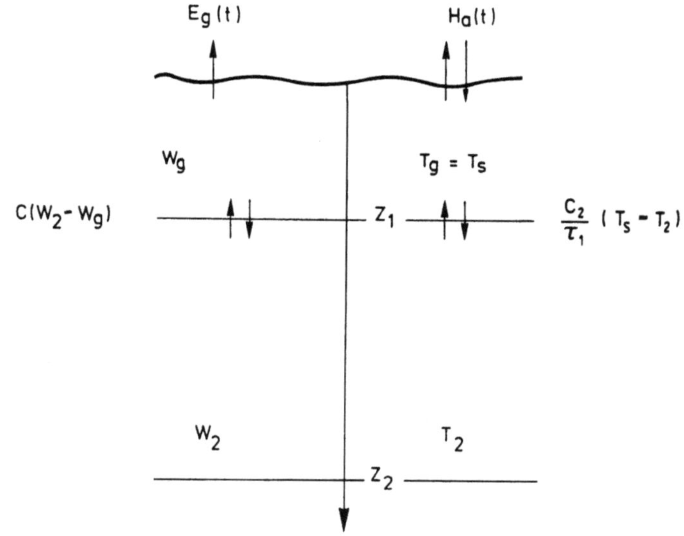

FIGURE 7.2. Surface parameterization introduced in the hydrological model from Soarès et al. (1988).

hydrological models. The fundamental quantity for these models is not only the surface water content but the content of a deeper layer (the layer concerned with the root water uptake). A realistic scheme must be able to take into account the water exchange between this second layer and the underlying soil and with the atmosphere through plant transpiration and diffusion to or from the soil surface.

In this study, the surface scheme used has been derived from two papers by Deardorff (1977) and (1978). It is fully described in Taconet et al. (1986) and Soarès et al. (1988).

3.1 Soil Parameterization

The soil is represented by a two-layer force-restored model for the temperature as well as for the water. Figure 7.2 describes the principle of the model. It depends on the soil specific heat, thermal conductivity, and hydraulic diffusivity. The heat and mass transfers are coupled by the dependency of the thermal conductivity on the water content.

The specific heat is nearly independent of the soil type. It is not the case for the thermal conductivity and the hydraulic diffusivity, which have to be assigned a value for each model grid point. This may be done following the soil classification used in the original surface scheme and the existing literature to obtain an order of magnitude for these parameters. The fine adjustment may then be done by a calibration of several years following a procedure identical to the one used for the calibration of a hydrological model.

3.2 The Vegetation Parameterization

To simulate realistically the exchanges between the land surface and the atmosphere, it is absolutely necessary to include the vegetation. The vegetation differs from the soil by taking water directly in the second reservoir and also by its transpiration control against atmospheric demand. But vegetation is a very complex physical and biological medium, far more complex than soil. It is not possible to describe it in a perfect way and simplifications must be done to preserve its essential properties. This has been done using a parameterization derived from Deardorff (1978). Figure 7.3 gives the general principle. The vegetation is represented by a single, semitransparent layer ventilated by a surface turbulent layer (10 m deep). The atmospheric forcing at the top of the surface layer may be either specified from data of a conventional meteorological network or predicted by a one-dimensional boundary layer model.

The second solution appears to be preferable (Taconet and Vidal-Madjar, 1988) because the atmospheric forcing is then relatively independent of local heterogeneities that always exist in network data.

As in Deardorff, the model is stratified in two layers, soil and vegetation. The general idea for

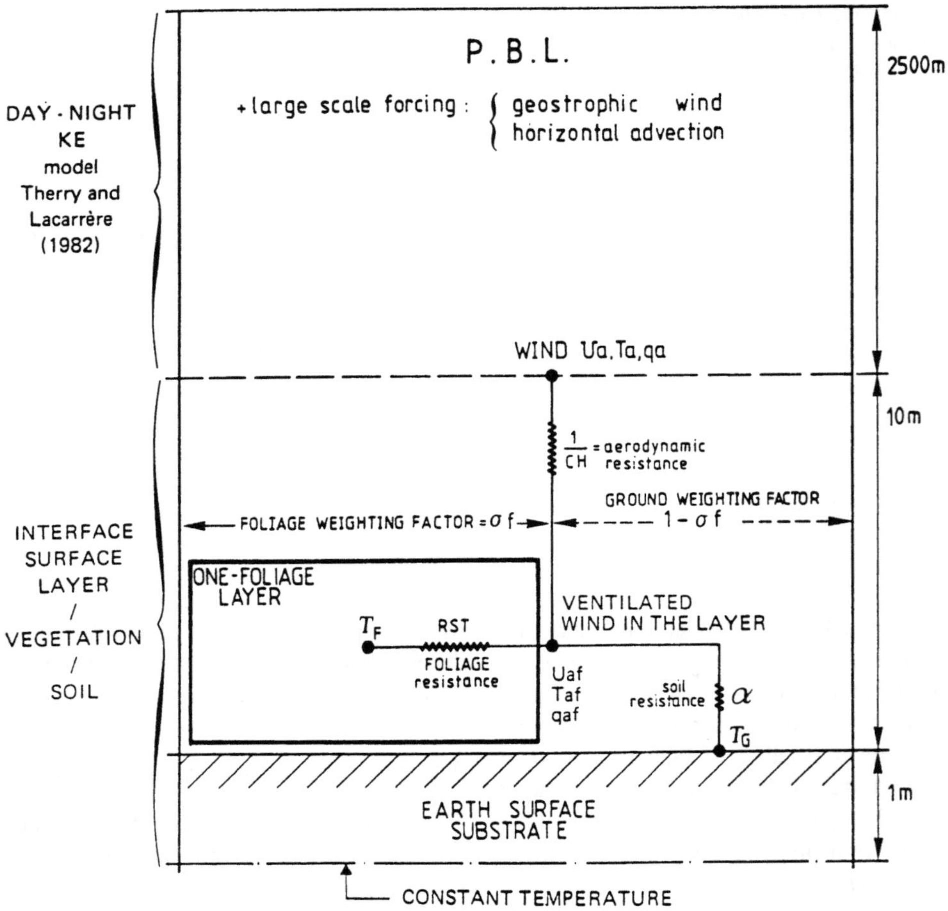

FIGURE 7.3. Basic framework of the model. Reprinted with permission from Taconet et al., 1986, *Journal of Climate and Applied Meteorology*, American Meteorological Society.

determining heat fluxes is to solve simultaneously the energy budget equations at the ground level and for the foliage, assuming an adequate partition of the available fluxes between vegetation and bare soil. Using this formulation, the vegetation canopy is then described by the following parameters where σ_f is a number measuring the percentage of the soil surface covered by the vegetation canopy, R_0 is the minimum value of the plant stomatal resistance, and h is the height above the ground and has a minor influence on the fluxes. R_0 depends strongly on the vegetation type. The stomatal resistance is calculated as a function of time using the mean water content of the two soil reservoirs.

They all depend on the location of the grid points and have to be changed with the season. There is some hope that they can be determined using visible, thermal infrared, and microwaves satellite remote sensing.

4. First Application to the HAPEX Hydrological Model

The scheme described above has not yet been implemented into the HAPEX hydrological model. The reason is that high costs of both time and work have resulted in a preference to first evaluate the ameliorations using a far more simplified algorithm.

4.1 The Simplified Scheme at Its Implementation

The soil is represented by two layers, a very superficial one (about 10 cm) and a bulk one of 1.2 m deep. The water contents of these reservoirs are calculated by the force restore method (Fig. 7.2).

The partition of the water between runoff and infiltration depends on the properties of the soil but

FIGURE 7.4. Cumulated evaporation simulated with the two parameterizations during the summer of 1985, compared with the precipitation.

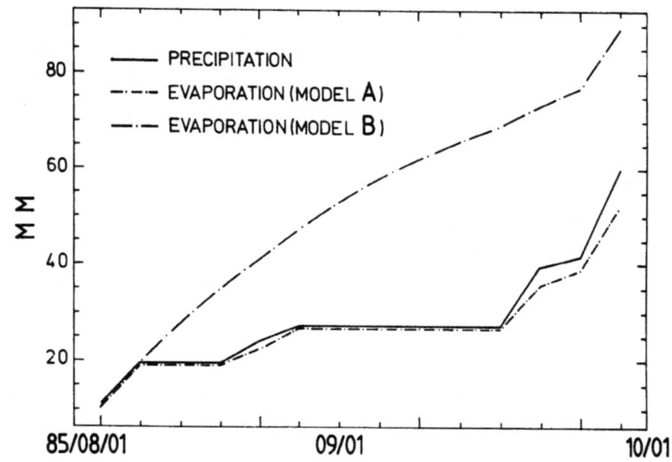

also strongly on the nature of the vegetation cover. For example, precipitation infiltrates more over vegetated areas than over bare soils. To take into account this effect in the model, runoff is calculated in the following way: over bare soils, surface runoff appears as soon as the first layer is saturated, while over vegetated areas, it starts only when the bulk layer is filled up. The remaining water is infiltrated according to the infiltration rate defined by the original scheme.

For the calculation of evapotranspiration, it is clear that real evaporation is lower than potential evaporation and that the properties of the vegetation cover have to be described precisely. To simplify the problem, the evaporation has been reduced from the potential one, by a factor β, depending on the soil water content of the bulk layer, following the parameterization of Manabé (1969).

$$E = \beta * ETP$$

$$\beta = \min\left(1.,\ 3.* \frac{W_2}{W_{2\max}}\right)$$

ETP is the potential evaporation and $W_{2\max}$ is the saturation value of the bulk layer.

First, a calibration was done on the soil parameters [maximum soil capacities of the two layers, soil hydraulic diffusivity (*c*), infiltration rates], to obtain the same monthly surface runoff and infiltration as the original model, already adjusted to the measurements made at the different outlets of the watershed. It was made for the 2 years 1984–1985, for which the meteorological and hydrological data (potential evaporation, precipitation, streamflows, and piezometric data) were available.

4.2 Comparisons Between the Two Schemes

Comparison of these two simulations has shown that although infiltration and runoff are the same on a monthly basis, due to the calibration procedure, surface layer behavior can be very different especially in summer. One drawback of the initial parameterization is that as long as there is some water left in the soil, it is evaporated at a potential rate. So as early as the beginning of summer, the model simulates the following situation: the upper reservoir is already at its minimum level and all the precipitation is immediately reevaporated.

The second parameterization simulates a much more realistic situation: the evaporation is limited as soon as the soil begins to dry and has more regular variations. Figure 7.4 clearly shows a plot of evaporation simulated by the two models during the summer 1985, averaged over a region of 10 × 10 km² in the South Western part of the HAPEX square (on the figure, model A is for the initial parameterization, model B is for the new one). On the other hand, because there is no minimum level for the reservoir (except 0), the amplitude of variation for soil moisture is greater and the drying of soils is more gradual. Figure 7.5 shows the variation of the bulk layer soil moisture, simulated by the two models during the summer of 1985 and the

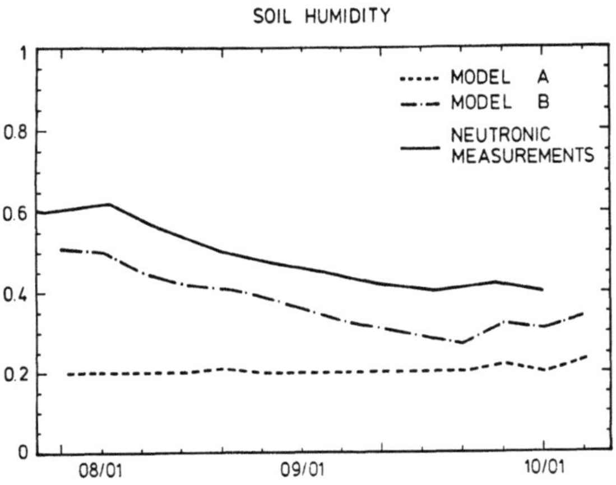

FIGURE 7.5. Comparison of the soil humidity simulated with the two parameterizations to the neutronic measurements.

measurements of a neutronic probe set up at the same location. The results of the second model are in very good agreement with the measurements (the variation and the values are realistic). Soils dried up by the end of September (especially dry in 1985), whereas with the first model (initial parameterization), the minimum value was already reached by the end of July and evaporation could occur only after precipitation.

For a further validation of the second parameterization, the results of the two simulations have been compared to the water deficit observed by satellite.

As a matter of reference, canopy temperature has long been recognized as an indicator of water availability. When the soil begins to dry, the plant reduces its evaporation to avoid drying out, and consequently leaf temperature increases.

Then, if one can follow this temperature day after day, measured always at the same time (for example, around the maximum heating), it may be an indicator of stress. Different indices to quantify plant water stress have been proposed, using canopy temperature measured by infrared radiometry (Jackson et al., 1977, 1981; Boatwright, 1986). Jackson et al. (1977) particularly proposed the use of the canopy–air temperature difference measured postnoon, called the stress degree day (SDD), and found, in many cases, a good correlation with water requirements.

The SDD has been computed all over the HAPEX square using infrared satellite data NOAA/AVHRR (channel 4, 10.5–11.5 μm) to determine surface temperatures, and *in situ* air temperature measurements, and compared with the water budget simulated by the two models.

In Figure 7.6, the map of the SDD over the HAPEX square on September 7 is compared with maps of the soil water content derived from the original model and the new one. The map of SDD shows high values in the southeastern part of the square (corresponding to gradients greater than 4°), which should point out a drying area. If we compare the soil moisture maps simulated by the two models, we can see that the original model simulates a very humid region (soil moisture greater than 40%, more humid than the "Landes" forest), and that the model shows a drying area that seems more realistic.

In the same way, with the second model, the gradients between the different zones are lower: the forest in the northern part is clearly delimited with values greater than 30% from the crop areas in the south, which dry out more rapidly. Generally, it appears that the new scheme (model B) simulates a distribution of humid and dry areas in better agreement with the distribution of SDD.

5. Conclusion

These preliminary results seem to show that the second model where the surface parameterization derived from Deardorff (1977) has been introduced simulates a more realistic water budget. The limitation of the evaporation and the introduction of the vegetation for the computation of surface runoff have certainly improved the simulation of soil water content, especially during summer months, where classical parameterizations give unrealistic results.

It is anticipated that when the complete surface scheme described in this paper will be implemented, the HAPEX hydrological model will be able to monitor the evapotranspiration flux on a 10-day basis.

Comparisons between derived fluxes from the model and other fluxes measured or calculated during the special observing period (SOP) of HAPEX will be done to validate this methodology.

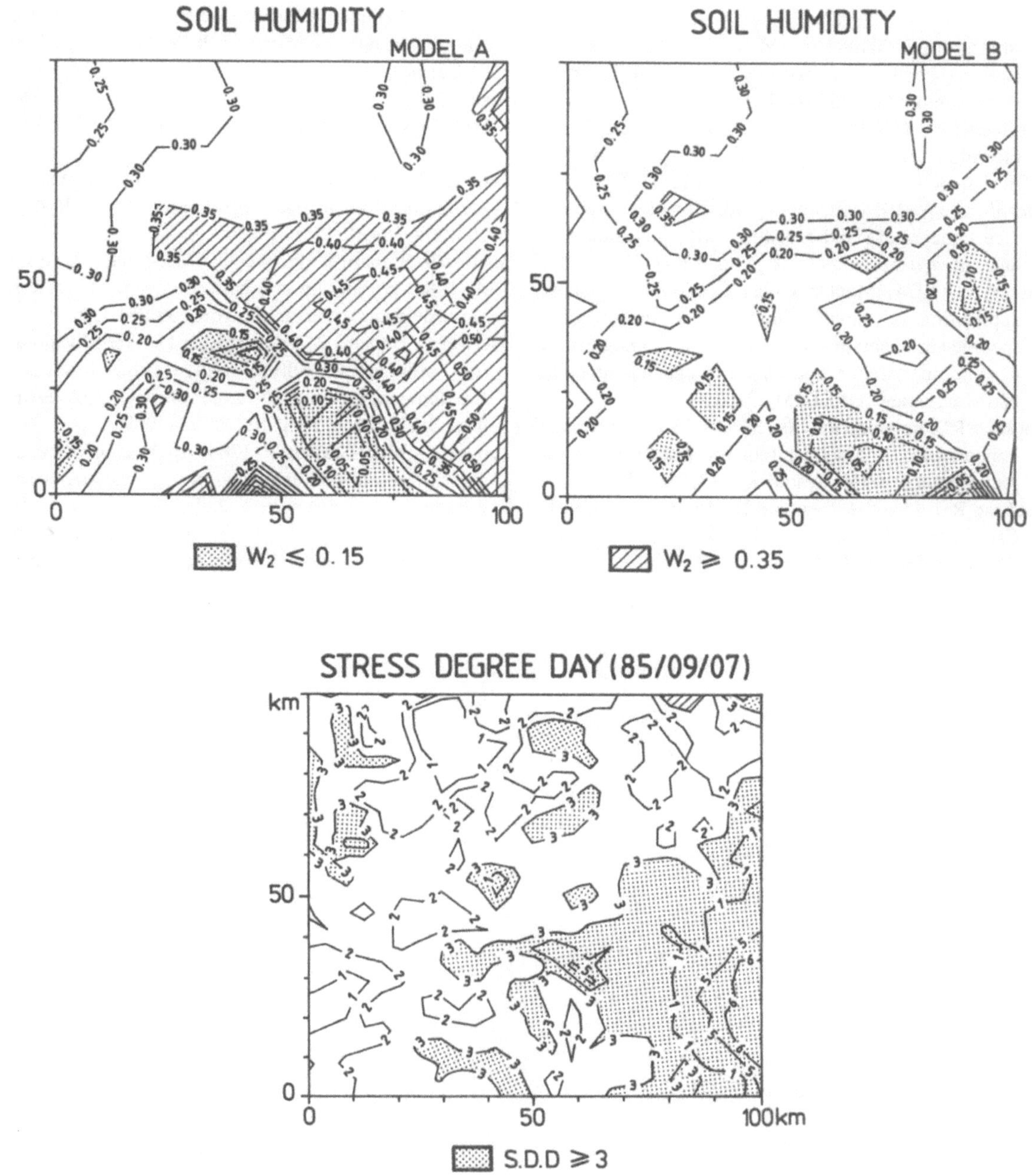

FIGURE 7.6. Comparison of the soil water budget simulated by the two models to the SDD computed with satellite data from Ottlé et al. (1989).

References

Boatwright GO, Whitehead VS (1986) Early warning and crop condition assessment research. IEEE Trans Geos Remote Sens GE-24(1):54–64.

Deardorff JW (1977) A parametrization of ground surface moisture content for use in atmospheric prediction models. J Appl Meteorol 16:1182–1185.

Deardorff JW (1978) Efficient prediction of ground surface temperature and moisture, with inclusion of a layer of vegetation. J Geophys Res 83(4):1889–1903.

Girard G (1974) Modèle global ORSTOM. Première application du modèle journalier à discretisation spatiale sur le bassin versant de la crique Grégoire en Guyane in "Atelier hydrologique sur les modèles mathématiques." ORSTOM, Paris.

Jackson RD, Reginato RJ, Idso SB (1977) Wheat canopy temperature: A practical tool for evaluating water requirements. Water Resour Res 13:651–656.

Jackson RD, Idso SB, Reginato RJ, Pinter PJ Jr (1981) Canopy temperature as a crop water stress indicator. Water Resour Res 17(4):1133–1138.

Ledoux E (1980) Modélisation intégrée des écoulements de surface et des écoulements souterrains sur un bassin hydrologique. Thèse de Docteur-Ingénieur, Ecole Nat. Supérieure des Mines de Paris et Université P et M Curie, Paris.

Manabé S (1969) The atmospheric circulation and the hydrology of the Earth's surface. Mon Weather Rev 97:739–774.

Ottlé C, Vidal-Madjar D, Girard G (1989) Remote sensing applications to hydrological modeling. J Hydrol 105:369–384.

Soarès JV, Bernard R, Taconet O, Vidal-Madjar D, Weill A (1988) Estimation of a bare soil evaporation from airborne measurements. J Hydrol 99:281–296.

Taconet O, Bernard R, Vidal-Madjar D (1986) Evaporation over an agricultural region using a surface flux/temperature model based on NOAA/AVHRR data. J Clim Appl Meteorol 25:284–307.

Taconet O, Vidal-Madjar D (1988) Application of fluxes algorithm to a field campaign over a vegetated area. Remote Sens Environ 26(3):227–239.

Therry G, Lacarrere P (1983) Improving the eddy kinetic energy model for planetary boundary layer description. Bound Layer Meteorol 25:63–88.

8
Land Surface Processes: Description, Theoretical Approaches, and Physical Laws Underlying Their Measurements

Alain Perrier and Andrée Tuzet

1. Introduction

For more than a decade, evidence has accumulated from climate model experiments that the exchanges of momentum, heat, and moisture at the surface are phenomena that strongly influence the dynamics and thermodynamics of the atmosphere.

Climate is particularly sensitive to variations in evaporation at the Earth's surface. Evaporation of water in the natural environment, either from free water surfaces or from land surfaces covered by vegetation, is one of the main components of the continental hydrological cycle and it is also the connecting link between energy budget and water budget.

Studies of regional interactions between vegetation and atmosphere have shown that there is a climatic feedback, as well as a long-term vegetation feedback (Charney, 1975). The heat and vapor released into the atmosphere from the soil or plant canopies change the temperature and humidity of the air above. In turn, these changes in temperature and humidity modulate the fluxes of heat and water vapor from soil and particularly vegetation, which are very sensitive to these parameters. Consequently, plant canopies modify their own microclimate and continuously adapt themselves to seasonal climatic changes to get the highest productivity. Plant productivity is very dependent on the influence of the microclimate on plant processes such as transpiration, carbon assimilation (photosynthesis), and respiration. There is a close relationship between crop evaporation and carbon assimilation. Carbon flows from the atmosphere into the leaves through the same critical point, i.e., the leaf stomata. The climatic dependence of stomata is their ability to open and close, in response to environmental factors, so that they act as regulators in the transfer of water vapor and CO_2 between plants and the atmosphere. Stomatal activity therefore makes the plant an active agent in the determination of its own microclimate. At the regional scale, changes in the relative proportions of vegetative surfaces induce spatial as well as temporal (crop development) variability of land surface characteristics and it is the former that we will discuss next.

2. Spatial Variability of Land Surface Characteristics

2.1 Description of Land Surface Processes

The land surface parameters that govern energy and water exchanges at the earth's surface are presented on the simplified scheme of Figure 8.1.

For a given set of boundary conditions corresponding to atmospheric air mass characteristics (T_a, the air temperature; T_d, the dew point temperature; U_a, the horizontal wind speed; and C_{CO_2}, the CO_2 concentration at a reference height z_R) and soil characteristics (T_G, the soil temperature; and Θ_G the soil moisture at a depth z_G), the response of the surface (defined by its properties: a, albedo; ε, emissivity; and z_0, aerodynamic roughness length) to the input of downward radiative fluxes (R_G, global incoming solar radiation; R_A, longwave atmospheric radiation) leads through

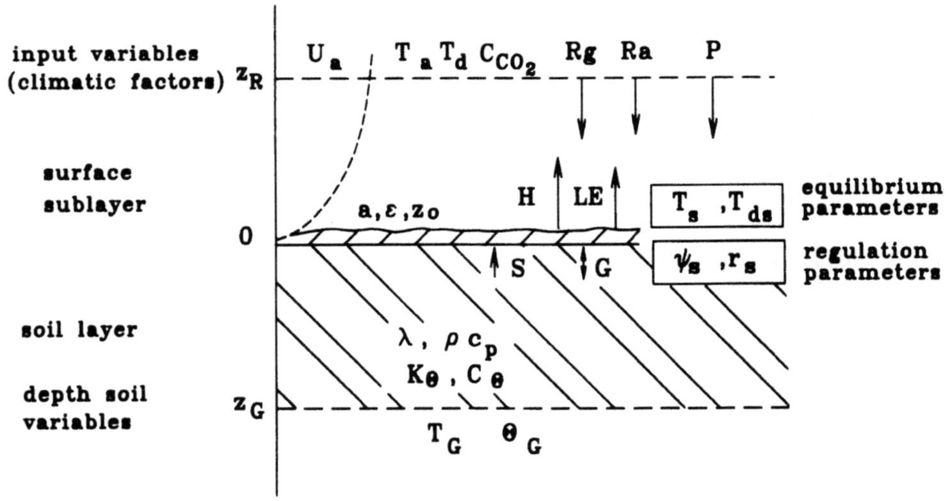

FIGURE 8.1. Land surface parameters that govern energy and water exchanges at the Earth's surface.

the energy and water balances to an instantaneous equilibrium. Here P is the rainfall input and S the water supply at the surface.

This state of equilibrium mainly depends on moisture availability at the surface. For a bare soil surface, moisture availability is determined to a large extent by the soil hydraulic parameters (K_Θ and C_Θ, respectively, hydraulic conductivity and water capacity of the soil), but for vegetation it is determined most strongly by plant morphology and their physiological properties (plant rooting, stomatal characteristics). Therefore, limitations to moisture availability are here generally expressed by global regulation parameters such as surface resistance, r_s or surface water potential, ψ_S.

At any time, a surface radiative imbalance results in a combination of convective exchanges with the atmosphere of either sensible (H) or latent (LE) heat, and conduction to or from the underlying soil (G, where G depends on soil characteristics: λ the thermal conductivity; and ρc_p the heat capacity). Thus the surface state is expressed by two equilibrium parameters: T_s the surface temperature, and T_{ds} the dew point temperature representing the energy and moisture balances, respectively.

Therefore, each surface will tend to regulate and partition the available energy and water in its own way, giving a unique energy and water balance. The dynamics of the flow tend to define for each surface an internal equilibrium sublayer in which air properties are fully adjusted to surface properties and where convective fluxes are roughly constant with height.

2.2 Effects of Spatial Inhomogeneity

Under natural conditions relatively few surfaces are sufficiently uniform and large, so that edge effects involving local horizontal advection by the mean wind become important. On a regional scale, the land surface appears as a patchwork of different fields. These differences may be due to different surface roughness, water availability, and microclimate in terms of temperature, humidity, and wind speed profiles. Thus there will be spatial discontinuity of microclimates, and horizontal gradients will exist. Near the surface, these gradients will be greatest and horizontal interactions will occur (Fig. 8.2a).

2.2.1 Edge Effect

As air passes from a surface to one with different characteristics, it must adjust to the new set of boundary conditions. The adjustment is not immediate throughout the thickness of the air layer; it is generated at the surface and diffuses upward. The layer of air whose properties have been affected by the new surface properties is referred to as an internal boundary layer (Fig. 8.2b). Its thickness

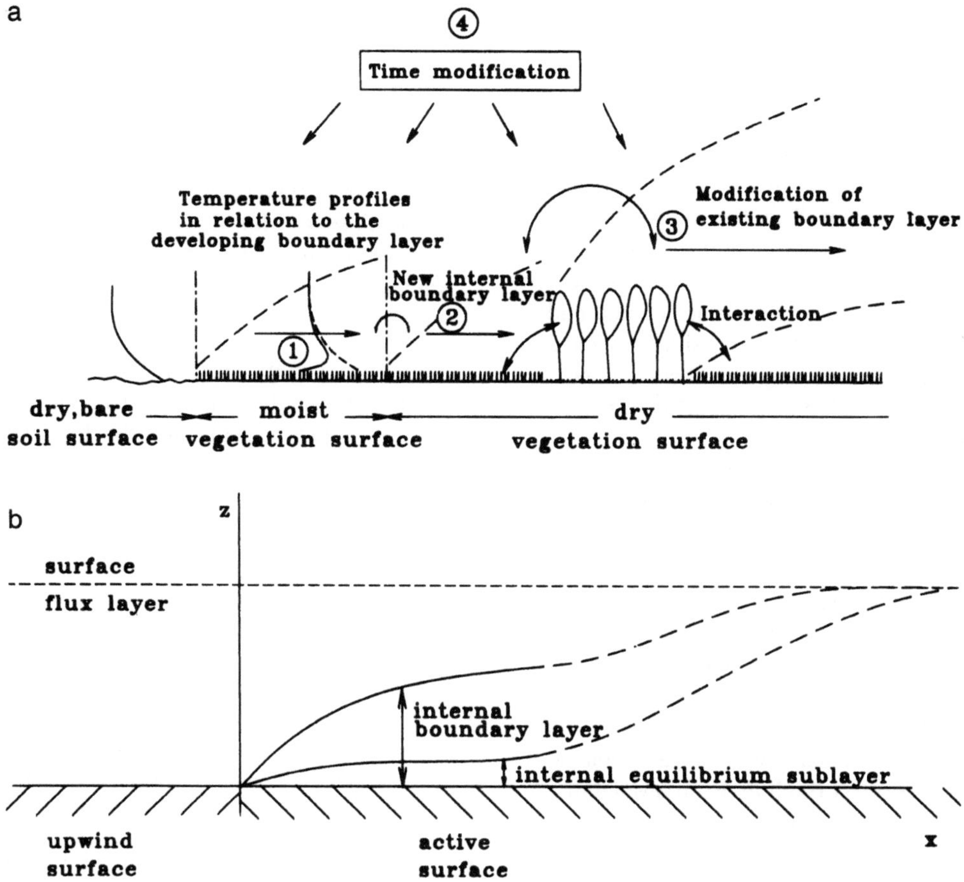

FIGURE 8.2. (a) The development of an internal boundary layer at a regional scale. (b) The development of an internal boundary layer as air flows from a surface to a new and climatically different surface.

grows with increasing distance, or fetch, downwind from the line of discontinuity, and the thickening rate depends on the relative change of roughness between the two surfaces and on the atmospheric stability. It is only in the lower part of this layer that the conditions are fully adjusted to the properties of the new surface. Above the internal boundary layer, flow characteristics are the same as in the upwind part flow at the same height above the surface; they are essentially determined by the mean upwind surface characteristics. The influence of the upstream surface may be expected to disappear at a sufficiently large distance from the line of discontinuity, where the internal boundary layer has grown to the equilibrium surface layer.

This equilibrium surface layer may in turn be defined as a fully turbulent zone where the vertical turbulent fluxes do not change appreciably from their mean values at the surface. The layer so defined is often referred to as the surface flux layer.

The surface flux layer is a zone where it is easiest to estimate turbulent fluxes; the sharpest variations with height in mean meteorological variables with height occur in this layer. As a good experimental field site should have a large enough upwind fetch, this is the prime requirement that determines the thickness of the constant flux layer. Itier and Perrier (1976) showed that the height-to-fetch ratio for the constant flux layer should be at most 1%. But one must keep in mind that local flux values calculated from measurements of mean wind speed, temperature, and humidity at two levels in the constant flux layer are different from those calculated from regional climatic data, the latter being more representative of the regional upwind flow.

2.2.2 *Oasis Effect*

Wetting of the ground surface by precipitation or artificial irrigation can greatly modify the surface energy balance, as well as the microclimate near the surface. The free availability of water allows evaporation to become an important, and even a dominant component of the surface energy budget, while the sensible heat flux is considerably reduced. The sensible heat flux may even become negative, particularly for small areas, due to regional advective effects.

Furthermore, in this case, the contrast between the wetter and dryer surfaces is so important that, even with a long fetch, the internal boundary layer never becomes thick enough to be really developed before the surface conditions change again. Then, it is more than an edge effect, it is the so-called oasis effect. As we shall see later, the relative importance of sensible and latent heat fluxes will depend on the surface moisture availability and the corresponding temperature at the surface; one can then define either an oasis effect (high moisture availability) or an island effect (low moisture availability).

2.2.3 *Regional Equilibrium*

On a regional scale the land surface appears as a patchwork of fields with different surface properties. These fields are associated with different internal boundary layers that are never completely developed most of the time. In these conditions, the overall energy and water balance induce a natural equilibrium that represents some mean integration of the different surfaces contribution, and the regional evapotranspiration tends naturally to an equilibrium value.

2.3 Effects of Surface Roughness Parameters

The aerodynamic roughness of a flat and uniform surface may be characterized from the average height of the various roughness elements, their areal density, characteristic shapes, and dynamic response properties (flexibility, mobility). All these characteristics are important because they exert a considerable influence on the position of the active surface with regard to the exchanges of heat, mass, and momentum. For exchange processes, it is convenient to define two parameters:

1. The height of the reference plane called displacement height (d), which is less than the height of the roughness elements (h). It increases with increasing roughness density and approaches a value close to h for very dense canopies.
2. The surface roughness length, z_0, which depends on the mean height and density of the roughness elements.

For a given canopy, Perrier (1982) expressed the displacement height d and the roughness parameter z_0 as a function of LAI, the total leaf area index,

$$\frac{d}{h} = 1 - \frac{2}{LAI}[1 - \exp(-LAI/2)]$$

$$\frac{z_0}{h} = [1 - \exp(-LAI/2)]\exp(-LAI/2)$$

These bulk expressions can be generalized to take into account vertical variations in the leaf density (see Fig. 8.3).

At the regional scale, d and z_0 are generally determined indirectly from a knowledge of the type and distribution of the surface roughness elements. Wieringa (1986) produced a table giving effective areal z_0 values in terms of terrain classes in the absence of significant orographic effects.

3. Energy Budget Near the Surface

For deriving a simplified equation for the energy budget near the surface, we assume the surface to be a finite-depth interfacial layer, which may include the small-scale surface inhomogeneities. This layer must have finite mass and heat capacity, which would allow the energy to be stored in or released from the layer over a given time interval. In the case of a soil we just consider, for example, the upper layer, so that energy storage is negligible. For vegetative surfaces, the energy budget of the whole canopy layer has to be considered. For such conditions, the rate of energy storage results from both the rate of physical heat storage and the rate of biochemical heat storage as a result of photosynthesis and carbon dioxide exchange. These two terms are usually negligible on micrometeorological time scales of a few hours to a day. Consequently,

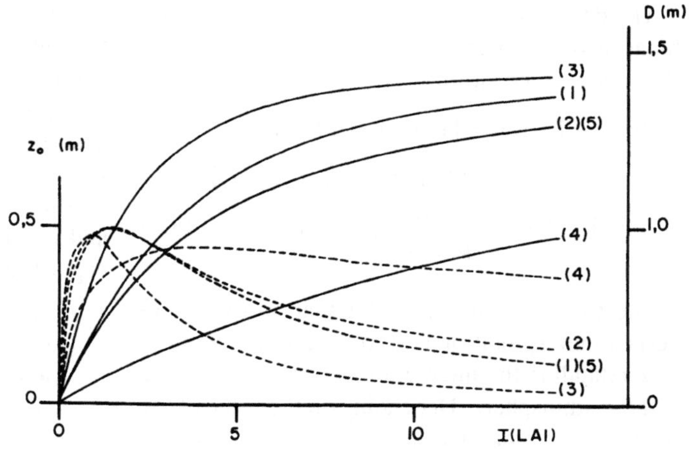

FIGURE 8.3. Theoretical variation of roughness parameter (z_0 dashed lines) and crop height displacement (d solid lines) as function of crop *LAI* for several leaf density profiles. (1) constant leaf density profile; (2) parabolic leaf density profile (maximum for $h/2$); (3) maximum leaf density at the top ($3/4h$); (4) maximum leaf density at the bottom ($h/4$); (5) measured maize leaf density.

most of the time it will be assumed that they are negligible, but not in the case of a forest energy budget. Then, a simplified one-dimensional energy budget for the layer can be expressed as

$$R_N = H + LE + G \tag{8.1}$$

where R_N is the net radiation, H and LE are the sensible and latent heat fluxes to or from the air, and G is the ground heat flux to or from the submedium.

Each term can be written as a function of atmospheric data and surface parameters.

3.1 Ground heat flux G

G depends on the soil thermal properties, which, in turn, depend on surface and soil type and on whether it is wet or dry and whether it is a bare or a vegetated surface.

3.2 Net radiation R_N

$$R_N = (1 - a)R_G + \varepsilon(R_A - \sigma T_S^4) \tag{8.2}$$

where R_G, global solar radiation, and R_A, longwave atmospheric radiation, are the two terms of incident radiation; a and ε define the radiative properties of the surface, the albedo, and emissivity; T_a and T_s are, respectively, the air temperature and radiative surface temperature, so that εR_A is the absorbed downwelling longwave radiation.

Net radiation R_N may be written as a climatic net radiation R_N^* multiplied by a correction term:

$$R_N = R_N^* + f(T_a - T_s) \tag{8.3}$$

with

$$R_N^* = (1 - a)R_G + \varepsilon(R_A - \sigma T_a^4) \tag{8.3a}$$

and

$$f(T_a - T_s) = 4\varepsilon\sigma T^3 (T_a - T_s) \tag{8.3b}$$

with T a mean value for either T_a or T_s.

3.3 Sensible heat flux H

$$H = \rho c_p [h]_{z_S}^{z_R} [T(z_s) - T(z_R)] \tag{8.4}$$

where ρ volumetric mass of air; c_p heat capacity of air; $[h]_{z_S}^{z_R}$ exchange coefficient between the surface (z_S) and the reference height (z_R); $T(z)$ air temperature at level z; $T(z_S)$ may be defined as ground surface temperature or canopy surface temperature.

3.4 Latent heat flux LE

$$LE = (LM/RT) [h]_{z_S}^{z_R} [P[T_d (z_S)] - P[T_d (z_R)]] \tag{8.5}$$

where L, latent heat of water; M, molar weight of water; R, constant of perfect gas; $P[T_d (z)]$, saturation vapor pressure at temperature $T_d (z)$; $T_d (z)$ dew point temperature at level.

A most commonly used expression of evapotranspiration (LE) is given by

$$LE = \frac{P'}{P' + \gamma} [(R_N - G) + \rho c_p [h]_{z_S}^{z_R} (Y_R - Y_S)] \tag{8.6}$$

where P' is the slope of the saturation water vapor pressure, γ the psychometric constant, and Y_z the air hygrometric temperature deficit at level $z[Y(z) = T(z) - T_d (z)]$.

The derivation of Equation (6) is based on combination of Equations (8.1), (8.4), and (8.5) and of the following equation:

$$\begin{aligned} T_d(z_R) - T_d(z_S) &= [T_d(z_R) - T(z_R)] \\ &\quad - [T_d(z_S) - T(z_S)] + [T(z_R) - T(z_S)] \\ &= Y_S - Y_R + \frac{H}{\rho c_p [h]_{z_S}^{z_R}} \end{aligned} \quad (8.7)$$

The two-term structure of Equation (8.6) suggests an interpretation that may serve as an aid in understanding the interaction between the surface and the atmosphere. Depending on the value of the relevant climatological parameters five different cases may occur:

1. $Y_S = 0$ defines potential evaporation (*EP*): evaporation from any large uniform surface that is sufficiently moist or wet, so that the air in contact with it is fully saturated. Such conditions prevail usually only after the occurrence of precipitation or dew.
2. $Y_S = Y_R$: there is an equilibrium between air hygrometric temperature deficit at reference level and at surface level. This particular value for evapotranspiration, which may be called climatic evapotranspiration, is also an equilibrium value (LE_0) toward which the regional evapotranspiration must naturally tend.
3. $Y_{S\max} = Y_R + (R_N - G)/\rho c_p [h]_{z_S}^{z_R}$: corresponds to a dry surface without evapotranspiration ($LE = 0$). In this case the surface temperature is a maximum.
4. $Y_S < Y_R$ is referred to as an "oasis effect," air hygrometric temperature deficit at the reference level being greater than that at the surface level, i.e., the air is drier at the reference level than it is at the surface. In the oasis the free availability of water sometimes allows evaporation to exceed energy supplied by radiation. This apparently anomalous situation is explained by the fact that the atmosphere supplies sensible heat to the surface because the oasis is cooler than the surrounding air in which it is embedded.
5. $Y_S > Y_R$ is referred to as an "island effect," air hygrometric temperature deficit at the reference level being lower than that at the surface level, i.e., air is wetter at the reference level than it is at the surface.

These different cases are schematically depicted in Figure 8.4.

Equation (8.6) has such a general form that it can be applied to describe the rate of evaporation above a surface having specified properties for water vapor transfer. Two main cases can be distinguished: a surface with constant water potential, and a surface with specified resistance to water vapor transfer.

1. Constant water potential at the surface: this situation is appropriate for a layer of uniformly wet porous material with relative humidity less than 1. In this case, the water potential at the surface ψ is related to water vapor pressure in the system, $P(T_{ds})$, by the expression:

$$\psi_S = \frac{RT}{M} \ln \frac{P(T_{ds})}{P(T_s)} \quad (8.8)$$

where $P(T_s)$ is the vapor pressure of pure, free water at the same temperature.

From this expression (8.8), the air hygrometric temperature deficit at the surface can be written as

$$Y_s = \frac{P(T_s)}{P'} (1 - \exp(\psi_S \cdot (M/RT))) \quad (8.9)$$

2. Specified surface resistance to vapor transfer: the surface having a temperature T_S and a water vapor pressure $P(T_{ds})$, the difference between $P(T_s)$ and $P(T_{ds})$ is proportional to the evaporation rate, and the constant of proportionality is described as a "surface resistance" r_s such that

$$P(T_{ds}) - P(T_s) = r_s \times LE \times \frac{RT}{LM} \quad (8.10a)$$

The most common example of this case is transpiration through the stomatal pores of a leaf. Extension of this analysis to crop canopies has been the subject of many experimental and theoretical works (Perrier, 1976; Monteith, 1965, 1981). This type of equation can also be used to study water vapor transfer from a drying soil to the atmosphere; in this case r_s is a function of the dry-soil layer thickness (mulch depth) at the surface (Sifaoui and Perrier, 1978).

From Equation (8.10a), an expression for the air hygrometric temperature deficit at the surface can be derived:

$$Y_s = r_s \times LE/(LMP'/\mathrm{RT}) \quad (8.10b)$$

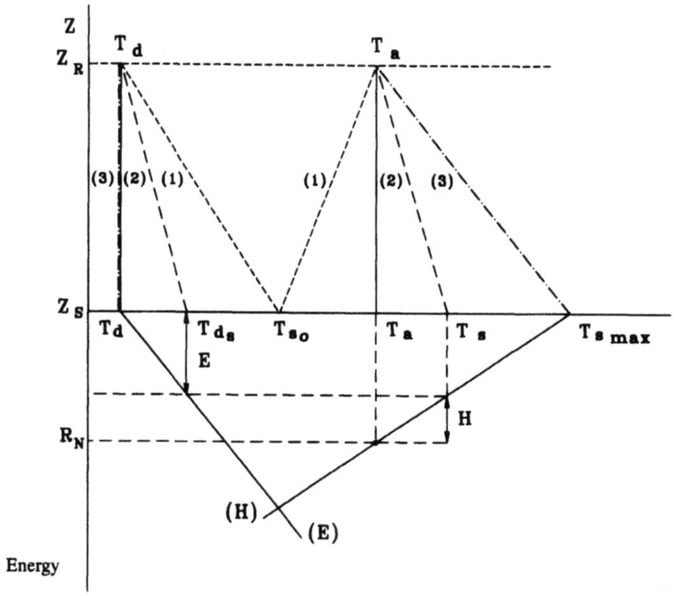

FIGURE 8.4. The upper graph, above the zero line, shows the surface layer and three schematic profiles: (1) saturation at the surface: $Y_S = 0$, $LE = EP$; (2) equilibrium situation: $Y_S = Y_R$, $LE = LE_0$; (3) completely dry situation: $Y_S = Y_{S\max} = T_{S\max} - T_d$, $LE = 0$. The lower graph, below the zero line, represents energy balance partition according to the value of Y_S ($Y_S = T_S - T_{ds}$). (*E*) represents the relationship between evapotranspiration and surface dew point temperature, the slope depending on the exchange coefficient; (*H*) represents the relationship between sensible heat flux and surface temperature, with the origin taken at the point defined by R_N and T_a (the sensible heat flux is positive for $T_S > T_a$ and negative for $T_S < T_a$). For any value T_{ds} of dew point temperature at the surface, an *LE* value is associated through the relationship (*E*). From this value, a horizontal projection on (*H*) gives the corresponding value of sensible heat flux. Then, from this value of *H* a vertical projection on the *x* axis gives the corresponding surface temperature T_S. For any given T_S value the reverse construction is valid.

This expression relates the air hygrometric temperature deficit to the surface resistance through the evaporation rate. It follows that substitution of this equation in Equation (8.6) gives an equation with the same general form as the original Penman–Monteith formula. In a same way, it is easy to show that the evaporation rate can be expressed as a function of climatic evapotranspiration:

$$LE = \frac{P'}{P' + \gamma} (R_N - G) \frac{1 + \gamma/(P'+\gamma)\, h\, r_c}{1 + \gamma/(P'+\gamma)\, h\, r_s} \quad (8.11a)$$

where r_c is a climatic resistance defined

$$r_c = \frac{P' + \gamma}{\gamma} \frac{\rho c_p Y_R}{R_N - G} \quad (8.11b)$$

4. Micrometeorological Methods of Determining Evaporation

4.1 Eddy Correlation Method

This method consists of determining the turbulent fluxes of momentum, water vapor, and sensible heat from covariances between fluctuating quantities u', w', q', and Θ'. Hence, over a uniform surface under steady conditions, the governing flow equations can be considerably simplified, as equations for horizontal mean motion, obtained from the Reynolds equations, simply become (e.g., Businger, 1973):

$$-\frac{1}{\rho}\frac{\partial \bar{p}}{\partial x} + f\bar{v} + \nu \frac{\partial^2 \bar{u}}{\partial z^2} - \frac{\partial}{\partial z}\overline{(w'u')} = 0 \quad (8.12)$$

$$-\frac{1}{\rho}\frac{\partial \bar{p}}{\partial y} - f\bar{u} + \nu \frac{\partial^2 \bar{v}}{\partial z^2} - \frac{\partial}{\partial z}\overline{(w'v')} = 0 \quad (8.13)$$

where the gradients of the mean pressure $\bar{p}$ are assumed to be constant along the horizontal. The Coriolis parameter f represents the influence of the earth's rotation, ν, is the viscosity of dry air and $\bar{u}$ and $\bar{v}$ are the latitudinal and meridional mean wind are the latitudinal and meridional mean wind components.

The analogous equation for mean specific humidity $\bar{q}$ can be expressed under the form

$$K_v \frac{\partial^2 \bar{q}}{\partial z^2} - \frac{\partial}{\partial z} \overline{(w'q')} = 0 \tag{8.14}$$

where K_v is the molecular diffusivity of water vapor in air.

Finally, the analogous equation for heat transport in a stationary and horizontally homogeneous boundary layer is

$$K_h \frac{\partial^2 \bar{\Theta}}{\partial z^2} - \frac{\partial}{\partial z} \overline{(w'\bar{\Theta}')} - \frac{1}{\rho c_p} \frac{\partial HR}{\partial z} = 0 \tag{8.15}$$

where K_h is the molecular diffusivity of dry air and $\partial HR/\partial z$ the radiative flux divergence.

In fully turbulent flow, the molecular transport terms depending on viscosity ν and molecular diffusivities K_v and K_h are orders of magnitude smaller than the Reynolds fluxes. With these assumptions and if the mean wind $\bar{u}$ near the surface is aligned on the x axis ($\bar{v} = 0$), Equations (8.12) and (8.13) can be simplified into (e.g., Businger, 1973)

$$u_* = \overline{w'u'} \tag{8.16}$$

and

$$\overline{w'v'} = 0 \tag{8.17}$$

When these assumptions are valid, Equation (8.14) shows that the evaporation rate, E, at the surface is constant:

$$E = \rho\, \overline{w'q'} = Cst \tag{8.18}$$

Similarly, for the sensible heat flux H, Equation (8.15) shows that in the absence of radiative flux divergence, we get

$$H = \rho c_p\, \overline{w'\theta'} = Cst \tag{8.19}$$

Consequently, theory predicts that fluxes from the surface can be measured by correlating the vertical wind fluctuations w' with fluctuations in temperature, water vapor, CO_2, etc. over the desired sampling or averaging time. A representative fetch is required since the measurements are made at some height above the surface.

As more extensively discussed by Wyngaard (this volume), the distribution of eddy sizes contributing to vertical transport creates a range of frequencies that has to be taken into account for eddy correlation measurements. The sensors must have a small enough response time to measure the frequencies at the higher end of the range, while covariance averaging times must be long enough to include frequencies at the lower end. A bandwidth combining these desired features and recommended by many investigators (McBean, 1972; Kaimal et al., 1972) is given by

$$10^{-3} \le f z/\bar{u} \le 10 \tag{8.20}$$

where z is the measurement height, f the frequency, and $\bar{u}$ the mean horizontal wind. As shown in Equation (8.20), the required sensor response time decreases, and averaging time increases as the measurement height increases.

4.2 Energy Balance Method (Bowen Ratio Method)

The energy balance method for flux measurement is derived from the energy balance of the underlying surface

$$R_N - G = H + LE \tag{8.21}$$

If the gradient transport theory, relating turbulent fluxes to mean quantities vertical gradients, is used, this equation becomes

$$R_N - G = -\rho c_p K \frac{dT}{dz} - L \rho K \frac{dq}{dz} \tag{8.22}$$

where K is the eddy diffusivity, assumed to be identical for both heat and water vapor turbulent transfer.

Using the relationship between specific humidity q and water vapor pressure $P(T_d)$, and measurements at two levels in the surface layer, the eddy diffusivity can be written as

$$K = \frac{R_N - G}{\rho c_p \left(\frac{\Delta T}{\Delta z} + \frac{1}{\gamma} \frac{\Delta P(T_d)}{\Delta z} \right)} \tag{8.23}$$

where ΔT and $\Delta P(T_d)$ are the differences in temperature and water vapor pressure at two levels in the surface flux layer. These measurements can be done through the use of dry and wet bulb thermometers. The wet bulb temperature is a measurable quantity that is commonly used for estimating water vapor pressure using the psychometric equation:

$$\Delta P(T_d) = P' \Delta T_w - \gamma(\Delta T - \Delta T_w) \qquad (8.24)$$

Then, from Equations (8.23) and (8.24), it is easy to show that sensible and latent heat fluxes can be written as (with $\Delta T_* = \Delta T - \Delta T_w$)

$$H = H_0\left(1 + \frac{\Delta T_*}{\Delta T_w}\right)$$

$$LE = E_0\left(1 - \frac{\gamma}{P'}\frac{\Delta T_*}{\Delta T_w}\right)$$

Here H_0 and E_0 are the equilibrium values defined in Section 3 for $Y_{z_S} = Y_{z_R}$.

The Bowen ratio concept thus provides simple expressions for the determination of sensible and latent heat fluxes.

4.3 Profile, or Aerodynamic, Method

Applying the Monin–Obukhov similarity relations to the horizontally homogeneous surface layer (e.g., Businger, 1973), the mean wind speed $\bar{u}$, the temperature $\bar{T}$, and the specific humidity $\bar{q}$ profiles can be represented by

$$\bar{u}(z) = \frac{u_*}{k}\left[\ln\frac{z-d}{z_0} - \psi_m\left(\frac{z-d}{L}\right)\right]$$

$$\bar{T}(z) - T_0 = \frac{H}{\rho c_p k u_*}\left[\ln\frac{z-d}{z_0} - \psi_h\left(\frac{z-d}{L}\right)\right]$$

$$\bar{q}(z) - q_0 = \frac{E}{\rho k u_*}\left[\ln\frac{z-d}{z_0} - \psi_{w'}\left(\frac{z-d}{L}\right)\right]$$

where T_0 and q_0 are the values at $z = d + z_0$, k is the von Karman's constant, and L is the Obukhov length scale defined as

$$L = -\frac{u_*^3}{k\frac{g}{T}\frac{H}{\rho c_p}}$$

The Monin–Obukhov similarity functions ψ_m, ψ_h, and $\psi_{w'}$ are fairly well known over a restricted range of stability conditions around neutral conditions, but their specification for extreme stability conditions (both stable and unstable) is much more uncertain (see, e.g., Stull, 1988).

For unstable and neutral conditions $[(z-d)/L \le 0]$ one can take

$$\psi_m\left(\frac{z-d}{L}\right) = 2\ln\frac{1+x}{2} + \ln\frac{1+x^2}{2} - 2\,\mathrm{arctg}\,x + \frac{\Pi}{2}$$

$$\psi_h\left(\frac{z-d}{L}\right) = \psi_w\left(\frac{z-d}{L}\right) = 2\ln\frac{1+x^2}{2}$$

where

$$x = \left(1 - 16\frac{z-d}{L}\right)^{0.25}$$

While for stable conditions $[(z-d)/L > 0]$ one has

$$\psi_m\left(\frac{z-d}{L}\right) = \psi_h\left(\frac{z-d}{L}\right) = \psi_{w'}\left(\frac{z-d}{L}\right)$$

$$= -5((z-d)/L) \qquad \text{if } 0 < ((z-d)/L) \le 1$$

$$= -5\left(1 + \ln\frac{z-d}{L}\right) \qquad \text{if } ((z-d)/L) > 1$$

(Brunet, 1984)

The above described flux-profile equations are the basis of the aerodynamic method; they allow the calculation of surface fluxes from measurements of mean wind speed, temperature, and humidity at two or more levels in the surface layer. The procedure for determining fluxes from such profiles measurements is to fit the appropriate flux relations to the observations; an iterative method is required to take into account the stability influence.

4.4 Combined Aerodynamic Method

Classical Method

To use the above described aerodynamic method, one needs measurements of mean wind speed, temperature, and humidity profiles. Reliable measurements of humidity profiles are, however, not easy to achieve. This difficulty can be avoided by combining the aerodynamic method with the energy balance method. In fact, the energy budget equation is then used to determine the latent heat flux from net radiation measurements and estimates of sensible heat flux.

Simplified Method

This method tries to combine the advantages of the classical combined method (no measurement of air humidity) with those of the Bowen ratio method (only two levels of measurement in the air). Besides net radiation measurements, the sensible heat flux is obtained by measurements of wind speed and air temperature at two levels above the canopy (z_1 and z_2). From these values, one can determine the Richardson number at the mean height $z_* = (z_1 z_2)^{1/2}$; it is approximately given by

$$R_i \simeq \frac{g}{T} z_* \left(\ln \frac{z_2}{z_1} \right) \frac{\Delta T}{(\Delta U)^2}$$

Then, Riou (1982) and Itier (1982) showed that currently used expressions of Monin–Obukhov similarity functions could be replaced in flux formulation by direct R_i functions such as

$$-1 < R_i < 0 \quad H = H_n (1 - 16 R_i)^{3/4}$$

$$0 < R_i < 0{,}14 \quad H = H_n (1 - 5 R_i)^2$$

where H_n is the sensible heat flux under neutral conditions.

The SAMER system is based on this simplified aerodynamic method (Itier, 1982). For larger instabilities ($R_i < -1$), the sensible heat flux is computed with the temperature difference alone, keeping the Priestley coefficient constant:

$$H = \alpha \Delta T^{3/2}$$

For larger stabilities ($R_i > 0{,}14$), the sensible heat flux is taken equal to

$$H = H_n/10$$

Comparisons between sensible heat fluxes from a SAMER system and from other systems based on the Bowen ratio method (BEARN system) or aerodynamic method are shown by Goutorbe (this volume).

5. Concluding Remarks

A primary objective of the various micrometeorological studies involving plant canopies has been to better understand the processes of momentum, heat, and mass exchanges between the lower atmosphere and the active canopy. These exchanges influence the local weather as well as the microclimate in which plants grow. A basic understanding of exchange mechanisms and development of practical methods to parameterize them in terms of more easily and routinely measured parameters is essential for many applications at the local scale, mesoscale, and also GCM scale. At the regional scale, changes in the relative proportion of land cover types in a region, or changes in the pattern of various land cover types, may also affect regional climate. In this case remotely sensed information on the pattern of landscapes and on its physical and biological properties may be necessary to infer changes in regional climate processes.

Acknowledgments. The authors gratefully acknowledge the helpful comments of Y. Brunet and the reviewers.

References

Brunet Y (1984) Modélisation des échanges sol nu–atmosphère. Essai de validation locale et influence de la variabilité spatiale du sol. Thèse de Docteur-Ingénieur, INPG.

Businger JA (1973) Turbulent transfer in the atmospheric surface layer. Workshop on Micrometeorology, pp 67–100. Amer. Meteor. Soc., Boston.

Charney JG (1975) Dynamics of deserts and drought in the Sahel. Quart J Ry Meteorol Soc 101:193–202.

Itier B, Perrier A (1976) Présentation d'une étude analytique de l'advection. I. Advection liée aux variations horizontales de concentration et de température. Ann Agron 27:111–140.

Itier B (1982) Révision d'une méthode simplifiée pour la mesure du flux de chaleur sensible. J Rech Atmos 16:85–90.

Kaimal JC, Wyngaard JC, Izumi Y, Cote OR (1972) Spectral characteristics of surface layer turbulence. Quart J Ry Meteorol Soc 98:563–589.

McBean GA (1972) Instrument requirements for eddy correlation measurements. J Appl Meteorol 11:1078–1084.

Monteith JL (1965) Evaporation and environment. Symp Soc Exp Biol XIX:205–234.

Monteith JL (1981) Evaporation and surface temperature. Quart J Ry Meteorol Soc 107:1–27.

Perrier A (1976) Etude et essai de modélisation des échanges de masse et d'énergie au niveau des couverts végétaux: profils microclimatiques, évapotranspiration et photosynthèse nette. Thèse Docteur d'Etat Es Sciences Physiques, Univ. Pierre et Marie Curie, Paris VI.

Perrier A (1982) Land surface processes: Vegetation. In "Land Surface Processes in Atmospheric General Circulation Models" PS Eagleson ed., pp. 395–448. Cambridge Univ. Press, Cambridge.

Riou Ch (1982) Une expression analytique du flux de chaleur sensible en conditions suradiabatiques à partir de mesures du vent et de la température à deux niveaux. J Rech Atmos 16:15–22.

Sifaoui MS, Perrier A (1978) Caractérisation de l'évaporation profonde. Int J Heat Mass Transfer 21:629–637.

Stull RB (1988) "An Introduction to Boundary-Layer Meteorology." Kluwer Academic Publishers, Dordrecht.

Wieringa J (1986) Roughness-dependent geographical interpolation of surface wind speed averages. Quart J Ry Meteorol Soc 112:867–889.

9
Fluxes in the Surface Layer Under Advective Conditions

H.A.R. de Bruin, N.J. Bink, and L.J.M. Kroon

1. Introduction

The surface fluxes of water vapor, sensible heat, and momentum are important in many meteorological, agricultural, and hydrological problems. In most applications it is (tacitly) assumed that the Earth's surface is, on the scale of interest, horizontally homogeneous. An example is the way in which the surface fluxes are parameterized in models for the prediction of weather and climate. Also, most measuring techniques used for the determination of the surface fluxes are based on the assumption of horizontal homogeneity.

Although the last word on the uniform case has certainly not been said, for most practical calculations the Monin-Obukhov similarity theory (Panofsky and Dutton 1984; Brutsaert 1982) provides a suitable description of the surface fluxes. Reliable estimates can be obtained from the vertical profiles of temperature, humidity, and wind speed or, alternatively, the variances of these quantities.

In reality the Earth's surface is seldom homogeneous, and even if a surface is uniform, the meteorological conditions are often such that horizontal differences exist. Examples are partly clouded skies or surfaces partly wetted by rain. Matters become much more complicated in the heterogeneous case, even if we exclude hilly or mountainous terrains. Several types of nonuniform surfaces can be distinguished, primarily depending on the scale being considered. For meteorological models the horizontal scale varies between, say, 10 km (mesoscale models) and 300 km (general circulation models). In agriculture the scale of interest is about 100 m to 1 km, while hydrologists deal with scales of 1 km to 100 km depending on the size of the watershed.

Small scale irregularities at the surface will affect only the atmospheric surface layer (ASL), but irregularities of the order of 10–50 km also influence the planetary boundary layer (PBL), while larger ones can affect the flow of the free atmosphere.

In this chapter we will consider only a surface that is nonuniform on a small scale; so only the atmospheric surface layer is affected. For this relatively simple case we distinguish two types of nonuniform surfaces:

Type a: A nonuniform terrain with a constant surface roughness, but different thermal properties described in the previous chapter as the Oasis effect. In this case, primarily the fluxes of heat and water vapor are affected by the irregularities and not the momentum flux. An example will be considered in this chapter.

Type b: A uniform terrain covered with isolated obstacles. Then, the momentum flux is affected, while the sensible and latent heat fluxes are hardly disturbed. An example is the typical Dutch landscape of meadows, where cows, measuring vans, isolated trees, and ditches are the obstacles.

Here we will consider an example of type a, viz. an irrigated grassland field, surrounded by dry land covered with a vegetation of about equal height. This field is located in the area of "La Crau" in the South of France, where the Mistral (i.e., a strong dry northerly wind) frequently blows. Some first results are shown from a field experiment car-

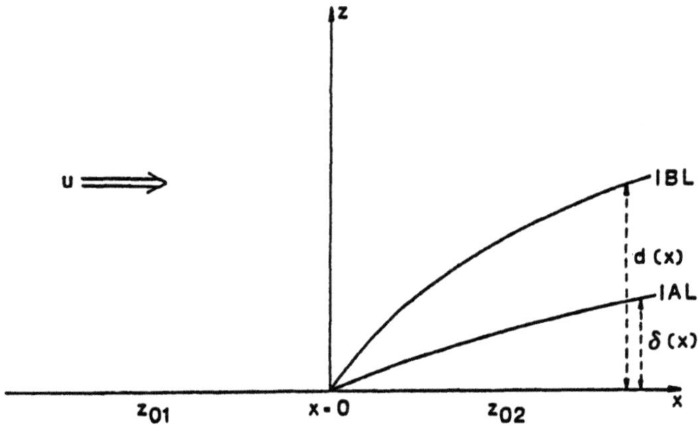

FIGURE 9.1. Definition sketch of a change in surface conditions.

ried out in June 1987. The turbulent fluxes of heat, water vapor, and momentum were measured with the eddy-correlation technique over the dry uniform upwind terrain and at several heights over the irrigated field downwind from a sudden step change in surface conditions. Also the vertical profiles of temperature, humidity, and wind speed were observed at different distances from the leading edge. It appears that the observed profiles are fairly well described by the second-order closure model of Rao et al. (1974). A first analysis of the collected data reveals that in the case where the Mistral occurs the fluxes of heat and water vapor over the irrigated field vary considerably with height in the first 15 m. Furthermore, it appears that the profiles of temperature and humidity change with the distance from the leading edge. Also, the applicability of the Monin–Obukhov similarity theory is considered. For this purpose, the fluxes are calculated with the variance technique (Tillman, 1972). Large deviations with observation above the irrigated field are found, but evidence is given that the variance method still can be applied to this surface type for water vapor.

Results for this field will be compared with those obtained earlier from a type b site in the Netherlands. Also results will be used from measurements over the dry land that surrounds the irrigated grassland. The dry area is an extensive uniform terrain.

It is noted that data of fluxes near the surface under advective conditions are rather scarce. Lang et al. (1983) did measurements in Australia above in irrigated rice field, thus also a type a field. These authors considered the differences between the exchange coefficients for heat and water vapor.

2. Theory

2.1 General

In most models that describe the structure of the turbulent flow over a sudden step-change in surface conditions, the region of the atmosphere affected by the step-change is referred to as the internal boundary layer (IBL). Furthermore, it is assumed that the flow in the lower portion of the IBL is locally adapted to the new surface conditions. This sublayer is called the internal adapted layer (IAL) (Kroon, 1985) or internal equilibrium sublayer (Brutsaert, 1982). Note that the surface fluxes within this adapted layer vary with the distance from the step change. The depth of the IBL and IAL, whose absolute value depends on their definition, grows with the distance, x, downwind from the step-change (Fig. 9.1).

Reviews of existing models dealing with this issue are given by Brutsaert (1982) and Kroon (1985). The majority of these models deal with a step-change in surface roughness and consider the behavior of the wind field and momentum flux only. It is concluded in both reviews that the second-order closure approach proposed by Rao et al. (1974) is the most promising. In Appendix 2 a brief description of this approach is given. Some applications of the model of Rao et al. (1974) will be presented in this chapter.

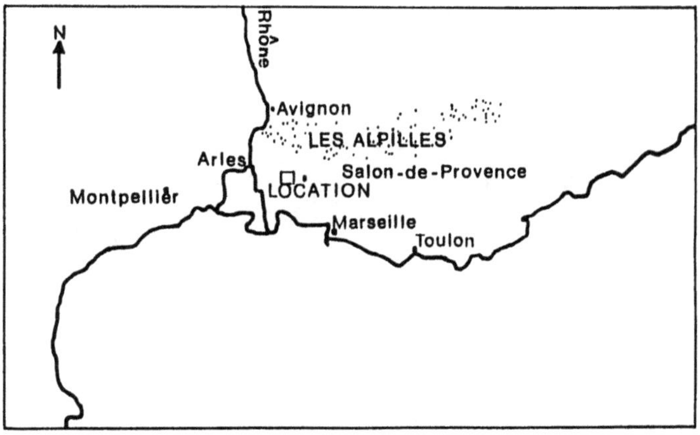

FIGURE 9.2. The location of the Crau experiment, part of the EC program "regional parameterization of the surface fluxes."

A drawback of this approach is that al first- and second-order moments such as $\overline{w'q'}$, $\overline{u'w'}$, $\overline{u'q'}$ etc. must be specified at the surface both upwind and downwind from the discontinuity at $x=0$. Rao et al. (1974) assumed that at the surface the relative humidity is a constant after the step-change. This is unrealistic (Brutsaert, 1982). For this reason we introduced the "big leaf" concept of Monteith (1981) as a lower boundary condition in the model.

2.2 Variance Method

To illustrate how a sudden dry-to-wet step-change at the surface affects the fluxes of heat and water vapor, we will consider the so-called "variance method." This method is based on the Monin-Obukhov similarity theory. For horizontally uniform surfaces it predicts that

$$\frac{\sigma_T}{|\theta_*|} = f\left(\frac{z}{L}\right) \quad (9.1)$$

For the explanation of the symbols see Appendix 1. Tillman (1972) has proposed:

$$\frac{\sigma_T}{|\theta_*|} = C_1\left(C_2 - \frac{z}{L}\right)^{-1/3} \quad (L < 0) \quad (9.2)$$

This leads to

$$H = \rho c_p\left(\frac{\sigma_T}{C_1}\right)^{3/2}\left(\frac{kgz}{\theta}\right)^{1/2}\left(\frac{C_2 - z/L}{-z/L}\right)^{1/2} \quad (L<0) \quad (9.3)$$

According to Equation (9.3) ($C_2=0.05$) for $z/L < -0.1$ the so-called free convection limit is reached, and H is approximately given by

$$H = \rho c_p\left(\frac{\sigma_T}{C_1}\right)^{3/2}\left(\frac{kgz}{\theta}\right)^{1/2} \quad (9.4)$$

In that case Equation (9.1) reads (Wyngaard et al., 1971)

$$\frac{\sigma_T}{|\theta_*|} = C_1\left(-\frac{z}{L}\right)^{-1/3} \quad (9.5)$$

This expression is very convenient for practical applications since it requires one fast responding thermometer only. Since, in most cases, H is relatively small if $z/L > -0.1$ this implies that H is primarily determined by σ_T while u_* plays a less important role. If this picture is correct, one expects that Equations (9.3) and (9.4) will be applicable for surfaces over which the temperature is not disturbed, whereas deviations of these equations must be expected in the case of the dry-to-wet transition we are considering.

3. Experimental Set-up and Data Processing

In June 1987 an experiment was carried out in which we studied the airflow over an extreme change of terrain conditions from a desert-like terrain to (occasionally) irrigated grass. The (local) surface roughness, z_0, of the dry terrain was about 0.01 m and the surface roughness of the grass terrain varied from 0.03 m at the beginning of the measurement period to 0.06 m at the end. The location was about 10 km east of Salon-de-Provence at Mas Capelan, France at 43.38°N and 4.59°E in the Crau plain (Fig. 9.2).

FIGURE 9.3. The experimental site and its direct surroundings.

The alluvial plain extends for 50,000 ha with the Rhone at its west side, the Alpilles mountains in the north, and the Mediterranean in the south. The experimental site (Fig. 9.3) was located on the border of the Petite or Wet Crau and the Grande or Dry Crau. The Petite Crau is mainly covered with orchards and fields with all kinds of crops, most of them surrounded by wind breaks. The Grande Crau is a stone desert covered with some herbs and isolated, low shrubs. At the northwest end of the

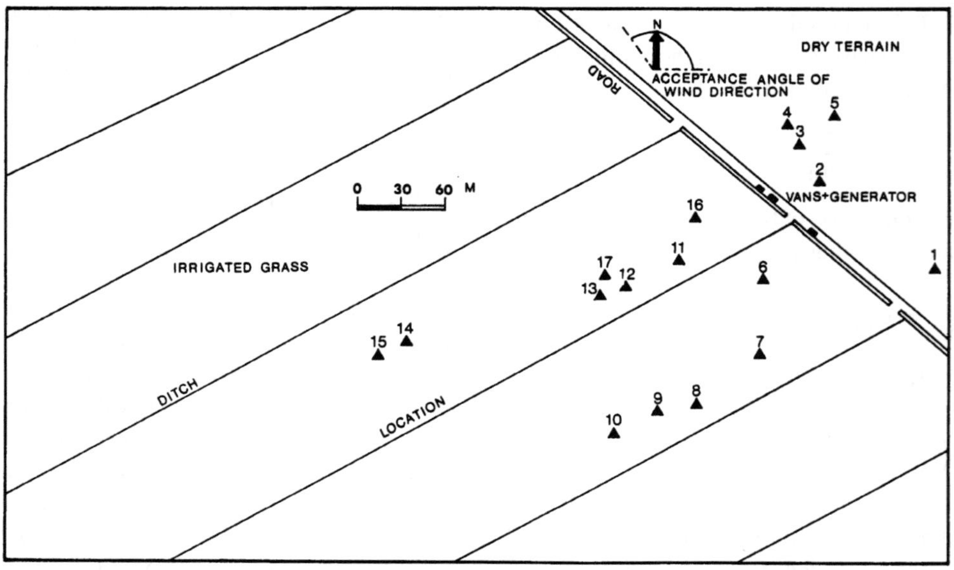

FIGURE 9.4. The experimental setup at La Crau, June 1987, of the Agricultural University Wageningen and the Free University of Amsterdam.

experimental site a small windbreak is present. The fetch for the dry terrain was at least 1 km.

In Table 9.1 and Figure 9.4 an overview of the experimental setup is given.

At no. 4 (Fig. 9.4), above the dry terrain a Kaijo–Denki one-component sonic anemometer, a Lyman-α humidity sensor and thermocouple were mounted at 6.10 m height. At the locations 8, 9, and 10 above the grass terrain at three different heights, 4.1, 6.7, and 13.0 m, three-component sonic anemometers (Kaijo–Denki) and Lyman-αs humidity fluctuation meters were installed. Fast responding thermocouples were mounted at the 4 and 6 m level.

The Lyman-αs are "self-build" except for the source and the detector; these are constructed by Glass Technologists, Maryland. The Lyman-αs were positioned downwind of the sonics and the separation between the w-sensor and the Lyman-α was 0.5 m (1 m at the 13 m level); all other components were nearer.

Corrections of the eddy correlation data due to distortion have not yet been applied. These errors can be significant, but do not affect our results, except those shown in Fig. 9.12. this is partly due to the fact that the relation between σ_T/Θ^* and z/L is not very sensitive to these distortion errors.

The profile measurements were carried out using shielded and aspirated psychrometers for temperature and humidity measurements and cup anemometers for the windprofiles. In the 10 m masts (11 and 12 in Fig. 9.4) psychrometers with Pt-100 resistance sensors are used (precision: 0.1 °C for the dry bulb temperature and 0.5 g/m³ for the humidity). The small cup anemometers have a mean starting speed of 0.2 m/s and the first-order response length is 0.9 m. The psychrometers have been calibrated both in the laboratory (absolute calibration of the Pt-100s) and the field (relative calibration of the psychrometer as a whole). The cup anemometers are calibrated in a wind tunnel. These instruments are self-developed and constructed at our laboratory.

In the 24 m masts the equipment of the department of Meteorology of the Free University of Amsterdam was installed (Vugts et al., 1988).

The turbulence signals were sampled at a rate of 10 Hz and 1 Hz was used for the profile measurements. All samples are written on magnetic tape. The sampling process was controlled with a PDP 11-24 minicomputer installed in a measuring van. The data-processing software allows us to collect samples digitally up to a total of 2000 Hz from a maximum of 200 channels. The system accepts analogue and digital signals. The frequency and the electrical range are separately adjustable for each channel such that the resolution is optimal and the natural range to be expected is covered. The frequency response of all the instruments used for the turbulence measurements ranges from at least 5 to 0.001 Hz. Within this range more than 95% of the flux-carrying fluctuations are covered (McBean and Miyake, 1972).

The data were further processed with the aid of a PDP 11-44 computer at the laboratory. The sampled signals were reduced to 30 min averages. The standard deviation and relevant correlation products of the signal with and without the removal of a linear trend were calculated. The

TABLE 9.1. The instruments.

Dry terrain
1. Psychrometers and cup anemometers at 3 and 0.75 m
2. Stevenson screen (thermohygrograph, minimum and maximum temperature
3. Net pyrradiometer and Heimann KT24 infrared-radiation thermometer (surface temperature)
4. Eddy correlation equipment at 6.10 m (1-component sonic, thermocouple, and Lyman-α humidity-fluctuation meter)
5. Net pyrradiometer and soil heat flux plates (2)
Wet terrain
6. Net pyrradiometer and soil heat flux plates (2×)
7. Net pyrradiometer, Stevenson screen
8. Eddy correlation equipment at 6.7 m (3-component sonic, Lyman-α thermocouple)
9. Cup anemometers (24, 18, 12, 6, 3, 1.5, 0.75 m); psychrometers (24, 12, 6, 3, 0.75 m); Eddy correlation equipment at 13 m (3-component sonic and Lyman-α)
10. Eddy correlation equipment (see 8) at 4.1 m
11. Cup anemometers (10, 7.5, 5, 3, 2, 1.5, 1, 0.75, 0.5, 0.25 m); psychrometers (10, 5, 3, 2, 1, 0.5) m
12. As 11
13. Net pyrradiometer and Heimann KT24 infrared-radiation thermometer (surface temperature)
14. Cup anemometers (24, 18, 12, 6, 3, 1.5, 0.75 m); psychrometers (24, 12, 6, 3, 0.75 m)
15. Stevenson screen
16. Soil heat flux plate and soil temperature (−2, −10, −25, −50 cm) with PT-100 resistance thermometers
17. Soil heat flux plate and soil temperature (−2, −10, −25, −50 cm) with PT-100 resistance thermometers

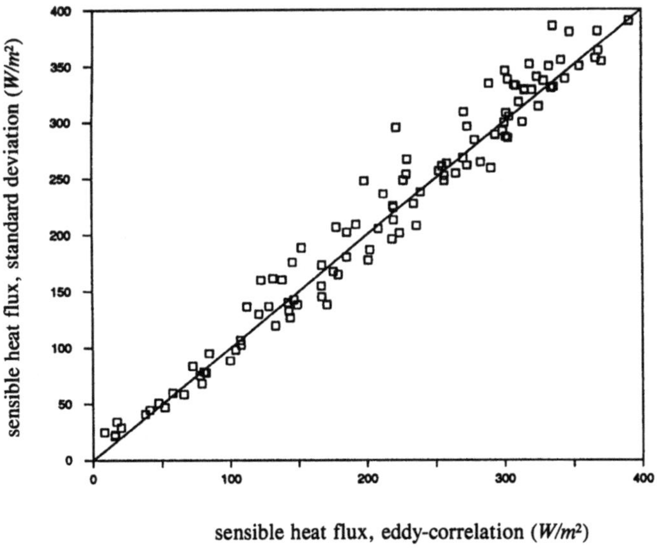

FIGURE 9.5. Uniform terrain. The sensible heat flux measured with the eddy-correlation method at location 4 (Fig. 9.4) compared with the sensible heat flux calculated with the variance technique.

reduced signals were converted to obtain engineering values. All raw data are also stored on tape.

The heat flux at 13 m is calculated from the sonic temperature using the method proposed by Schotanus et al. (1983) since no thermocouple was installed at that height. The same procedure is also adopted to calculate other products like σ_T and $\overline{Tq'}$ at the 13 m level. The method could be checked with the measurements at 4 and 6 m and provides accurate results.

The friction velocity, necessary for Equation (9.3), at the dry terrain for z/L negative is obtained from (Panofsky and Dutton, 1984)

$$\frac{\sigma_w}{u_*} = 1.25\left(1 - 3\frac{z}{L}\right)^{1/3} \tag{9.6}$$

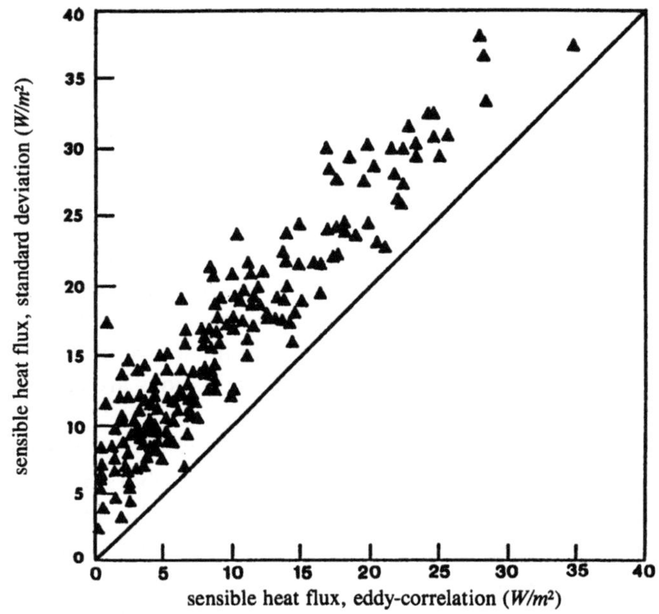

FIGURE 9.6. Nonuniform terrain. The sensible heat flux measured (eddy-correlation) at different locations and levels after the step change (locations 9, 8, and 10 in Fig. 9.4) compared with the sensible heat flux calculated with the variance technique [Equation (3)].

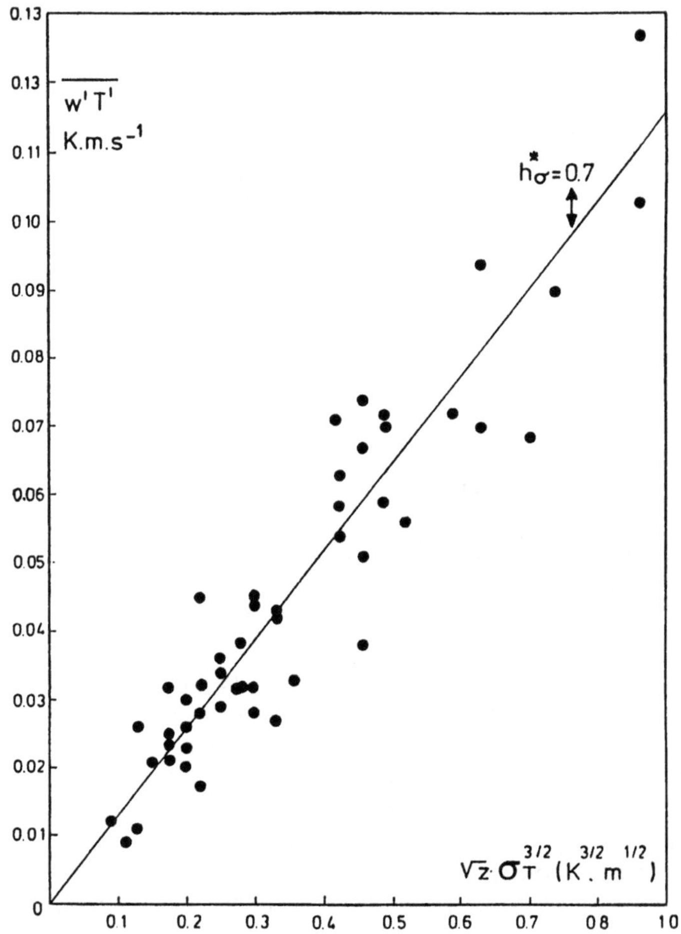

FIGURE 9.7. Nonuniform terrain at Cabauw. The sensible heat flux at z=22.5 m measured with the eddy-correlation method compared with the sensible heat flux derived from the variance technique [Equation (4) from de Bruin 1982]. Note that the axes differ from those in Figures 9.5 and 9.6 ($\overline{w'T'}=u^*\theta^*$ versus $z^{1/2}\sigma_T^{3/2}$).

and the measured heat flux using an iterative procedure.

For the data selection the following criteria were used:

1. Only unstable conditions are considered, $z/L<0$.
2. Air had to be advected from the dry terrain to the wet terrain during a period of days with no rain. The range of wind directions of interest to us is indicated in Figure 9.4. Days with wind directions within this range are selected.

For a period of 5 consecutive days the wind direction was within this range. The angle of incidence of the flow at the leading edge was about 30° with a wind direction of 340° during the selected period, June 20 to 24, 1987. During that period the weather was fair with midday temperatures (measured in a Stevenson screen above the dry terrain) of 23°C on June 20 up to 28°C on June 24. The Mistral was blowing with 10 to 15 m/s at 10 m height. On June 19 there was some rain.

4. Results and Discussion

In Figures 9.5, 9.6, and 9.7 the observed (eddy-correlation) sensible heat flux density is plotted against the corresponding estimated values obtained with the variance method [Equation (3)] for three terrains: (1) the upwind dry field in the Crau, (2) the downwind irrigated field in the Crau, and (3) Cabauw, the Netherlands; respectively a uniform terrain, a nonuniform of type a and a nonuniform type b terrain (see Section 1). Figure 9.7, referring to the latter site, is taken from de Bruin (1982).

It can be clearly seen that the variance method yields good results for the uniform site as well as

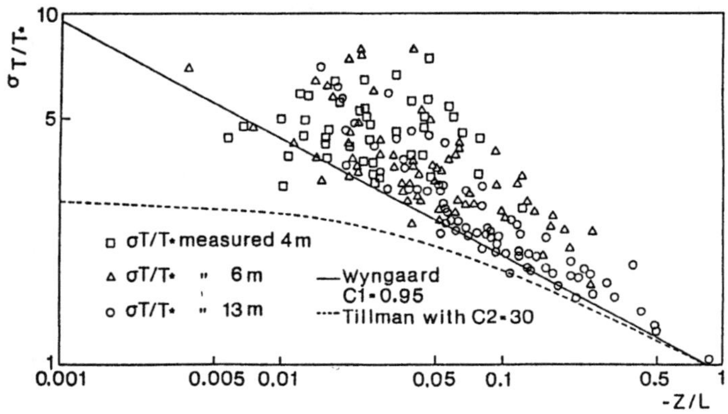

FIGURE 9.8. The measured σ_T/θ^* versus z/L compared with the free convection formulations of Wyngaard et al. (1971) and Tillman (1972) [Equations (5) and (3)].

for the terrain at Cabauw, which is nonuniform of type b. Over the irrigated field the variance method overestimates the sensible heat flux significantly under the present conditions.

This example shows that it makes sense to distinguish between type a and type b. In the first case the temperature and sensible heat flux are highly affected by the step-change in surface conditions, while in the second case only the wind speed and the momentum flux are influenced. Note that the flux-profile relationships found at Cabauw deviate significantly from those over uniform terrain (Beljaars et al., 1983).

In Figure 9.8 σ_T/θ_* is plotted versus the *local* value of z/L; these data were gathered over the irrigated field at a height of 4, 6, and 13 m. For comparison, the curves corresponding to Equation (9.2) (Tillman curve) and Equation (9.5) (free convection) are drawn. It is seen that almost all points lie above both curves. In Figure 9.8 we used the *local* values of z/L because the fluxes vary significantly with height (see later). It is questionable whether this local z/L is a suitable scaling parameter (Lang et al., 1983). This is yet an unsolved problem. Note that it is one of the objectives of this study to show the limitations of the M–O similarity theory in real applications. These limitations are clearly demonstrated in Figure 9.8.

A similar plot for water vapor, σ_q/q^* versus z/L, is presented in Figure 9.9. This can be done since it is expected that Equations (9.1) to (9.5) are valid for all scalars.

It appears that in this case the deviation from the Tillman curve is almost absent and falls within the experimental error. Apparently, the *local* water vapor flux scales with the *local* z/L similarly to the homogeneous case. Note that Figure 9.9 does not contain information of the surface evaporation, because the water vapor flux varies so much with height.

In Figure 9.10 some measured profiles of temperature and humidity for increasing distances from the step-change at midday are presented.

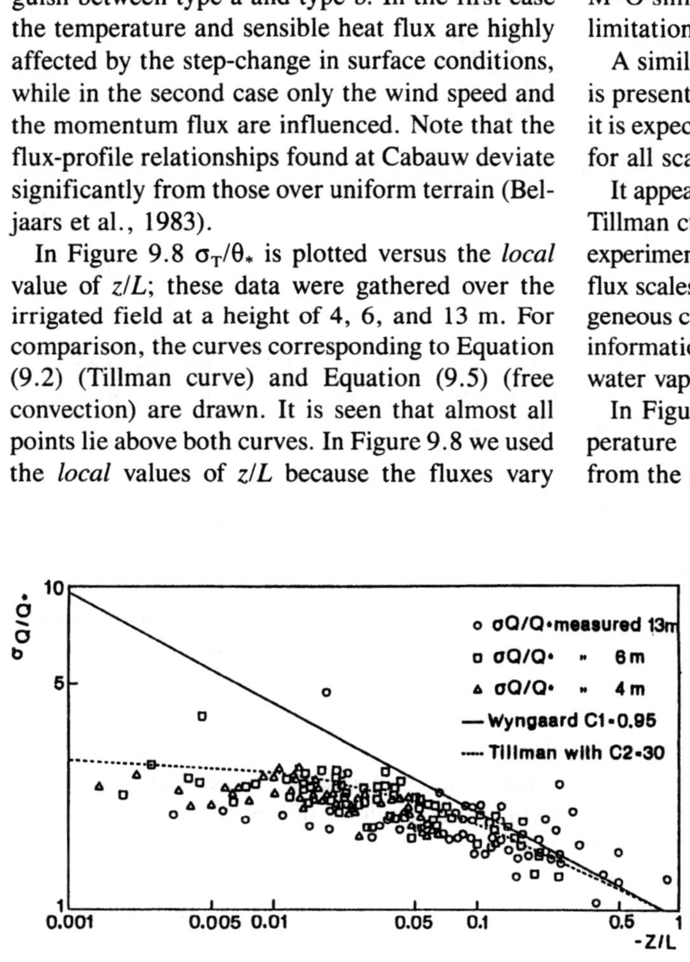

FIGURE 9.9. The measured σ_q/q^* versus z/L compared with the free convection formulation of Wyngaard et al. (1971) and Tillman (1972) [Equations (5) and (3), but for moisture].

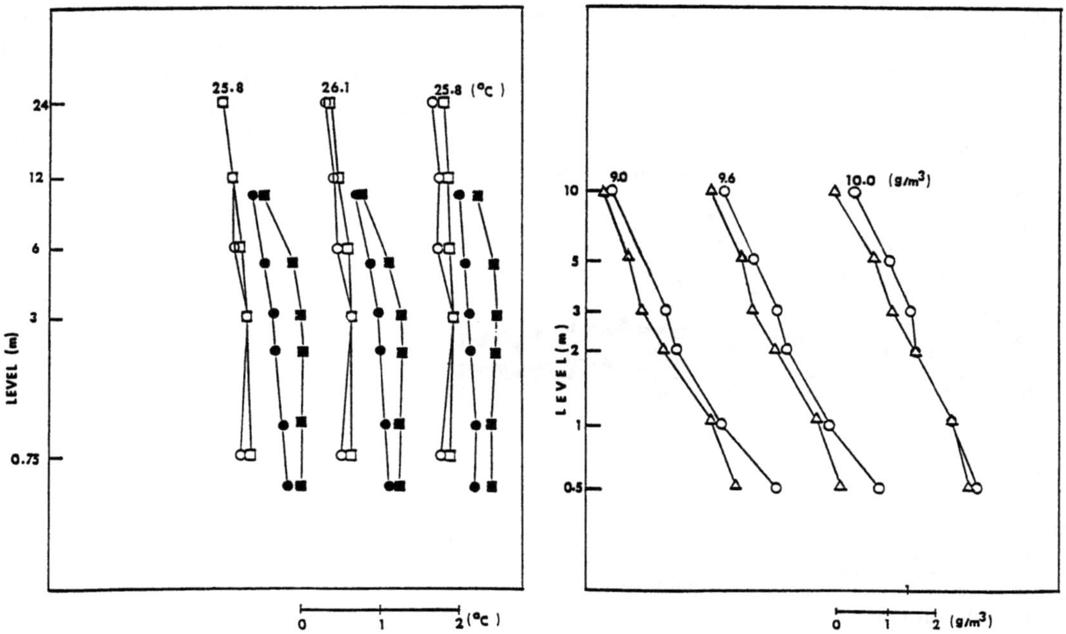

FIGURE 9.10. Profiles of temperature (80, 120, 150, and 200 m after the step) and for humidity (80 and 120 m after the step).

The temperatures measured at about 80 m from the step are the highest. Cooling of the air with distance from the step change is clearly observed. In the first 120 m the most rapid changes in the profile can be noticed where the IBL is still thin. At 150 m and 200 m the profiles show a so-called advective inversion (Lang et al., 1983).

The variation of the profiles in the horizontal direction is reflected in the local divergence of the fluxes with height. The latter is clearly shown in Figure 9.11, in which the observed flux densities of latent and sensible heat are depicted over, respectively, the dry terrain at 6 m height and for the 4, 6, and 13 m levels at almost the same distance from the step change over the irrigated field.

In Figure 9.12 a comparison is given between the calculated profiles of temperature and humidity at 80 m from the leading edge, using the second-order closure model of Rao et al. (1974) and the corresponding measured profiles. As noted before we included the Monteith's "big leaf" concept as lower boundary condition in this version of the model. We carried out the calculations for different values of the surface resistance r_s. It appears that for r_s = 30–45 s/m the observed profiles are reasonably described by the model. In a next study the calculated fluxes will be compared with the measured ones.

5. Concluding Remarks

This chapter deals with the surface fluxes of heat and water vapor near a nonuniform surface, in particular an irrigated field surrounded by dry terrain. A first analysis of data obtained from an extensive micrometeorological field experiment reveals that the fluxes can vary significantly with height and that (therefore) strong deviations from the Monin–Obukhov similarity theory can occur over nonuniform terrain.

Also, it is shown that it makes sense to distinguish between two types of nonuniform terrains:

1. Type a, where the temperature and humidity field are highly affected and the wind field is not, and
2. Type b, where the wind field is disturbed and the temperature and humidity field is not.

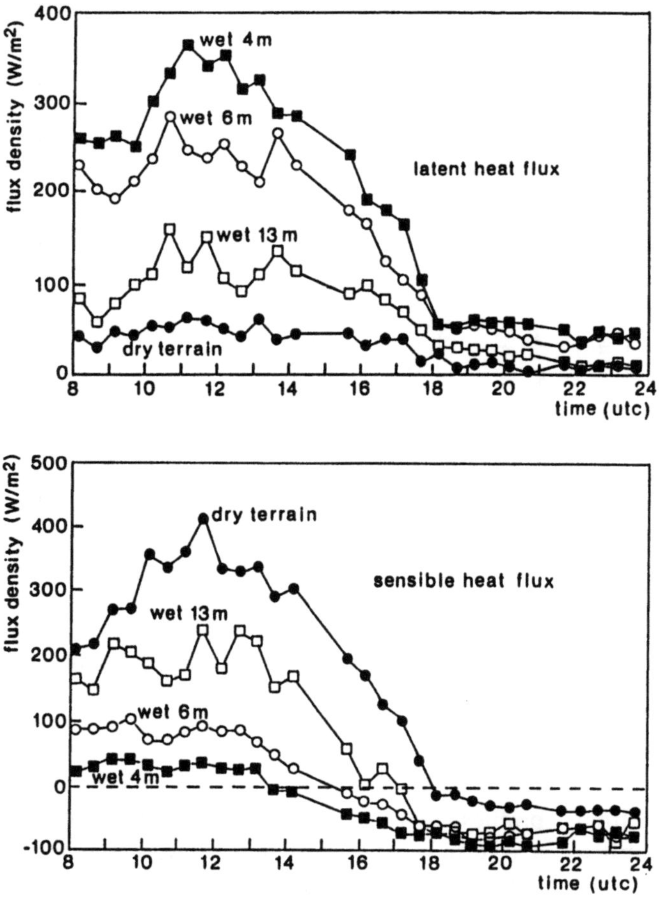

FIGURE 9.11. The flux densities for heat and water vapour at Crau, France, June 23, 1987, above the dry terrain and above the wet terrain at different levels (locations 4, 8, 9, and 10 in Fig. 9.4). Not corrected for distortion errors.

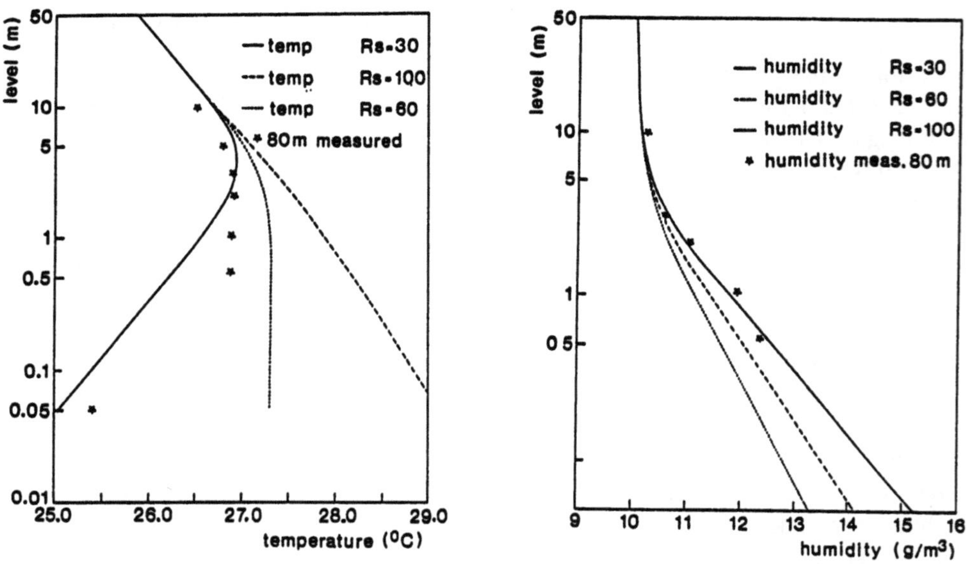

FIGURE 9.12. Model calculations and measurements of temperature and humidity profiles behind a dry-to-wet step change.

Evidence is presented that the variance method as used by Tillman (1972), by which fluxes can be obtained from simple single-level observations is still applicable over nonuniform fields of type b.

The first comparison between data and calculations done with the second-order closure model of Rao et al. (1974), in which we changed the surface boundary conditions by using the big-leaf concept of Monteith, is promising. It appears that the results depend critically on the surface resistance.

Acknowledgments. Mr. Py allowed us to use his farm and crop for the experiment. The data used in this paper were collected in close cooperation with Dr. H. F. Vugts, Dr. F. Cannemeijer, and F. Tesselaar from the Department of Meteorology of the Free University of Amsterdam. Dr. B. Sequin and Dr. F. Verbrugge from I.N.R.A, Avignon are acknowledged for their hospitality and support in France. Both the technical staffs of our institute and of the Free University of Amsterdam did indispensable work before and after the measuring period. The software for the data processing was skillfully developed by Dr. Ir. J. van Boxel. One of the authors, Ir. N. J. Bink, is supported by the Working Group on Meteorology and Physical Oceanography (MFO) with financial aid from the Dutch Organization for Scientific Research (NWO). This study is partly sponsored by the EC under contract number EV4C/0015NL. We are grateful for the valuable comments of Bart v.d. Hurk.

Appendix 1: List of Symbols

Symbols of quantities	Dimension	Description
a_t		Constant for the parameterization of the third order moments
b		Constant for the parameterization of ε_M
C_1		Constant in free convection scaling formulation of Wyngaard et al. (1971)
C_2		Constant in scaling formulation of Tillman (1972)
c		Constant for the parameterization of ε
c_p	J kg^{-1} K^{-1}	Specific heat of air at constant pressure
$d_{i=1,2,3}$		Constants for the parameterization of the pressure-gradient covariance term
g	m s^{-1}	Acceleration of gravity
H	W m^{-2}	Sensible heat flux density
k		von Karman constant, taken as 0.4
L	m	Monin–Obukhov length
p'	Pa	Pressure fluctuations
q'		Moisture fluctuations
q^*		Characteristic specific humidity
r_s	s m^{-1}	Surface resistance
$U_{i=1,2,3}$	m s^{-1}	Wind speed vector (U, V, W) in the x, y, z or x_1, x_2, x_3 directions
u^*	m s^{-1}	Friction velocity
$u'_{i1,2,3}$	m s^{-1}	Vector of the velocity fluctuations (u, v', w') in the x, y, z or x_1, x_2, x_3 directions
x	m	Horizontal distance along the wind direction
$x_{i1,2,3}$	m	Spatial coordinates x, y, and z
y	m	Horizontal distance perpendicular to the wind direction
z	m	Height
z_0	m	Surface roughness length
δ_{ij}	–	Kronecker's delta: if $i = j$ then $\delta = 1$, else $\delta = 0$
ε	m^2 s^{-3}	Rate of dissipation of the turbulent energy $\overline{u'_i u'_i}$ into heat
ε_M	m^2 s^{-3}	Rate at which $\overline{\theta'^2}$, $\overline{q'^2}$, or $\overline{\theta' q'}$ are smoothed out by molecular conduction
ρ	g/m^3	Air density
σ_s	(depends	Standard deviation of the quantity s
τ	s on s)	Turbulent relaxation time
θ	K	Potential temperature
θ'	K	Potential temperature fluctuations
$\theta_* = H/\rho\, c_p\, u^*)$	K	Characteristic potential temperature

Appendix 2: Second-Order Closure Modeling

The conservation equations of momentum, mass, specific entropy, and water vapor are used as a starting point for the derivation of the transport equations for various meteorological variables that will be affected by a change in surface conditions.

If the proper simplifying assumptions are used, and if the restriction is made of a flow at right angles to the surface discontinuity (a two-dimensional case), then it is possible to derive a set of transport equations for the mean quantities $U_i=(U, O, W)$, θ, and Q (see, e.g., Brutsaert, 1982; Kroon, 1985). The corresponding fluctuating components are indicated by $u'_i=(u', v', w')$. Note that repeated indices in one term denote summation (Einstein's convention).

By deriving the equations for the four mean quantities, additional unknowns are created, i.e., $\overline{u'w'}$, $\overline{w'\theta'}$, and $\overline{w'q'}$. Hence the four equations do not form a closed system, i.e., a system where the number of equations equals the number of unknowns. To close this system several solutions are proposed. In the second-order closure models a set of transport equations for the turbulence quantities is derived from the governing equations.

These transport equations, however, also contain a number of unknown variables (third-order terms, pressure terms etc.). In second-order closure models (e.g., Rao et al., 1974) the third-order and pressure terms are closed by expressing these quantities as a function of first- and second-order moments using (semi)empirical relations.

Generally speaking, second-order closure models contain three different higher order terms, which have to be parameterized to close the system of equations. These terms will be briefly discussed.

A2.1 Turbulent Transport Terms

These terms all have the form $\partial\overline{Mu'}/\partial x_j$, where M can be any second moment combination of u', w', θ', and q' (e.g., $\overline{u'q'}$). Using Gauss' integral theorem it is possible to show that these terms merely tend to redistribute the quantity M from one point inside the flow to another. In the model of Rao et al. (1974) a simple ad hoc gradient diffusion modeling of these terms is used viz.

$$\overline{Mu'_i} = a_t \frac{\partial M}{\partial x_j} \overline{u'_i u'_j} \tau \qquad \text{(A2.1)}$$

where a_t is a constant and τ is a turbulent relaxation time.

A2.2 Pressure Covariance Terms

In the equation of the Reynolds' stress $\overline{u_i' u_k'}$ the pressure covariance term reads

$$\overline{u'_k \frac{\partial p'}{\partial x_i} + u'_i \frac{\partial p'}{\partial x_k}} \qquad \text{(A2.2)}$$

Analysis of homogeneous two-dimensional flows (e.g., Hinze, 1959) shows that this term redistributes the turbulent kinetic energy in that it forces the $(\overline{u_i'})^2$ distributions toward isotropy and destroys the off-diagonal components of the $\overline{u_i' u_k'}$ tensor.

A pressure term is also generated in the $\overline{u_i'\theta'}$ and $\overline{u_i'q'}$ transport equations. In Rao's model they are modeled as follows:

$$-\overline{M\frac{\partial p'}{\partial x_i}} = d_i \frac{\overline{Mu'_i}}{\tau} \qquad \text{(A2.3)}$$

where M is either q' or θ', and d_i is a constant.

No general agreement on the modeling of the pressure covariance terms exists. In Rao's model no buoyancy and mean strain effects are incorporated in the pressure covariance term. Expressions incorporating these effects were presented by Wyngaard (1975).

A.2.3 Dissipation Terms

The dissipation terms indicate the molecular action on the turbulent correlations. It appears that viscous dissipation is the major loss term for turbulent kinetic energy. It can be demonstrated that in a Newtonian fluid the dissipation terms only affect the variances $\overline{u_i'^2}$, $\overline{q'^2}$, $\overline{\theta' q'}$. The off-diagonal elements of the Reynolds stress tensor $\overline{u_i' u_k'}$ as well as the turbulent fluxes $\overline{u_i'\theta'}$ and $\overline{u_i'q'}$ are not affected. In Rao's model the dissipation terms were modeled by

$$\varepsilon = c\frac{\overline{u'_i u'_k}}{\tau} \delta_{ik} \qquad \text{(A2.4)}$$

$$\varepsilon_{M} = b\frac{M}{\tau} \quad \text{(A2.5)}$$

where δ_{ik} is Kronecker's delta ($\delta = 1$ if the indices are equal else $\delta = 0$), M is either $\overline{\theta'^2}$, $\overline{q'^2}$, or $\overline{\theta' q'}$, and c and b are constants.

References

Beljaars ACM, Schotanus P, Nieuwstadt FTM (1983) Surface layer similarity under non-uniform fetch conditions. J Climatol Appl Meteorol 22:1800–1810.

Bruin HAR de (1982) The energy balance of the Earth's surface. Ph.D. dissertation, Agricultural University, Wageningen.

Brutsaert WH (1982) Evaporation into the atmosphere. "Environmental Fluid Mechanics." D. Reidel, Dordrecht, Holland.

Hinze JO (1959) "Turbulence." McGraw-Hill, New York.

Kroon LJM (1985) Profile derived fluxes above inhomogeneous terrain: A numerical approach. Ph.D. dissertation, Agricultural University, Wageningen.

Lang ARG, McNaughton KG, Fazu C, Bradley EF, Othaki E (1983) Inequality of eddy tranfer coefficients for vertical transports of sensible heat and latent heat under advective inversions. Bound Layer Meteorol 25:25–41.

McBean GA, Miyake M (1972) Turbulent transfer mechanisms in the atmospheric surface layer. Quart J R Meteorol Soc 98:383–398.

Monteith JL (1981) Evaporation and surface temperature. Quart J R Meteorol Soc 107:1–27.

Panofsky HA, Dutton JA (1984) "Atmospheric Turbulence, Models and Methods for Engineering Applications." Wiley, New York.

Rao KS, Wyngaard JS, Coté OR (1974) Local advection of momentum, heat and moisture in micrometeorology. Bound Layer Meteorol 7:331–348.

Schotanus P Nieuwstadt FTM, Bruin HAR de (1983) Temperature measurement with a sonic anemometer and its application to heat and moisture fluxes. Bound Layer Meteorol 26:81–93.

Tillman JE (1972) The indirect determination of stability, heat and momentum fluxes in the atmospheric boundary layer from simple scalar variables during dry unstable conditions. J Appl Meteorol 11:783–792.

Vugts HF, Cannemeijer F, Tesselaar F (1988) The Crau experiment. Parameterization of the surface fluxes. Internal report Free University Amsterdam, Department of Meteorology.

Wyngaard JC (1975) Modelling the planetary boundary layer-extension to the stable case. Bound Layer Meteorol 9:441–460.

Wyngaard JC, Coté OR, Izumi Y (1971) Local free convection, similarity, and the budgets of shear stress and heat flux. J Atmos Sci 28:1171–1182.

10
A Critical Assessment of the Samer Network Accuracy

J.-P. Goutorbe

During the HAPEX-MOBILHY experiment a network of 12 stations was set up to monitor the surface energy balance over agricultural land. These so-called SAMER stations were an essential component of the experiment. This paper is a summary of validation studies for this station. A more extensive description of the station and data processing can be found in Bessemoulin et al. (1987). A description of the sites is given by Goutorbe and Tarrieu (this volume).

1. Physical Principles

The chapter by Perrier and Tuzet in this volume gives a description of methods used for flux measurements including the one used here. The station operates on the principle of the difference method. A stationary energy balance is considered at the surface:

$$R_n + H + LE + G = 0$$

where R_n = net radiation
H = sensible heat flux
LE = Latent heat flux
G = ground heat flux
In this equation net radiation, sensible heat flux, and ground heat flux are measured and evaporation is obtained as a residual.

The sensible heat flux is computed using a simplified aerodynamic method described by Riou (1982) and Itier (1982). The inputs for the computation are vertical gradients of temperature $D\Theta$ and wind velocity DV, measured between two levels Z_2 and Z_1.

$$H = AD\Theta^{3/2} \qquad R_i < -1$$
$$H = H_0(1 - 16R_i)^{3/4} \qquad -1 < R_i < 0$$
$$H = H_0(1 - 5R_i)^2 \qquad 0 < R_i < 0.14$$
$$H = H_0/10 \qquad 0.14 < R_i$$

R_i is the Richardson number,

$$R_i = Z_*(g/T) \log(Z_2/Z_1)\ (D\Theta/DV^2)$$

$D\Theta$ is the temperature difference between the lower and upper probes and DV is the wind speed difference between these levels.

$$Z_* = (Z_1\ Z_2)^{1/2}$$
$$A = 1.3\rho C_p\ (g/T)^{1/2}/(3(-Z_1^{-1/3} - Z_2^{-1/3}))^{3/2}$$
$$H_0 = KD\Theta DV$$
$$K = -\ \rho C_p k^2/[\log\ (Z_2/Z_1)]^2$$

k is the von Karman constant and T the air temperature at 2 m above the surface.

2. Description of the Station

The following basic measurements are necessary for surface energy balance measurement:

1. Net radiation: a Crouzet net radiometer is used, the wavelength interval is 0.3–50 μm.
2. Ground heat flux: Thornwaite heat flux plates are used, buried 3 cm below the surface.
3. Temperature difference: the two probes are 1.5 m apart. Each probe is compared to a common reference body and temperature differences are measured with a set of six copper-constantan thermocouples. The probes are well ventilated.

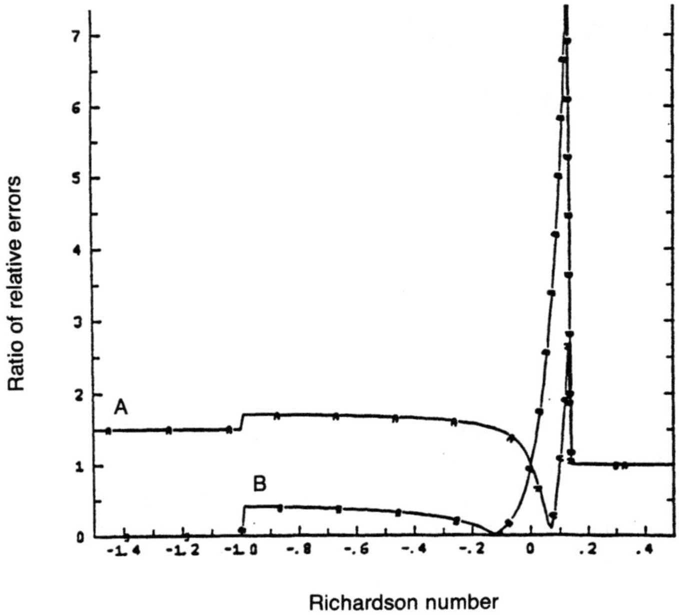

FIGURE 10.1. Ratio of relative errors as function of Richardson number; curve A: $(dH/H) / (d D\theta / D\theta)$; curve B: $(dH/H) / (dDV/DV)$.

4. Wind speed: the measurement levels are the same as for temperature. MCB anemometers with low speed threshold are used.

The system can move up and down to adjust for vegetation height. The following other measurements are performed by this station:

1. Air temperature using a platinium wire and air humidity (SPSI capacitive sensor). These two sensors are in a small shelter and are not ventilated.
2. Rainfall is recorded with a tipping bucket Précis Mécanique.
3. Shortwave and total (0.3–60 μm) upward and downward radiations are measured with a set of four Schenck sensors. Total components can be computed from these measurements using the temperature of the core of the instrument, which is measured with a 100-Ω platinium probe.

3. Data Processing

All parameters are averaged over a 15-min interval. The sensible heat flux is computed by the station processor using the simplified formula and the evaporation flux retrieved by balancing the fluxes at the surface. However, the original gradients $D\Theta$ and DV have been stored as well, so that the sensible heat flux can be recomputed using other similarity formulations and taking into account the actual vegetation/sensors heights. In the final HAPEX-MOBILHY data set, for instance, the formulation of Dyer and Hicks (1970) has been used.

4. Error Analysis for Sensible Heat Flux Retrieval

Considering again the simplified aerodynamic formulas and differentiating with respect to $D\Theta$ and DV

$R_i < -1$

$$dH/H = 1.5\, d(D\Theta)/D\Theta$$

$-1 < R_i < 0$

$$dH/H = [(1-28R_i)/(1-16\text{-}R_i)]\, d(D\Theta)/D\Theta + [(1+8R_i)/(1-16R_i)]\, d(DV)/DV$$

$0 < R_i < 0.14$

$$dH/H = [(1-15R_i)/(1-5R_i)]\, d(D\Theta)/D\Theta + [(1+15R_i)/(1-5R_i)]\, d(DV)/DV$$

$0.14 < R_i$

$$dH/H = d(D\Theta)/D\Theta + d(DV)/DV$$

The contribution of $D\Theta$ and DV to the error is plotted on Figure 10.1. The unstable part of the graph

($R_i < 0$) is of interest here, for which the contribution of $D\Theta$ is dominant. For a large range of stabilities we have approximately

$$dH/H = 1.5\, d(D\Theta/D\Theta \quad R_i < -0.2$$

The value of H is also sensitive to the reference level. A displacement height is introduced as

$$Z_d = 0.66 \text{ (height of canopy)}$$

Z_1 and Z_2 are then measured from this reference level. The height of the canopy is known with an uncertainty of 5 to 10 cm. More uncertain is the specification of Z_d as a fraction of the vegetation height.

Differentiating with respect to Z yields

$$\frac{1}{H}\frac{dH}{dZ} = 2\frac{Z_2 - Z_1}{Z_1 Z_2 \log\left(\frac{Z_2}{Z_1}\right)\left[\frac{(1-10R_i)}{(1-16R_i)}\right]} - \frac{Z_1 + Z_2}{Z_1 Z_2\left[\frac{6R_i}{(1\text{-}16R_i)}\right]}$$

For usual values

$$\frac{1}{H}\frac{dH}{dZ} = 1.7 - \frac{28R_i}{1\text{-}16R_i} \quad \text{or} \quad 1.7 \quad \text{m}^{-1} \text{ for } R_i = -1$$

An error of 10 cm for Z leads to a relative error of 17% for the sensible heat flux.

5. Intercomparison of Stations in the Same Field

When the SAMER stations are in actual operation they experience different conditions and cannot be easily compared. To get a better idea of the measurement scatter, two of these stations have been installed over the same field of grass after the experiment. By operating over short grass, the problem of displacement height determination was avoided.

The two stations have been operated for 16 clear days in April and May 1987. The total data set, on which statistical analysis has been performed, consists of 800 samples. The results of the intercomparison are summarized in Table 10.1.

For conditions of small sensible heat flux and large evaporation this gives a lower bound of the scatter to be expected for individual estimates (15 min averages) of key SAMER output. For daily estimates, the scatter is reduced to 1.5, 12, and 3.5% for net radiation, sensible heat flux, and latent heat flux, respectively.

TABLE 10.1. Intercomparison results.

Parameter	Mean	Bias (%)	Scatter (%)
Net radiation	290 W m^{-2}	2.5	3.5
Wind speed	2.8 m s^{-1}	2	8
DV	0.58 m s^{-1}	13	10
$D\Theta$	−0.28°C	0	18
H	53 W m^{-2}	0	20
LE	259 W m^{-2}	3.4	6

6. Validation for Net Radiation

Net radiation, as already said, is directly measured at each SAMER site by a Crouzet net radiometer. It can also be inferred from the two Schenck units measuring total upward and downward radiations.

On clear days, and flux density exceeding 400 W m^{-2} a large discrepancy can be noticed between the two estimates at some sites (Fig. 10.2). Station 12 is recording very low values of maximum net radiation for both systems on June 16. This low value cannot be accounted for by specific characteristics of the surface, which is a field of maize as for eight others experimental sites. On the same day, measurement of net radiation flux by the NCAR KING AIR aircraft (Hildebrand, 1988) indicates very homogeneous conditions over the farmland area, and in particular over the area of station 12. At flight height (100 m), the net radiation flux recorded by the aircraft is 750 W m^{-2}. Later in the season, on July 4, a similar although less extreme situation is found. In Figure 10.3, the nine stations located over a maize field have been plotted and again estimates are low for station 12 and for the Crouzet net radiometer of station 2. Such large differences occur in fact on any clear day and the stations always rank in the same order.

For overcast days and for long time periods these differences are greatly reduced, some compensation occurring at night. For a period of 34 days, for instance, the overall difference between the Schenk and Crouzet networks is only 10 W m^{-2}.

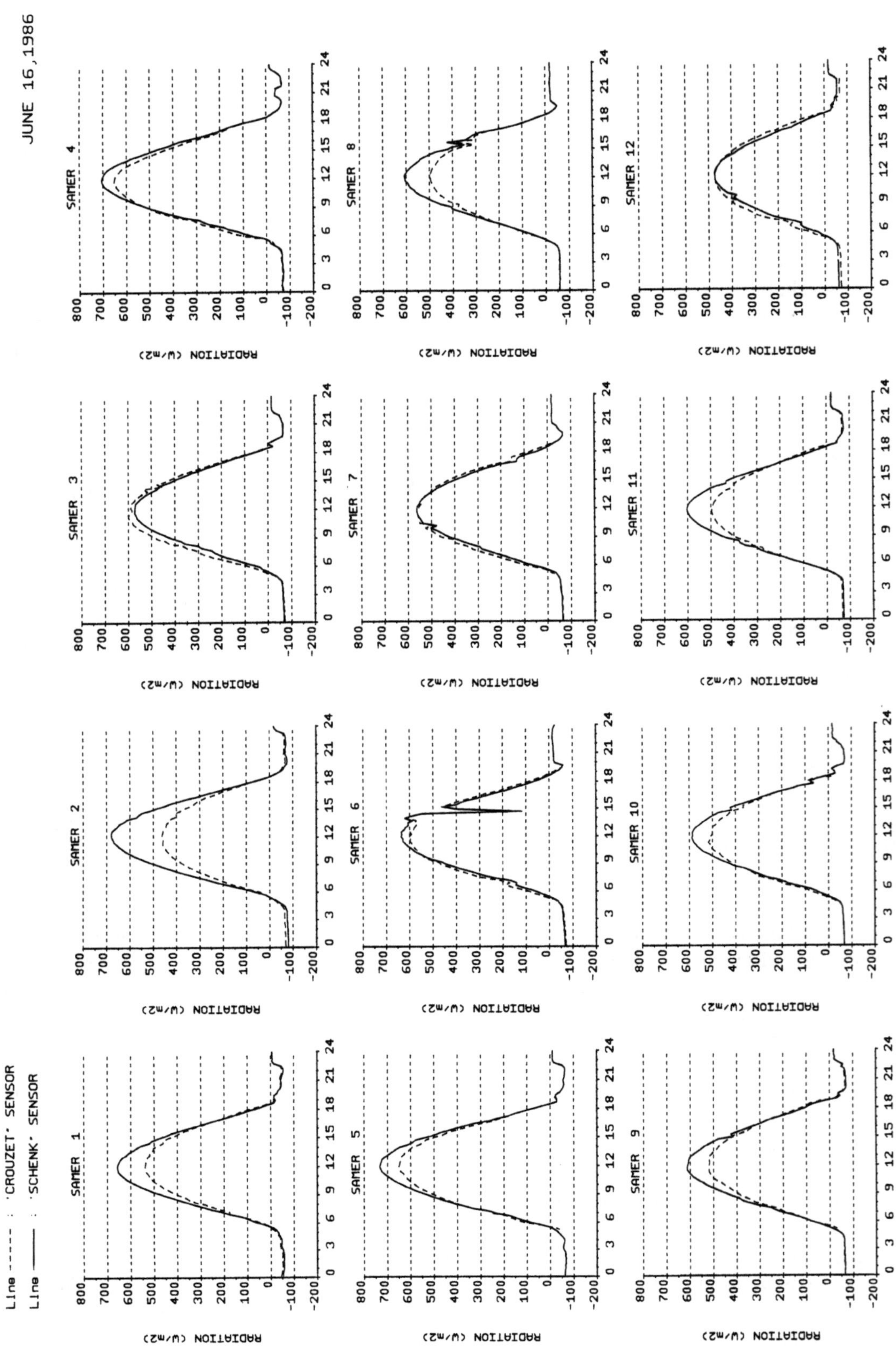

FIGURE 10.2. Net radiation intercomparison for June 16.

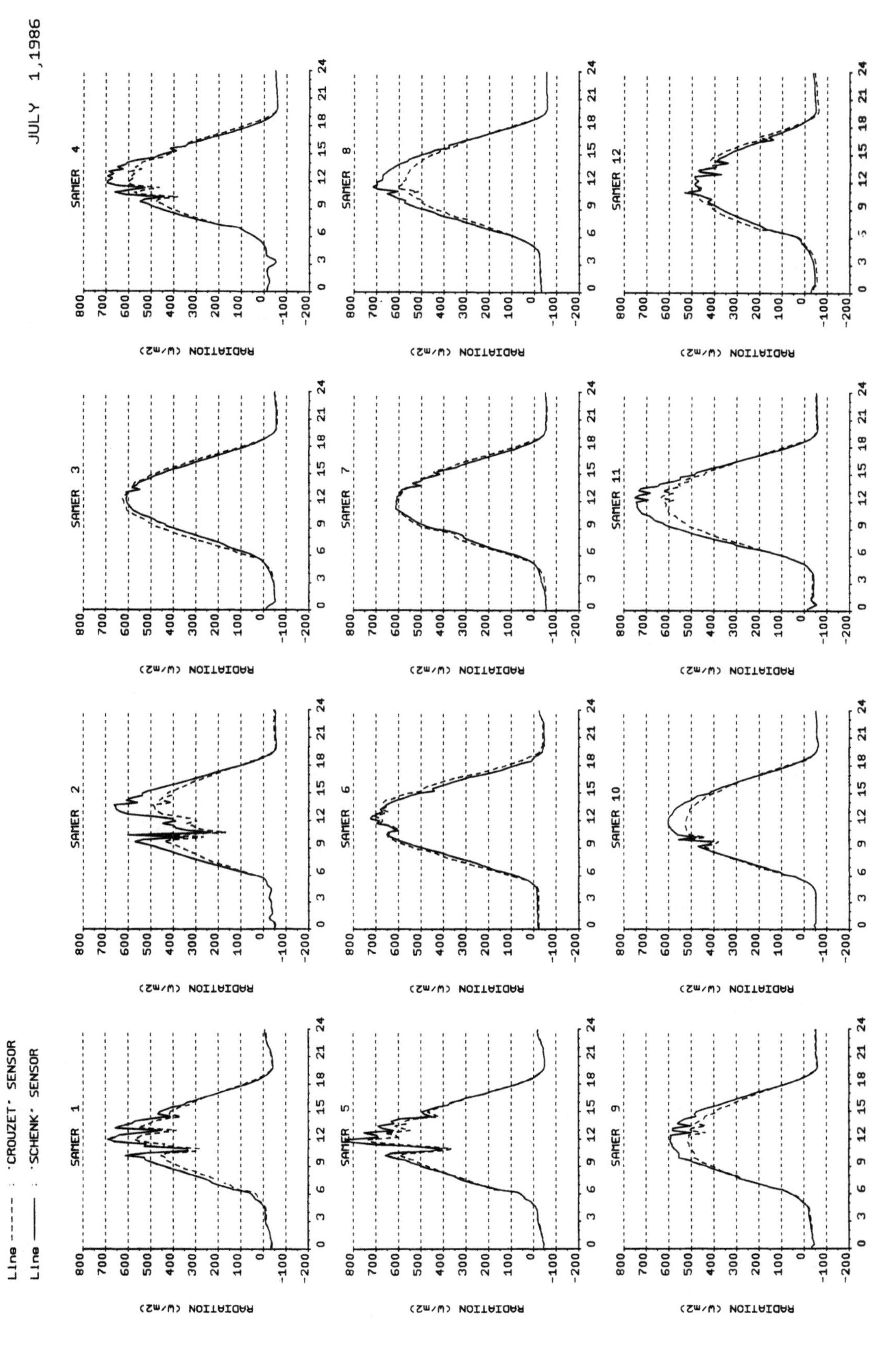

FIGURE 10.3. Net radiation intercomparison for July 1.

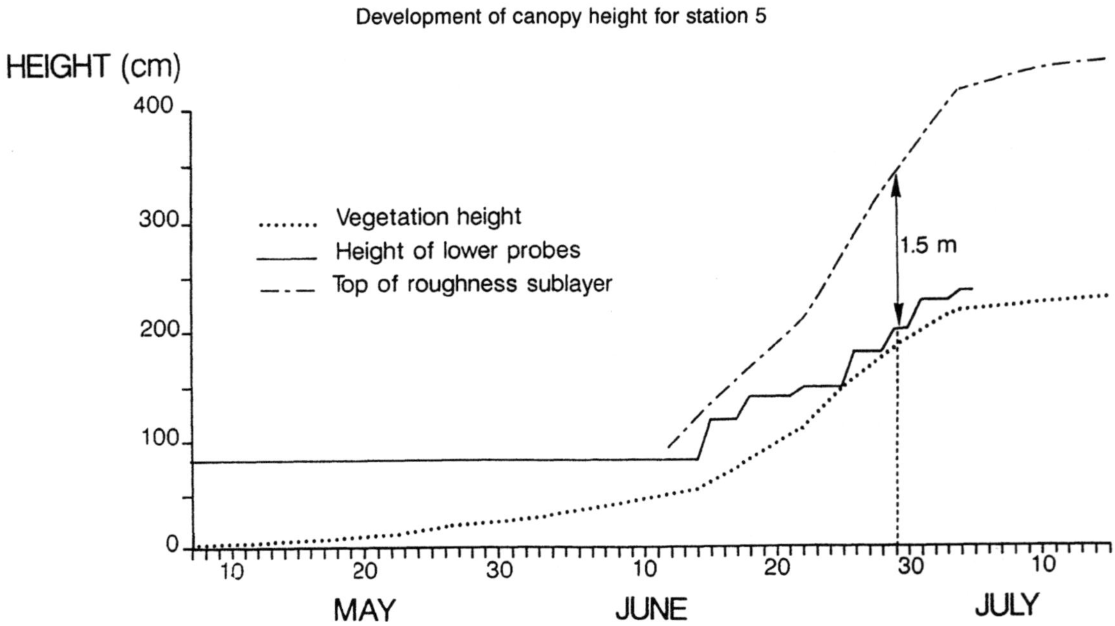

FIGURE 10.4. Growth of a maize field at station 5.

The analysis has shown, however, that 6 of the 12 stations yielded very coherent measurement on clear days. This set of stations is large enough for radiation sampling on fair weather. The other set of six stations has been recalibrated for clear days using the mean of the six good stations.

As a second step, all radiation sensors have been sent back to the laboratory for calibration; analysis is in progress.

7. Possible Inadequate Formulation of Similarity Functions

It has been argued by several authors (Thom et al., 1975; Raupach et al., 1980) that the commonly used flux-gradient relationships may be inadequate over rough canopies, close to the canopy. Some of the HAPEX-MOBILHY measurements were performed over well developed maize with the sensors very close to the canopy (see Fig. 10.4). The problem was investigated by Calvet (1988) in the context of HAPEX-MOBILHY. The argument is that in a so-called roughness sublayer, turbulence is not governed only by the usual parameters of similarity theory but also by geometric parameters describing the canopy and scaling the big eddies. In this layer, similarity theory does not apply or at least should be modified. According to Cellier (1984), this is the case over maize for the layer up to twice the canopy height.

Considering the situation at station 5 (Fig. 10.4), one can see that the lower probes are usually in the roughness sublayer. Following a suggestion by Cellier (1984), who ran an experiment over maize, a correction has been tried, based on exchange coefficients proposed by Dyer and Hicks (1970):

$$K_m = ku_*(Z-Z_d)\{1-16[(Z-Z_d)/L]\}^{1/4}$$
$$K_h = ku_*(Z-Z_d)\{1-16[(Z-Z_d)/L]\}^{1/2}$$

K_m and K_h are exchange coefficients for momentum and heat respectively, U_* is the friction velocity, $Z - Z_d$ the height over the zero-displacement plane, and L the Monin–Oboukhov length scale.

The proposal by Cellier (1984), backed by experimental data, is to modify the exchange coefficient for heat in the roughness sublayer:

$$K_h = ku_*(Z_r-Z_d)\{1-16[(Z-Z_d)/L]\}^{1/2}$$

where Z_r is the depth of the roughness sublayer.

The result of this modification is for a given temperature gradient to increase the sensible heat flux and decrease evaporation (Fig. 10.5). In many

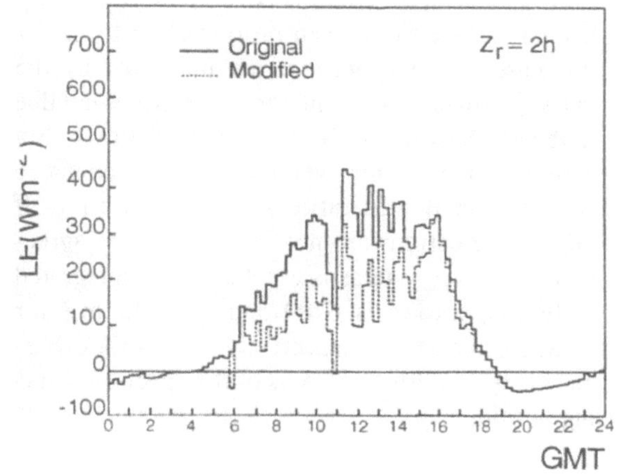

FIGURE 10.5. Modification of the evaporation flux over maize at station 5 using Cellier's hypothesis.

circumstances, however, the correction is too large and the enhanced sensible heat flux exceeds net radiation in the middle of the day, a very unlikely result. The correction also proves to be sensitive to the specification of Z_r (Fig. 10.5). These difficulties could have been avoided by moving the sensors upward. But gradients of temperature decrease with height, fetch requirements are more stringent and accuracy on sensible heat flux estimate would deteriorate. For these reasons a height of only 75 cm over the zero-displacement height was considered as the best trade-off, provided the lower sensors were above the canopy.

8. Validation of the Diurnal Cycle

At the central site of the experiment, over a field of oats, a SAMER station was located near a Bowen ratio system called BEARN. The intercomparison shown (Fig. 10.6) is for June 28, over the maturing crop: André et al., (1988). The sensible heat flux estimated by both systems agree fairly well until 1 PM, but later in the day estimates start diverging. A large sensible heat flux, almost exceeding net radiation is computed by the SAMER station. The discrepancy, as noticed by John Gash (private communication) starts when wind direction turns from 300° (excellent fetch) to 200°, in which direction the border of the field is only 200 m away. Further upwind are farm buildings and cooler, evaporating maize fields. Between 1 PM and 3 PM, the vertical gradient of temperature recorded by the SAMER is close to −1.4°C. To match the BEARN sensible heat flux measurement this gradient should be only −0.9°C. The difference is larger than the 0.1°C

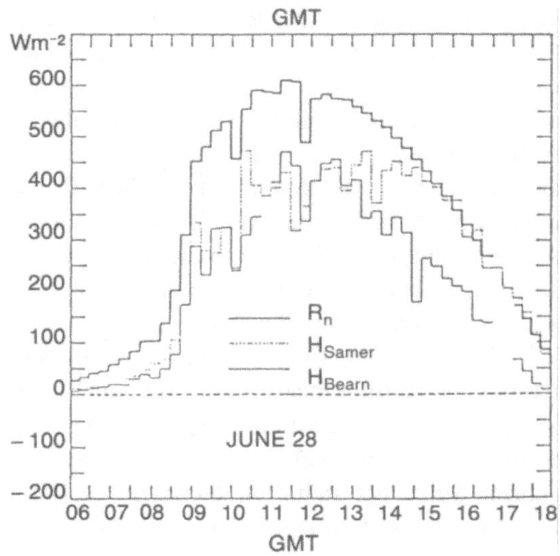

FIGURE 10.6. Comparison of fluxes as observed from Bowen ratio method (BEARN system) and from SAMER station 1.

error expected for the vertical temperature gradient. It is therefore likely that the disagreement is due to the advection of cooler air from the maize field, distorting the temperature profile over the oat field. At the same time a weighing lysimeter located in the oat field recorded a sizable evaporation flux. The conclusion of this first intercomparison is that under good fetch conditions the SAMER system performs satisfactorily, when compared with a Bowen ratio system, although the principles of measurement are different. It is, however extremely vulnerable to advection effects.

At another location, over a large field of maize, a SAMER station was installed next to an aerodynamic mast. The mast was equipped with temperature and wind sensors at six levels, from 0.50–11 m above the canopy. The ground heat flux was carefully measured by using flux plates and also a temperature profile in the soil (seven levels of measurement down to 1 m). The ground heat flux can be computed from the rate of change of temperature in the ground. The heat capacity of a soil layer is estimated by adding the heat capacity of the dry soil and the heat capacity of the water in the layer, known from neutron probe soundings. The heat stored in the first 2 cm of soil above the upper probe is estimated from the radiative temperature rate of change; this is adequate for a bare surface, but unrealistic over vegetation. A few intercomparisons of ground heat flux as measured from the flux plates (SAMER) and from an average of flux plates and profile method (mast) are shown in Figure 10.7. On June 3, the surface is almost bare and heating moderate; the agreement between the two methods is excellent. For the two other days shown, the mast system ground heat flux is larger by 50 W m^{-2} at midday; also, there is a different phasing of the flux due to the neglect of the surface layer in the heat plates method. On July 4, the maize canopy is well developed and the profile method overestimates the soil heat flux. In both methods the sensible heat flux is computed using the same similarity formulation and evaporation retrieved as a residual of the surface energy balance. Since net radiation compares well for these two stations, one can proceed directly to evaporation fluxes intercomparison (Fig. 10.8). For June 3, evaporation is three to four times larger than the sensible heat flux and the agreement for the evaporation flux excellent. Two weeks later, over a short (20 cm) maize crop and strongly heated surface the Bowen ratio is close to 1. For this case, the evaporation flux measured by the mast system is smaller and the difference is not due to the soil heat flux only. The last case, July 4, is a case of strong heating over a well-developed maize canopy; the Bowen ratio is again close to 1. If minor fluctuations are not considered, the agreement between systems is excellent. The integrated daily evaporation is 6700 and 7800 kJ m^{-2} for SAMER and mast, respectively. This 15% difference is typical for most days of the special observing period.

9. Validation at Monthly Time Scale

A good general consistency check is the computation of a local water balance using precipitation and evaporation from the SAMER station, and soil moisture variation from neutron probe measurements. This budget has been computed for periods of 3 weeks or more for the drying period. For 9 of the 12 SAMER sites the budgets proved to be very consistent (see Cuenca and Noilhan, this volume; Goutorbe et al., 1989). For the remaining three stations, the discrepancies can be explained in terms of "runin" or "runoff." This study supports the statement that SAMER measurement of the evaporation flux is reliable at the monthly time scale and does not exhibit systematic bias.

10. Summary of Validation Studies

The HAPEX-MOBILHY program is essentially a study of evaporation. As a residual, the evaporation flux incorporates errors for the three other fluxes. Differentiating the surface energy budget yields

$$dLE/LE = (R_n/LE)dR_n/R_n + (H/LE)dH/H + (G/LE)dG/G$$

The error on the sensible heat flux in conditions typical of the intensive period of HAPEX-MOBILHY can be estimated from Table 10.2 and Table 10.3. Table 10.2 indicates, for the time interval 10–14 h GMT, the distribution of the Richardson number

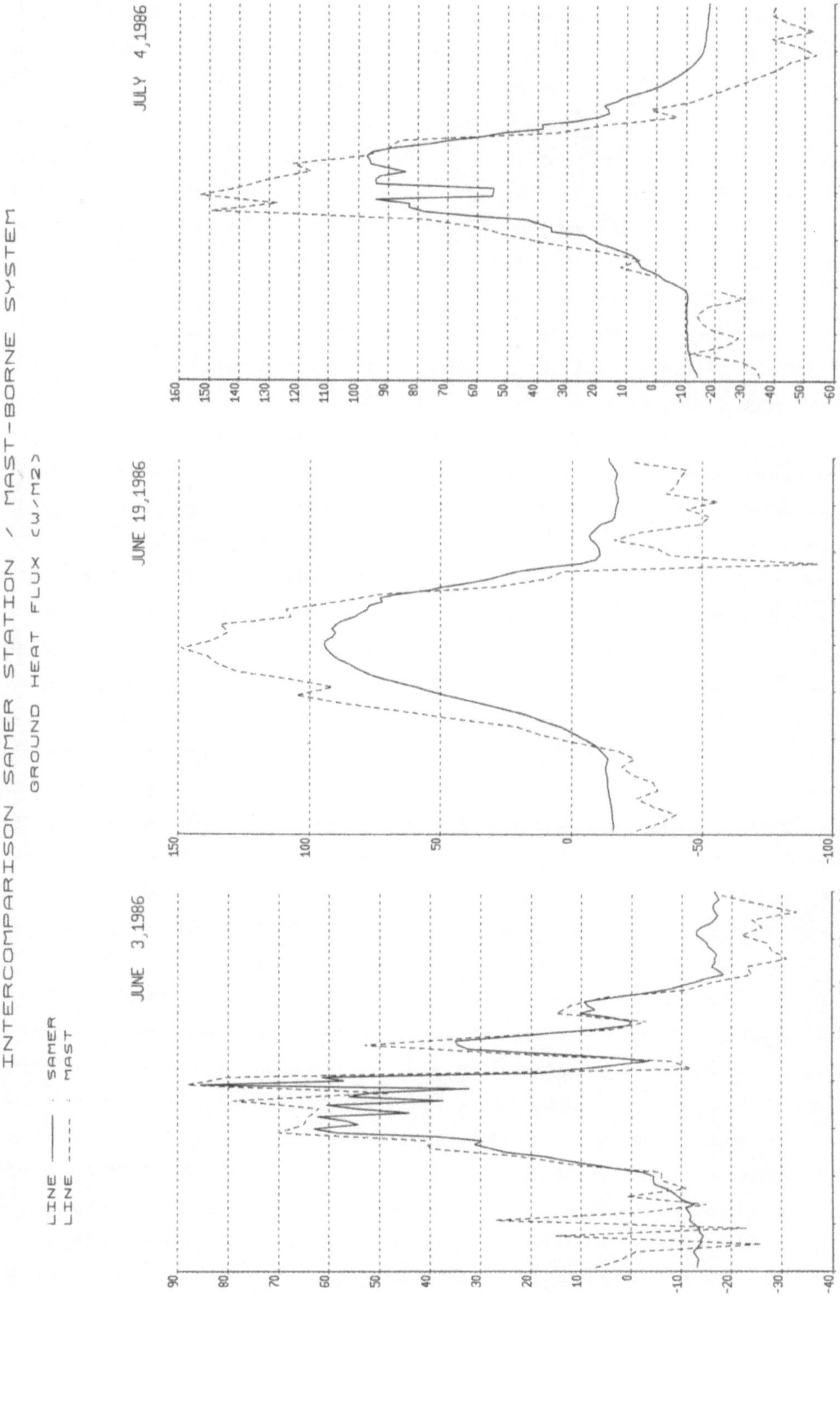

FIGURE 10.7. Intercomparison of ground heat flux at Castelnau.

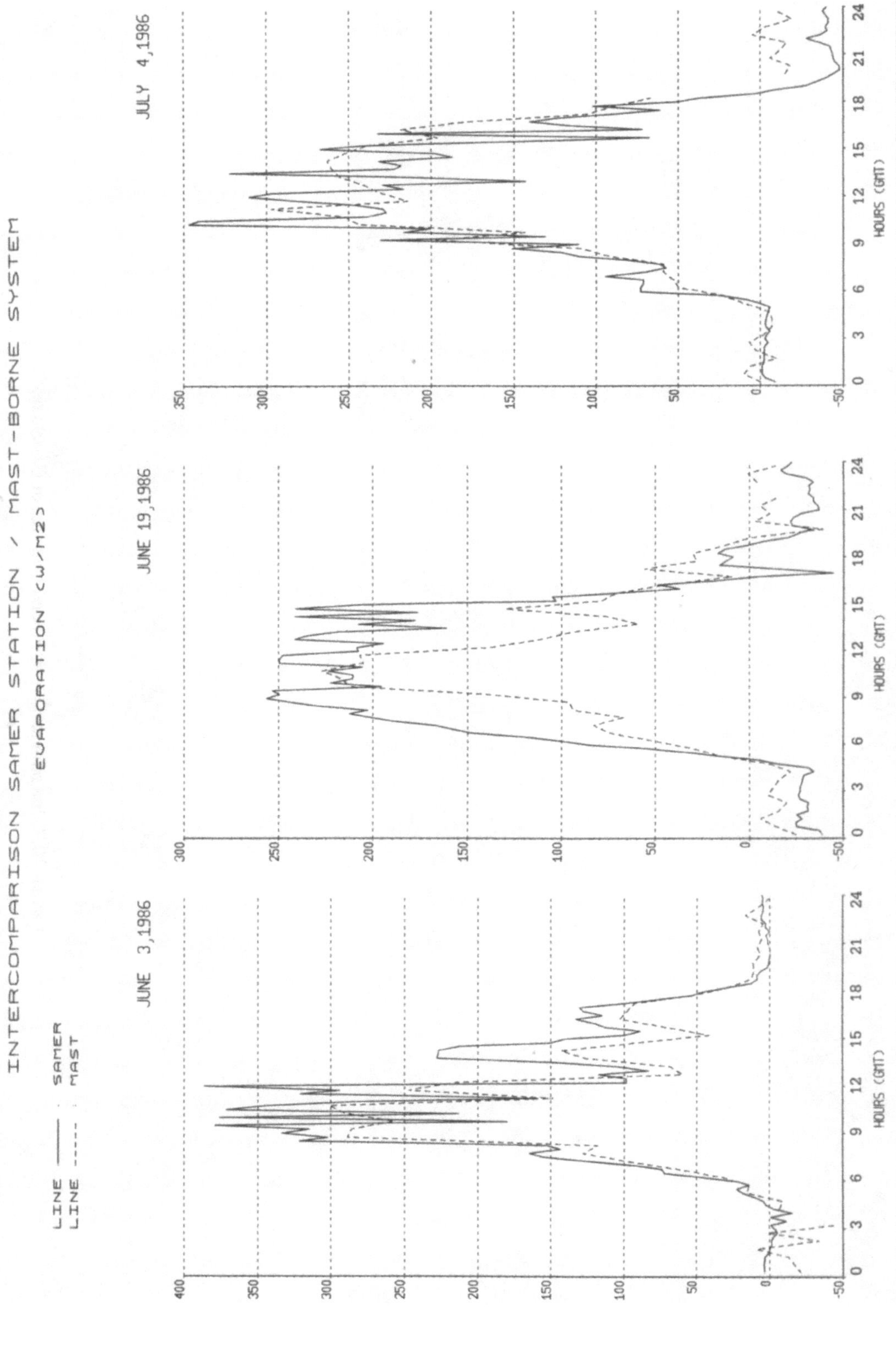

FIGURE 10.8. Intercomparison of evaporation.

TABLE 10.2. Richardson number distribution for May 8 to July 14.

Station	$R_i < -1$ (%)	$-1 < R_i < 0$ (%)	$0 < R_i < 0.14$ (%)	$D\Theta$ (°C)	DV m s^{-1}
1	2.3	85.9	11.6	−0.55	0.62
2	5.2	92.0	0.9	−0.56	0.62
3	2.5	95.9	1.5	−0.61	0.57
4	0.0	99.0	0.7	−0.79	0.84
5	4.8	94.3	0.1	−0.58	0.66
6	9.6	89.4	0.1	−0.83	0.75
7	5.5	94.4	0.0	−0.65	0.56
8	1.7	88.7	7.9	−0.39	0.56
9	2.9	96.3	0.1	−0.44	0.53
10	3.9	95.7	0.0	−0.61	0.49
11	0.8	99.1	0.0	−0.68	0.65
12	3.4	96.5	0.0	−0.59	0.68

during the special observing period of the experiment (May 8–July 14). The average values of $D\Theta$ and DV are also given. A histogram of $D\Theta$ for station 5 is presented on Table 10.3.

One can see that except for station 1, which suffered from severe advection effects, mixed convection ($-1 < R_i < 0$) was the rule. According to the error analysis for the sensible heat flux the following approximate relation holds:

$$dH/H = 1.5\ d(D\Theta/D\Theta$$

For most stations, according to Tables 10.2 and 10.3, the mode of $D\Theta$ is about 0.7 °C. The error on the temperature gradient being 0.1 °C, the error on the sensible heat flux should be about 20% for an individual estimate (15 min). Considering that part of this error is random and decreases with time averaging, this estimate is an upper bound. In the control experiment over grass described above, measurement scatter for the heat flux was reduced from 20 to 12% by moving from 15 min to daily averages of sensible heat flux.

Applying the same rate of decrease would mean a 12% uncertainty for the sensible heat flux over short vegetation.

The other large error on the sensible heat flux stems from the determination of the zero-plane displacement, estimated to 16% for an error of 10 cm on the zero-plane displacement level. A typical error for the sensible heat flux over tall crops would then be, after adding randomly

$$DH/H = [(0.16)^2 + (0.20)^2]^{1/2}$$
$$= 0.25 \text{ (15-min estimates)}$$

Other typical errors are 5% for net radiation and 30% for ground heat flux. Typical ratios are

$$H/LE = 0.5$$
$$G/R_n = 0.1$$

After collecting all the terms, the estimation of error for evaporation is

$$dLE/LE = 25\%$$

This again applies for 15-min estimates. At daily scale the error for temperature gradient is reduced to 12% and the contribution of the ground heat flux almost negligible. At this time scale the error is close to 18%. Further averaging would decrease the likely error for temperature gradients and for zero plane displacement to 10% and to a small, negligible value, respectively. If after a new calibration the radiation error is reduced to 3%, the final result for evaporation may meet the 10% accuracy at monthly scale.

TABLE 10.3. $D\Theta$ for station 5.

Interval for $D\Theta$	Percentage of values
−1.43 to −1.05	5.7
−1.05 to −0.65	34.4
−0.65 to −0.26	41.8
−0.26 to 0.12	16.0
0.12 to 0.52	0.2

11. Conclusion

The SAMER station, backbone of the HAPEX-MOBILHY network, has been tested in three ways: through analytical error analysis, through intercomparison of two stations in the same field, and by comparison with other systems. Various limitations have appeared. Some problems can be solved, at least partially, such as the error in net radiation or the underestimation of the ground heat flux in the morning and early afternoon.

Other shortcomings, such as the extreme sensitivity of the sensible heat flux estimate to the zero-plane displacement height are more fundamental and the price to pay for using so simple a system. No real improvement can be expected because the virtual sources for heat, for moisture, and for momentum can be located at different levels, and there is no way to know with two measurement levels only.

The system, at short time scale, will not reliably differentiate surfaces with fluxes differing by less than 15 or 20%.

At the longer period, for instance monthly time scale, measurements, in particular of the temperature gradient, are largely bias free.

Pending some improvement of radiation measurement, the accuracy at this time scale can be as good as 10%, which is adequate for the hydrological part of the program.

Acknowledgments. The SAMER station concept is due to Institut National de la Recherche Agronomique. It was built by Societe Flambeau, Pau, France. Many people at Centre National de Recherche Meteorologique were involved in this program, especially the 4M team that was in charge of the SAMER network and of data processing. The intercomparison study between two SAMER stations is borrowed from work by G. Fortune. The B.E.A.R.N. data have been processed and analyzed by M. Payen. The aerodynamic mast in Castelnau was operated and data processed by J-P. Lagouarde.

References

André J C, Goutorbe J P, Perrier A, Becker F, Bessemoulin P, Bougeault P, Brunet Y, Brutsaert W, Carlson T, Cuenca R, Gash J, Gelpe J, Hildebrand P, Lagouarde J P, Lloyd C, Mahrt L, Mascart P, Mazaudier C, Noilhan J, Ottlé C, Payen M, Phulpin T, Stull R, Shuttleworth J, Schmugge T, Taconnet O, Tarrieu C, Thepenier R-M, Valencogne C, Vidal-Madjar D, Weill A (1988) Evaporation over land-surfaces. First results from HAPEX-MOBILHY special observing period. Ann Geophys 6:477–492.

Bessemoulin P, Desroziers G, Payen M, Tarrieu C (1987) Atlas des données SAMER, 256 pp. Internal report available at CNRM, 42 avenue G. Coriolis, 31057 Toulouse cedex, France.

Calvet, J C (1988) Synthèse des recherches ayant trait à la détermination des flux de chaleur sensible et latente sur couvert très rugueux. 37 pp. Internal report available at CNRM, 42 avenue G. Coriolis, 31057 Toulouse cedex, France.

Cellier P (1984) Tentative de correction des relations flux-gradient au voisinage des surfaces très rugueuses. J Rech Atmos 18(4):253–261.

Dyer A J, Hicks BB (1970) Flux-gradient relationships in the constant flux layer. Quart J R Meteorol Soc 96:715–721.

Goutorbe JP, Noilhan J, Valancogne C, Cuenca RH (1989) Soil moisture variations during HAPEX-MOBILHY. Ann Geophys 7:415–426.

Hildebrand, PH (1988) Flux and sounding data from the NCAR King Air Aircraft during HAPEX. National Center for Atmospheric Research, Boulder, CO. NCAR Tech Note-319.

Itier B (1982) Révision d'une méthode simplifiée pour la mesure du flux de chaleur sensible. J Rech Atmos 16:85–90.

Raupach MR, Thom AS, Edwards I (1980) A wind tunnel study of turbulent flow close to regularly arrayed rough surfaces. Bound Layer Meteorol 18(4):373–397.

Riou C (1982) Une expression analytique du flux de chaleur sensible en conditions suradiabatiques à partir de mesures du vent et de la température à deux niveaux. J Rech Atmos 16:15–22.

Thom AS, Stewart JB, Olivier HR, Gash JHC (1975) Comparison of aerodynamic and energy budget estimates of fluxes over a pine forest. Quart J R Meteorol Soc 101:93–105.

11
Using One- or Two-Layer Models for Evaporation Estimation with Remotely Sensed Data

Peter J. Camillo

1. Introduction

Remotely sensed radiation in various wavebands has been used to calibrate models to estimate surface energy balances over relatively homogeneous areas. Theoretical work on the precise structure of these models and the mathematical formulations of the various submodels is continuing. In many cases the models have been tested by comparisons with data, but rarely have the tests included data from many days over the same site, or data from different sites. This chapter addresses the issue of whether the energy balances of soil and vegetation need to be included separately. One- and two-layer energy balance models were compared using the HAPEX data set, which contains long time periods of measurements over different types of vegetation.

Remote sensing in the hydrological community is mainly used to estimate energy and moisture fluxes at the Earth's surface. The wave bands found to be most useful are visible (0.3–0.7 μm), near-infrared (0.8–1.1 μm), thermal infrared (8.0–12.0 μm), and microwave (1–50 cm). The first two bands may be used to estimate canopy density and possibly the solar albedo. The thermal infrared band provides estimates of surface temperature, and parts of the microwave band correlate well with near surface soil moisture. These remotely sensed data may be used in combination to estimate state variables such as temperature, moisture, and albedo. To estimate the fluxes throughout the entire day, state variable measurements may be used to calibrate time-dependent numerical models of the surface hydrological processes. No combination of all the possible remotely sensed data in a model has yet been made.

This chapter describes a comparison of a one-layer energy balance model (canopy and soil at the same temperature, one surface energy balance) and a two-layer energy balance model (canopy and soil at different temperatures, with two energy balances and interactions between the two surfaces). The longer term goal is to calibrate a new model with all possible remotely sensed data, including surface temperature from thermal infrared radiances, soil moisture from L-band microwave radiation, and solar albedo from visible and near-infrared radiances. It is necessary first to determine if the more complex two-layer model is needed to model the fluxes accurately or if the single-layer model is sufficient, and to see how the calibrations vary over space and time. It is then possible to determine the minimum complexity needed in a more operational model and how often it must be recalibrated. In this chapter remotely sensed surface temperatures are used for calibration and ground measurements of canopy density and solar albedo were used to partition the solar energy between the canopy and soil.

2. Approach

2.1 Data

The HAPEX experiment took place in southwestern France in the summer of 1986. Intensive ground and remote measurements were made

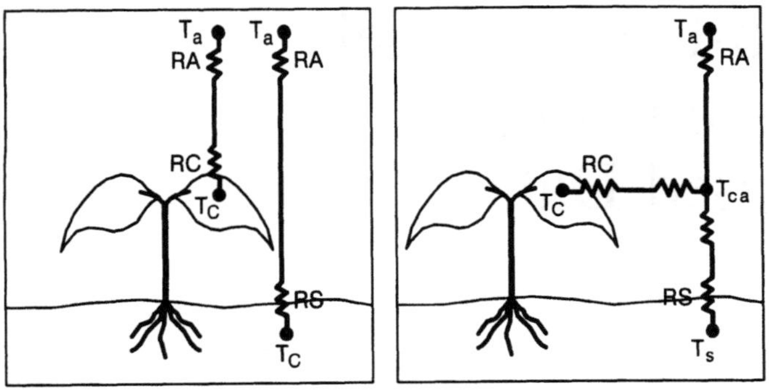

FIGURE 11.1. Resistance networks for the one- and two-layer models.

during a 6-week period in June and July, called the special observing period (SOP). The experiment encompassed a 100 × 100 km area. For testing their models data were used from a north–south swath through the central portion, roughly 5 × 50 km, for which all suitable data were available. There were 13 ground flux stations in this area—one in a forest and 12 (called SAMER stations) on managed agricultural plots scattered throughout the area. An area called the central site had two SAMER stations in adjacent oat (#1) and corn (#5) fields, and some ground data (volumetric soil moisture, soil surface temperatures, leaf area index, etc.) were also taken there. A NASA C-130 aircraft with a scanning thermal infrared radiometry, TIMS (six thermal infrared bands), a visible and near-infrared scanning radiometer NS001 (four visible and three near-infrared channels), and a microwave radiometer (21 cm) were also taking data during the special observing period.

2.2 Models

To evaluate the efficiency of a single- versus a multiple-layer model, an existing energy and moisture balance model was modified to simulate either a canopy and soil surface at the same temperature, or a canopy and soil surface at different temperatures. The former is referred to as the one-layer model and has one energy balance equation

$$G = R_{\mathrm{NET}} + LE + H \qquad (11.1)$$

where the fluxes are, respectively, soil heat, net radiation, latent heat, and sensible heat. Camillo et al. (1983) and Gurney and Camillo (1984) describe how these are coupled to the soil surface boundary conditions and may be solved for the surface temperature.

The two-layer model has separate energy balances for the soil surface and the vegetation canopy and includes interactions between the two. With subscripts s and c referring to the soil and canopy, respectively, these are

$$G_s = R_{\mathrm{NET,S}} + LE_s + H_s \qquad (11.2a)$$
$$O = R_{\mathrm{NET,C}} + LE_c + H_c \qquad (11.2b)$$

Two continuity equations couple these to the fluxes between the canopy and the height above the canopy where the micrometeorological data are measured:

$$H = H_s + H_c \qquad (11.2c)$$
$$LE = LE_s + LE_c \qquad (11.2d)$$

Figure 11.1 shows the resistance networks for the one- and two-layer models. In the one-layer model the canopy and soil are each connected directly to the atmosphere, with no interaction between the two. In the two-layer model the canopy and soil are each connected to the canopy air space, which is then connected to the atmospheric layer above the canopy. A generic equation for the latent heat flux is

$$LE = -\frac{\rho c_p}{\gamma} \frac{(e_1 - e_2)}{(RS_1 + RA_{12})} \qquad (11.3)$$

where e_1 and e_2 are, respectively, the vapor pressures of the evaporating surface and the point to which it is connected. RA_{12} is the appropriate aerodynamic resistance. For example, in the one layer model e_1 and RS_1 are, respectively, the vapor pressure (e_c) and stomatal resistance (*RC*) of the bulk canopy, and e_2 and RA_{12} are the measured atmospheric vapor pressure (e_a) and the Paulson (1970) model for the aerodynamic resistance between the canopy and atmosphere, with roughness and displacement lengths appropriate for the canopy.

The sensible heat flux is modeled as

$$H = -\rho c_p (T_1 - T_2)/RA_{12} \qquad (11.4)$$

where T_1 and T_2 are the temperatures of the two surfaces that exchange sensible heat.

Sensitivity studies (Taconet et al., 1986; Choudhury and Monteith, 1988) have shown that under most environmental conditions the largest resistance in the network is the bulk canopy resistance. Its value is almost always larger than any of the aerodynamic resistances by a factor of two or more, and it is therefore the most important unknown parameter. That is, varying the bulk canopy resistance, *RC*, changes the values of *LE*, *H*, and the surface temperature more than do changes on any other single-model parameter. Therefore the bulk canopy resistance, *RC*, is our primary unknown parameter.

The value of the bulk canopy resistance, *RC*, may not be taken to be a constant. It is a function of the amount of intercepted photosynthetically active radiation (*PAR*) absorbed by the canopy, and of environmental conditions such as temperature, vapor pressure, and leaf water potential.

A well-known model for the resistance to water transfer in a bulk canopy is (Sellers et al., 1986)

$$RST = \frac{a}{b\,L_T + PAR_l} \frac{1}{f(T_c,\, VPD,\, \psi_c)} \qquad (11.5)$$

where a and b are constants that differ among plant species, L_T is the total leaf area index, and PAR_l is the *PAR* incident normal to the leaves integrated over the canopy. The function f models canopy stress factors that increase the resistance and is a function of canopy temperature (T_c), vapor pressure deficit (*VPD*) in the atmosphere, and leaf water potential (ψ_c). The bulk canopy resistance is therefore itself controlled by many unknown variables, which can complicate the modeling. In this work we take a, the overall constant, as an unknown fitting parameter and assume no stress factors, that is $f=1$.

If more than one energy balance is used, the radiation that supplies the energy input must be apportioned between vegetation and soil. Many models exist for apportioning the incoming solar radiation among canopy absorption, soil absorption, and reflection (albedo). They can be as simple as an exponential decay of the direct beam through the canopy, the two stream approximation to the radiative transfer equation, or a more exact solution that allows for directional reflectance. However, they all depend on approximately the same set of parameters: single scattering albedos for the visible and near-infrared wavebands, soil albedo, total leaf area index, leaf angle distributions, and the fraction of the incoming solar radiation that is diffuse. Ideally these models should be calibrated with measured visible and near-infrared reflectances at various sun angles. The integrated *PAR* should then be calculated for the canopy resistance model and the canopy albedo, canopy absorption, and soil absorption. However, the measurement of such reflectances was insufficient during the HAPEX experiment. The measured total leaf area index was therefore used in a canopy reflectance model (Camillo, 1987). Values were assumed for the single scattering albedos, and the other parameters were varied until the predicted and measured canopy albedos agreed. For full canopies (leaf area index greater than 4.0) the procedure is trivial; soil absorption is zero and canopy absorption is one minus the measured solar albedo. For less dense canopies the procedure is more complex but the model of Camillo (1987) can be used.

In the two-layer model, longwave radiation is calculated as follows (Ross, 1981):

$$RL_c = (1-\tau)\,(\sigma\varepsilon_a T_a{}^4 + \sigma\varepsilon_s \mathrm{T_s}{}^4 - 2\sigma\varepsilon_c \mathrm{T_c}{}^4) \qquad (11.6a)$$

$$RL_s = \tau\,\sigma\varepsilon_a \mathrm{T_a}{}^4 - \sigma\varepsilon_s \mathrm{T_s}{}^4 + (1-\tau)\sigma\varepsilon_c \mathrm{T_{-c}}{}^4 \qquad (11.6b)$$

where subscripts c, s, and a refer to the canopy, soil, and atmosphere, respectively, *RL* is the net longwave radiation, T is the temperature, σ is the Stefan–Boltzmann constant, ε is the emissivity, and τ is the transmission coefficient for the canopy and is a function of the total leaf area index and leaf angle distribution (Ross, 1981).

Air temperature [sensible heat flux, Equation (11.4)], vapor pressure [latent heat flux, Equation (11.3)], and wind speed (aerodynamic resistance models) measured 2 m above the canopy, that is

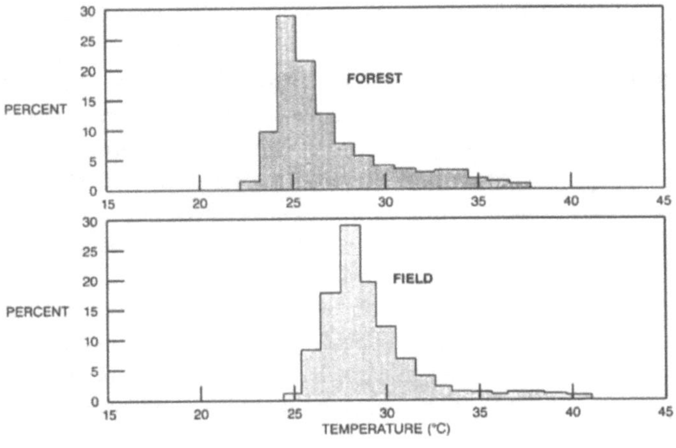

FIGURE 11.2. Remotely sensed surface temperature histograms for the SAMER 5 cornfield and adjacent forest on June 27.

from the top level of the SAMER stations, were used to give upper boundary conditions on the models. The one layer model uses net radiation directly in the energy balance equation [Equation (11.1)], since the soil and canopy are assumed to have the same temperature. The two-layer model uses incoming solar radiation and apportions it between the soil and canopy as already described.

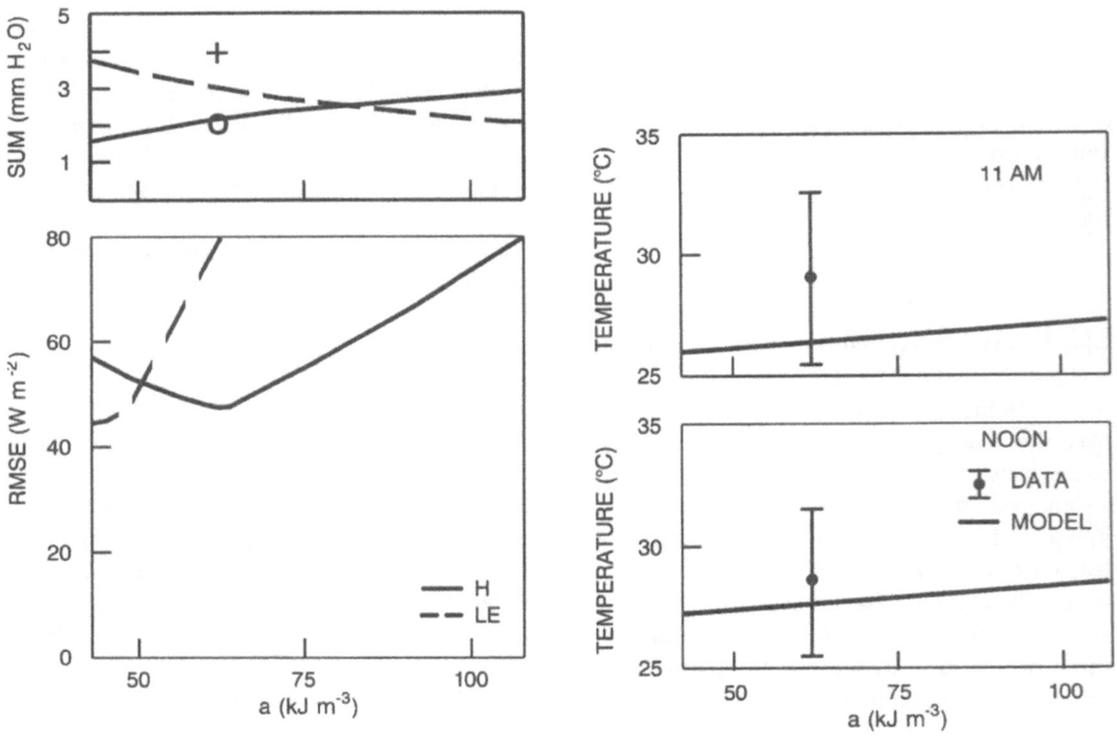

FIGURE 11.3. Results of fitting the one-layer model to the June 27 SAMER 5 data. Clockwise from the lower left graphs are latent and sensible heat flux root mean square errors, daytime integrated latent and sensible heat fluxes, 11 AM temperature comparison, and noon temperature comparison.

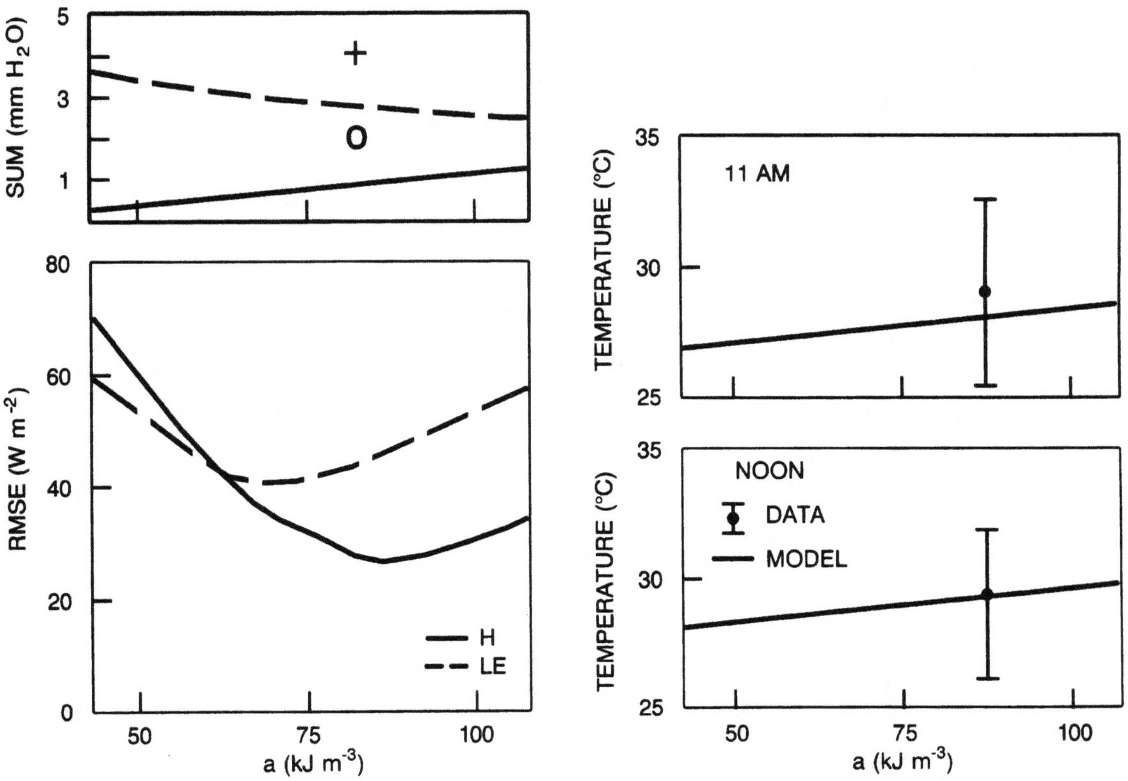

FIGURE 11.4. Results of fitting the two-layer model to the June 27 SAMER 5 data. Clockwise from the lower left graphs are latent and sensible heat flux root mean square errors, daytime integrated latent and sensible heat fluxes, 11 AM temperature comparison, and noon temperature comparison.

3. Fitting

Operationally, one would like to vary the canopy resistance model parameter [a, Equation, 11.5)] until the measured and predicted state variables, such as surface temperature at some time near the daily maximum, agree. The model then estimates the fluxes, both instantaneous and integrated over the day. However, for the purpose of testing the models we fitted to the measured sensible heat flux and then compared model and data for the remaining fluxes and surface temperature. This allowed a decision to be taken more easily on the complexity of the model required.

We now present the results of fitting the model to the central site corn field (SAMER 5) on June 27. Histograms of the remotely sensed surface temperatures are shown in Figure 11.2. Although the SAMER 5 was in the corn field, it was just 150 m northeast of a small forested area, and on some days the temperatures of these areas differed by as much as 15 °C, with the forested area always being cooler. If the wind was blowing from the southwest, then the effective fetch for the SAMER 5 station is not the corn but the forest, and the modeled temperature should match that of the forest, not the corn. However, on June 27 the two temperatures were very similar, the difference being less than half the standard deviation of the measurements in each area. The temperatures were all measured using channel 5 (10.3–11.1 μm) of the TIMS instrument. The temperatures at 11 AM were measured at an altitude of 300 m and the noon temperatures from 1500 m. The noon histograms are not shown but have the same skewed distribution as the 11 AM histograms. No atmospheric corrections were applied.

Some of the model output as a function of a are plotted in Figures 11.3 and 11.4 for the one- and two-layer models, respectively. The root mean

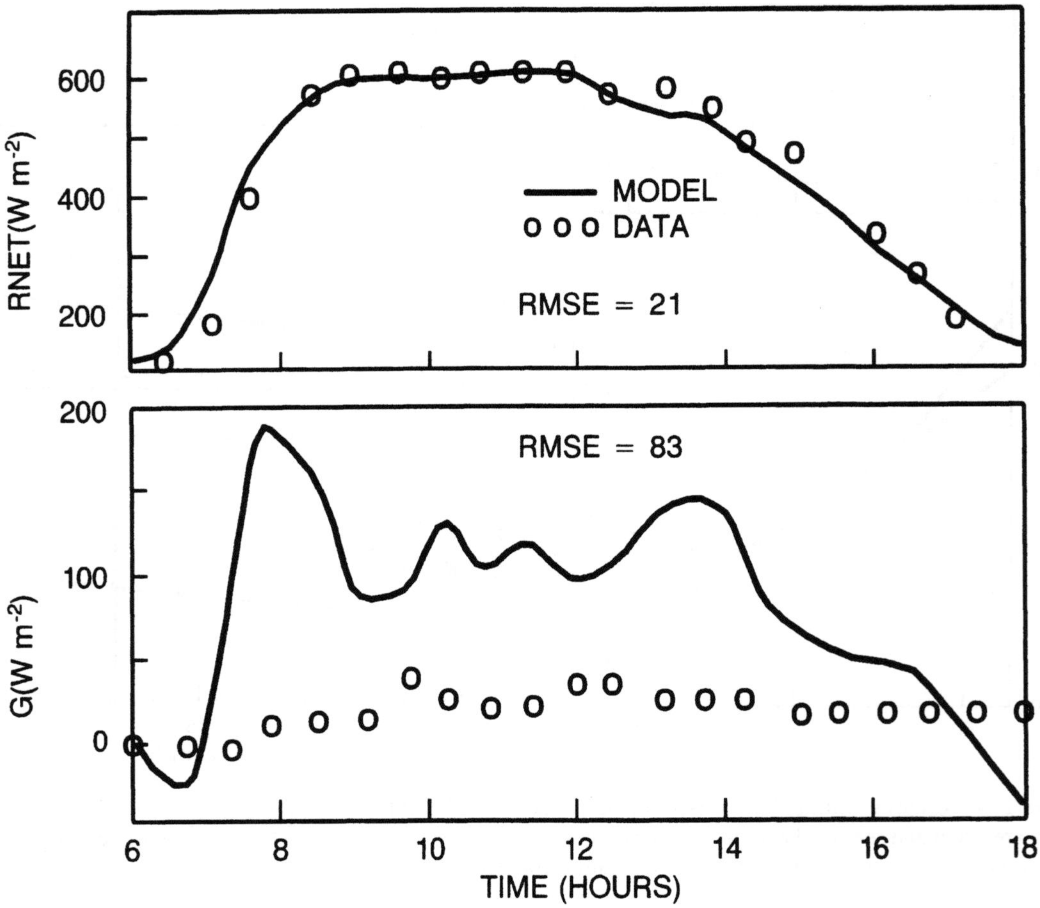

FIGURE 11.5. Comparison of the measured and modeled (one-layer model) surface energy balance fluxes for the June 27 SAMER 5 data. (a) Net radiation and soil heat flux.

square errors (RMSEs) between measured and modeled latent and sensible heat fluxes are plotted in the bottom left graph. Note that the minimum is smaller for the two-layer model, indicating a better fit. The top left graph in each figure shows the total sensible and latent heat flux during the day in millimeters of water. The data points are plotted at the value of a corresponding to the minimum RMSE for the sensible heat flux. Note that the cumulative sensible heats for both the one- and two-layer models agree with the data, which is not surprising since a minimum RMSE is roughly the same as an equal integral. However, if one uses this graph to estimate the cumulative latent heat flux, the one-layer model underestimates it by more than 1 mm (25% of the measurement), whereas the two-layer model estimate is within one third of a millimeter, which is about an 8% error. The two-layer model gives a better estimate for cumulative evaporation.

The two right-hand graphs in each figure compare the measured and modeled surface temperature at 11 AM (top) and noon (bottom). If the spatial variability of the surface temperature is as large as it is here, the model may not be calibrated by fitting to surface temperature alone. The error bars are so large that there is no hope of restricting the range of the fitting parameter to get a reasonable evaporation estimate. This does not mean that surface temperature cannot be used as it can bound parameter values. If the modeled temperature is well outside the error bars, this can point to errors in other model variables.

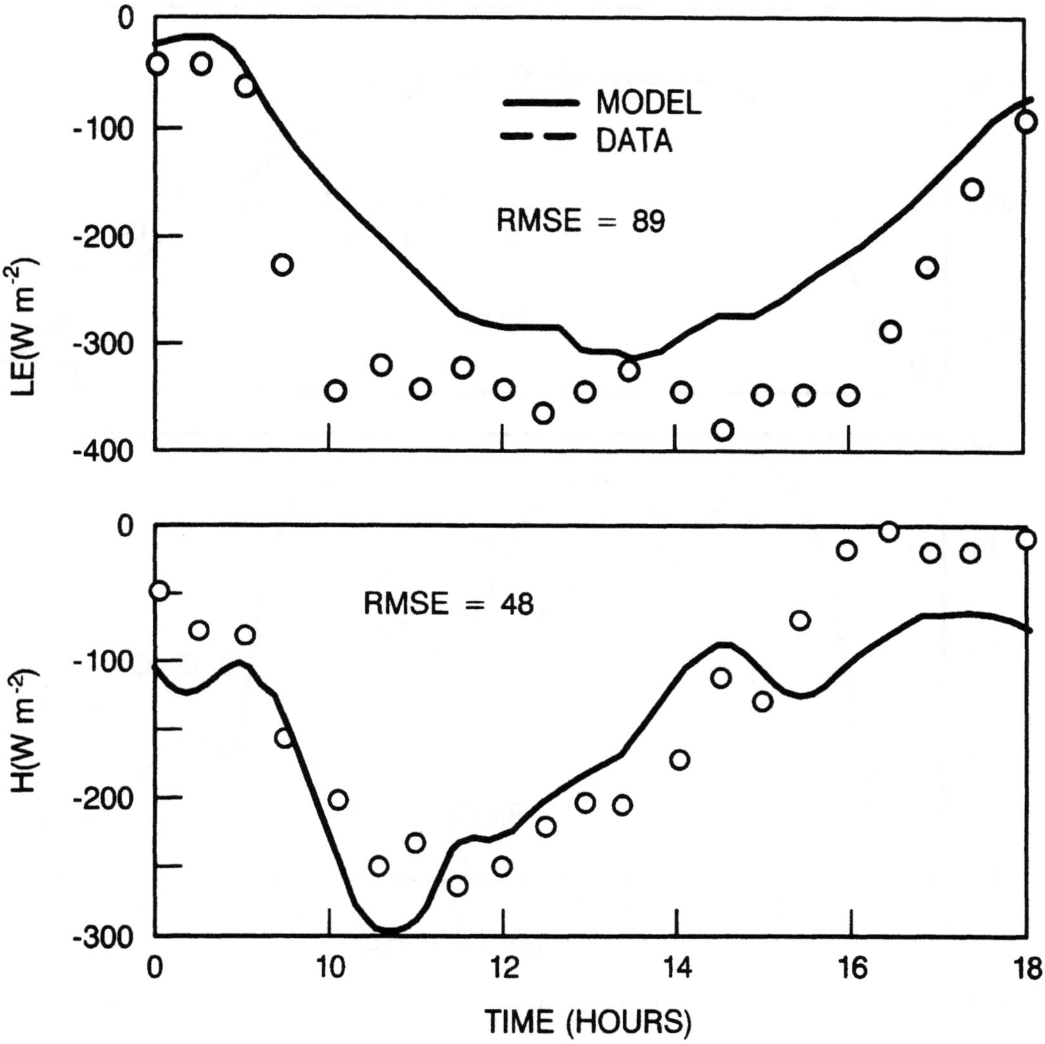

FIGURE 11.5. (b) latent and sensible heat fluxes.

The measured and modeled fluxes are compared in Figure 11.5 (one-layer model) and Figure 11.6 (two-layer model). One would expect the RMSE for net radiation to be zero for the one-layer model because the data are used directly in the model. This is not the case because the data represent half hourly averages of nearly continuous measurements, whereas the model half hourly average to which the data are compared is actually the average of only two points, one each at the start and stop of the half hour. However, an RMSE of about 20 W m^{-2} is well within the expected measurement error. The net radiation RMSE for the two-layer model (Fig. 11.6a) is about 10% of the maximum, an acceptable figure. Again, some of this discrepancy may be attributed to the differences in the averaging periods.

The one-layer model greatly overestimates the soil heat flux (Fig. 11.5a). This is not surprising, as the soil surface temperature used for the calculation is certainly much larger than the actual temperature at the heat flux plate. The two-layer model with its lower soil surface temperature gives a better estimate of the soil heat flux, although the RMSE is still about half the peak value. Figure 11.6b shows that the two-layer model provides much

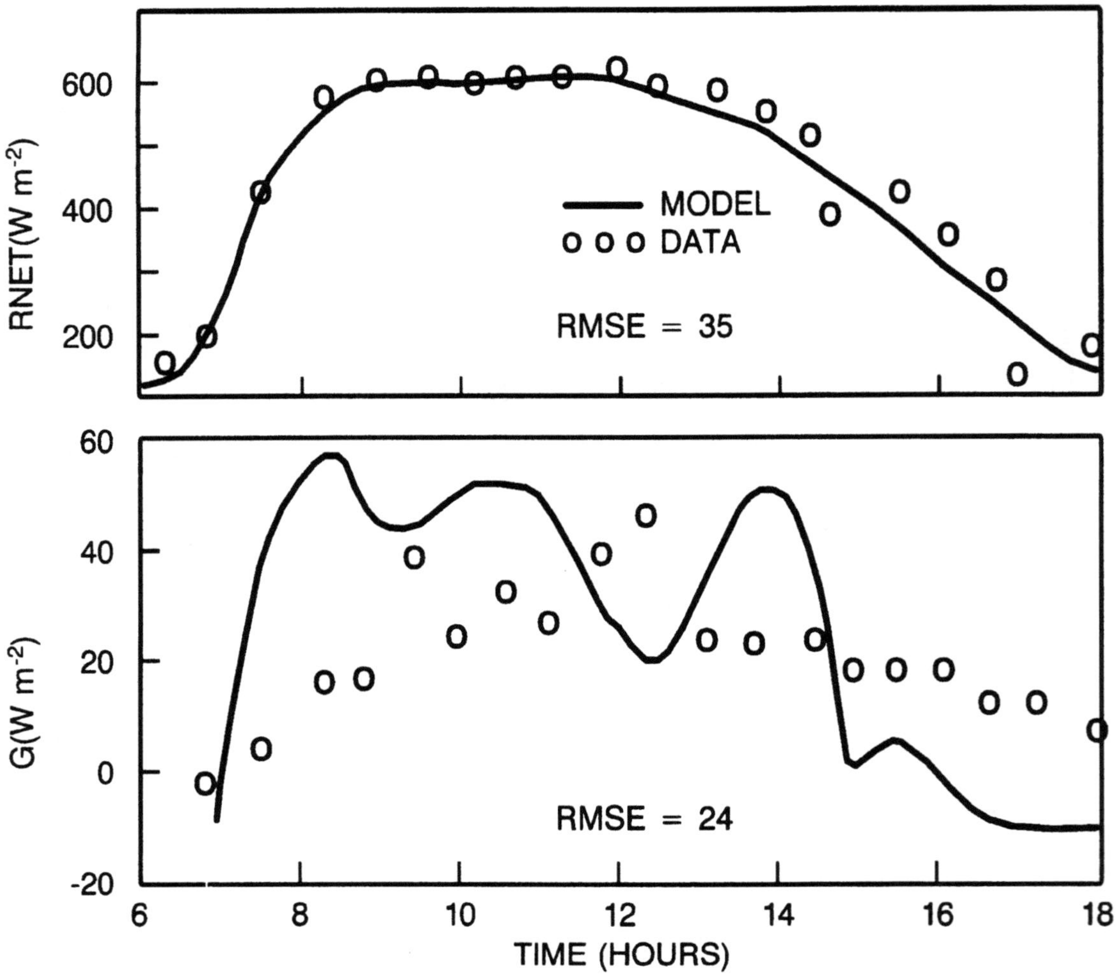

FIGURE 11.6. Comparison of the measured and modeled (two-layer model) surface energy balance fluxes for the June 27 SAMER 5 data.

better agreement with measured latent and sensible heat fluxes than does the one layer model.

This analysis was repeated for the same site on June 16. Figure 11.7 shows the temperature histograms for the field and adjacent forest. The mean temperature of the forest canopy is 30°C, but the field is much hotter, about 43°C. The Bowen ratio estimated from the data was about 0.5 (Fig. 11.11b), which represents a cool, freely evaporating surface. It therefore appears that a component of the sensible heat flux actually measured by the SAMER 5 station is from the forest (i.e., the effective fetch is from the forest, not the field). The model was therefore fitted to the forest surface temperature. Figures 11.8 through 11.11 show the fitting results and flux comparisons for the one- and two-layer models. The basic conclusion from this data set is the same as that from the June 27 data set; the two-layer model is significantly better for modeling the fluxes.

The values of the fitting parameter for the two-layer model are significantly different for the two days (144 and 82 kJ m^{-3} for June 16 and June 27, respectively), which means that the canopy resistance model RC as used here is not physically realistic. It should be possible to make the parameter *a* the same for both data sets and still provide realistic fluxes by including other parameters in the stress factor *f*.

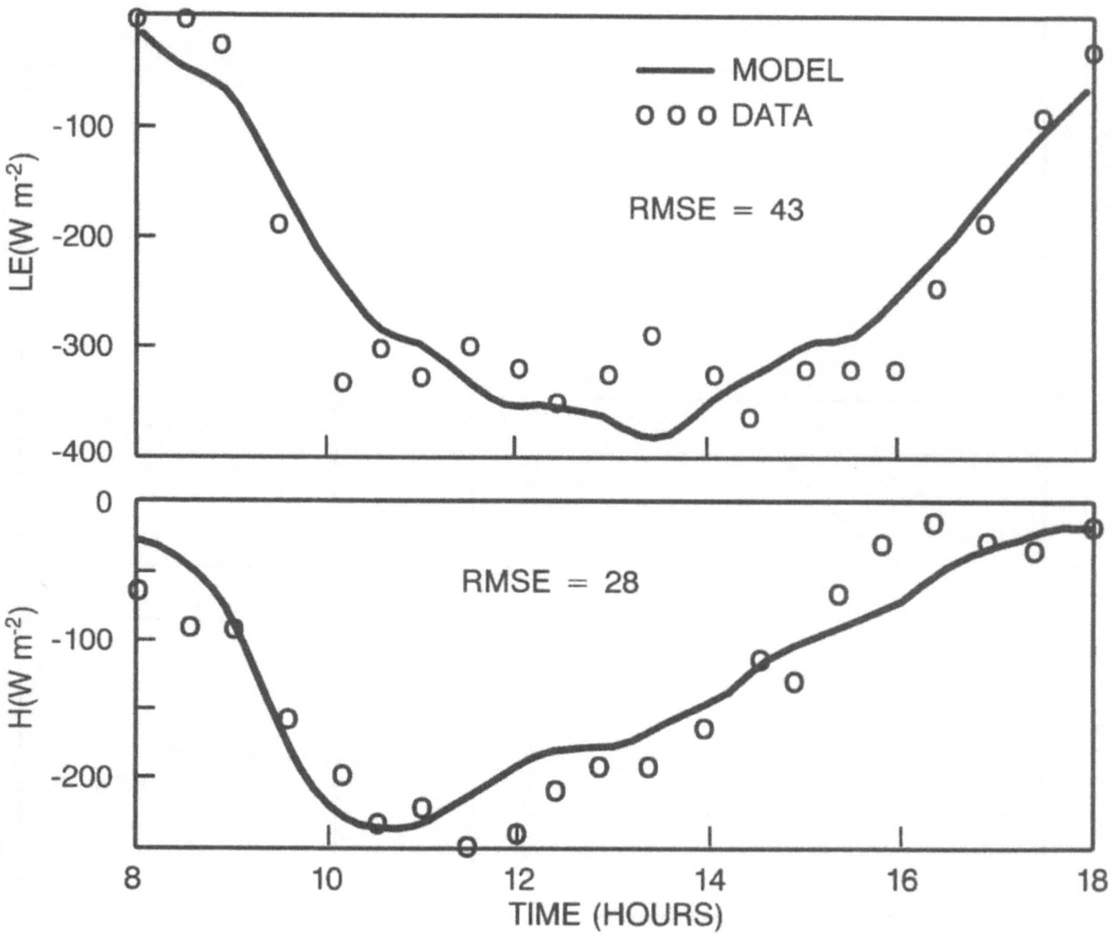

FIGURE 11.6. (b) *Continued.*

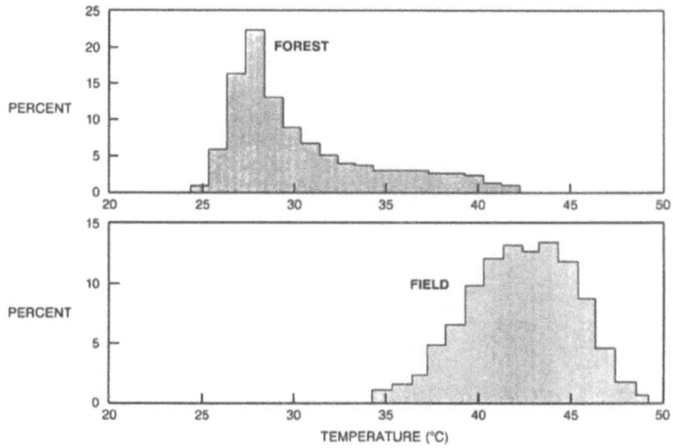

FIGURE 11.7. Remotely sensed surface temperature histograms for the SAMER 5 forest and cornfield on June 16.

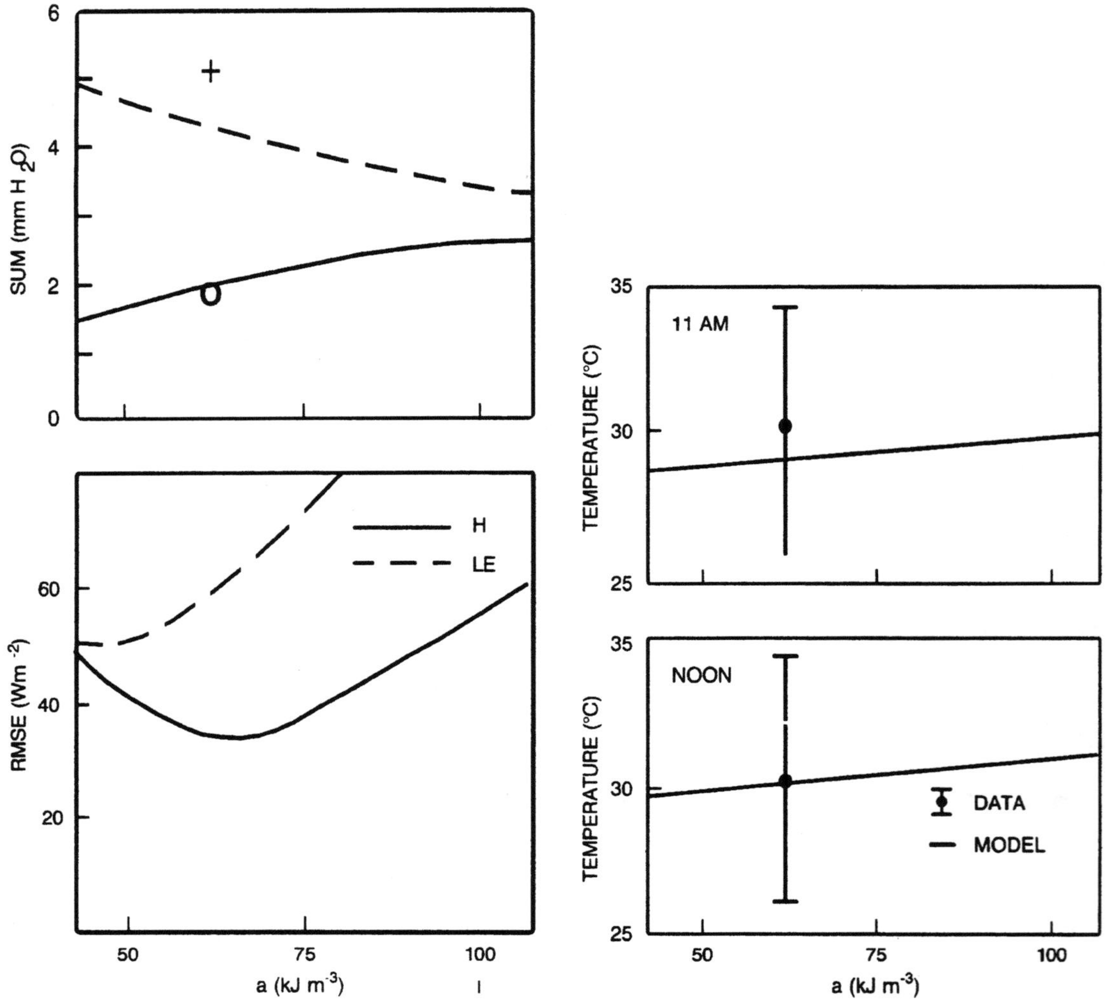

FIGURE 11.8. Results of fitting the one layer model to the June 16 SAMER 5 data. Clockwise from the lower left graphs are latent and sensible heat flux root mean square errors, daytime integrated latent and sensible heat fluxes, 11 AM temperature comparison, and noon temperature comparison.

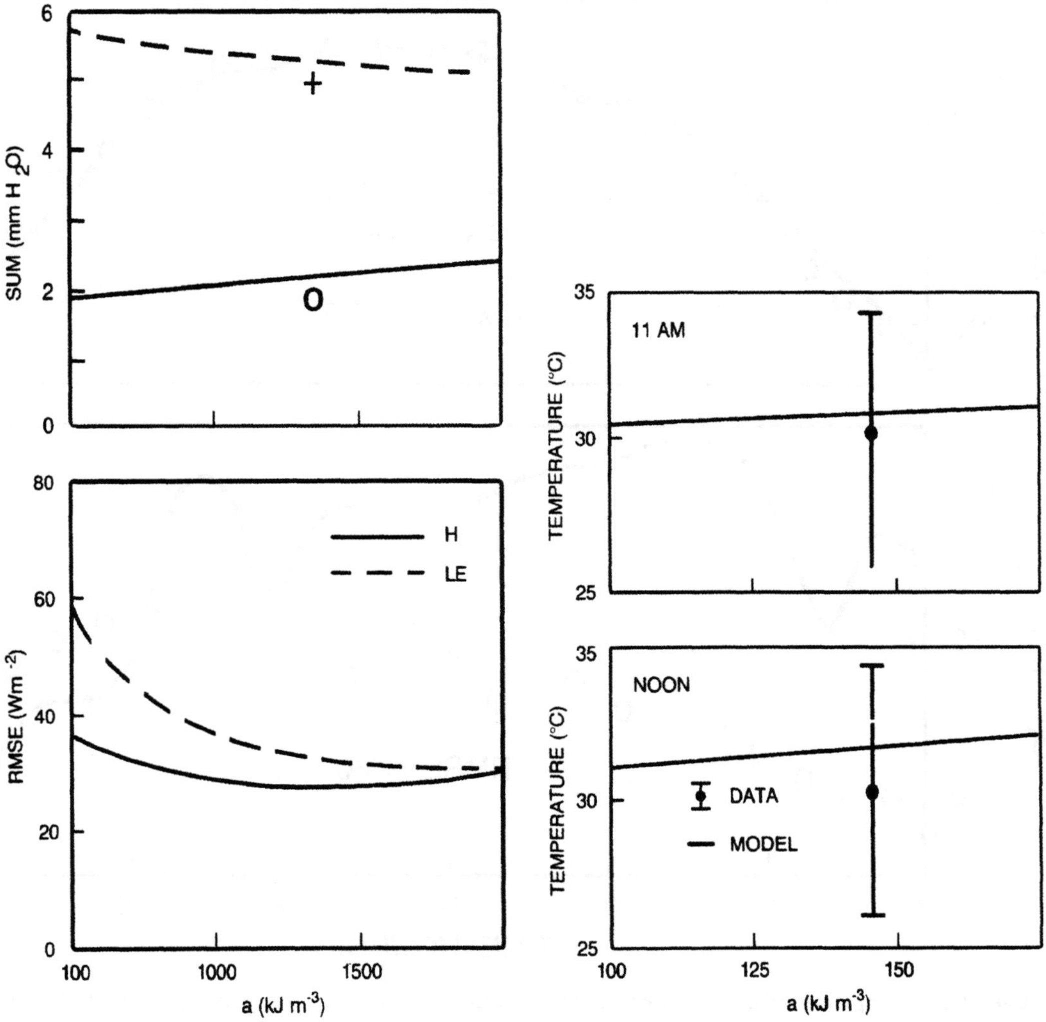

FIGURE 11.9. Results of fitting the two-layer model to the June 16 SAMER 5 data. clockwise from the lower left graphs are latent and sensible heat flux root mean square errors, daytime integrated latent and sensible heat fluxes, 11 AM temperature comparison, and noon temperature comparison.

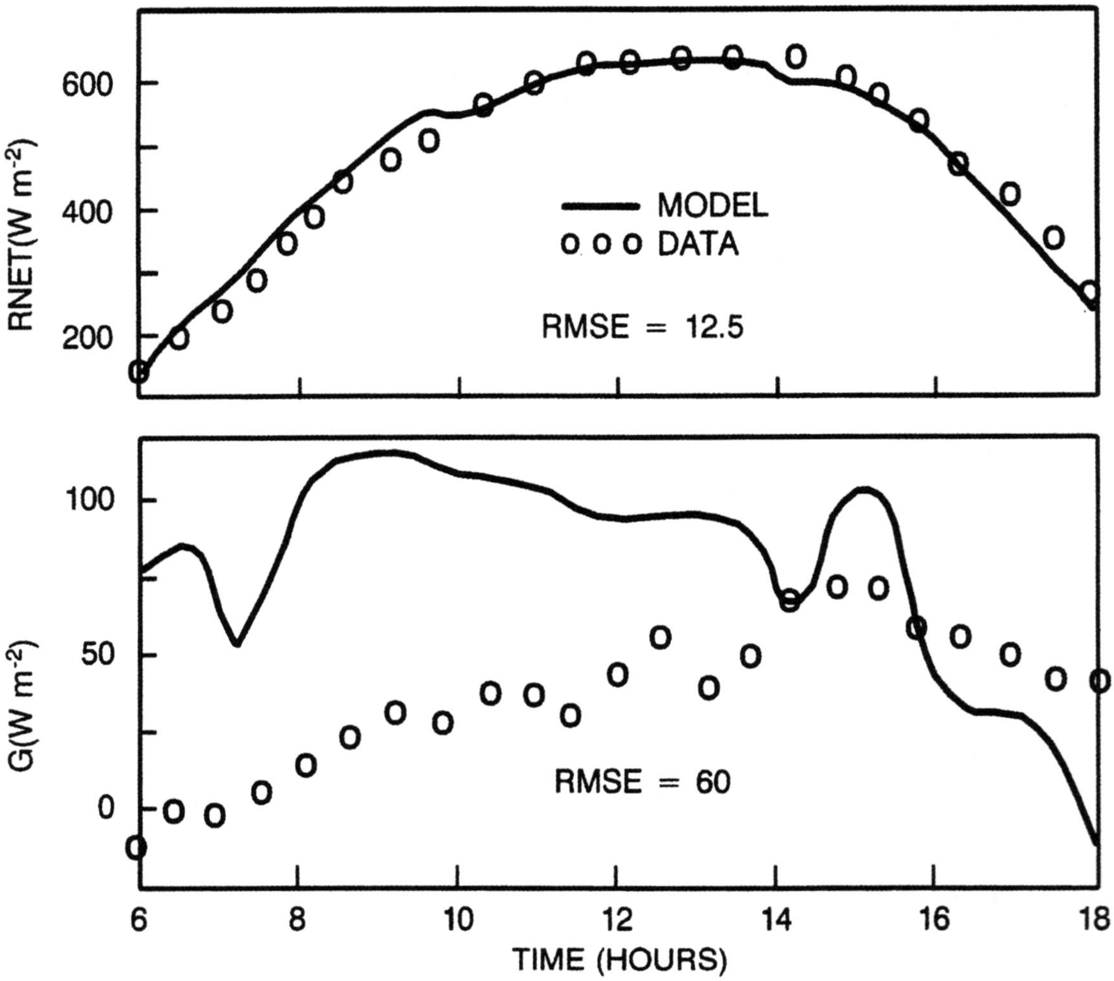

FIGURE 11.10. Comparison of the measured and modeled (one-layer model) surface energy balance fluxes for the June 16 SAMER 5 data.

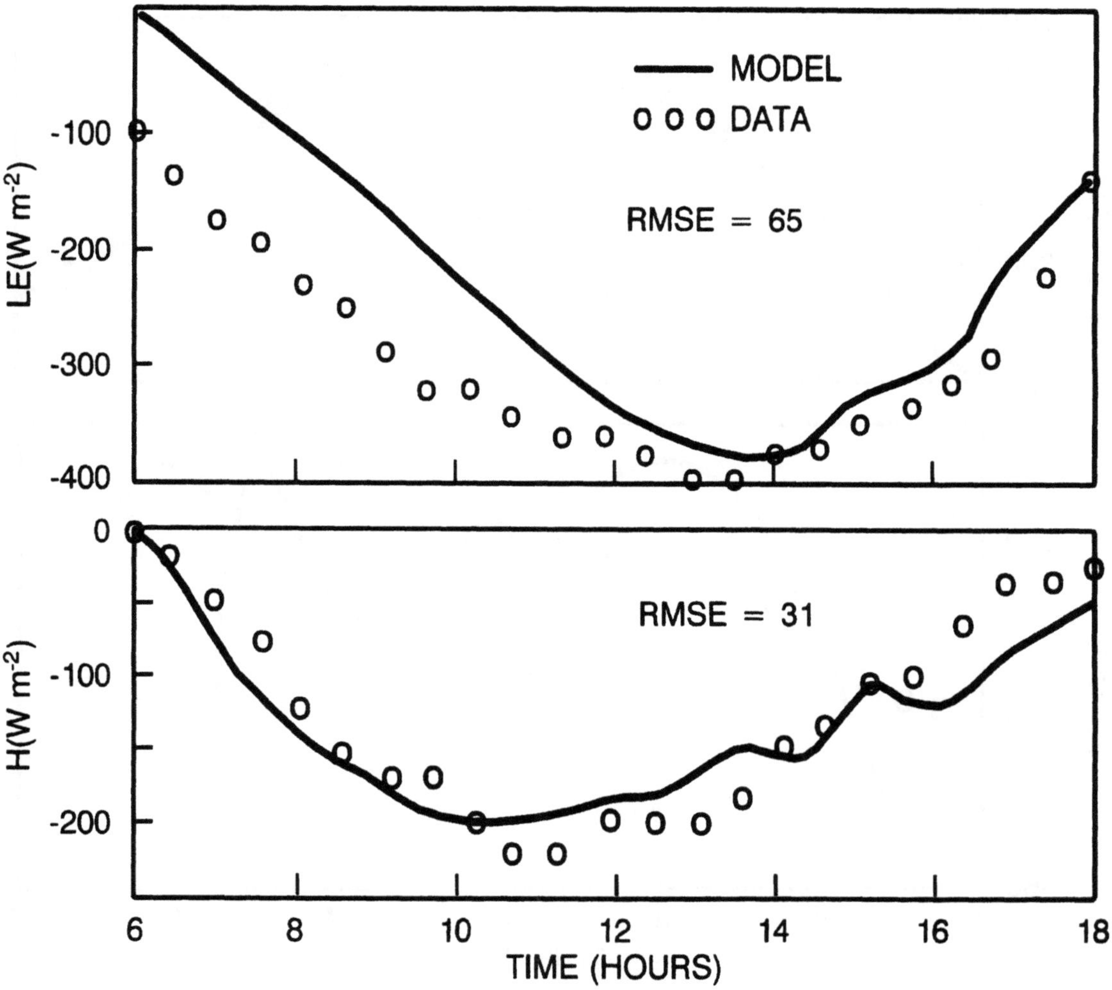

FIGURE 11.10. (b) *Continued.*

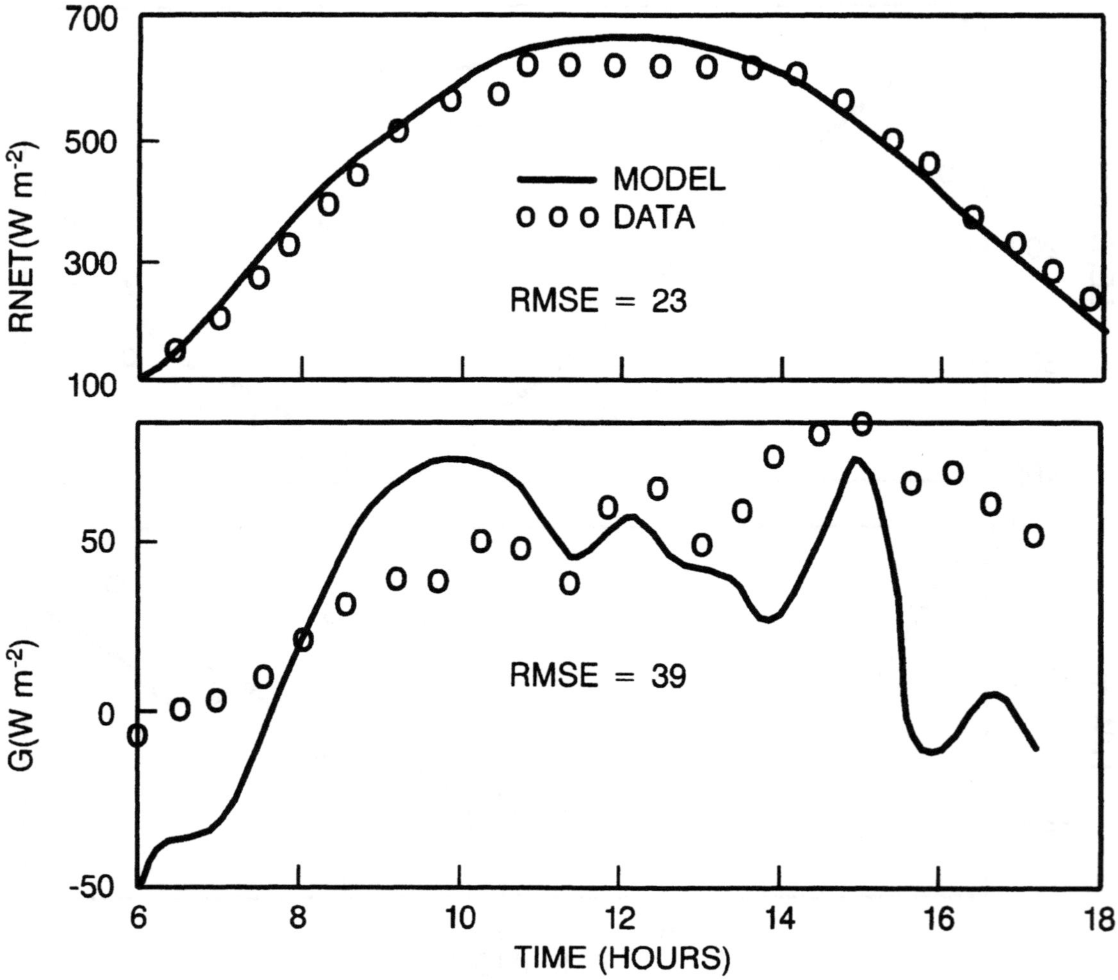

FIGURE 11.11. Comparison of the measured and modeled (two-layer model) surface energy balance fluxes for the June 16 SAMER 5 data.

4. Summary

When fitted to the sensible heat flux, the two-layer model provides much better estimates of daytime integrated evaporation. It was found also that the main fitting parameter, the constant multiplying the bulk canopy resistance model, had to be reevaluated on a daily basis. We speculate that this is because we ignored the canopy stress corrections, which are known to be important. However, in general the two-layer model was found to be adequate and extra layers are not needed.

Acknowledgments. This work would not have been possible without the help of Tom Schmugge, who provided the TIMS data and guidance on how to interpret it. Anne Hsu provided technical support for processing these data. Robert Gurney and Bhaskar Choudhury are to be recognized for helpful discussions about the models.

References

Camillo PJ (1987) A canopy reflectance model based on an analytical solution to the multiple scattering equation. Remote Sens Environ 23:453–477.

Camillo, PJ, Gurney RJ, Schmugge TJ (1983) A soil and atmospheric boundary layer model for evaporation and soil moisture studies. Water Resour Res 19:371–380.

Choudhury BJ, Monteith JL (1988) A four-layer model

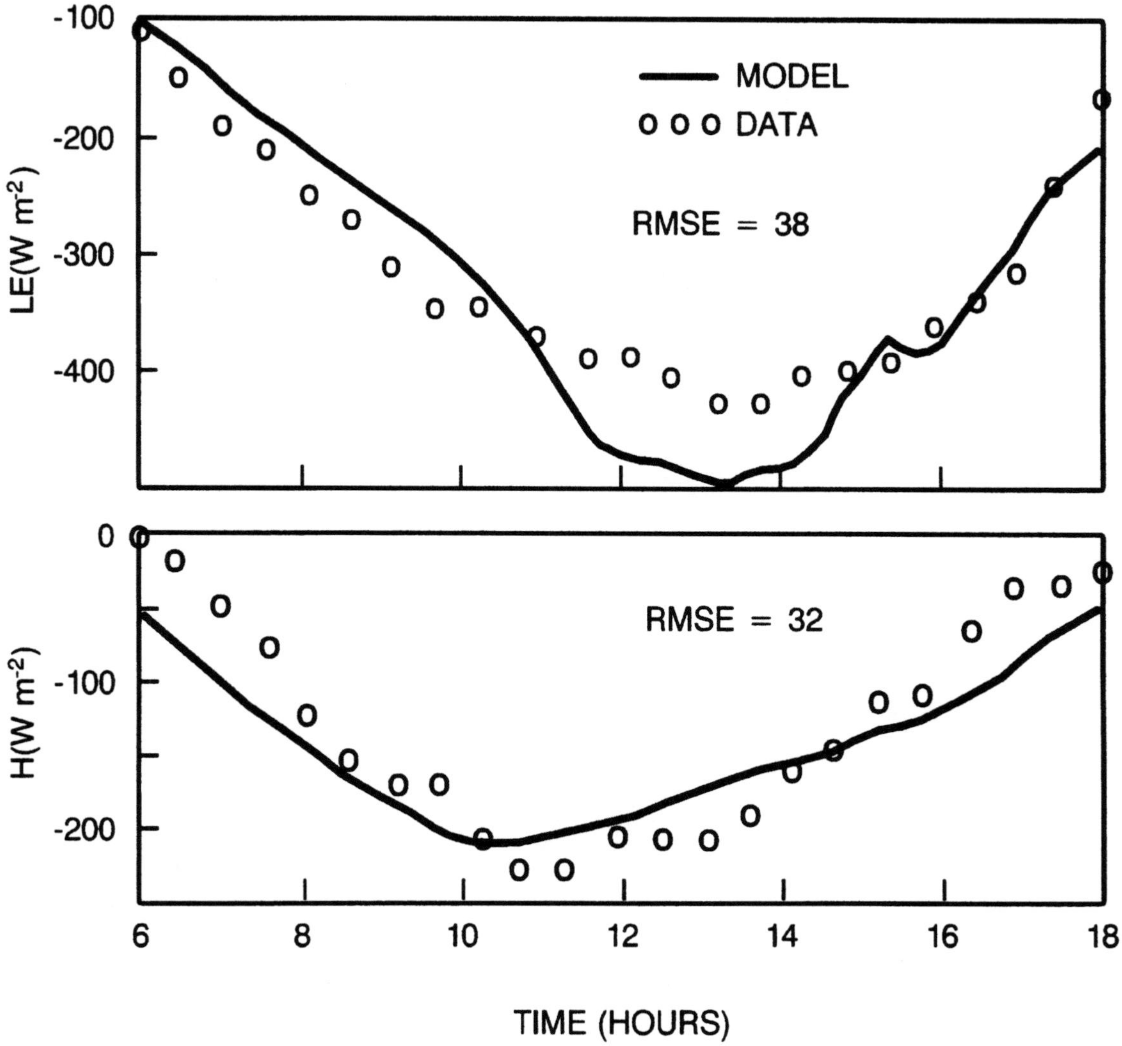

FIGURE 11.11. (b) *Continued.*

for the heat budget of homogeneous land surfaces. Quart J R Meteorol Soc 114:373–398.

Gurney RJ, Camillo PJ (1984) Modelling daily evapotranspiration using remotely sensed data. J Hydrol 69: 305–325.

Paulson CA (1970) The mathematical representation of wind speed and temperature profiles in the unstable atmospheric surface layer. J Appl Meteorol 9:851–861.

Ross J (1981) "The Radiation Regime and Architecture of Plant Stands." Dr. W. Junk Publ., Boston.

Sellers PJ, Mintz Y, Sud YC, Dalcher A (1986) A simple biosphere model (SIB) for use within general circulation models. J Atmos Sci 43:505–531.

Taconet O, Bernard R, Vidal-Madjar D (1986) Evapotranspiration over an agricultural region using a flux/temperature model based on NOAA-AVHRR Data. J Clim Appl Meteorol 25:284–307.

12
On the Maintenance and Measurement of Scalar Fluxes

John C. Wyngaard

1. Introduction

I will address the topic of land-surface evaporation fluxes from my perspective as a researcher concerned with the structure of the atmospheric boundary layer. That boundary layer is turbulent, and as with any turbulent flow what we know about its structure and transport properties has been gained from both theoretical and experimental approaches.

The material in this chapter is divided into three sections. In the first, Section 2, I survey the behavior of scalar fluxes in the planetary boundary layer (PBL) from a theoretical standpoint and discuss some salient features of the PBL turbulence that carries these fluxes—including its stochastic nature. The next, titled Measurement Physics, deals with a number of fundamental aspects of the flux-measurement process. The last section covers some of the important measurement physics of a key component of the modern flux-measuring system, the sonic anemometer.

More details about turbulent fluxes, their measurement, and PBL turbulence can be found in Lumley and Panofsky (1964), Tennekes and Lumley (1972), Haugen (1973), Panofsky and Dutton (1984), and Lenschow (1986).

2. Scalar Fluxes in the PBL

In this section we apply some of the recent advances in understanding of boundary-layer turbulence to the theory and measurement of scalar fluxes in the lower atmosphere. We will focus on the gross features of turbulent transfer across the boundary layer, including the diffusive aspects of PBL turbulence. These features have become much more distinct over the past two decades, as a result of advances made through field programs and numerical modeling. We will not dwell on the details of the transfer processes in canopies or near the surface, although these are quite important as well and have seen comparable advances recently. These areas are discussed by Brutsaert (1982) and Raupach (1988).

2.1 Scalar Density Flux Profiles

The mass density $\mathcal{C}$ of a scalar in the lower atmosphere satisfies the conservation equation

$$\frac{\partial \mathcal{C}}{\partial t} + \frac{\partial}{\partial x_i} \mathcal{C}\mathcal{U}_i = D \frac{\partial^2 \mathcal{C}}{\partial x_i^2} + \mathbf{S} \qquad (12.1)$$

where $\mathcal{U}_i$ is the velocity, D is the molecular diffusivity for $\mathcal{C}$, and $\mathbf{S}$ is the net source of $\mathcal{C}$. We use the convention that repeated indices are summed. Since the flow near the Earth's surface is turbulent, we traditionally represent all dependent variables as the sum of ensemble-mean and fluctuating parts. We distinguish these by upper- and lower-case symbols, respectively:

$$\mathcal{C} = C + c, \quad \mathcal{U}_i = U_i + u_i, \ \mathbf{S} = S + s \quad (12.2)$$

The equation for C is obtained by substituting the decomposition (12.2) into (12.1) and averaging,

$$\frac{\partial C}{\partial t} + \frac{\partial}{\partial x_i}(U_i C + \overline{u_i c}) = D\frac{\partial^2 C}{\partial x_i^2} + S \quad (12.3)$$

where the overbar represents an average. Since $U_i C$ is the advective flux due to the mean motion,

$\overline{u_i c}$ is the turbulent flux of C, and $-D\partial C/\partial x_i$ is the flux of C due to molecular diffusion, we can write Equation (12.3) in "flux form":

$$\frac{\partial C}{\partial t} + \frac{\partial}{\partial x_i}\left(U_i C + \overline{u_i c} - D\frac{\partial C}{\partial x_i}\right) = S \qquad (12.4)$$

This says the local change in C is caused by the divergence of the total C-flux (i.e., the sum of the advective flux, the turbulent flux, and the molecular flux) and the net source. Therefore, time changes of a conservative scalar (i.e., one with $S = 0$) are due solely to flux divergence.

The molecular and turbulent fluxes are of order Dc_*/ℓ and qc_*, respectively, where c_* and ℓ are characteristic intensity and length scales for the C-field and q is a turbulent velocity scale. Their ratio $D/(q\ell)$, an inverse Peclet number, is very small except very close to the surface. [For example, if $D = 10^{-5}$ m^2 s^{-1}, $q = 1$ m s^{-1}, and $\ell = 1$ m, $D/(q\ell)$ is 10^{-5}.] Thus, except at and immediately above the surface the molecular flux is negligible and the vertical flux of a scalar constituent is $F = U_3 C + \overline{u_3 c}$. As we discuss in the next section, the mean vertical flux $U_3 C$ is not always negligible.

Consider some simple examples where $\mathcal{C}$ is the water vapor density in the absence of phase change so $S = 0$. Over a horizontally homogeneous surface with $U_3 \sim 0$ (12.4) becomes

$$\frac{\partial C}{\partial t} = -\frac{\partial F}{\partial x_3} \qquad (12.5)$$

and time changes are due entirely to turbulent flux divergence. In steady-state flow over an inhomogeneous surface (this is often called "local advection") the balance is between horizontal advection and flux divergence:

$$U_1\frac{\partial C}{\partial x_1} + U_2\frac{\partial C}{\partial x_2} + \frac{\partial F}{\partial x_3} = 0 \qquad (12.6)$$

Over the sea, slight subsidence can cause a steady balance between vertical advection and flux divergence:

$$U_3\frac{\partial C}{\partial x_3} + \frac{\partial F}{\partial x_3} = 0 \qquad (12.7)$$

In each of these cases the flux divergence is important, so let us now consider it more closely.

In a quasi-steady situation over a horizontally homogeneous surface we can differentiate Equation (12.5) with respect to x_3 and require that the time derivative vanish, so that

$$\frac{\partial}{\partial t}\frac{\partial C}{\partial x_3} = 0 = -\frac{\partial}{\partial x_3}\frac{\partial F}{\partial x_3} \qquad (12.8)$$

This implies that F varies linearly between values of F_0 at the surface and F_h at the boundary-layer top at $x_3 = h$; i.e.,

$$F = F_0 + \frac{(F_h - F_0)}{h}x_3 \qquad (12.9)$$

The surface flux F_0 is maintained by molecular diffusion, but the transition to turbulent flux is achieved at very small heights above the surface.

If there is minimal entrainment at the boundary layer top (e.g., in a stably stratified case or in some convective boundary layers) we can have $F_h \ll F_0$, so that

$$F = F_0\left(1 - \frac{x_3}{h}\right) \qquad (12.10)$$

If, on the other hand, there is no surface flux but there is a flux at h due to entrainment, the profile is

$$F = F_h\frac{x_3}{h} \qquad (12.11)$$

although this is more likely to be found for certain nondepositing trace species than for water vapor.

In practice, the flux of water vapor at h is apt to be significant (for example, it is typically nearly equal to the surface flux in the marine PBL). Thus, a conservative estimate for the flux divergence in such cases is

$$\frac{\partial F}{\partial x_3} \sim -\frac{F_0}{h} \qquad (12.12)$$

Thus, the fractional difference between the flux at height x_3 and the surface flux F_0 does not exceed

$$\frac{F(x_3) - F_0}{F_0} = \frac{x_3}{h} \qquad (12.13)$$

so at $x_3 = 10$ m in a horizontally homogeneous, 1-km-deep boundary layer the scalar flux is likely to be within 1% of its surface value—so we typically need not measure the "surface" flux right at the surface.

Now consider a *conservative* scalar attribute $\mathcal{A}$—i.e., one that (neglecting molecular diffusion) satisfies the conservation equation

$$\frac{D\mathcal{A}}{Dt} = \frac{\partial \mathcal{A}}{\partial t} + \mathcal{U}_j\frac{\partial \mathcal{A}}{\partial x_j} = 0 \qquad (12.14)$$

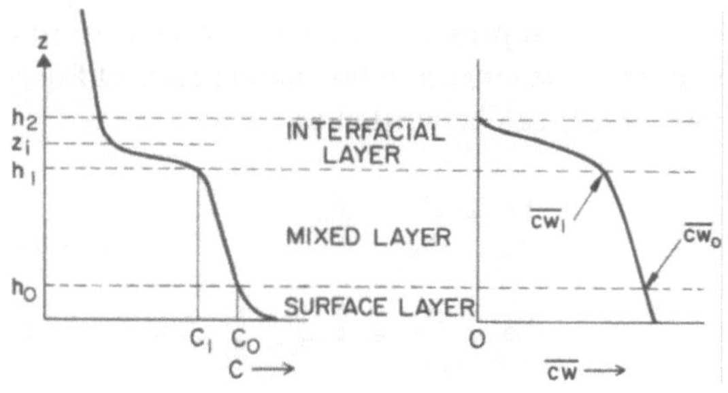

FIGURE 12.1. Typical vertical profiles of the mean value (left panel) and vertical turbulent flux (right panel) of a conservative scalar constituent in the convective boundary layer. Reprinted with permission from Wyngaard and Brost, 1984, *Journal of Atmospheric Sciences*, American Meteorological Society.

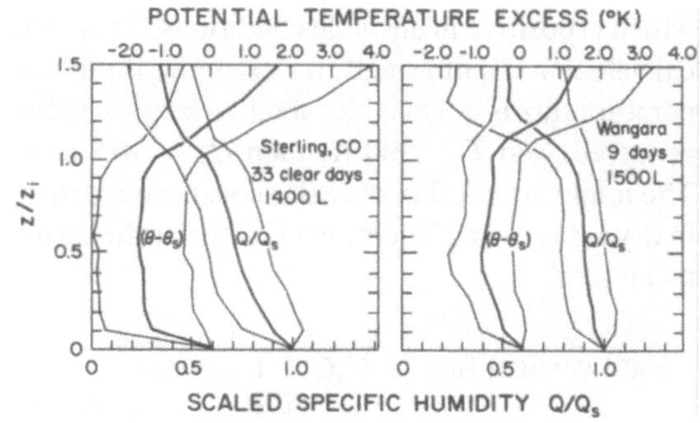

FIGURE 12.2. Composite profiles of potential temperature and specific humidity from the NHRE (left panel) and Wangara (right panel) experiments. Reprinted with permission from Mahrt, 1976, *Monthly Weather Review*, American Meteorological Society.

The physical interpretation of Equation (12.14) is that $\mathcal{A}$ does not change—i.e., is conserved—following the fluid motion. One expects such a variable to have a "strong-mixing" limit in which the mean vertical profile $\overline{\mathcal{A}} = A$ is flat, or well mixed in the vertical. One expects the convective boundary layer to approach this limit; potential temperature is one example of a conservative variable that displays this behavior there. Scalar constituent density, however, is not conservative; its conservation equation (12.1) in the nondiffusive, zero-source limit is subtly different from Equation (12.14). Since both $\mathcal{C}$ and air density ρ_a satisfy Equation (12.1), one can show easily that the "mixing ratio" $\mathcal{C}/\rho_a$ is conservative, however. Thus, in the well-mixed limit we expect flat profiles of scalar mixing ratio, but not of scalar density.

Figure 12.1 sketches typical vertical profiles of the mean value and vertical flux of a conservative scalar in a convective boundary layer. We have taken the flux at mixed-layer top to be positive, the usual case for water vapor but not for potential temperature. This difference often causes a noticeable difference in their mean profiles. Even in strongly convective conditions mean water vapor mixing ratio can be distinctly "unmixed" because the positive fluxes at the surface and at the top both require negative mean gradients, which add to give a noticeable mean mixing-ratio change across the boundary layer. The different signs of surface and entrainment fluxes of temperature tend to give equal and opposite mean gradients, by contrast, yielding a much more nearly well-mixed potential temperature profile. Figure 12.2 shows an example of this.

2.2 The "Webb Correction"

We just showed that the total vertical flux of a scalar constituent density is $U_3C + \overline{u_3c}$, the sum of mean–mean and turbulent–turbulent contributions. Webb et al. (1980) found that the mean–mean contribution cannot always be neglected, because even over a horizontally homogeneous surface a vertical heat flux causes a nonzero mean vertical velocity. This is a consequence of the constraint that the total vertical flux of air density $\rho_a = \bar{\rho}_a + \rho'_a$ must vanish over a level surface:

$$U_3\bar{\rho}_a + \overline{u_3\rho'_a} = 0 \qquad (12.15)$$

Thus, the mean vertical velocity is

$$U_3 = -\frac{\overline{u_3\rho'_a}}{\bar{\rho}_a} \qquad (12.16)$$

For dry air $\rho'_a/\bar{\rho}_a \simeq -T'/\bar{T}$, where $\bar{T}$ and T' are mean and fluctuating air temperature; at the low Mach numbers of lower atmospheric motions we can neglect pressure fluctuations here (Lumley and Panofsky, 1964). Thus, the mean vertical velocity is, in more familiar terms,

$$U_3 \simeq \frac{1}{T}\overline{u_3T'} \qquad (12.17)$$

which is positive in unstable conditions. This vertical velocity is quite small; for example, if the temperature flux is 0.3 m s^{-1} K, about as large as is ever observed, and $\bar{T} \doteq 300$ K, then $U_3 = 10^{-3}$ m s^{-1}. The total vertical flux of scalar constituent density in dry air is, using Equation (12.16) for the mean–mean part,

$$\text{total vertical flux} = U_3 C + \overline{u_3 c} = -\frac{\overline{u_3 \rho'_a}}{\bar{\rho}_a} C + \overline{u_3 c} \qquad (12.18)$$

An interesting sidelight is that the mean-velocity contribution to the total scalar flux vanishes if we deal with scalar mixing ratio $\mathcal{C}/\rho_a$ rather than density $\mathcal{C}$. To first order in fluctuating quantities,

$$m'_c = \left(\frac{\mathcal{C}}{\rho_a}\right)' = \frac{c}{\bar{\rho}_a} - \frac{C}{\bar{\rho}_a}\frac{\rho'_a}{\bar{\rho}_a} \qquad (12.19)$$

It follows from Equation (12.19) that the product of ρ_a and turbulent vertical flux of scalar mixing ratio is

$$\rho_a \overline{u_3 m'_c} = -\frac{\overline{u_3 \rho'_a}}{\bar{\rho}_a} C + \overline{u_3 c} \qquad (12.20)$$

which from Equation (12.18) is the total flux of scalar constituent density. Thus, should a scalar constituent sensor measure mixing ratio rather than density (which is usually not the case) then its turbulent flux needs no Webb correction.

In general, then, we need to use the Webb correction, and Equation (12.18) must therefore be generalized to moist conditions. Let us quickly review the full derivation of Webb et al. (1980). Consider the atmosphere to be a mixture of air and water vapor. Each satisfies the gas law

$$p_a = \frac{R}{m_a}\rho_a T$$
$$p_v = \frac{R}{m_v}\rho_v T \qquad (12.21)$$

where p is pressure, R is the gas constant, m is molecular weight, and subscripts a and v refer to air and vapor, respectively. From Dalton's law of partial pressures we have

$$\frac{\rho_a}{m_a} + \frac{\rho_v}{m_v} = \frac{p}{RT} \qquad (12.22)$$

Assuming that pressure remains constant, to first order in T' the mean and fluctuating parts of Equation (12.22) can be written as

$$\frac{\bar{\rho}_a}{m_a} + \frac{\bar{\rho}_v}{m_v} = \frac{p}{R\bar{T}} \qquad (12.23)$$
$$\frac{\rho'_a}{m_a} + \frac{\rho'_v}{m_v} = -\frac{p}{R\bar{T}}\frac{T'}{\bar{T}}$$

Solving for ρ'_a gives

$$\rho'_a = -\frac{m_a}{m_v}\rho'_v - \bar{\rho}_a\left(1 + \frac{\bar{\rho}_v m_a}{\bar{\rho}_a m_v}\right)\frac{T'}{\bar{T}} \qquad (12.24)$$

Combining Equations 12.16 and 12.23 then gives

$$U_3 = \frac{m_a}{m_v}\frac{\overline{u_3\rho'_v}}{\bar{\rho}_a} + \left(1 + \frac{\bar{\rho}_v m_a}{\bar{\rho}_a m_v}\right)\frac{\overline{u_3 T'}}{\bar{T}} \qquad (12.25)$$

Thus, the total vertical flux of a trace constituent density is

$$\text{total flux} = U_3 C + \overline{u_3 c} = \frac{m_a}{m_v}\frac{\overline{u_3\rho'_v}}{\bar{\rho}_a} C + \left(1 + \frac{\bar{\rho}_v}{\bar{\rho}_a}\frac{m_a}{m_v}\right)\frac{\overline{u_3 T'}}{\bar{T}} C + \overline{u_3 c} \qquad (12.26)$$

For constituents whose mean value C is large compared to the fluctuation $(\overline{c^2})^{1/2}$, the first two terms on the far right side of Equation (12.26) [which had been traditionally neglected before the Webb et al. (1980) analysis] can be significant. This is discussed also by Businger (1986).

Let us rewrite (12.26) for the particular case of water vapor density flux:

$$\text{total flux} = \left(1 + \frac{29\bar{\rho}_v}{18\bar{\rho}_a}\right)\left(1 + \frac{\bar{\rho}_v \overline{u_3 T'}}{\bar{T}\,\overline{u_3\rho'_v}}\right)\overline{u_3\rho'_v} \qquad (12.27)$$

Since $\bar{\rho}_v/\bar{\rho}_a$ typically does not exceed a few percent, the first factor on the right side of Equation (12.27) ranges from 1.0 to about 1.05. The second factor can be larger than that under conditions of large temperature flux and small water vapor flux. In fact, Equation (12.27) shows that when the turbulent water vapor density flux is identically zero, the true vertical flux (due entirely to the mean advection) is $\bar{\rho}_v\overline{u_3 T'}/\bar{T}$.

2.3 Boundary-Layer Turbulence

The wind velocity in the boundary layer is a random, three-dimensional, time-dependent vector field. Its evolution in space and time is formally governed by the Navier–Stokes and continuity equations, which express momentum and mass conservation, respectively. However, a given realization of any turbulent velocity field is unpredictable in detail because of the complex physics embodied in these equations and their sensitivity to initial and boundary conditions.

In spite of these formal difficulties with turbulence, we do have the benefit of the "folk wisdom" gained from several generations of experience in trying to deal with it experimentally, theoretically, and numerically. Thus, although we still have no rational scheme for solving the turbulence equations analytically, we do have an effective physical understanding of turbulence; we can deal with it quite successfully in many applications, including those in the boundary layer.

Let us discuss several concepts that are useful when dealing with PBL turbulence. We call turbulence *stationary* if its statistical properties are independent of time. In everyday terms, this means that any piece of a time record of a flow property looks like any other piece; there are detail differences, of course, but the statistics are the same. Thus stationarity implies statistical invariance with respect to translation of the time axis. In the PBL this is often the case near mid-day in fair weather.

If the field is statistically invariant to translation of the spatial axes we call it *homogeneous*. Although PBL turbulence is never homogeneous in the vertical (it is strongly affected by the lower surface and the capping inversion) it can be horizontally homogeneous, to a good approximation.

An *isotropic* field is statistically independent of translation, rotation, and reflection of the spatial axes. This is clearly not the case in the PBL, since the upper and lower boundary conditions and the effects of buoyancy make properties different in the horizontal and vertical. However, experimental data and theoretical arguments suggest that the small-scale (meters and smaller) structure of the PBL is effectively isotropic. This is called *local isotropy*, meaning isotropy confined, or localized, to the smallest-scale structure.

If we had a detailed air velocity record from a fast moving aircraft in the PBL we could interpret it as a spatial record, since the velocity fluctuations are small compared to the flight speed. By forming the correlation function as a function of separation along the flight path we could then determine the integral scale. This is a measure of the distance over which the signal is correlated; we interpret this length as the scale (in the flight direction) of the dominant eddies. This application of *Taylor's hypothesis* is nearly universal, even with time series from *in situ* sensors.

The eddies carrying the bulk of the turbulent kinetic energy and doing most of the turbulent transport are (in order of magnitude) integral-scale sized. In principle there is an integral scale for every flow property; in general it depends on distance from the surface, stability, and the direction of the measurement path (lateral or streamwise). For example, some integral scales (e.g., that for vertical velocity) scale with height, x_3, in the surface layer and with PBL depth in the middle and upper regions. Others (e.g., that for horizontal velocity under convective conditions) scale with PBL depth, even in the surface layer. Stability can strongly restrict integral scales, particularly in the nocturnal PBL. We will deal with these variations later, and for now will simply denote the integral scale by l.

Another important scale is λ, the Taylor microscale. It is defined by the turbulent dissipation rate per unit mass (ε), the turbulent kinetic energy per unit mass (q^2), and the molecular kinematic viscosity $\nu : \lambda = (\nu q^2/\varepsilon)^{1/2}$.

The scale λ does not mark the size of the dissipative eddies; rather, we can think of it as the eddy size where viscous effects begin to become significant. As such, it roughly marks the small-scale end of the inertial subrange, that broad range of scales between the anisotropic, energy-containing eddies and the isotropic, dissipative eddies that convert kinetic energy into internal energy through their viscous friction.

A third length scale, the Kolmogoroff scale η, does indicate the size of the viscous eddies. A turbulent fluid, having a kinematic viscosity ν and needing to dissipate kinetic energy at a rate ε, establishes a dissipative range of eddies scaling in size with $\eta = (\nu^3/\varepsilon)^{1/4}$.

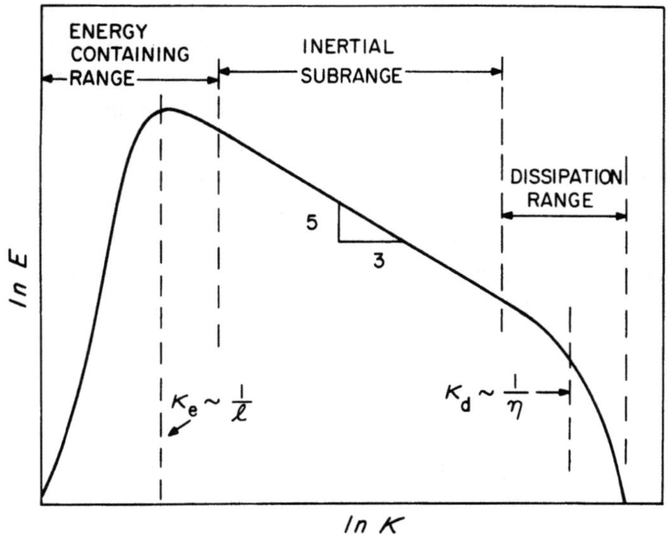

FIGURE 12.3. A schematic of the spectral energy density (spectrum) of turbulence as a function of wave number magnitude κ. A wave number (radians/length) characteristic of the energy-containing range is $\kappa_e \sim 1/\ell$, where ℓ is the integral scale; a dissipative-range wave number is $\kappa_d \sim 1/\eta$, where η is the Kolmogoroff microscale. Reprinted with permission from Wyngaard, 1986, American Meteorological Society.

We can now quickly sketch the distribution of turbulent kinetic energy over wave number magnitude, κ. The spectral density E (kinetic energy/wave number) has a peak at wave numbers on the order of $1/\ell$. At larger κ (smaller scales), Kolmogoroff argued that E depends only on ε and k; on dimensional grounds this gives the well-known inertial range form $E \sim \varepsilon^{2/3}\kappa^{-5/3}$. This extends to wave numbers roughly on the order of $1/\lambda$, where viscous effects begin to be felt. There E begins to fall more sharply, and it cuts off very steeply near the viscous wave number $1/\eta$. Figure 12.3 shows a schematic of the turbulence spectrum.

In a convective PBL, z_i and ℓ can range from several hundred meters to a few kilometers. If the surface temperature flux H_0 is 0.2 m K/s, typical over land in fair weather, then the mid-PBL buoyant production rate of turbulent kinetic energy, which is on the order of $gH_0/(2T)$, is about 3×10^{-3} m^2/s^3. If $q^2 = 2$ m^2/s^2, and using $\nu = 1.5 \times 10^{-5}$ m^2/s for air, we find that $\lambda = (\nu q^2/\varepsilon)^{1/2} \sim 0.1$ m. Very roughly, the spectral region between $1/\ell$ and $1/\lambda$—three decades, if $\ell = 100$ m and $\lambda = 0.1$ m—contains the inertial range. The dissipative eddy size is $\eta = (\nu^3/\varepsilon)^{1/4} = 0.001$ m, so the eddy size range covers roughly five decades! An artist's conception of the convective boundary layer is sketched in Figure 12.4.

The nocturnal PBL over land is usually stably stratified as a result of surface radiative cooling. Its turbulence is driven by shear production and suppressed by buoyancy, with the result that it is weaker, has smaller energy-containing eddies, and carries smaller fluxes than its convective counterpart. The stably stratified boundary layer limits its own depth by its stable stratification; this depth is typically not more than a few hundred meters. Figure 12.5 is an artist's conception.

The PBL is important in part because the turbulent fluxes it carries are an essential link between the earth's surface and the free atmosphere. We are specifically interested here in the turbulent flux of water vapor—the covariance between u_3 and q. The cospectrum of two variables represents the density of contributions to the covariance as a function of spatial wave number $\boldsymbol{\kappa}$. (As we discuss in Section 3, the three dimensionality of turbulence in physical space implies that it is also three dimensional in wave number space.) Thus, we can write the scalar flux of any constituent c schematically as

$$\overline{u_3 c} = \int Co_{u_3 c}(\kappa)\, d\boldsymbol{\kappa} \qquad (12.28)$$

In the inertial subrange the flux cospectrum is observed to fall as $\kappa^{-7/3}$, where $\kappa = |\boldsymbol{\kappa}|$, slightly faster than the $\kappa^{-5/3}$ decay of the energy spectrum. This means that, broadly speaking, the eddies containing most of the turbulent kinetic energy also carry most of the flux. Since an isotropic field has zero scalar flux, this more rapid decay of the flux cospectrum is consistent with the notion of an

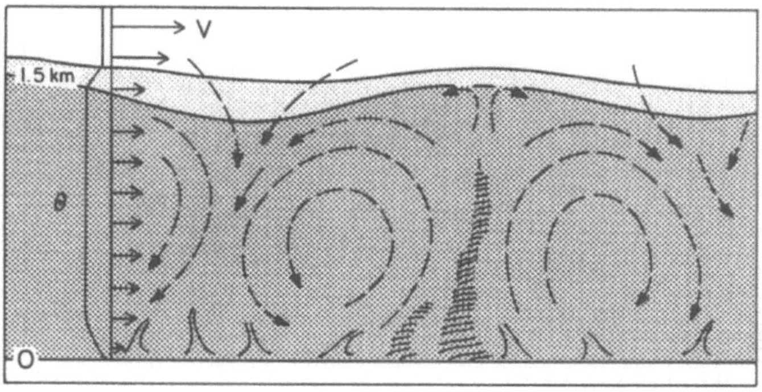

FIGURE 12.4. An artist's conception of the convective boundary layer, showing its large eddies, capping inversion, and well-mixed mean profiles of wind and potential temperature. Reprinted with permission from Wyngaard, 1986, American Meteorological Society.

approach to local isotropy of turbulence structure at larger wave numbers.

2.4 Flux Measurements–Two Approaches

The traditional (i.e., analog) way to calculate the scalar flux $\overline{u_3c}$ from time series $u_3(t)$ and $c(t)$ is to multiply the time signals and average. This is equivalent to integrating the frequency cospectrum:

$$\overline{u_3c} = \int_0^\infty Co(\omega)\,d\omega \qquad (12.29)$$

Thus, accurate determination of $\overline{u_3c}$ by analog techniques requires that the $u_3(t)$ and $c(t)$ signals have amplitude and phase fidelity into the inertial range.

Today we are more likely to use digital signal processing. Although the underlying analog signals must still be reliable into the inertial range, there is no restriction on the digital sampling rate except that it be effectively simultaneous for the two signals. In effect, with digital sampling we determine the joint probability density $\beta(u_3,c)$ and find the flux from the property

$$\overline{u_3c} = \int_{-\infty}^{\infty} \int_0^{\infty} \beta(u'_3,c')u'_3c'\,du'_3dc' \qquad (12.30)$$

With digital signal processing the flux calculation is most efficient when the sampling rate is slow enough that successive samples are statistically independent (i.e., uncorrelated), for then each sample has maximum "weight." Put another way, if the sampling rate is so fast that adjacent samples are highly correlated then there is a high fraction of "redundant" samples. Researchers often want not only the flux but also its cospectrum, in which case achieving the required frequency bandwidth mandates a sampling rate higher than required for the flux alone. One can think of the samples that are redundant for the flux value as serving to define its frequency distribution.

A variation on this second approach is the "eddy accumulation" technique. Equation (12.30) can be rewritten as

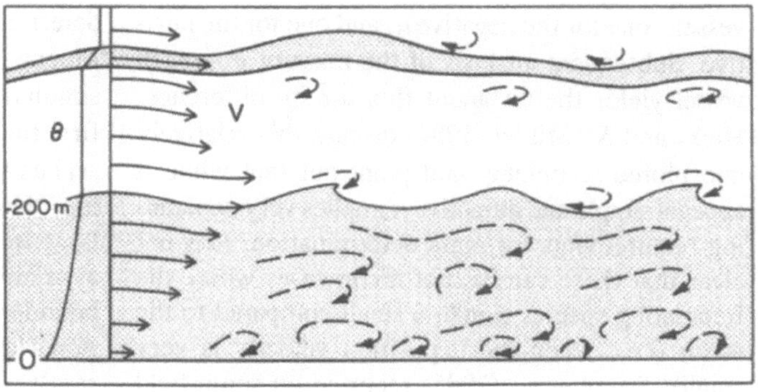

FIGURE 12.5. An artist's conception of the stably stratified boundary layer, showing its small eddies, shallow depth, overlying jet, and strong gradients of mean wind and temperature. Reprinted with permission from Wyngaard, 1986, American Meteorological Society.

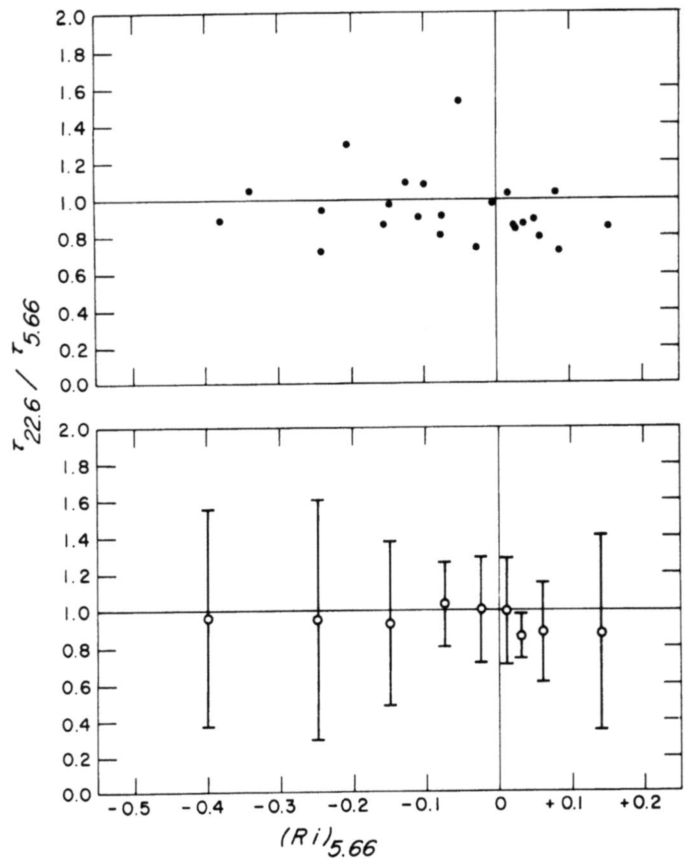

FIGURE 12.6. Ratios of shearing stress measured at 22.6 and 5.7 m in the 1968 Kansas experiments. Upper panel, 15-min data in 1-hr blocks. Lower panel, data grouped into Richardson number intervals, with circle indicating mean and bars plus and minus one standard deviation. (From Haugen et al., 1971.)

$$\overline{u_3c} = \int_0^\infty \int_0^\infty \beta(u'_3, c') u'_3 c' \, du'_3 dc' - \int_{-\infty}^0 \int_0^\infty \beta(u'_3, c') |u'_3| c' \, du'_3 dc' \tag{12.31}$$

which is simply the difference of the fluxes due to upward and downward motions. In principle one can measure each of these separately by sampling air at a flow rate proportional to u_3 and storing it in two vessels, one for the negative u_3 and one for the positive. Subsequent analysis of the mass of c in each vessel yields the turbulent flux $\overline{u_3c}$ by difference. Hicks and McMillen (1984) discuss this relatively unexplored technique, and point out that while it appears simple and attractive it implies very demanding requirements for mass determination. It is not clear that these can be met for in cases where the fluctuating species signal is small compared to the mean (i.e., when the deposition velocity is very small). Speer et al. (1985) reported on some field trials of a prototype device, but their results suggest that substantial development work will be needed to produce a practical eddy-accumulation instrument.

More recently, Businger and Oncley (1990) proposed a "relaxed" eddy-accumulation technique in which the air-flow rate would be constant, not proportional to u_3. The final mass difference would be multiplied by a constant b times $(\overline{w^2})^{1/2}$ to obtain the flux. Through numerical experiments with actual time series from the surface layer they determined that $b \sim 0.6$, independent of stability. Moeng (personal communication) used large-eddy simulation of the convective boundary layer to confirm that $b \sim 0.6$ in bottom-up diffusion (i.e., for scalars with a surface flux but zero flux at the PBL top). However, she found that b was about 3/4 that large in the top-down case, a manifestation of the asymmetry of scalar transport in the convective boundary layer. In this variant of eddy accumulation a key limitation still remains—the necessity to resolve small mass differences.

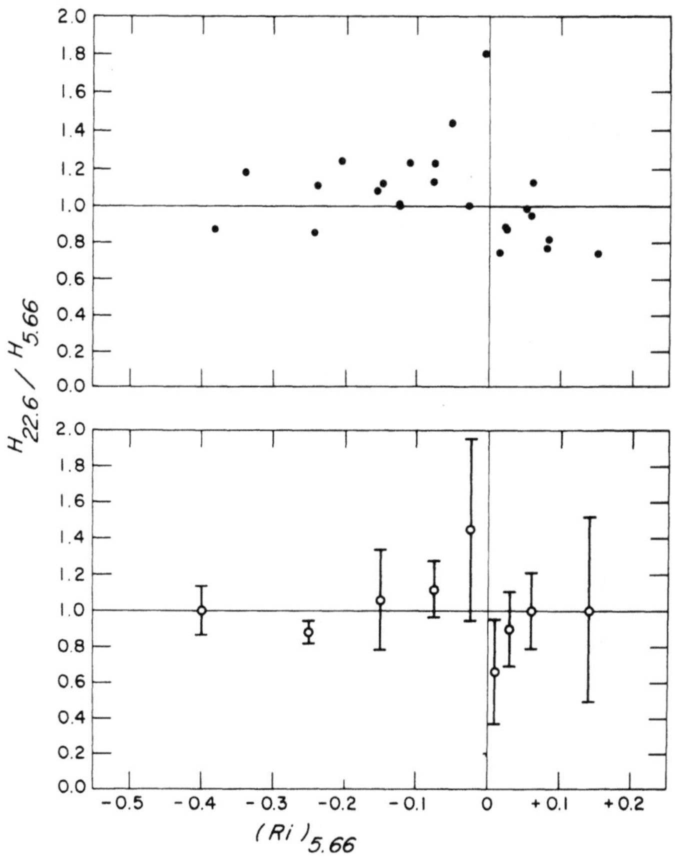

FIGURE 12.7. As in Figure 12.6 but for heat flux.

In practice there are a number of pitfalls in measuring turbulent scalar fluxes via eddy-correlation techniques, including spatial and temporal resolution requirements and flow-distortion effects. I survey these in Section 3.

2.5 Averaging Times

Turbulence theory is formally posed in terms of the ensemble average—the average over an infinite number of realizations of the "same" experiment. In the atmosphere we invariably must use time averages, however. In a statistically stationary field the two averages converge in the limit, but that is little solace to the experimenter who does not have the luxury of generating arbitrarily long time series; in practice, he must deal with time averages over short records. Even in "perfect" conditions, therefore, the experimenter can only *estimate* the ensemble average through a time average that carries statistical uncertainty due to the finite averaging time. Unfortunately, laboratory turbulence experience does not always prepare one adequately for the challenge of measurements outdoors.

In the summer of 1968 a group from the Air Force Cambridge Research Laboratories, of which I was a new member, conducted a surface-layer field program at a beautifully flat, uniform site in southwest Kansas. Our objective was to study the "constant-flux" layer immediately above the surface.

The first results from this experiment, summarized in Figures 12.6 and 12.7, seemed to show that the fluxes were far from constant in the first 20 m for the usual averaging periods—say 15 min. Even over an hour, the fluxes were constant only within 20%, and over the 24 hr of carefully chosen, research-quality runs the fluxes were constant only to within 5–10%.

The root problem here, of course, is that our theories deal with ensemble-average properties, while we most often measure time averages. The prediction of a constant-stress layer, for example, comes from analysis of the ensemble-mean horizontal momentum equation; that analysis shows that a quasi-steady, horizontally homogeneous flow cannot support a significant surface-layer stress divergence in the ensemble mean.

These results from Kansas and other experiments focused attention on the statistical scatter in time-averaged measurements. A convenient framework for interpreting this scatter had been provided in the Lumley-Panofsky (1964) monograph, where it was shown that the averaging time T required to determine a time mean $\bar{f}$ to an accuracy a compared to its expected (ensemble average) value is

$$T = 2\tau \frac{\overline{f'^2}}{a^2\bar{f}^2} \tag{12.32}$$

where τ is the integral scale of $f(t)$ and $\overline{f'^2}$ is the ensemble variance of f about its ensemble mean.

I used this equation (Wyngaard, 1973) to explore required averaging times for various statistics in the surface layer. In general, the required averaging time (for a given accuracy) increases with the order of the moment, so that fluxes, for example, require longer averaging times than mean winds. Of the second moments, momentum fluxes require the longest averaging times, variances the least, with scalar fluxes intermediate. The Kansas results in Figures 12.6 and 12.7 are consistent with this.

In practice, mean profiles averaged over 15 min are smooth enough for many analysis purposes; variance profiles are somewhat rougher. Fluxes require longer averaging times, 1 hr being often used. One-hour momentum fluxes can have large scatter, however, and the usual practice is to do some ensemble averaging by compositing a number of runs. Third moments, such as those whose vertical gradients enter the second-moment conservation equations, can require even longer averaging times. Here averages over an entire experiment—over several tens of hours of runs—are usually not as smooth as 15-min profiles of means or variances. Some of this scatter is due also to the differing conditions in the various members of the ensemble, undoubtedly.

2.6 Direct versus Indirect Techniques

Direct measurements of turbulent fluxes are now fairly routine, but not long ago they were a central research goal in micrometeorology. The turnaround has come through improvements in sensor and data-handling technology and also better understanding of the physics of flux measurement. The paper by Högström (1988) discusses a recent surface-layer experiment and gives some insight into the contemporary approach to flux measurements.

However, the last 20 years have taught us that while flux measurements are possible, they can also be notoriously difficult. For example, at the conclusion of the 1968 Kansas experiments the investigators mounted their three 3-component sonic anemometers less than 1 m apart at the same height (5.6 m) to compare their flux outputs. Although the means and standard deviations of the wind components agreed well, the covariance $\overline{u_1u_3}$ did not; the 1-hr averages were 1.50, 0.93, and 0.57 times the overall mean. They traced the problem to small (of the order of 1 degree) errors in sonic anemometer alignment (Kaimal and Haugen, 1969). Subsequent work has shown that flow distortion due to the sensor array itself can cause comparably large errors in flux measurements, as I discuss in detail in Section 3. Thus, it is very difficult to make *accurate* measurements of turbulent fluxes.

It is also difficult to make *representative* measurements of turbulent fluxes. In an experiment comparing 30-min measurements of temperature fluxes from two instruments separated horizontally by 5 m at 4 m height, Businger et al. (1967) found that the two fluxes agreed only within a factor of 2. Averaged over all runs, the agreement was much better, leading them to conclude that "the Eulerian point average does not provide an adequate statistical sample of the heat flux."

Similar results were found in the 1968 Kansas experiments. Figure 12.7 shows the ratio of 1-hr temperature fluxes at 22.6 and 5.66 m. The alignment errors mentioned earlier were accounted for, so the ratio averaged over the experiment is indeed close to 1.0 as expected. However, there is a good deal of scatter within a given hour due to the statistical variability of time averages, as previously discussed. This means that direct measurements of flux, however accurate, may not be representative.

For these reasons—the expense, difficulty, and nonrepresentativeness of direct flux measurements—indirect techniques for inferring fluxes in the surface layer are of great interest in both research and applications.

The framework of relationships that makes indirect techniques possible in the surface layer is Monin–Obukhov (M–O) similarity (Haugen, 1973; Panofsky and Dutton, 1984). This relates most mean and turbulent properties through length, temperature, constituent, and velocity scaling parameters plus a stability index. We can briefly illustrate. The length scale is z (respecting convention, we now change notation from x_3 to z); the velocity scale is $u_* = (\tau_0/\rho)^{1/2}$, where τ_0 is surface stress; the temperature scale is $-H_0/u_*$, where H_0 is surface temperature flux; the constituent scale is C_0/u_* where C_0 is the surface flux of constituent; and the stability index is z/L, where L is the M–O length, $u^3_*T/(kgH_0)$. Since in general there is water vapor present, it is traditional (see, for example, Lumley and Panofsky, 1964) to consider the mixture of air and water vapor as a perfect gas having the same gas constant as dry air but a "virtual" temperature $T_v = T(1 + 0.61q)$, where q is specific humidity (the fraction of water vapor in the air). Thus, we should interpret temperature in the M–O hypothesis as virtual temperature, temperature flux as virtual temperature flux, and so forth. The similarity hypothesis is that nondimensionalized properties are universal (that is, the same in all surface layers) functions of z/L. For example, the mean wind and temperature gradients are

$$\frac{kz}{u_*}\frac{\partial U_1}{\partial z} = \phi_m\left(\frac{z}{L}\right) \tag{12.33}$$

$$\frac{kz}{T_*}\frac{\partial \Theta}{\partial z} = \phi_h\left(\frac{z}{L}\right) \tag{12.34}$$

where $k \sim 0.4$, the von Karman constant, is traditionally used for scaling purposes. Similarly, there are universal functions for dimensionless variances, structure functions, and other statistics.

In general the M–O functions are found by experiment. For some of the reasons I will point out in later sections, such experiments are difficult; many obstacles must be overcome to determine the universal functions accurately. Probably the best known of the M–O functions are ϕ_m and ϕ_h, although there is continuing controversy over the details of their shapes. It is generally assumed that ϕ_h also represents the mean gradient function of any conservative scalar in the surface layer. There is some evidence that the gradient function for water vapor is different, but this evidence is inconclusive and even contradictory (e.g., Verma et al., 1978, and Lang et al., 1983, have opposite findings).

Since ϕ_m and ϕ_h are known empirically, measurements of $\partial U_1/\partial z$ and $\partial\Theta/\partial z$ at height z can be used through Equations (12.33) and (12.34) and the definition of L to give the surface fluxes τ_0 and H_0. As we saw earlier, the flux of water vapor contributes to the surface virtual temperature flux H_0; this contribution can be quite important over strongly evaporating surfaces. In fact, the flow over such surfaces can have a downward heat flux and, thus, appear stably stratified, while due to a strong moisture flux the virtual temperature flux can be positive and the layer actually unstably stratified. Thus, in general, one measures the profiles of both temperature and water vapor, converts to virtual temperature, solves for the surface virtual temperature flux, and separates it into moisture and temperature fluxes.

The Monin–Obukhov similarity hypothesis works very well for some quantities (e.g., vertical wind fluctuations), but poorly for others (e.g., horizontal wind fluctuations). Figure 12.8 illustrates this. Its failure for horizontal wind fluctuations is due to the large convective eddies that have velocities of order w_* and cause horizontal wind fluctuations in the surface layer.

There are also some subtler deficiencies of M–O similarity. Hill (1989) proved that M–O similarity implies that the correlation coefficient between any two conservative scalars—e.g., water vapor mixing ratio and potential temperature—is 1.0 in magnitude. Although there is theoretical and experimental evidence for strong correlation between two scalars (Wyngaard et al., 1978), the correlation coefficient in the surface layer, while typically large, is often distinctly less than 1.0. If two conservative scalars diffusing through the surface layer had precisely the same surface physics (i.e., transfer mechanisms, boundary conditions), it is reasonable to expect that they would indeed be perfectly correlated. The troublesome point seems to be the physics of the surface transfer. This differs for water vapor and temperature; they do not have the same thermodynamic roles in the surface energy balance, and their detailed transfer mechanisms need not be the same. This suggests that to cover scalars adequately, the M–O hypothesis needs to be extended to include some consideration of surface-transfer physics.

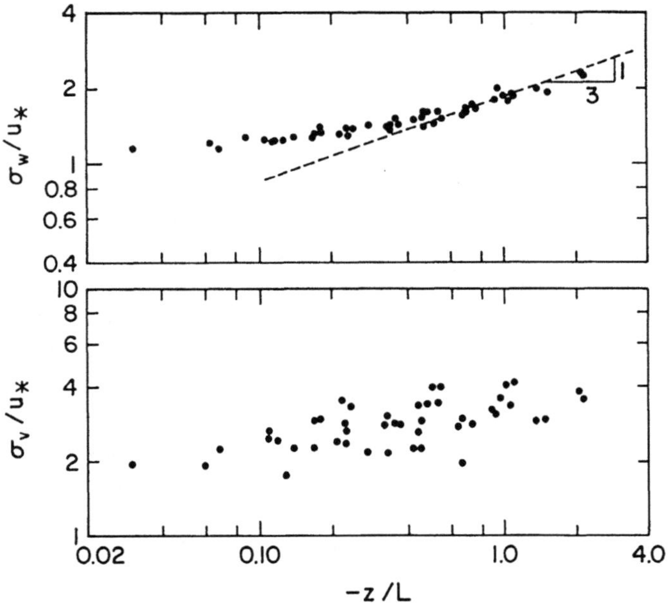

FIGURE 12.8. The standard deviations of vertical (upper panel) and lateral (lower panel) wind fluctuations at 5.7, 11.3, and 22.6 m height in the 1968 Kansas experiments. The scatter in the σ_v plot is due to the failure of M–O similarity, but σ_w follows it quite well, even showing indications of the $(-z/L)^{1/3}$ regime of local free convection. Reprinted with permission from Wyngaard, 1988c, *Lectures on Air Pollution Modeling*, American Meteorological Society.

In routine applications an even simpler method is used to infer evaporation (Businger, 1986). The surface energy balance can be used to deduce the total surface heat flux, the sum of the "sensible" and "latent" fluxes. The sensible heat flux is $\rho_a C_p \overline{u_3\theta}$; the latent heat flux is $\rho_a C_p L_v \overline{u_3 \rho'_v}$. Here C_p is the specific heat of air at constant pressure and L_v is the latent heat of evaporation. If, in addition, the mean gradients of temperature and water vapor are measured at the same height, then (by M–O similarity and the assumed equivalence of ϕ_h and that for water vapor) their ratio is also the ratio of their surface fluxes. These two conditions then allow both the surface moisture and temperature fluxes to be inferred. This is called the "Bowen ratio" technique because the ratio of the sensible and latent heat fluxes is the Bowen ratio.

Many practical applications deal with fluxes to vegetation canopies. Here it is necessary to replace distance from the surface, z, with $z - d$, where d is the "zero-plane displacement." This is not a well-defined parameter, and indeed it is not clear whether d is different for heat, momentum, and transported scalars.

Wesely (1988) discusses the use of variance techniques to infer surface fluxes. Coulter and Wesely (1980) present a method for determining the surface temperature flux from acoustic backscatter and from laser scintillation, and show that it can give good agreement with direct measurements. Wyngaard and Clifford (1978) suggested obtaining momentum and moisture flues as well as the temperature flux from structure parameters of velocity, humidity, and temperature inferred from scintillations. Such techniques have recently shown promise (e.g., Hill et al., 1988). Scintillation measurements can also be used to infer stability (Andreas, 1988) and the Kolmogorov microscale (Hill, 1988).

In this way one can use simple, statistically stable measurements to infer the surface fluxes, which can be difficult to measure directly. The underlying M–O similarity relationships are now fairly well established for stable, neutral, and unstable conditions (Haugen, 1973; Panofsky and Dutton, 1984) although as just mentioned we now know that some parameters (notably statistics of the horizontal wind components) do not follow M–O similarity under unstable conditions (Panofsky et al., 1977).

There have been some innovative indirect techniques proposed. For example, Tillman (1972) examined the hypothesis that the skewness S of temperature fluctuations [defined as $S = \overline{\theta^3}/(\overline{\theta^2})^{3/2}$] is a unique function of z/L in the unstable surface layer. He showed data from several sites that support this. He then demonstrated a technique whereby S and the temperature variance are used to infer the surface heat and momentum fluxes.

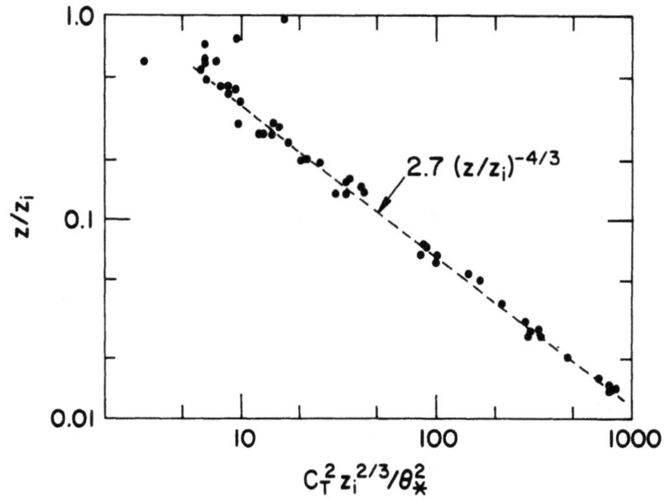

FIGURE 12.9. The temperature structure parameter C_T^2 from the 1973 Minnesota experiments closely follows mixed-layer scaling. Reprinted with permission from Wyngaard and LeMone, 1980, *Journal of Atmospheric Sciences*, American Meteorological Society.

There is also a fairly well-established similarity structure for the middle and upper portions of the convective PBL (Caughey, 1982), and therefore indirect measurement techniques could be feasible there as well. For example, the turbulent velocity scale is $w_* = (gH_0z_i/T)^{-1/3}$ and in mid-PBL the vertical velocity variance is about $0.4w_*^2$. Thus, in principle a Doppler acoustic sounder could measure both z_i and $\overline{w^2}$, and one could infer H_0. As another example, above the surface layer Figure 12.9 shows that the temperature structure parameter behaves as

$$\begin{aligned} C_T^2 &= 2.7H_0^{4/3}\,(g/T)^{-2/3}z^{-4/3} \\ &= 2.7\theta_*^2 z_i^{-2/3}(z/z_i)^{-4/3} \end{aligned} \qquad (12.35)$$

(Wyngaard and LeMone, 1980), and acoustic sounder measurements of C_T^2 could be used to give H_0; this is one of the techniques explored by Coulter and Wesely (1980).

In summary, indirect measurements of PBL parameters can be more representative than direct measurements, because the latter can be very sensitive to local conditions, particularly in the surface layer. On the other hand, the results can only be as accurate as the underlying similarity relation, and generally these are not known to better than 20%.

A particularly challenging task is to develop indirect techniques for obtaining PBL parameters from routine meteorological data. This is frequently required in air quality modeling, for example. Van Ulden and Holtslag (1985) present methods for obtaining H_0, u_*, U_1, and z_i under convective conditions from routine measurements, and find good agreement between their indirect values and direct measurements in field experiments.

3. An Introduction to Measurement Physics

In this section I expand my earlier discussion of some of the salient features of surface-layer turbulence. My focus here will be on basic aspects of its measurement, a topic I call "measurement physics." I hope to show that by dealing with the measurement process on a mathematical level we can obtain better data and better interpretations of them. Unfortunately, in my limited space here it is not possible to give an adequate introduction to the important turbulence concepts and also cover the measurement physics. I must limit the discussion on turbulence, therefore. Good general references on boundary-layer turbulence are Lumley and Panofsky (1964), Tennekes and Lumley (1972), Haugen (1973), Panofsky and Dutton (1984), and Lenschow (1986).

3.1 Spatial and Temporal Resolution

Turbulence near the earth's surface has a tremendous range of spatial scales—from millimeters to kilometers. Fortunately, we seldom need information spanning this entire range; instead, we can usually use sensors and data acquisition techniques that attenuate, or filter, some spatial scales and temporal frequencies.

To illustrate the effects of this filtering, let us consider first a one-dimensional problem. Take the simplest case, where the measurement process can be characterized as a first-order, linear system obeying

$$\tau \frac{de_o}{dt} + e_o = e_i \qquad (12.36)$$

where $e_o(t)$ and $e_i(t)$ are output and input signals (that is, the "measured" and "true" variables), and τ is the time constant. This is the familiar first-order, low pass RC filter; it is a good representation of the effects of thermal lag in many temperature sensors and inertia in many windspeed sensors. As in linear circuit theory, we can represent the signals by a combination of harmonics of frequency ω:

$$\begin{bmatrix} e_i(t) \\ e_o(t) \end{bmatrix} = \int_{-\infty}^{\infty} e^{i\omega t} \begin{bmatrix} E_i(\omega) \\ E_o(\omega) \end{bmatrix} d\omega \qquad (12.37)$$

Strictly speaking, we should use stochastic integrals here (Lumley and Panofsky, 1964). Substituting Equation (12.37) into Equation (12.36) gives the relation between input and output amplitudes:

$$\frac{E_o}{E_i} = \frac{1 - i\omega\tau}{1 + (\omega\tau)^2} \qquad (12.38)$$

At very low frequencies ($\omega\tau \ll 1$) Equation (12.38) shows that the input and output signals are the same, while at high frequencies ($\omega\tau \gg 1$) the output is strongly attenuated. The "break frequency" is $1/\tau$. Multiplying Equation (12.38) by its complex conjugate gives the relation between power spectral densities, which we denote by ϕ:

$$\frac{\phi_o}{\phi_i} = \frac{1}{1 + (\omega\tau)^2} \qquad (12.39)$$

We can extend this analysis to the spatial averaging of the three-dimensional turbulence fields we meet in the PBL. Here the Fourier representation is

$$\begin{bmatrix} e_i(\mathbf{x},t) \\ e_o(\mathbf{x},t) \end{bmatrix} = \int_{-\infty}^{\infty} e^{i\boldsymbol{\kappa}\cdot\mathbf{x}} \begin{bmatrix} E_i(\boldsymbol{\kappa},t) \\ E_o(\boldsymbol{\kappa},t) \end{bmatrix} d\boldsymbol{\kappa} \qquad (12.40)$$

The coordinate in three-dimensional physical space is $\mathbf{x}$, that in wave number space $\boldsymbol{\kappa}$. Now assume the sensor does a centered average over a path vector $2\mathbf{L}$—for example, over the wire filament of a temperature sensor, or over the propagation path of a sonic aneometer or Lyman-α humidity sensor. Then the input–output relation is

$$E_o(\boldsymbol{\kappa},t;\mathbf{L}) = \frac{\sin(\boldsymbol{\kappa}\cdot\mathbf{L})}{(\boldsymbol{\kappa}\cdot\mathbf{L})} E_i(\boldsymbol{\kappa},t) \qquad (12.41)$$

The relation between the input and output spectra then follows directly:

$$\phi_o(\mathbf{x},t;\mathbf{L}) = \frac{\sin^2(\boldsymbol{\kappa}\cdot\mathbf{L})}{(\boldsymbol{\kappa}\cdot\mathbf{L})^2} \phi_i(\boldsymbol{\kappa},t) \qquad (12.42)$$

We see that the wave number spectrum begins to be seriously affected by spatial averaging at wave numbers such that $\boldsymbol{\kappa} \cdot \mathbf{L} \simeq 1$.

These relations are not directly applicable to data, however, because even in a statistically steady flow we cannot measure the three-dimensional wave number spectrum $\phi(\boldsymbol{\kappa})$. (We call it three-dimensional because the argument $\boldsymbol{\kappa}$ is a three-dimensional vector.) A simpler, one-dimensional spectrum F is defined as an integral of the three-dimensional spectrum over two wave number components. This is (suppressing the time argument)

$$F(\kappa_1) = \int_{-\infty}^{\infty} \int_{-\infty}^{\infty} \phi(\boldsymbol{\kappa}) d\kappa_2 d\kappa_3 \qquad (12.43)$$

This has a simple interpretation: $F(\kappa_1)$ is the wave number spectrum one would find from a one-dimensional spatial record along the x_1 direction. If we take this direction to be along the mean wind in a flow measured with a fixed sensor, or along the flight path of an aircraft, then we see intuitively that $F(\kappa_1)$ is connected with the one-dimensional frequency spectrum we would measure from the time series. With "Taylor's hypothesis" (next section) we can, in fact, make this connection between frequency and one-dimensional wave number spectra. Thus, researchers routinely use time series measured from aircraft to find the one-dimensional wave number spectrum along the flight path, and time series measured from towers to find the one-dimensional wave number spectrum in the mean wind direction.

Because of the integration in Equation (12.43) over two wave numbers, we generally do not have the simple, multiplicative "transfer functions" for one-dimensional wave number spectra such as we find in Equation (12.39) for frequency spectra. This generally complicates the interpretation of

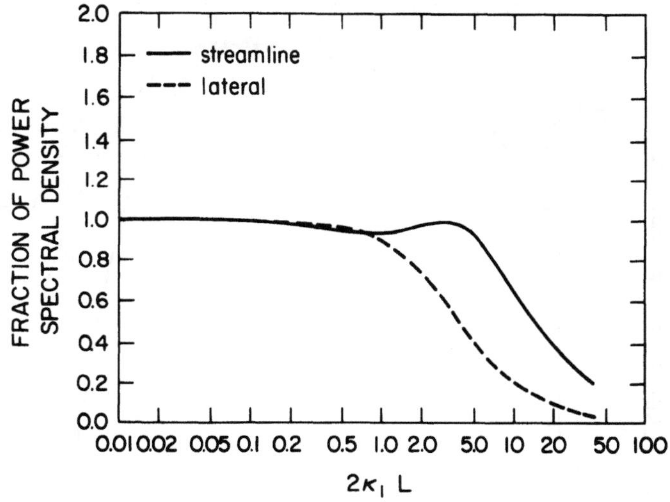

FIGURE 12.10. The one-dimensional spectral response of a typical three-component sonic anemometer. Solid curve, streamwise velocity; dashed curve, lateral velocity. L is the acoustic path length. Reprinted with permission from Kaimal et al., 1968, *Journal of Applied Meteorology*, American Meteorological Society.

spatial-averaging effects. For example, to determine the effect of the finite length of a hot-wire or temperature-sensing filament on the one-dimensional spectrum, we must integrate the product of the filter function in Equation (12.42) and the spectrum over two wave numbers. Generally, this cannot be done analytically. The only exceptions are spatial effects that depend only on the streamwise (one-direction) coordinate and, hence, can be taken outside the integral in Equation (12.43).

Turbulent flux and other covariance measurements involve an additional spatial filtering effect. If the two variables are not measured at the same spatial point but are instead separated by a vector $\mathbf{d}$, then one can see from Equation (12.40) that the spectral transfer function has the additional factor $\exp(i\boldsymbol{\kappa}\cdot\mathbf{d})$, which is complex. The cross spectrum of two variables is in general also complex: the real part (cospectrum) is even in $\boldsymbol{\kappa}$ and integrates to the covariance; the imaginary part (quadrature spectrum) is odd and integrates to zero. Thus, the inevitable spatial separation between the two sensors makes the measured cospectrum a combination of the true cospectrum and the quadrature spectrum; in principle, then, it can induce a covariance when none actually exists. These effects enter at spatial wave numbers $\boldsymbol{\kappa}$ such that $\boldsymbol{\kappa}\cdot\mathbf{d} \sim 1$, so we should be wary of data in that range.

We can calculate spatial-averaging effects reliably when they involve only the smaller scales, which can be considered to be isotropic and, hence, have tractable analytical forms. Wyngaard (1971) has calculated the response of single and multiple arrays of hot and cold wires. Kaimal et al. (1968) and Horst (1973) have calculated the spatial-averaging properties of sonic anemometers, and Andreas (1981) has done so for the Lyman-α hygrometer. Koshiek (1984) has calculated the effect of instrument separation on the cospectrum of two scalars. Figure 12.10 illustrates the calculated one-dimensional spectral response of a sonic anemometer.

We have traditionally relied heavily on our experience and intuition when assessing the effects of spatial averaging on the energy-containing range of turbulence (i.e., the range of substantial contributions to the energy spectrum) including the effects on turbulent fluxes. Our spectral models tend not to be applicable there, so it has been difficult to calculate the effects numerically. However, Kristensen and Fitzjarrald (1984) developed a cospectral model and used it to calculate the attenuation of scalar flux due to sonic path length. Their results indicate that one of the old rules of thumb—that measurement height should be 20 times the path length—is very conservative.

3.2 Taylor's Hypothesis

We have focused on spatial structure (e.g., wavenumber spectra), but most often we measure temporal structure; for example, we mount sensors on towers and observe the temporal variations as the turbulent flow passes by. Even when we use a moving probe, as with aircraft, we measure time series.

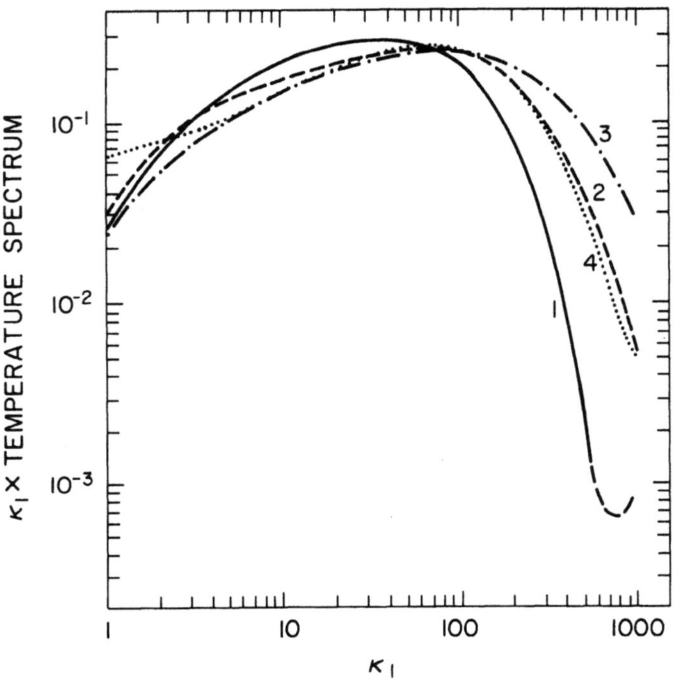

FIGURE 12.11. Temperature spectra, measured by a probe moving horizontally at speed U_0 in laboratory free convection of velocity scale w_*, plotted against horizontal wave number κ_1. These were obtained from frequency spectra through Taylor's hypothesis. Curve 1, $U_0/w_* = 4.27$; curve 2, 0.51; curve 3, 0.27; curve 4, 0. Note the aliasing to larger κ_1 due to the fluctuating convection velocity. (From Deardorff and Willis, 1982.)

Nonetheless, we traditionally interpret these temporal measurements in terms of streamwise structure—that is, as if its structure were "frozen" in space and convected past the probe by the mean flow. We can trace the origin of this interpretation to Taylor (1938), who wrote "If the velocity of the air stream which carries the eddies is very much greater than the turbulent velocity, one may assume that the sequence of changes in u at the fixed point is simply due to the passage of an unchanging pattern of turbulent motion over the point; i.e., one may assume that $u = f(t) = f(x/U)$, where x is measured upstream at time $t = 0$ from the fixed point where u is measured. In the limit when $u/U \to 0$ [this] is certainly true." This is now called "Taylor's hypothesis," or sometimes the "frozen-field hypothesis."

We can interpret Taylor's limit $u/U \to 0$ as follows. An eddy of size l and velocity u has a lifetime on the order of l/u and is convected past the probe in a time l/U. Taylor's hypothesis should be valid if the eddy lifetime is much greater than its convection time, or $l/u \gg l/U$; this is equivalent to $u/U \ll 1$.

Unfortunately, $u/U \ll 1$ is usually clearly satisfied only for aircraft measurements. For tower measurements, Taylor's hypothesis can fail in different wave-number regions for a variety of reasons. I know of no way of correcting for this failure within the energy-containing spectral range, so in practice we live with "wave number" spectra that are actually distorted versions of the real ones.

At small scales there is hope for understanding, however. Lumley (1965) considered the effects of mean wind shear and turbulence level, and found that if certain criteria were satisfied, "the spectrum beyond roughly $30/l$ may be described adequately by isotropic frozen regions convected by a spatially uniform fluctuating velocity having the characteristics of the energy-containing eddies."

Lumley presented an analytical model in which the structure at inertial range and smaller scales is effectively frozen but is convected with a velocity which fluctuates about a mean U. He found that this fluctuating convection velocity causes the high-wave-number spectrum to be overestimated. The complete explanation for this overestimation is complicated, particularly for the velocity field, because one must consider the three-dimensional, tensor geometry of the problem. However, we can explain the essence of the mechanism in very simple terms. Turbulent spectral energy appears at any $\kappa_1 = \omega/U$ from three sources: from eddies of smaller κ_1 convected at a speed less than U, from eddies of wave number κ_1 convected at U, and from

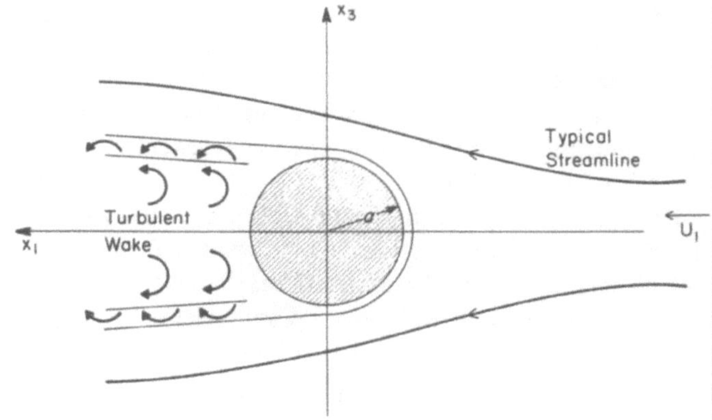

FIGURE 12.12. A schematic of flow near a circular cylinder. The turbulent flow ahead of the cylinder can easily be calculated if the integral scale ℓ in the free stream is much larger than the cylinder radius a. Reprinted with permission from Wyngaard, 1981a, *Journal of Applied Meteorology*, American Meteorological Society.

eddies of larger κ_1 convected faster than U. A crude model of this process is

$$\begin{aligned}\phi^m(\kappa_1) \sim\ & \varepsilon\phi(\kappa_1 - \Delta\kappa_1) && \text{(leakage from smaller } \kappa_1\text{)} \\ & + (1 - 2\varepsilon)\phi(\kappa_1) && \text{(attenuated contribution from } \kappa_1\text{)} \\ & + \varepsilon\phi(\kappa_1 + \Delta\kappa_1) && \text{(leakage from larger } \kappa_1\text{)} \\ \sim\ & \phi(\kappa_1) + \varepsilon(\Delta\kappa_1/\kappa_1)^2\kappa_1^2(d^2\phi/d\kappa_1^2) \end{aligned} \tag{12.44}$$

Here a superscript m denotes measured and ε is the fractional "leakage" from adjacent wave numbers. We expect $\varepsilon(\Delta k_1/k_1)^2$ to scale with u^2/U^2, where u^2 is the variance of the convection velocity. For a power-law spectrum the second derivative term in Equation (12.44) is positive, causing the measured spectrum to be overestimated.

Lumley's full calculations are more complicated because the full problem is three dimensional. Wyngaard and Clifford (1977) have extended them and shown how convection velocity fluctuations influence the applicability of Taylor's hypothesis to both small-scale velocity and scalar statistics.

Deardorff and Willis (1982) confirmed, through measurements of temperature spectra in their convection tank, that a fluctuating convection velocity aliases energy to higher wave numbers. This energy is borrowed from the energy-containing range, and their frequency spectra (interpreted as wave number spectra, Fig. 12.11) show the distortion in that region as well.

3.3 Flow Distortion and Tower-Mounted Sensors

Traditional (i.e., "*in situ*" or "immersion") sensors inevitably distort the flow in their immediate vicinity and thereby also at the measurement point. Perhaps this is most obvious in the case of velocity sensors, but it affects measurements of scalars as well.

Micrometeorologists have for years been sensitive to the influence of masts, booms, towers, and probe bodies on mean wind measurements. The literature contains an abundance of papers concerning "tower shadow effects" on mean winds. Less attention seems to have been given to flow distortion in other situations (on aircraft measurements, for example) or to the effect on turbulence measurements. Part of the reason for this neglect, certainly, is that turbulent flow around obstacles is very complicated and not generally amenable to theoretical analysis. Good progress has been made recently, however. Here I will introduce the problem, make its broad outlines clear, and review some recent work.

3.3.1 Velocity Statistics

Let us consider a simple example of probe-induced flow distortion. Figure 12.12 shows a turbulent flow approaching a probe in the shape of a two-dimensional circular cylinder of radius a. Let the approach flow have a velocity $(U_1 + u_1, u_2, u_3)$ with U_1 the mean. We assume the Reynolds number U_1a/ν is large enough to produce a turbulent wake behind the probe, while upstream of the cylinder, where we measure the velocity, the flow remains smooth but is distorted. We wish to know how this distortion affects the measured turbulent velocity components there. The following analysis is taken from Wyngaard (1981a).

The approach flow has an integral scale ℓ, and an important simplification occurs if $\ell \gg a$, where a is now a characteristic probe scale. If that is true, then as judged by the spatial and temporal scales of the flow near the probe, the approach flow appears spatially uniform and very slowly varying in time.

In the case where $\ell \gg a$ one can write a potential flow solution for the distorted turbulence components near the probe. The distorted flow velocity, the sum of mean ($\tilde{U}_i$) and turbulent ($\tilde{u}_i$) parts, is taken to be a function of the position of measurement x_m, the free-stream speed (which we assume is along the x_1 axis) U_1, and the free-stream turbulent velocity $u_i(t) = [u_1(t), u_2(t), u_3(t)]$:

$$\tilde{U}_i + \tilde{u}_i = f_i[\mathbf{x}_m, U_1 + u_1(t), u_2(t), u_3(t)] \quad (12.45)$$

Writing the distorted velocity field as a Taylor series expansion about nonturbulent flow down the x_1 axis then gives

$$\tilde{U}_i + \tilde{u}_i = \tilde{U}_i(\mathbf{x}_m, U_1, 0, 0) + a_{i1}(\mathbf{x}_m)u_1(t) + a_{i2}(\mathbf{x}_m)u_2(t) + a_{i3}(\mathbf{x}_m)u_3(t) \quad (12.46)$$

where

$$a_{ij}(\mathbf{x}_m) = \left. \frac{\partial \tilde{U}_i(\mathbf{x}_m)}{\partial U_j} \right|_0 \quad (12.47)$$

is a matrix of flow-distortion coefficients. We are able to write the free-stream turbulent velocity in Equations (12.45) and (12.46) as a function of t only, rather than a function of position $\mathbf{x}_0$ in the free stream and t, because we have assumed that the body is small compared to the spatial scale of u_i (i.e., $a \ll \ell$). It follows from subtracting the average of Equation (12.46) from Equation (12.46) that the distorted turbulent velocity field is, to first order,

$$\tilde{u}_i(\mathbf{x}_m,t) = a_{i,j}(\mathbf{x}_m)u_j(t) = a_{i1}(\mathbf{x}_m)u_1(t) + a_{i2}(\mathbf{x}_m)u_2(t) + a_{i3}(\mathbf{x}_m)u_3(t) \quad (12.48)$$

We use the convention that repeated indices are summed. Equation (12.48) says that the distorted turbulence components are linear combinations of the free-stream components. This means, for example, that a streamwise fluctuation $u_1(t)$ in the free stream can induce lateral components $\tilde{u}_2(t)$ and $\tilde{u}_3(t)$ at position $\mathbf{x}_m$ near the probe. As we will see, this can cause serious measurement errors in fluxes.

The distortion coefficients a_{ij}, which depend on the body shape and on position $\mathbf{x}_m$, can be written as

$$a_{ij}(\mathbf{x}_m) = \delta_{ij} + \varepsilon_{ij}(\mathbf{x}_m) \quad (12.49)$$

where $\varepsilon_{ij} = 0$ in the free stream. Thus, in the undistorted flow far ahead of the body Equation (12.48) becomes

$$\tilde{u}_i(\mathbf{x}_m,t) = \delta_{ij}u_j(t) = u_i(t) \quad (12.50)$$

as required.

The restriction to a linear expansion is no problem in aircraft applications, since the high flight speeds typically make turbulence levels very low. It should also not be a problem in typical tower applications, although in low mean wind speed, convective turbulence the higher order terms in Equation (12.46) might be significant.

For simple bodies one can calculate the a_{ij} analytically using potential flow theory. For the special case of a two-dimensional cylinder

$$a_{13}(\mathbf{x}_m) = a_{31}(\mathbf{x}_m) = \varepsilon_{13}(\mathbf{x}_m) = \frac{\tilde{U}_3(\mathbf{x}_m)}{U_1}$$
$$a_{33}(\mathbf{x}_m) = 1 + \varepsilon_{33}(\mathbf{x}_m) = 2 - \frac{\tilde{U}_1(\mathbf{x}_m)}{U_1}$$
$$a_{11}(\mathbf{x}_m) = 1 + \varepsilon_{11}(\mathbf{x}_m) = \frac{\tilde{U}_1(\mathbf{x}_m)}{U_1} \quad (12.51)$$

so that the distorted turbulence components are

$$\tilde{u}_1(\mathbf{x}_m,t) = [1 + \varepsilon_{11}(\mathbf{x}_m)]u_1(t) + \varepsilon_{13}(\mathbf{x}_m)u_3(t) \quad (12.52)$$

$$\tilde{u}_3(\mathbf{x}_m,t) = \varepsilon_{13}(\mathbf{x}_m)u_1(t) + [1 - \varepsilon_{11}(\mathbf{x}_m)]u_3(t) \quad (12.53)$$

Equations (12.52) and (12.53) indicate that there is "cross-talk" between the turbulence components; free stream u_1 fluctuations affect $\tilde{u}_3$, and free stream u_3 fluctuations affect $\tilde{u}_1$. We can immediately calculate the effect on turbulent velocity statistics near the cylinder. For example, multiplying Equations (12.52) and (12.53) and averaging gives

$$\widetilde{\overline{u_1 u_3}}(\mathbf{x}_m) \simeq \overline{u_1 u_3} + \varepsilon_{13}(\mathbf{x}_m)(\overline{u_1^2} + \overline{u_3^2}) \quad (12.54)$$

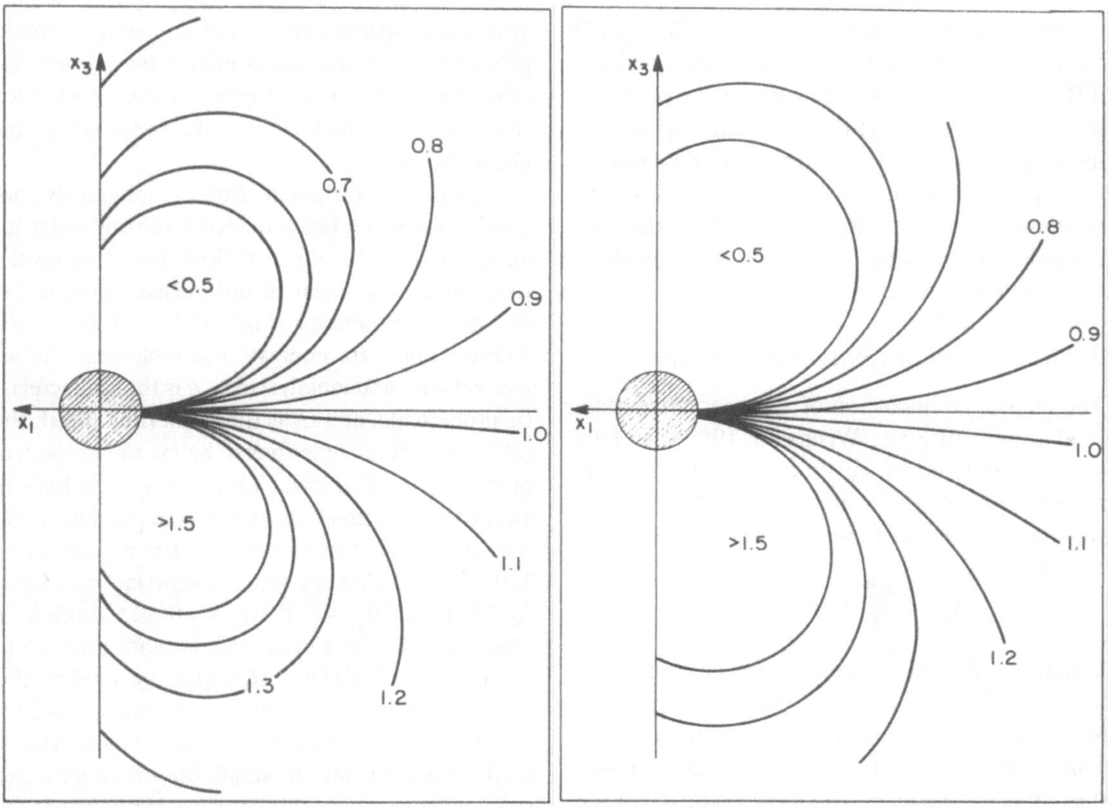

FIGURE 12.13. Contours of the ratio of distorted and undistorted shearing stress $\overline{u_1u_3}$ upstream of a circular cylinder in turbulent flow typical of the surface layer. Left panel is for flow with a narrow turbulent wake; right panel has a wider wake. Reprinted with permission from Wyngaard, 1981a, *Journal of Applied Meteorology*, American Meteorological Society.

Thus the shearing stress measured in the distorted flow is contaminated by the variances. Figure 12.13 shows examples of the serious errors this can cause.

Some of the most conspicuous effects of probe-induced flow distortion are seen in the mean flow; for example, ahead of the cylinder the mean flow has a local inclination angle θ, where $\cos\theta = \tilde{U}_3/\tilde{U}_1$. Some researchers have suggested that the effects on the turbulent velocities are equivalent to "tilt errors" that can be removed by rotating the axes through an angle θ. In fact, if we rewrite Equations (12.52) and (12.53) in terms of θ we find, suppressing the arguments,

$$\tilde{u}_1 = (1 + \varepsilon_{11})u_1 + (1 + \varepsilon_{11})u_3 \sin\theta \qquad (12.55)$$

$$\tilde{u}_3 = (1 + \varepsilon_{11})u_1 \sin\theta + (1 - \varepsilon_{11})u_3 \qquad (12.56)$$

which are different from the tilt equations. However, the "tilt correction" for flow distortion violates the conservation of vorticity and, hence, is incorrect. In practice, the errors it introduces are sometimes not serious, and it is sometimes a simple and quick way to make approximate corrections.

More fundamentally, however, this development indicates that flow distortion can seriously degrade turbulent velocity statistics; in fact, they can be affected much more seriously than the mean velocity. The key to this approach is determining the distortion coefficients $a_{i,j}$, for knowledge of them allows a simple, quantitative estimate of flow-distortion effects. For simple bodies one can calculate them analytically; for more complicated bodies one can measure them. One can also use them to set design criteria for turbulence sensors (Wyngaard, 1988a).

The full flow-distortion problem is even more complicated than we have shown here. Hunt (1973) found that the effects on turbulence are completely different at scales small compared to the body size. In that case the distortion is no longer effectively instantaneous, and the nonlinear turbulent interactions play a role in the distortion physics. That problem needs a good deal more work, particularly from the experimental side, since it seems all but intractable theoretically.

3.3.2 Statistics of Conservative Scalars

The physics of distortion of a conservative scalar field is quite different (Wyngaard, 1988b), as I will summarize here. As we showed in Section 2, we define a conservative scalar attribute $\mathcal{C}$ as one that satisfies the conservation equation

$$\frac{D\mathcal{C}}{Dt} = \frac{\partial \mathcal{C}}{\partial t} + \mathcal{U}_j \frac{\partial \mathcal{C}}{\partial x_j} = 0 \qquad (12.57)$$

Equation (12.57) indicates that $\mathcal{C}$ does not change following the fluid motion, which means we neglect molecular diffusion and that there are no sources or sinks of $\mathcal{C}$. To a good approximation, this holds for the mass density of nonreactive species at low speeds (i.e., at typical wind speeds, but not at aircraft speeds; we will consider the latter separately).

A key implication of Equation (12.57) is that in a steady velocity field

$$\frac{\partial \mathcal{C}}{\partial t}(\mathbf{x}_0, t) = \frac{\partial \tilde{\mathcal{C}}}{\partial t}(\mathbf{x}_m, t + \Delta t) \qquad (12.58)$$

where $\mathbf{x}_0$ and $\mathbf{x}_m$ are points in the freestream and near the body, respectively, that are connected by a fluid particle trajectory; Δt is the transit time from $\mathbf{x}_0$ to $\mathbf{x}_m$. This means that given a steady velocity field, the time series of a conservative scalar in the free stream and on a downstream trajectory in a region of flow distortion are identical except for the time delay. Thus, their frequency spectra are identical.

This has a simple physical interpretation. In a steady flow the velocity gradients distort the spatial gradients of the convected scalar field; flow deceleration ($\partial \tilde{U}_1/\partial x_1 < 0$) causes increased scalar gradients, for example. However, the velocity changes exactly compensate the gradient changes, leaving the time signal at a point unchanged. Thus, for example, a steady flow carrying a sinusoidal streamwise spatial variation of temperature would produce the same sinusoidal time series for observers in the free stream, in the decelerated flow ahead of a body, and in the accelerated flow above the body.

Although a turbulent flow is obviously not steady, this result has provided a starting point for an analysis of the effect of flow distortion on the time series of a scalar in turbulence. If the scalar distortion parameter $(aq^{1/3})/U_1\ell^{1/3})$ is small, where q and ℓ are energy-containing-range turbulent velocity and length scales, a is the characteristic probe scale, and U_1 is the mean flow speed, the scalar spectrum is unaffected below the probe frequency $f_p = U_1/(2\pi a)$. This is easy to satisfy in aircraft applications; there, for example, one might have $q \sim 1$ m s^{-1}, $a \sim 1$ m, $\ell \sim 100$ m, and $U_1 \sim 100$ m s^{-1}, so that the scalar distortion parameter $(qa^{1/3})/(U_1\ell^{1/3}) \sim 10^{-8/3} \sim 0.002$, which is clearly small. On a tower this is more difficult to meet because both ℓ and U_1 tend to be smaller. For example, one might have the same q but $a \sim 0.2$ m, $\ell \sim 10$ m, $U_1 \sim 5$ m s^{-1}, so that the parameter is 0.05. This might still be small, but for larger a and smaller ℓ there could be distortion of the frequency spectrum.

3.3.3 Conservative Scalar Fluxes

Let us assume our scalar distortion parameter $(qa^{1/3})/(U_1\ell^{1/3})$ is sufficiently small that the time series of the conservative scalar, say $c(t)$, is unaffected by flow distortion. If the turbulent velocity signal measured in a region of flow distortion is given by Equation (12.48) and the scalar is measured at the same location, it follows that the measured vertical ($i = 3$) component of scalar flux is

$$\tilde{\overline{u_3c}}(\mathbf{x}_m) = a_{31}(\mathbf{x}_m)\,\overline{u_1c} + a_{32}(\mathbf{x}_m)\,\overline{u_2c} + a_{33}(\mathbf{x}_m)\,\overline{u_3c} \qquad (12.59)$$

In a horizontally homogeneous surface layer the lateral flux $\overline{u_2c}$ is normally zero, but the streamwise flux $\overline{u_1c}$ is not; it is generated by the interaction between the vertical profiles of mean wind and mean scalar (Wyngaard et al., 1971). In fact, in the case of temperature the horizontal flux can exceed the vertical one by a factor of 3!

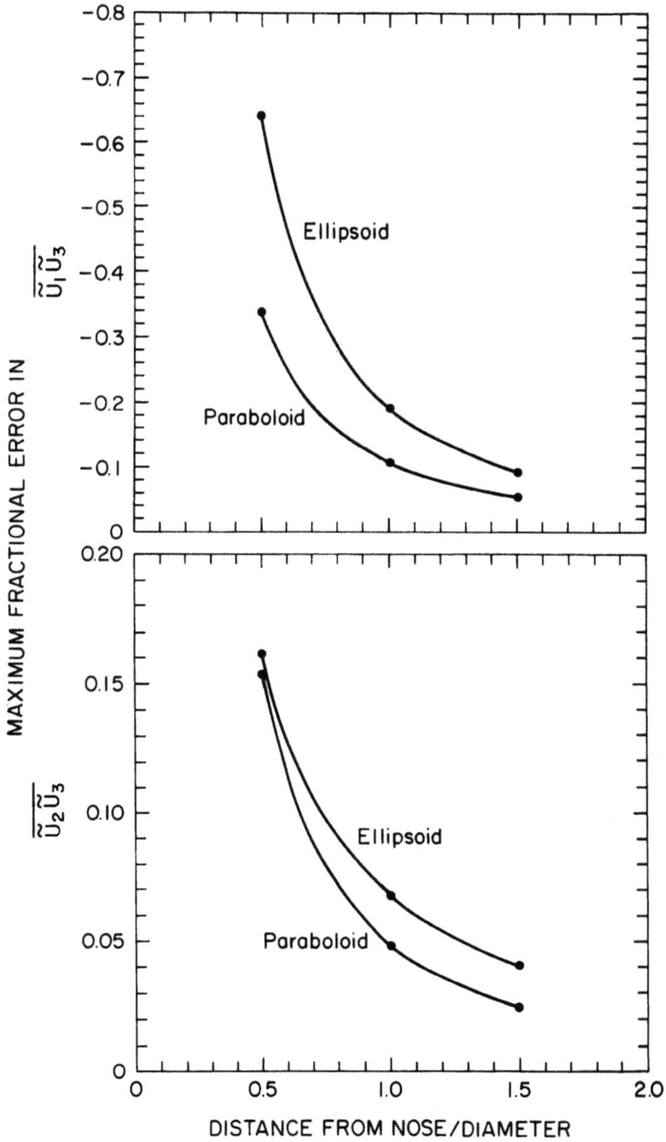

FIGURE 12.14. Errors in measured streamwise stress $\overline{u_1u_3}$ and lateral stress $\overline{u_2u_3}$ ahead of bodies of revolution. Reprinted with permission from Wyngaard et al., 1985, *Journal of Atmospheric & Oceanic Technology*, American Meteorological Society.

I proposed two design criteria in order to minimize the flow-distortion errors in measured scalar flux (Wyngaard, 1988a). The first is that the sensor array (the combination of u_3 and c sensors) should be symmetric about the horizontal plane passing through the point of measurement of u_3. This makes a_{31} zero and eliminates the contamination from the $\overline{u_1c}$ term in Equation (12.59). The second criterion is to minimize the bulk of the sensor array so that u_3 is measured at a point where stagnation effects are minimal, thus making $a_{33} \to 1.0$.

3.4 Flow Distortion and Aircraft-Mounted Sensors

The effects of flow distortion on turbulence measurements from aircraft nose booms can be even more serious than with towers, since nose booms must be short. Flow-distortion calculations for bodies of revolution (Wyngaard et al., 1985) indicate that errors in Reynolds stress can be severe (Fig. 12.14), although it appears that such calculations for real aircraft have not yet been published.

In principle the nose-radome technique (Brown et al., 1983) for sensing air motion can eliminate flow-distortion errors in the velocity field since the radome is calibrated in situ. (In principle, in situ calibration of a boom-mounted velocity sensor will also remove any errors due to flow distortion; good experimental practice dictates such calibration.) However, in eliminating the nose boom in favor of the nose radome one is forced to locate scalar sensors in the strongly distorted flow near the radome to have access to scalar fluxes. In this case the time series of a conservative scalar is unchanged from that in the free stream if the scalar distortion parameter of Section 3.3.2 is small, which is invariably the case.

However, some sensors do not measure scalar *mixing ratio*, a conservative variable, but instead measure scalar *density*. At aircraft speeds scalar density is appreciably nonconservative, which can cause serious difficulties in the presence of flow distortion. I will briefly summarize those arguments here.

From conservation of scalar mixing ratio it follows that the mixing ratio in the flow-distortion region is that which existed in the undisturbed flow along the trajectory upstream. Using this, one can show that the fluctuating scalar density in the flow-distortion region is

$$\tilde{\rho}'_s = \frac{\tilde{\rho}_a}{\rho_a}\rho'_s + \frac{\rho_s}{\rho_a}\left(\tilde{\rho}^1_a - \frac{\tilde{\rho}_a}{\rho_a}\rho'_a\right) \quad (12.60)$$

where for brevity I have suppressed the arguments. They should now be obvious: distorted fluctuating variables (i.e., those with a prime and a tilde) are functions of measurement point $\mathbf{x}_m$ and t; free-stream fluctuations are functions of only t because again we assume that $a \ll l$, so that spatial variations in the free stream do not enter the problem. Here ρ_a is the mass density of air, ρ_s is the mass density of the scalar species, unprimed symbols are means, and primes are fluctuations. Equation (12.60) says that scalar density fluctuations in a region of flow distortion are amplified by the ratio $\tilde{\rho}_a/\rho_a$, which at a stagnation point and an aircraft speed of 100 m s^{-1} is about 1.04; furthermore, they are contaminated by cross-talk from air density fluctuations. In fact, Equation (12.60) shows that the latter effect induces scalar density fluctuations in a flow-distortion region even if there are none in the free stream.

If the flow-distortion process is isentropic, one finds for the second group on the right side of Equation (12.60)

$$\tilde{\rho}^1_a - \frac{\tilde{\rho}_a}{\rho_a}\rho'_a = \left(\frac{\rho_a}{T}\right)\frac{1}{(k-1)}\left(\frac{\tilde{\rho}_a}{\rho_a}\right)^{2-k} \left[\tilde{T}' - \left(\frac{\tilde{\rho}_a}{\rho_a}\right)^{k-1} T'\right] \quad (12.61)$$

where $k = 1.4$ is the specific-heat ratio for air and T is mean temperature. If we use energy conservation and neglect compressibility effects on the velocity we can write

$$\tilde{T}' = T' + \frac{u_i}{C_p}(U_i - a_{ji}U_j) \quad (12.62)$$

where C_p is the specific heat at constant pressure and a_{ij} is the matrix of flow-distortion coefficients; we use the summation convention. If we combine Equations (12.60)–(12.62) we have

$$\tilde{\rho}^1_s = \frac{\tilde{\rho}_a}{\rho_a}\rho'_s + \rho_s\frac{U_1^2}{C_pT(k-1)} \left[-\frac{(1-\tilde{U}^2/U_1^2)T'}{2T} + \frac{(\delta_{1i}-a_{ji}\tilde{U}_j/U_1)u_i}{U_1}\right] \quad (12.63)$$

Thus, we see that the air-density cross-talk [the second term on the right side of Equation (12.60)] introduces free stream fluctuations of temperature and velocity into the fluctuating scalar density signal. These crosstalk terms vanish when $\tilde{U} = U_1$ and $a_{ij} = \delta_{ij}$, so they are nonzero only in a region of flow distortion. Furthermore, they are multiplied by the aerodynamic factor $U_1^2/C_pT(k-1)$, which increases as the square of the mean wind speed and, hence, tends to be important only in aircraft applications. At $U_1 = 100$ m s^{-1} and $T = 300$ K, for example, this factor is about 0.08; thus, it is only 0.0008 at $U_1 = 10$ m s^{-1}. We conclude that if high wind speeds, flow distortion at the measurement point, and free-stream velocity or temperature fluctuations are present then there is the potential for spurious scalar density signals.

For simplicity let us assume the turbulent velocity component u_3 is measured without flow-distortion error. Then multiplying (3.28) by u_3 and averaging gives an expression for the error in measured scalar density flux,

$$\overline{\rho'_s u_3} = \frac{\tilde{\rho}_a}{\rho_a}\overline{\rho'_s u_3} + \rho_s \frac{U_1^2}{C_p T(k-1)} \left[-\frac{(1-\tilde{U}^2/U_1^2)\overline{T'u_3}}{2T} + \frac{(\delta_{1i} - a_{ji}\tilde{U}_j/U_1)\overline{u_i u_3}}{U_1} \right] \quad (12.64)$$

which says that the scalar density flux measured in a region of flow distortion is contaminated by the temperature flux as well as by the Reynolds stress components $\overline{u_1 u_3}$, $\overline{u_2 u_3}$, and $\overline{u_3^2}$.

We see from Equation (12.64) that the principal error in scalar flux (the term in square brackets) depends on the mean scalar density, which is not generally related to the scalar flux itself. Thus, this is an error that becomes particularly serious at low scalar flux values; in effect, it sets a threshold for scalar flux measurements.

Assessing the effects of this scalar-flux contamination quantitatively is not simple: we must know the aerodynamic factor $U_1^2/C_p T(k-1)$; the free-stream turbulent covariances $\overline{T'u_3}$ and $\overline{u_i u_3}$, $i = 1, 2, 3$; and complete details of the flow distortion, i.e., the distorted mean velocity $\tilde{U}_i$ and the flow-distortion coefficients a_{ij}. However, some simple, analytical calculations for flow around a sphere (the simplest abstraction of an aircraft nose radome) show that the flux errors are potentially quite serious. In the case of CO_2 fluxes, for example, it seems that the normally low values over the ocean may fall below this threshold. Water vapor fluxes are affected somewhat less, because of its typically higher ratio of fluctuation and mean.

This is a problem that has been recognized only fairly recently and needs a good deal more study before we clearly understand how to deal with it. One approach is to minimize the flux error. This occurs naturally if speeds are sufficiently low that the aerodynamic factor is negligible, as on towers, or if we use a mixing ratio sensor and keep the flow-distortion parameter small. We can also minimize flow distortion effects by designing aerodynamically "clean" sensors and mounting them in the undisturbed flow well away from the aircraft. Finally, we can strategically locate sensors in error-free positions on the aircraft if we know in detail its flow-distortion characteristics. As a last resort, we can try to correct for the errors by removing them from the time series or from the statistics.

Although the physics discussed here bears a certain similarity to that discussed in Section 2, one should not confuse them. There we saw that the total scalar density flux can have a significant contribution from the product of mean vertical velocity and mean scalar density; thus, in general one must add this "Webb correction" to the measured turbulent scalar density flux to get the total flux. However, if one measures the scalar mixing ratio rather than density, then its turbulent flux includes the mean–mean contribution and no Webb correction is needed. In case of aircraft flow distortion, in measuring a scalar density flux one can face severe errors caused by the combination of slight density changes (i.e., non-conservation effects) and flow distortion. If the flow-distortion parameter is sufficiently small, these errors can be eliminated by using a sensor that measures the mixing ratio of the scalar, rather than its density.

Thus, if we measure a scalar density flux from aircraft we need to account for flow-distortion errors and also make the Webb correction for the mean–mean contribution to the flux. If we use a mixing-ratio sensor we can eliminate the flow-distortion errors and the need for the Webb correction as well. Perhaps this recognition will generate interest in developing mixing-ratio sensors.

4. Sonic Anemometers

Sonic (also called acoustic) anemometers, in conjunction with modern developments in data processing and recording, have revolutionized surface-layer turbulence research over the past 20 years. Kaimal (1979) writes:

> The sonic anemometer with its rapid response, linear output and stable calibration has become a valuable tool for atmospheric research. It has provided high-quality turbulence data in a number of field experiments conducted during the last decade. A new three-axis configuration is currently being used for continuously monitoring the wind at eight levels on the Boulder Atmospheric Observatory's 300 m tower.
>
> Because of its complexity and high cost, the sonic anemometer is likely to remain a research instrument. Attempts are being made to reduce cost and to simplify its operation. Even if these attempts are successful, the sonic anemometer has an inherent limitation which precludes its use in adverse weather conditions. The

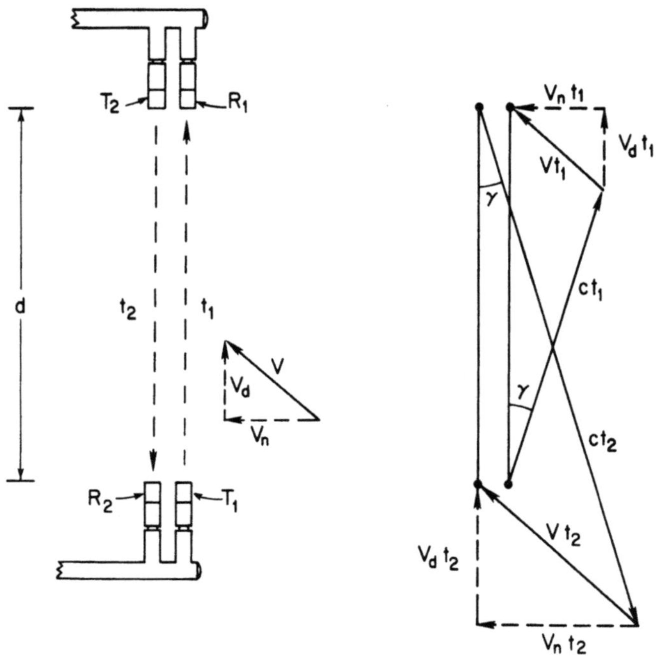

FIGURE 12.15. Kaimal's portrayal of sound-ray vectors for a single-axis sonic anemometer, showing principle of operation. T and R represent transmitter and receiver, respectively. (From Wyngaard, 1981b.) Reproduced, with permission, from the *Annual Review of Fluid Mechanics*, vol. 13. © 1981 by Annual Reviews Inc.

instrument fails to respond in rain, wet snow, and heavy fog. Formation of water drops on the transducer temporarily affects its operation. Thus the sonic anemometer's main contribution will be in fair-weather observations of atmospheric turbulence, where it is still the instrument of choice among boundary-layer physicists.

Although those words were written 10 years ago, they remain substantially true today. The "sonic," as it is colloquially called, remains the standard anemometer for tower-based micrometeorological research; it still has no serious competition when the ultimate in performance is needed. Kaimal (1986) has provided a modern view of its use in tower micrometeorology. In principle, a well-designed sonic anemometer measures the projection of the wind-velocity vector on the acoustic path and is free of the nonlinearities, time lag, and most other deficiencies of cup, propeller, and vane anemometers. As Kaimal indicates in the quote above, a principal disadvantage is its high cost; a three-component unit is at least an order of magnitude more expensive than a three-component propeller anemometer, for example.

In this section I will cover some of the interesting measurement physics of the sonic anemometer. I will not discuss sonic circuitry, signal processing, or transducer technology, although each is important in the total picture. Instead, I will focus on the dynamic performance of the sonic. There are some interesting developments to report; its performance is not as impeccable as we perhaps once thought.

4.1 Some History

Suomi (1957) was a pioneer in sonic anemometry. His device had a 1-m path; the transit-time difference of acoustic pulses traveling in opposite directions along this path gave the average velocity in the path direction. Later workers used continuous acoustic signals (Gurvich, 1959; Bovsheverov and Voronov, 1960; Kaimal and Businger, 1963). The 1968 Kansas experiments (Haugen et al., 1971; Haugen, 1973) saw the first use of sonic anemometry in a large-scale field program. Three-component, pulse-type sonic anemometers with 20-cm paths were operated at heights of 5.7, 11.3, and 22.6 m in a study of the structure and dynamics of the surface layer.

Kaimal et al. (1974) describe an improved device, an updated version of which is commercially available from Applied Technologies Corporation of Boulder, Colorado. The Coppin-Taylor anemometer (Coppin and Taylor, 1983) is now also commercially available from Dobbie Brothers in Australia. Kaijo–Denki also markets a three-component unit (Hanafusa et al., 1982).

4.2 Theory of Operation

Figure 12.15, from Kaimal (1979), shows the basic principle of the sonic anemometer. The wind vector **V** has components V_d an V_n along and normal to the acoustic path (we assume that all velocities are averages along the paths). The transit times for two simultaneous pulses traveling from T_1 to R_1 and from T_2 to R_2 are, from the vector diagram in Figure 12.1,

$$t_1 = \frac{d}{c \cos \gamma + V_d} \qquad (12.65)$$

$$t_2 = \frac{d}{c \cos \gamma - V_d} \qquad (12.66)$$

Here t_1 and t_2 are the pulse transit times, d is the path length, c is the velocity of the sound in the air, and $\gamma = \sin^{-1}(V_n/c)$. Thus for $V^2 \ll c^2$ the transit-time difference is, from Equation (12.65),

$$\Delta t = \frac{2d}{c^2} V_d \qquad (12.67)$$

Knowing the temperature T gives c^2 from the relation $c^2 = kRT$, where k is the ratio of specific heats and R the gas constant. Thus, when T is known, measuring V_d becomes a matter of measuring the transit-time difference. For $d = 20$ cm this typically can be done with a resolution of about 10^{-7} s in commercial units, giving a velocity resolution of the order of 3 cm s^{-1}. Kaimal (1979) has discussed the influence of temperature and humidity fluctuations, which affect the sound speed c, on the velocity measurements; they are generally negligible. The sensitivity of c to humidity and temperature creates the possibility of measuring temperature and humidity with a sonic (Schotanus et al., 1983.)

The newest units utilize an improvement on this scheme. They measure the transit times t_1 and t_2, invert them, and subtract; this allows V_d to be obtained directly from

$$\left(\frac{1}{t_1} - \frac{1}{t_2}\right) = \frac{d}{2V_d} \qquad (12.68)$$

which eliminates the need to know the sound speed c.

4.3 Three-Dimensional Units

Designing a three-dimensional sonic array with unobstructed exposure for all the axes is difficult. Some models use a pair of horizontal axes separated by 120° and one vertical axis; the three orthogonal velocity components are then retrieved from these outputs. However, the outputs must be monitored to ensure that the natural variations in wind direction do not exceed the allowable range of this array, which is of the order of 90° in the horizontal.

Units with a pair of acoustic paths per axis typically have pulse rates on the order of 400 s^{-1}. Kaimal (1979) points out that 10 or 20 successive measurements can be block-averaged to improve resolution and reduce spectral-aliasing effects, while leaving a signal with a rate of the order of 20 samples s^{-1}. Some units use only one transducer at each end of the path, acting alternately as a transmitter and a receiver; here the sample rate must be less (typically no greater than 10 s^{-1}) to accommodate settling times in the transducers.

The spatial-resolution properties of three-component sonic arrays have been calculated in some detail by Kaimal et al. (1968) and by Horst (1973). They determined the effects of path length and path separation on measured horizontal and vertical velocity spectra, assuming the true spectra are inertial and isotropic. As we discussed in Section 3, the results indicate that the measured streamwise wave number spectra (inferred from frequency spectra through Taylor's hypothesis) of horizontal and vertical velocity components become significantly distorted at streamwise wave numbers κ_1 of the order of d^{-1}; thus the "distance constant" is effectively the path length, which at (typically) 20 cm is less than the distance constant of most cup or propeller anemometers. This gives sufficient spatial resolution to recover the vertical turbulent fluxes at heights of the order of 5 m and greater.

Sonic-anemometer velocity measurements can be systematically in error because of the velocity defects in the transducer wakes along the acoustic paths. The severity of this effect, which was relatively unexplored until fairly recently, depends on the array design.

4.4 Recent Developments

Although sonic anemometers have been used in micrometeorological research for at least 20 years,

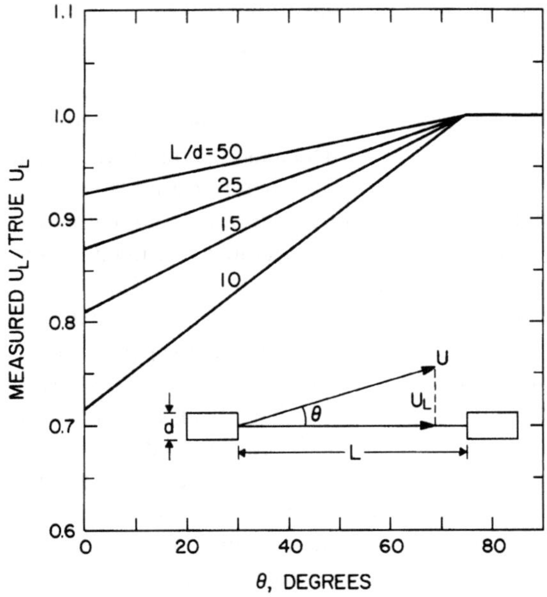

FIGURE 12.16. Kaimal's (1979) curves on axial velocity attenuation in an array with BAO-type transducers. Reprinted with permission from Wyngaard and Zhang, 1985, *Journal of Atmospheric & Oceanic Technology*, American Meteorological Society.

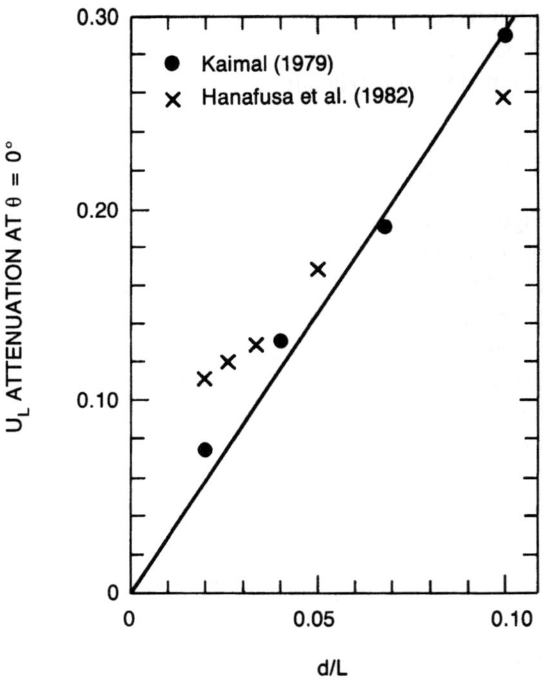

FIGURE 12.17. Axial velocity attenuation at θ = 0°. Reprinted with permission from Wyngaard and Zhang, 1985, *Journal of Atmospheric & Oceanic Technology*, American Meteorological Society.

it is only fairly recently that we have uncovered some of the subtleties of their performance. These subtleties can have profound effects on turbulence statistics, however. Let us consider these next.

4.4.1 The Transducer-Shadow Effect

"Transducer shadow" refers to the velocity deficit in the transducer wakes along the acoustic paths. It seems to have been discussed first by Kaimal (1979), who presented some measurements of the resulting path-averaged velocity errors in a simple array. Kaimal and Gaynor (1983) show how it is being accounted for in the sonic anemometers at the Boulder Atmospheric Observatory. Hanafusa et al. (1982) and Coppin and Taylor (1983) discuss improved sonic anemometers that, nonetheless, suffer from transducer-shadow effects.

Wyngaard and Zhang (1985) analyzed the impact of transducer shadow effects on turbulence statistics measured by sonic anemometers. Their results were rather surprising, indicating that serious errors could result if shadow effects were not minimized or accounted for in data analysis. Several other studies have corroborated that finding. Let us review this recent work, using the Wyngaard-Zhang study as a point of departure.

Kaimal's (1979) measurements in the wind tunnel and atmosphere gave the results in Figure 12.16, which shows axial velocity attenuation as a function of the inclination θ of the flow vector to the acoustic path. The attenuation is greatest when the flow is along the path, decreases as θ increases, and also increases as L/d (path length/transducer diameter) increases. The latter point is emphasized by Figure 12.17, which shows the axial velocity attenuation at $\theta = 0°$ as a function of d/L.

The details of the velocity attenuation depend on the shape of the transducer; for example, the right circular cylinder (BAO-type) and torpedo-shaped (Kaijo–Denki type) transducers have different axial attenuation curves. Wyngaard and Zhang suggested the parameterizations

$$\frac{U_L^m}{U_L} = C + (1 - C)\sin\theta, \quad \text{BAO–type} \tag{12.69}$$

$$\frac{U_L^m}{U_L} = 1 - (1-C)e^{-a\sin^2\theta}, \text{ Kaijo–Denki–type} \tag{12.70}$$

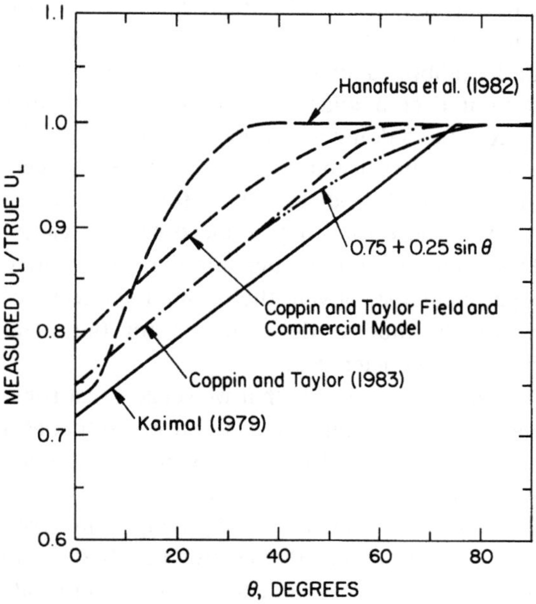

FIGURE 12.18. Axial velocity attenuation at $L/d = 10$, contrasting the behavior of Kaijo–Denki-type and BAO-type transducers. Reprinted with permission from Wyngaard and Zhang, 1985, *Journal of Atmospheric & Oceanic Technology*, American Meteorological Society.

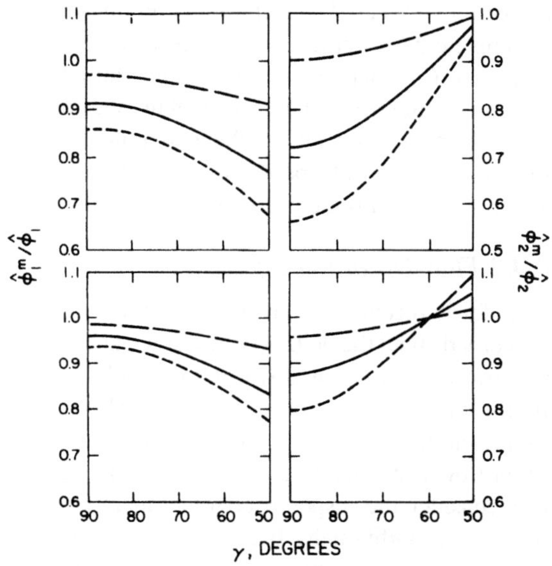

FIGURE 12.19. Calculated spectral response of a 3-component sonic anemometer with BAO transducer-wake effects included. The coordinates are aligned with the mean wind. Long-dash curves, $L/d = 60$; solid, $L/d = 20$; short-dash, $L/d = 10$. Top panel, 90° array; bottom, 120° array. Reprinted with permission from Wyngaard and Zhang, 1985, *Journal of Atmospheric & Oceanic Technology*, American Meteorological Society.

where the superscript m means measured value and C is the response at $\theta = 0°$. Figure 12.18 shows that Equation (12.69) with $C = 0.75$ is a fairly good fit to the Coppin-Taylor (1983) data; a slightly smaller C would fit the Kaimal data. Equation (70) with $C = 0.74$ and $a = 11.5$ fits the Hanafusa et al. data very well.

Using the parameterizations of transducer shadow effects in Equations (12.69) and (12.70), Wyngaard and Zhang developed a set of equations for the linearized response of a three-component sonic anemometer in the presence of a turbulent wind field. Figure 12.19 shows representative results for the streamwise and lateral wind spectra measured with BAO-type transducers of various L/d values with 90 and 120° arrays. The results show that the transducer wake effects can have serious consequences.

Zhang and Wyngaard (1986) later showed that the "U.W." sonic anemometer, designed by J. A. Businger and S. P. Oncley of the University of Washington, suffers less from transducer shadow effects than most 3-component units because of its geometry.

Very recently, Grant and Watkins (1989) used an analytical technique slightly different from that of Wyngaard and Zhang (1985) to assess the impact of transducer shadow effects on turbulence statistics. Rather than linearizing the response, they posed the problem in terms of the probability density of wind fluctuations and then solved for the response after assuming a form for this density. Their results thus depend on the turbulence level and reduce in the small-amplitude, linear limit to those of Wyngaard and Zhang.

Grant and Watkins (1989) also made field measurements with two 120°, 3-component Kaijo–Denki TR 61A sonic anemometers oriented in different directions and therefore having different errors due to shadow effects. The axis of one sonic was kept within 20° of the mean wind direction; the other sonic was oriented between 20 and 90° to the wind direction. The measured error was as large as 10% when the wind was down one axis of the array, in good agreement with their predictions. These results confirm that transducer shadow effects cannot, in general, be neglected.

Research practice today is to correct the "raw" sonic anemometer outputs for transducer shadow effects, generating new time series with the shadow effects removed. This is somewhat risky, in that the corrections are based on somewhat preliminary and incomplete data, but it seems the best we can do at this point. Coppin (personal communication) developed a fast, iterative scheme for making the corrections. Further research on the magnitude of the shadow effects would be desirable, and of course ideally we would minimize them through design.

4.4.2 Flow-Distortion Effects

Until fairly recently researchers were relatively unconcerned about the bulk of typical sonic anemometer arrays and the flow distortion this bulk causes. As the subtler aspects of measurement physics became clearer, however, and concerns were expressed about flow distortion in general, researchers began to reconsider the response of sonic anemometers. Sonics had traditionally been held to measure the projection of the free-stream velocity along the acoustic path. The first inkling that all was not well with that assumption came with the discovery of transducer wake effects. It was found that the velocity averaged along the acoustic path could differ appreciably from the projection of the free-stream value, and this was shown to be capable of causing substantial errors in turbulence statistics.

Subsequently, researchers noticed that the sonics themselves could cause measureable flow distortion—e.g., induce an upward mean vertical velocity. Dyer (1981) discusses some of the flow-distortion effects observed in the International Turbulence Comparison Experiment. The mathematical analysis of flow distortion discussed in Section 3 shows that such effects can be particularly serious for Reynolds stresses.

I believe the principal use of that flow-distortion analysis is not in correcting data, but rather as a tool for studying its effects so that better instruments can be developed. For example, they give two simple design criteria for scalar flux sensors, as discussed in Section 3.3.3.

4.4.3 Anemometer Calibration

Even the newest three-dimensional sonic anemometer design can have response that differs considerably from the ideal because flow distortion from the array bulk and transducer shadow effects can be minimized but not eliminated. Reliable, accurate measurements of turbulence statistics can be obtained only by accounting for both of these. The best way to do this, I believe, is through calibration.

Kraan and Oost (1988) proposed that the entire sonic anemometer unit, mounting assembly and all, be calibrated directly in a wind tunnel by operating it over a wide range of orientations and speeds. In so doing a "calibration surface" can be generated that accounts directly for transducer-shadow and flow-distortion effects and, hence, removes the errors they would otherwise cause. This seems a prudent step for the conservative experimentalist, given the insidious effects of seemingly modest departures from the ideal response characteristics.

Perhaps it seems odd that we suggest that sonic anemometers be calibrated, when we do that as a matter of course with any other anemometer. If so, recall that one of the long-standing attractions of the sonic has been that it is, in principle, an "absolute" device; i.e., one that needs no calibration. In the early years users came to rely on this attribute rather than doing calibrations, which could be difficult to carry out. Because of the bulk of typical arrays, either a very large wind tunnel was needed or the tunnel blockage effects had to be accounted for very carefully. Even after the transducer-wake effect was detected, it continued to be more practical to correct for it than to absorb it in a whole-anemometer calibration. Today, we know that flow distortion due to the array bulk can cause comparable errors. Quantifying these bulk flow-distortion effects (e.g., determining their flow-distortion coefficients) would require direct tests of the whole unit in a wind tunnel, however. Given that, it makes more sense, in my opinion, to adopt the Kraan–Oost approach of using the anemometer calibration surface than it does to use transducer-shadow corrections and/or flow-distortion coefficients to correct the data.

Acknowledgments. This paper is based on lectures given at the Workshop on Measurement and Paramaterization of Land-Surface Evaporation Fluxes, Banyuls, France, October 10–21, 1988. The National Center for Atmospheric Research is sponsored by the National Science Foundation.

References

Andreas EL (1981) The effects of volume averaging on spectra measured with a Lyman-α hygrometer. J Appl Meteorol 20:467–475.

Andreas EL (1988) Atmospheric stability from scintillation measurements. Appl Optics 27:2241-2246.

Bovsheverov VM, Voronov VP (1960) Acoustic anemometer. Izv Geophys Ser 6:882-885.

Brown EN, Lenschow DH, Friehe CA (1983) The use of pressure fluctuations on the nose of an aircraft for measuring air motion. J Clim Appl Meteorol 22: 171-180.

Brutsaert WH (1982) "Evaporation into the Atmosphere." Reidel, Dordrecht.

Businger JA (1986) Evaluation of the accuracy with which dry deposition can be measured with current micrometeorological techniques. J Clim Appl Meteorol 25:1100-1124.

Businger JA, Miyake M, Dyer AJ, Bradley EF (1967) On the direct determination of the turbulent heat flux near the ground. J Appl Meteorol 6:1025-1032.

Businger JA, Oncley SP (1990) Flux measurement with conditional sampling. J Atmos Ocean Tech 7:349-352.

Caughey SJ (1982) Observed characteristics of the atmospheric boundary layer. In "Atmospheric Turbulence and Air Pollution Modelling" (FTM Nieuwstadt and H van Dop eds.), pp 107-158. Reidel, Dordrecht.

Coppin PA, Taylor KJ (1983) A three-component sonic anemometer/thermometer system for general micrometeorological research. Bound Layer Meteorol 27: 27-42.

Coulter, RL, Wesely ML (1980) Estimates of surface heat flux from sodar and laser scintillation measurements in the unstable boundary layer. J Appl Meteorol 19:1209-1222.

Deardorff JW, Wilis GE (1982) Investigation of the frozen-turbulence hypothesis for temperature spectra in a convectively mixed layer. Phys Fluids 25:21-28.

Dyer AJ (1981) Flow distortion by supporting structures. Bound Layer Meteorol 20:243-252.

Grant ALM, Watkins RD (1989) Errors in turbulence measurements with a sonic anemometer. Bound Layer Meteorol 46:181-194.

Gurvich AS (1959) Acoustic microanemometer for investigation of the microstructure of turbulence. Acoust J (USSR) 5:368-369.

Hanafusa T, Fujitani T, Kobori Y, Mitsuta Y (1982) A new type sonic anemometer-thermometer for field operation. Pap Meteorol Geophys 33:1-19.

Haugen DA (ed) (1973) "Workshop on Micrometeorology." American Meteorological Society, Boston.

Haugen DA, Kaimal JC, Bradley EF (1971) An experimental study of Reynolds stress and heat flux in the atmospheric surface layer. Quart J R Meteorol Soc 97:168-180.

Hicks BB, McMillen RT (1984) A simulation of the eddy-accumulation method for measuring pollutant fluxes. J Clim Appl Meteorol 23:637-643.

Hill RJ (1988) Comparison of scintillation methods for measuring the inner scale of turbulence. Appl Optics 27:2187-2193.

Hill RJ (1989) Implications of M-O similarity theory for scalar quantities. J Atmos Sci 46:2236-2244.

Hill RJ, Bohlander RA, Clifford SF, McMillan RW, Priestley JT, Schoenfeld WP (1988) Turbulence-induced millimeter-wave scintillation compared with micrometeorological measurements. IEEE Trans Geo Rem Sensing 26:330-342.

Högström U (1988) Non-dimensional wind and temperature profiles in the atmospheric surface layer: A reevaluation. Bound Layer Meteorol 42:55-78.

Horst TW (1973) Spectral transfer functions for a three-component sonic anemometer. J Appl Meteorol 12: 1072-1075.

Hunt JCR (1973) A theory of two-dimensional flow round two-dimensional bluff bodies. J Fluid Mech 61:625-706.

Kaimal JC (1979) Sonic anemometer measurement of atmospheric turbulence. Proc Dynamic Flow Conf 1978, pp. 551-565. P.O. Box 121, DK 2740 Skovlunde, Denmark.

Kaimal JC (1986) Flux and profile measurements from towers in the boundary layer. In "Probing the Atmospheric Boundary Layer" (DH Lenschow, ed.), pp. 19-28. American Meteorology Society, Boston.

Kaimal JC, Businger JA (1963) A continuous wave sonic anemometer thermometer. J Appl Meteorol 2:156-165.

Kaimal JC, Gaynor JE (1983) The Boulder Atmospheric Observatory. J Appl Meteorol 22:863-880.

Kaimal JC, Haugen DA (1969) Some errors in the measurement of Reynolds stress. J Appl Meteorol 8:460-462.

Kaimal JC, Wyngaard JC, Haugen DA (1968) Deriving power spectra from a three-component sonic anemometer. J Appl Meteorol 7:827-837.

Kaimal JC, Newman JT, Bisberg A, Cole K (1974) An improved three component sonic anemometer for investigation of atmospheric turbulence. In "Flow—Its Measurement and Control in Science and Industry," Vol 1, pp. 349-359. Instrum. Soc. Amer.

Koshiek W (1984) Inertial subrange correlation between temperature and humidity fluctuations in the unstable surface layer above vegetated terrains. Bound Layer Meteorol 29:211-224.

Kraan C, Oost WA (1988) A new way of anemometer calibration and its application to a sonic anemometer. J Atmos Ocean Tech 6:516-524.

Kristensen L, Fitzjarrald DR (1984) The effect of line averaging on scalar flux measurements with a sonic anemometer near the surface. J Atmos Ocean Tech 1:138-146.

Lang, ARG, McNaughton KG, Chen Fazu, Bradley EF, Ohtaki E (1983) Inequality of eddy transfer coefficients for vertical transport of sensible and latent heats during advective conditions. Bound Layer Meteorol 25:25–41.

Lenschow DH Ed (1986) "Probing the Atmospheric Boundary Layer." American Meteorological Society, Boston.

Lumley JL (1965) Interpretation of time spectra measured in high-intensity shear flows. Phys Fluids 8: 1056–1062.

Lumley JL, Panofsky HA (1964) "The Structure of Atmospheric Turbulence." Interscience, New York.

Mahrt L (1976) Mixed-layer moisture structure. Mon Weather Rev 104:1403–1418.

Panofsky HA, Dutton JA (1984) "Atmospheric Turbulence." Wiley, New York.

Panofsky HA, Tennekes H, Lenschow DH, Wyngaard JC (1977) The characteristics of turbulent velocity components in the surface layer under convective conditions. Bound Layer Meteorol 11:355–361.

Raupach MR (1988) Canopy transport processes. In "Flow and Transport in the Natural Environment: Advances and Applications" (WL Steffen and OT Denmead, eds.). Springer-Verlag, Berlin.

Schotanus P, Nieuwstadt FTM, De Bruin HAR (1983) Temperature measurement with a sonic anemometer and its application to heat and moisture fluxes. Bound Layer Meteorol 26:81–93.

Speer RE, Peterson KA, Ellestad TG, Durham JL (1985) Test of a prototype eddy accumulator for measuring atmospheric vertical fluxes of water vapor and particulate sulfate. J Geophys Res 90:2119–2122.

Suomi VE (1957) Sonic anemometer. In "Exploring the Atmosphere's First Mile," Vol 1, pp 356–366. Pergamon, New York.

Taylor GI (1938) The spectrum of turbulence. Proc R Soc London A Ser 164:476–490.

Tennekes H, Lumley JL (1972) "A First Course in Turbulence." MIT Press, Cambridge.

Tillman JE (1972) The indirect determination of stability, heat and momentum fluxes in the atmospheric boundary layer from simple scalar variables during dry unstable conditions. J Appl Meteorol 11:783–792.

Van Ulden AP, Holtslag AAM (1985) Estimation of atmospheric boundary layer parameters for diffusion applications. J Clim Appl Meteorol 24:1196–1207.

Verma S, Rosenberg NJ, Blad BL (1978) Turbulent exchange coefficients for sensible heat and water vapor under advective conditions. J Appl Meteorol 17:330–338.

Webb EK, Pearman GI, Leuning R (1980) Correction of flux measurements for density effects due to heat and water vapor transfer. Quart J R Meteorol Soc 106: 85–100.

Wesely ML (1988) Use of variance techniques to measure dry air-surface exchange rates. Bound Layer Meteorol 44:13–31.

Wyngaard JC (1971) Spatial resolution of a resistance wire temperature sensor. Phys Fluids 14:2052–2054.

Wyngaard JC (1973) On surface-layer turbulence. In "Workshop on Micrometeorology" (DA Haugen, ed.), pp 101–149. American Meteorological Society, Boston.

Wyngaard JC (1981a) The effects of probe-induced flow distortion on atmospheric turbulence measurements. J Appl Meteorol 20:784–794.

Wyngaard JC (1981b) Cup, propeller, vane, and sonic anemometers in turbulence research. Annu Rev Fluid Mech 13:399–423.

Wyngaard JC (1986) Measurement physics. In "Probing the Atmospheric Boundary Layer" (DH Lenschow, ed.), pp 5–18. American Meteorology Society, Boston.

Wyngaard JC (1988a) Flow-distortion effects on scalar flux measurements in the surface layer: Implications for sensor design. Bound Layer Meteorol 42:19–26.

Wyngaard JC (1988b) The effects of probe-induced flow distortion on atmospheric turbulence measurements: Extension to scalars. J Atmos Sci 45:3400–3412.

Wyngaard JC (1988c) Structure of the PBL. In "Lectures on Air Pollution Modeling" (A Venkatram and JC Wyngaard, eds.), pp 9–61. American Meteorology Society, Boston.

Wyngaard JC, Brost RA (1984) Top-down and bottom-up diffusion of a scalar in the convective boundary layer. J Atmos Sci 41:102–112.

Wyngaard JC, Clifford SF (1977) Taylor's hypothesis and high-frequency turbulence spectra. J Atmos Sci 34: 922–929.

Wyngaard JC, Clifford SF (1978) Estimating momentum, heat, and moisture fluxes from structure parameters. J Atmos Sci 35:1204–1211.

Wyngaard JC, LeMone MA (1980) Behavior of the refractive index structure parameter in the entraining convective boundary layer. J Atmos Sci 37:1573–1585.

Wyngaard JC, Zhang SF (1985) Transducer-shadow effects on turbulence spectra measured by sonic anemometers. J Atmos Ocean Tech 2:548–558.

Wyngaard JC, Coté OR, Izumi Y (1971) Local free convection, similarity, and the budgets of shear stress and heat flux. J Atmos Sci 28:1171–1182.

Wyngaard JC, Pennell WT, Lenschow DH, LeMone MA (1978) The temperature-humidity covariance budget in the convective boundary layer. J Atmos Sci 35: 47–58.

Wyngaard JC, Rockwell L, Friehe CA (1985) Errors in the measurement of turbulence upstream of an axisymmetric body. J Atmos Ocean Tech 2:605–614.

Zhang SF, Wyngaard JC (1986) Response characteristics of the U.W. sonic anemometer. J Atmos Ocean Tech 3:315–323.

13
Errors in Eddy Correlation Turbulence Measurements from Aircraft: Application to HAPEX-MOBILHY

Peter H. Hildebrand

1. Introduction

Aircraft are excellent platforms for measuring atmospheric turbulence. Due to their mobility, aircraft can make a wide variety of measurements over a large domain in a short length of time and the aircraft flight pattern can be easily adjusted to accommodate to changes in the atmosphere and to accommodate new measurement techniques.

This chapter will discuss the measurement of planetary boundary layer fluxes and turbulence from research aircraft. The discussion will cover aircraft instrumentation and accuracies, aircraft flight patterns, and the eddy correlation data analysis technique for measurement of atmospheric turbulence. The discussion will be directed toward the problem of measuring the variability of the atmospheric turbulence and fluxes over complex terrain. For inhomogeneous atmospheres, such as over complex terrain, the modulation of boundary layer turbulence structure must be measured as well as the mean state of the atmosphere. The data collection and data processing strategies must be adjusted accordingly.

The techniques of eddy correlation turbulence measurement will be discussed using examples from the HAPEX experiment (Andre et al., 1986). The NCAR King Air aircraft was used to measure boundary layer fluxes during HAPEX. The operations of the NCAR King Air are described by Stull (1986) and Hildebrand (1988).

2. Aircraft Instrumentation for Eddy Correlation Measurements from Aircraft

Aircraft instrumentation for turbulence measurements must include sensors that are sampled at a high rate with respect to the important turbulent scales of motion. The instrumentation is designed to be as free as possible from distortions due to airflow around the aircraft (see Wyngaard, 1990 this volume). The basic turbulence instrumentation includes high rate measurements of winds, temperatures, humidity, pressure, altitude, and aircraft location. For convective turbulence, these measurements are typically made at 20–50 Hz, which corresponds to about 1–4 m at a flight speed of 80 m/s. The measurements thus include a good portion of the turbulent spectrum, from the energy-containing spectral peak down to the inertial subrange. It is useful to also make low rate (≈ 1 Hz or ≈ 80 m scale) measurements of solar and terrestrial radiation and surface radiative temperature. An overview of aircraft instrumentation for turbulence measurements can be seen in Lenschow (1986) and Desjardins and MacPherson (this volume). Details of NCAR King Air turbulence instrumentation can be seen in Resaerch Aviation Facility (1988a,b,c).

For all instrumentation, major concerns relate to the correct frequency and amplitude (gain) response throughout the range of frequencies to be

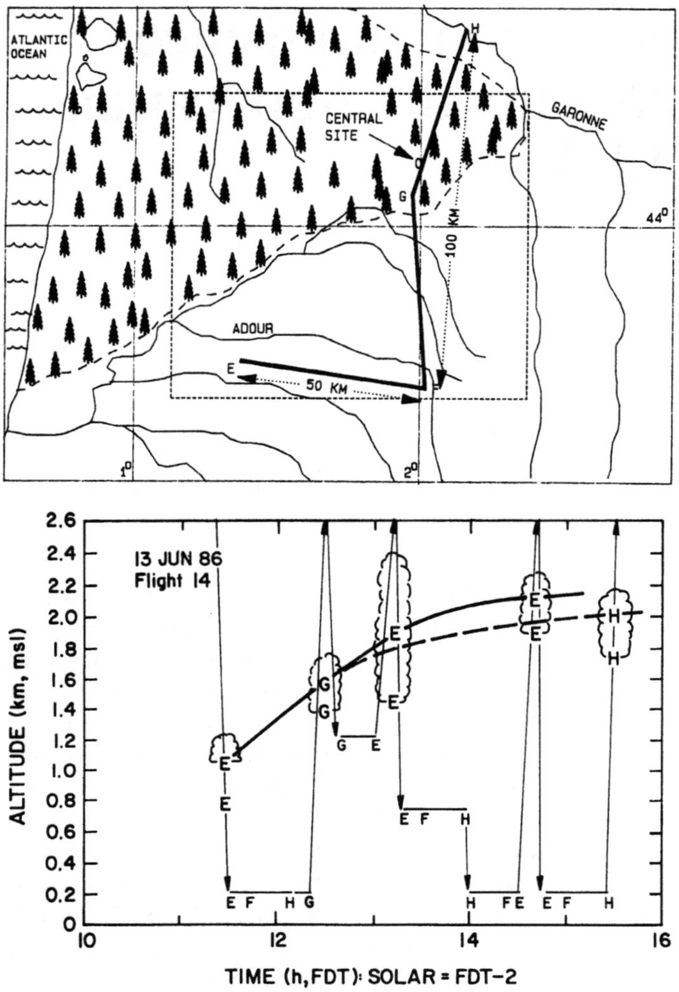

FIGURE 13.1. Flight plan showing the 50 and 100 km flight legs used for HAPEX (top) and sequence of flight legs flown on a typical day (bottom). The 100-km north–south flight leg was subdivided into 21 segments and the 50-km east-west flight leg was subdivided into 9 segments. These segments were used to evaluate the small scale variability of boundary layer turbulence.

sampled. Typical instrumentation calibration techniques evaluate only the low frequency gain and biases of the instruments. Evaluation of the high-frequency response of the instruments is typically evaluated using in-flight evaluations and comparisons between different systems. For measurement of convective boundary layer turbulence, frequency response up to about 50 Hz are needed, and for measurement of stable boundary layer turbulence, up to 250 Hz may be necessary.

Turbulent air motions are typically measured using a gust probe system that combines measurements from an Inertial Navigation System, static and dynamic pressure measurements, and a wind gust sensing system (See Research Aviation Facility, 1987b). Common wind sensing systems include vane-type strain gauges and pressure-sensing systems. Typical vane systems can be troublesome if extreme care is not taken in the design, maintenance, and protection of the system. Boom pressure systems frequently suffer from attenuation of high-frequency response due to pressure line length, and from noise due to boom vibration. An improved pressure-type gust probe (Brown et al., 1983) makes use of the pressure fluctuations around the aircraft's radome to measure the turbulent gusts. This system provides improved frequency response and more reliable operation and is being installed on numerous research aircraft.

Temperature is typically measured using a resistance element temperature probe. As mentioned above, good measurements of the turbulent spectrum require frequency response up to about 50 Hz. Many commercially available temperature sensors have inadequate frequency response for turbulence

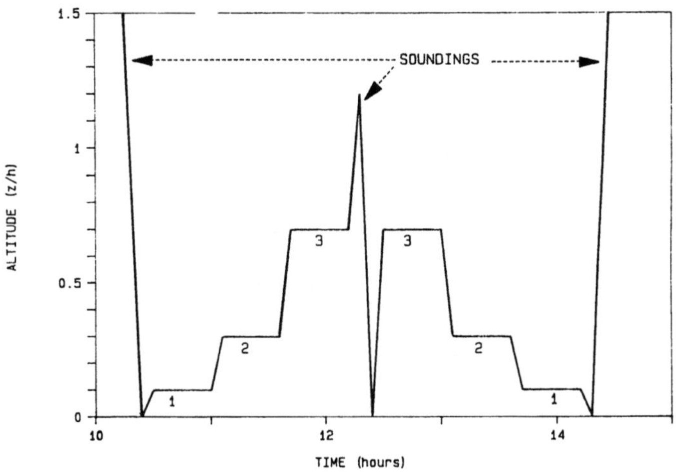

FIGURE 13.2. Conceptual flight plan which accommodates linear diurnal variability in the measurements (after the idea of FIFE Boundary Layer Group).

measurements (Spyers-Duran and Baumgardner, 1983). Adequate frequency response and calibration plus good exposure to the unperturbed airstream are the major concerns relating to temperature measurement (e.g., see Wyngaard, this volume).

Humidity is typically measured using Lyman-alpha hygrometers and thermoelectric hygrometers such as the EG&G 137 hygrometer (see Research Aviation Facility, 1987a). Both units have problems. The Lyman-alpha hygrometers typically suffer from drift in the output gain and bias and have to be calibrated against other, more stable humidity measurements. The dew point hygrometers, such as the EG&G 137, are typically of slow response and can also drift. Consequently, two units are frequently flown. Friehe et al. (1986) and others have discussed approaches to calibration of the Lyman-alpha hygrometer using other reference hygrometers.

3. Aircraft Flight Track Design

The major considerations for the design of an aircraft flight track for eddy correlation flux measurements are (1) representativeness of the flight track with respect to the terrain being studied, (2) length of the flight track with respect to the large eddy size, and (3) the ability to eliminate diurnal effects in the data using information contained in the data. Less important, but also critical are the efficiency of the use of the flight time, the ability to interject sounding at important locations, etc.

The representativeness of the flight track to the terrain being studied can be a significantly complicating factor for flux measurements over complex terrain. For example, the HAPEX research area was a 100 × 100 km area in the southwest of France. The land surface consisted of about 40% forest, 45% mixed agricultural use, and 15% wine growing. The HAPEX flux flight track (Fig. 13.1) was selected to cover most types of land use and terrain type. The flight track length should be selected keeping in mind the sampling guidelines for separation of mean and turbulent portions of the flow. For HAPEX, the E–W and N–S flight track lengths were 50 and 100 km. In the vicinity of the flight track, the typical land use scales were ≥ 20 km. The track was designed to be short enough that repeated flight tracks could be flown in a single flight.

Diurnal effects can easily produce biased data. This is particularly true for aircraft measurements, where the need to fly at different altitudes and over different terrain can produce the need to fly for 2–4 hr. During that time, considerable diurnal variation can occur. A variety of corrections for the diurnal trends can be made. In the HAPEX example, at the 80 m/s aircraft flight speed, it took about 30–35 min to fly the full 150 km flight track. Since repeated low level flights were made (Fig. 13.1, bottom), solar noon values could be reliably estimated using an interpolation technique.

The FIFE experiment made use of a novel flight plan (Desjardins and MacPherson, this volume) in which the aircraft made the measurements twice, with the first and second sets of measurements being made in opposite order (Fig. 13.2). Using

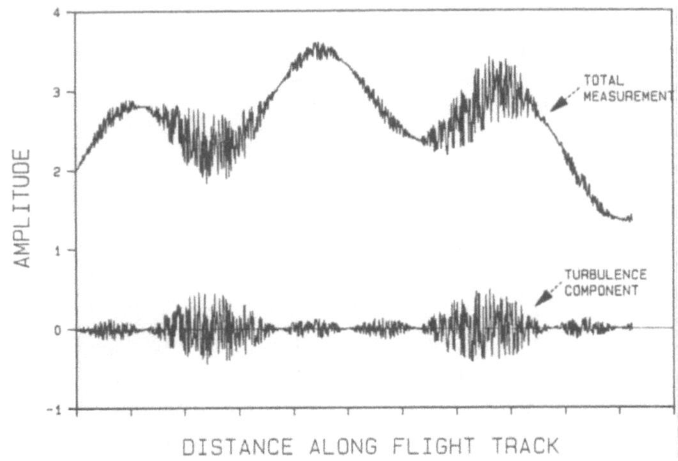

FIGURE 13.3. Simulation of turbulent flow showing decomposition of the total flow through filtering into a variable mean value, plus a turbulent component having only small scale deviations from the mean.

this type of flight pattern, simple averages of the data could be used to produce time-centered averages. This technique deserves much attention for it can produce time-centered data with minimal computational effort and almost no added burden on the flight plan. However, its success depends on the assumption of a linear time variation of fluxes with time. Grossman (this volume) explores the time variability of fluxes by using many short runs between the same two points.

4. Eddy Correlation Measurements of Turbulence over Complex Terrain

4.1 Data Processing

The King Air data system recorded measurements from the instruments listed above at a sample rate of 50 samples per second, and selected data were also plotted in real time (at one sample per second) for assistance in guiding the flight. After the field project, the data were processed to produce calibrated data at a rate of 20 samples per second. This data processing includes calibration, despiking, filtering, and recording of the data on magnetic tape and microfilm. The filtering process reduces possible aliasing problems due to high-frequency oscillations in the data by low-pass filtering the data using a recursive, 4-pole Butterworth filter having a cutoff at 10 Hz. The resultant data are recorded at 20 samples per second and are the standard NCAR RAF data (Research Aviation Facility, 1988c).

4.2 Decomposition into Mean and Turbulent Components

The decomposition of turbulence measurements into mean and turbulent components is based on the assumption that a representative mean value (p) exists, and that any total measured value (P) can be decomposed into mean (p) and turbulent (p') portions using the relation $P = p + p'$. This decomposition assumes there is a wavelength that accurately can be used to separate the long wavelength mean values (p) from the short wavelength turbulent (p') portions of the flow. Classical use of this decomposition assumes that the atmosphere is horizontally homogeneous and that a single mean value or linear trend will be representative of the mean state of the atmosphere.

This assumption implies that the flight path length or measurement domain is short with respect to the large scale atmospheric variability and at the same time is long with respect to the turbulence scales. This is equivalent to assuming that a representative mean exists and that a representative average of the turbulence can be made. We shall see that this is the crux of the sampling problem: in many cases, if the flight track is long enough to provide an accurate estimate of a mean value, then the mean state may have changed by the time the flight track is completed. The change in the mean state can come from diurnal changes or from flying from one mesoregion into another.

In carefully selected undisturbed atmospheres over homogeneous terrain, these conditions can

TABLE 13.1. HAPEX King Air Instrumentation for turbulence measurements.

Variable	Instrument	Manfacturer and model	Absolute accuracy	Resolution
Position	Inertial navigation	Litton LTN-51	≤1.0 N mi/flt hour	0.0014°
Statis pressure	Oscillation frequency	Rosemont 1501	± 1 mb	0.034 mb
Airspeed	Variable capacitance	Rosemont 1221	± 0.7 mb	0.006 mb
Air temperature	Platinum resistance	NCAR Fast Response	± 0.05°C	0.006°C
Air temperature	Platinum resistance	Rosemont 102	± 0.5°C	0.006°C
Humidity	Lyman-alpha hygrometer	NCAR LA-3	± 5%	0.2%
Humidity	Thermoelectric hygrometer	EG&G 137-C3-S3	± 0.5°C if T>0°C ± 1.0°C if T<0°C	0.006°C
Horizontal winds	Gust probe and INS	NCAR	± 0.1 m/s, t<0.6 hr ±(1.0+0.5t) t in hr	0.012 m/s
Vertical winds	Gust probe and INS	NCAR	± 0.1 m/s	0.012 m/s
Geometric altitude	Radio altimeter	Collins ALT-55	± 2–3%	0.1 m
Infrared radiation	Pyrgeometer 4–45 μm	Eppley PIR	–	0.4 W/m²
Visible radiation	Pyranometer 0.285–2.8 μm	Eppley PSP	–	0.4 W/m²
Surface temperature	Bolometric radiometer	Barnes PRT-5	± 1°C	0.005°C

hold well. Extensive measurements of convective boundary layer fluxes have been made using this technique, and typical temperature and moisture flux profile measurements illustrate a scatter of 10–30% in the temperature and moisture flux measurements (e.g., AMTEX, Wyngaard et al., 1978). Clearly, the major problem with this technique lies with the definition of the mean value and the technique is only as good as the initial assumption of a homogeneous atmosphere.

An improved method of decomposition of the data into turbulent and mean components can be through the use of spectral decomposition or filtering of the data. The techniques of spectral separation or bandpass filtering should be equivalent, provided the data are handled properly. In the HAPEX data processing (Hildebrand, 1988) bandpass filtering was used for computational convenience. Using these techniques, the data can be decomposed into short wavelength "turbulent" components and long wavelength "mean" components. This technique is schematically illustrated in Figure 13.3. In this decomposition, the variance of the total sample (0.27) is distributed between the variance of the mean (0.25) and the variance of the turbulent component (0.02). If this is not done, the total variance (0.27) would erroneously be attributed to turbulence and the decomposition wavelength would effectively be placed at the scale of the total measurement flight path.

For the flux calculations, the winds (u, v, w), temperature, humidity, and all other data for each flight leg (see Table 13.1), are input and processed as a single unit. The data are screened again for erroneous data values, then are decomposed into mean and turbulent values using a harmonic filter (Lanczos, 1956). For all HAPEX flights, the mean-turbulence cutoff filter wavelength was set to about 15 km (180 sec). This cutoff filter length was selected because the HAPEX boundary layer typically grew to about 1.5 km depth. About 10 times this typical maximum boundary layer depth was selected as the cutoff filter length as a compromise between adequate sampling of boundary layer turbulence and adequate resolution of the horizontal variability of the turbulence. The effects of the selection of filter length are more fully discussed in Section 5.2.

The result of this filtering process was to produce high-pass turbulence values and low-pass mean values for all variables. The high-pass data were used to calculate variances, covariances, and other quantities that were measures of boundary layer turbulence. The low-pass data were used to describe the average background state of the boundary layer.

Clearly, very different estimates of the mean and turbulent state of the measurements can result from use of the different analysis techniques. In the mean value technique, the decomposition wavelength is arbitrarily set to be the full measurement flight path length. In the spectral decomposition technique, the wavelength can be selected based on some physical reasoning. The major advantages of use of the filtering technique are the ability to view the spatial modulation of the mean and turbulent flow and the reasonably solid basis for the decomposition.

4.3 Calculation of Flux and Turbulence Values

The filtering process separates the total set of measurements into mean and turbulent components, retaining the original data sampling rate of 20 samples per second. From these data, individual mean values, variances, and covariances are calculated. The mean value of any measurement, for example, temperature or wind speed, is calculated as the average of the low-pass output for that variable, averaged over the spatial domain of interest. Similarly, the turbulence and flux values are calculated as the mean value of the instantaneous high-pass products for that variable, again averaged over the spatial domain of interest. Thus, vertical temperature flux would be the covariance of the T' and w' values, $T'w'$, averaged over the domain of interest.

For the HAPEX data processing, we calculated the mean and variance of u, v, w, T, and q; the vertical fluxes of u, v, T, and q; and horizontal fluxes of U, v, w, T, and q. Additionally, the turbulent kinetic energy (e) and vertical flux of e were calculated. Mean values of aircraft location, altitude above ground, pressure, radiative surface temperature, up- and downward long- and shortwave radiation were calculated, as well as total up- and downward radiation and net radiation. Boundary layer scaling parameters included L, $w*$, $u*$, estimates of surface values of Q_s and Q_l, and inversion height.

4.4 Measurement of Turbulence on Small Scales over Complex Terrain

Since the data rate is not affected by the decomposition of the total data set into turbulent and mean components, the decomposed data can be sampled over any domain of interest. The filter decomposition technique is thus ideally suited to studying the effects of inhomogeneous terrain on the atmosphere.

Unlike selection of the wavelength for decomposing the total flow into turbulent and mean components, selection of the averaging domain is not highly critical, for no turbulence energy will be artificially "lost" through partitioning into the "mean" state. Selection of too short or long an averaging domain will simply increase the effect of small scale inhomogeneities; selection of too long an averaging domain will attenuate the small scale signals. Using this technique, the selection of the averaging domain is an analysis decision that can be made on the basis of the scientific questions being asked. For studies of mesoscale variability of the boundary layer, a scale that encompasses several large eddies (i.e., on the order of $10h$: see Section 2.3) should adequately represent the average turbulence intensity.

In the case of HAPEX, the turbulence data were averaged over 10-km segments, with the segments located every 5 km along the flight track. This meant that the 150-km HAPEX flight track was broken into 30, 10-km-long subsegments located every 5 km along the flight track (Fig. 13.1). Successive 5-km observations were thus 50% correlated with each other. The effect of the averaging interval was tested with scales of 2 to 15 km. The 10-km averaging interval was selected on the basis of the relatively stable turbulence values that resulted and the need to provide adequate spatial resolution of the Landes Forest, which was about 45 km across.

5. Analysis of Errors Resulting from the Eddy Correlation Data Processing Technique

The accuracy of the aircraft measurements of turbulence is dependent on sensor measurement accuracy, the accuracy of the decomposition into turbulence and mean flow, and any atmospheric sampling errors. This section will address these three sources of error.

5.1 Input Sensor Accuracy

The basic measurements of winds, temperature, and humidity each has measurement accuracies and resolutions (Table 13.1) that are based on calibration procedures. These calibration procedures generally concentrate on calibration of the low-frequency response capabilities of the sensor. As noted in the last paragraph of this section, this can be a problem. The accuracy of individual products of the measured variables (Table 13.2) can be calculated from the sensor measurement resolution values.

The accuracy of mean turbulence values, averaged over many points, will be considerably

TABLE 13.2. Accuracy for turbulence measurements.[a]

Parameter	Units	Typical value	Error	Percent error
$\sigma^2_{u,u,w}$	$m^2\ s^{-2}$	1.0	0.07	7
σ^2_T	deg^2	0.04	0.008	21
σ^2_q	$(g\ m^{-3})^2$	0.04	0.028	71
$u'w'$	$m^2\ s^{-2}$	0.0	0.07	14–23
$v'w'$				
$T'w'$	$deg\ m\ s^{-1}$	0.12	0.013	11
$q'w'$	$g\ m\ s^{-1}$	0.09	0.023	26

[a] These accuracies are the accuracies of the individual products which are used to compose the measurements of turbulence.

enhanced over the accuracy of the individual sample values. If the sampling error has a Gaussian distribution, then the resultant error of the mean values (i.e., the numbers in Table 13.2) are reduced by a factor of $N^{0.5}$, where N is the sample size. For example in the HAPEX case, the individual 10-km data segments include about 3600 samples. Thus for those 10-km segments, the random instrumentation sampling errors listed in Table 13.2 would be reduced by a factor of about 60. Because of these sampling effects, with any reasonable sampling interval, the instrumentation measurement inaccuracy would be reduced to the point of insignificance. This conclusion is empirically supported by comparisons of measurements made on aircraft having multiple sensors or by comparisons of measurements made by aircraft flying in close formation.

The notable exception in HAPEX is the humidity measurements made using the Lyman-alpha instrument. The mean HAPEX humidity flux values from two different instruments typically differed by 10–30% and the humidity variance values by 20–70%. These differences are considerably larger than would be expected from any colocated instrument that is making accurate measurements. The best explanation of these differences is that the two Lyman-alpha sensors could have had different gains for high-frequency measurements. Unfortunately, a more definitive explanation has been elusive.

5.2 Errors Resulting from Decomposition of the Data into Mean and Turbulent Quantities

Decomposition of the measurements into mean and turbulent components requires analysis decisions concerning the definition of the mean value or, using the technique of this chapter, the cutoff frequency. These analysis decisions introduce errors or uncertainties into the turbulence estimates. Although it is not clear how to accurately estimate such errors, they can potentially be large. For example, since the large-scale variability of the wind speed near the surface is of the same order of magnitude as the wind speed, then assumptions that determine the mean/turbulent decomposition could change the turbulence values by as much as 50% of the magnitude of the winds.

As discussed by Wyngaard (1986, 1990) this volume), the spectrum of turbulence can be divided into components or ranges, the largest of which is the energy containing range. Kaimal et al. (1976) found that the spectral peaks scale as $1.5h(1\text{-}e^{-5z/h})$ for $0.1h<z<h$, where h is boundary layer depth and z is the measurement altitude. These large-scale, energy-containing eddies perform the majority of the turbulent transport in the convective boundary layer. These large eddies are highly variable in intensity, space, and time. Therefore, accurate measurement of boundary layer turbulence requires sampling a reasonably large number of these large eddies to attain an accurate ensemble average of boundary layer turbulence. Due to the size of the large eddies and the need to sample a number of them, flight tracks need to be fairly long if accurate turbulence measurements are to be made.

The selection of the filter cutoff wavelength was tested using HAPEX data. Turbulence values for $T'w'$, $q'w'$, σ^2_T, σ^2_q, stress, and turbulent kinetic energy, from 16 June 1986 in HAPEX are presented in Figures 13.4–13.6. These data consist of one pass of the aircraft over the 100-km, north–south flight leg (Fig. 13.1). The three curves in each figure correspond to filter lengths of about 10, 15, and 20 km (120, 180, and 240 sec of flight time). For each curve, the data were filtered using the chosen filter, then final 10 km mean data values were produced at intervals along the flight track.

These curves illustrate that as expected, the smoothness of the final data increased as the filter length increased. However, the changes were not large, and the smoothness typically changed by $<10\%$ as the filter length was changed between 10 and 20 km. In addition, the 15- and 20-km filter lengths produced data that were generally more similar than the 10 and 15 km data. It thus appeared that the effectiveness of the filter was decreasing for scales longer than 15 km. On this

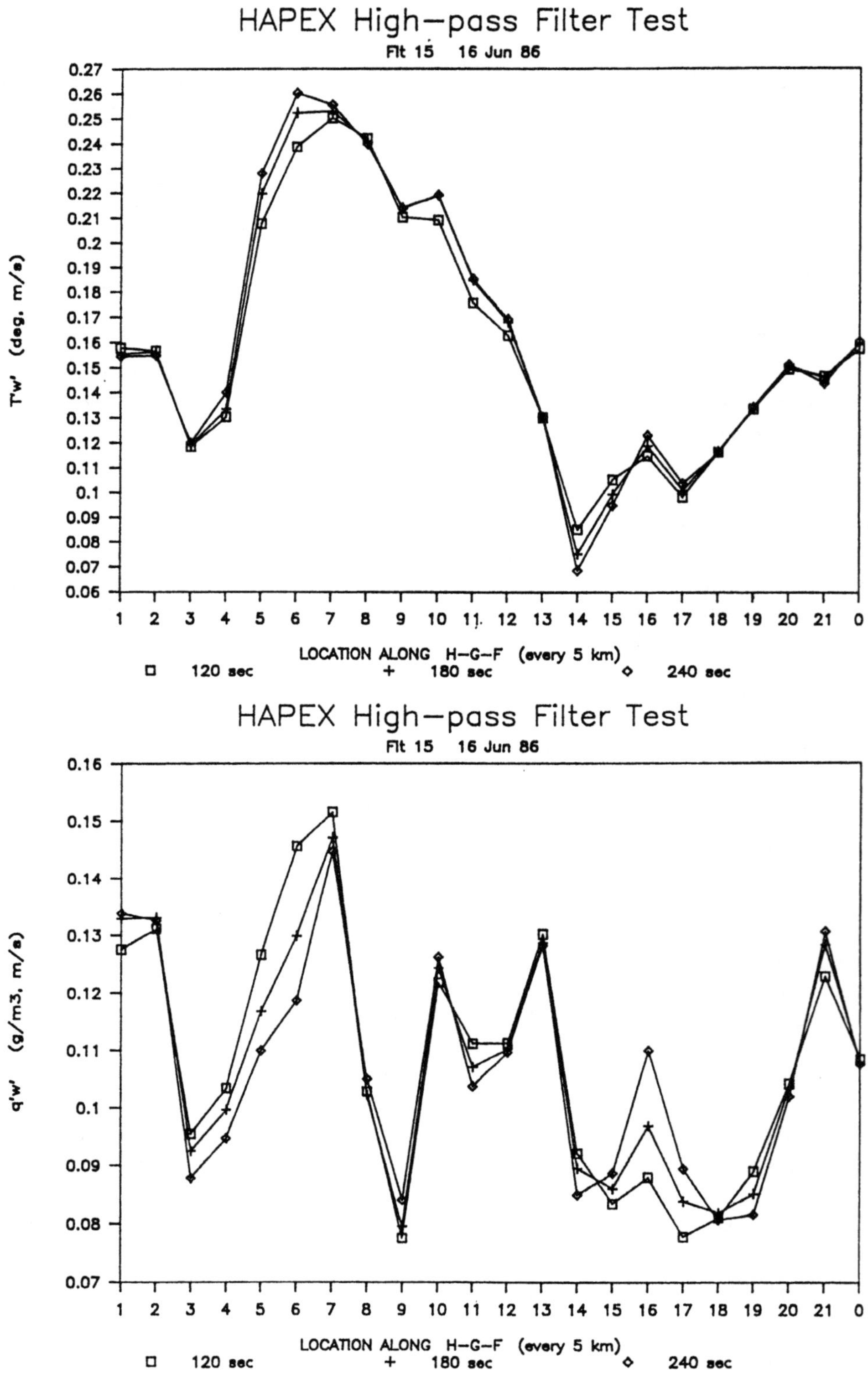

FIGURE 13.4. Measurements of $T'w'$ and $q'w'$ from the NCAR King Air aircraft during HAPEX. The numbers 30–10 correspond to the 21 segments along the north-south HAPEX flight leg, with 30 being at the north end and with segments 20–26 being over the forested area. The three curves are for filter cutoff wavelengths of about 10, 15, and 20 km (120, 180, and 240 sec of flight time).

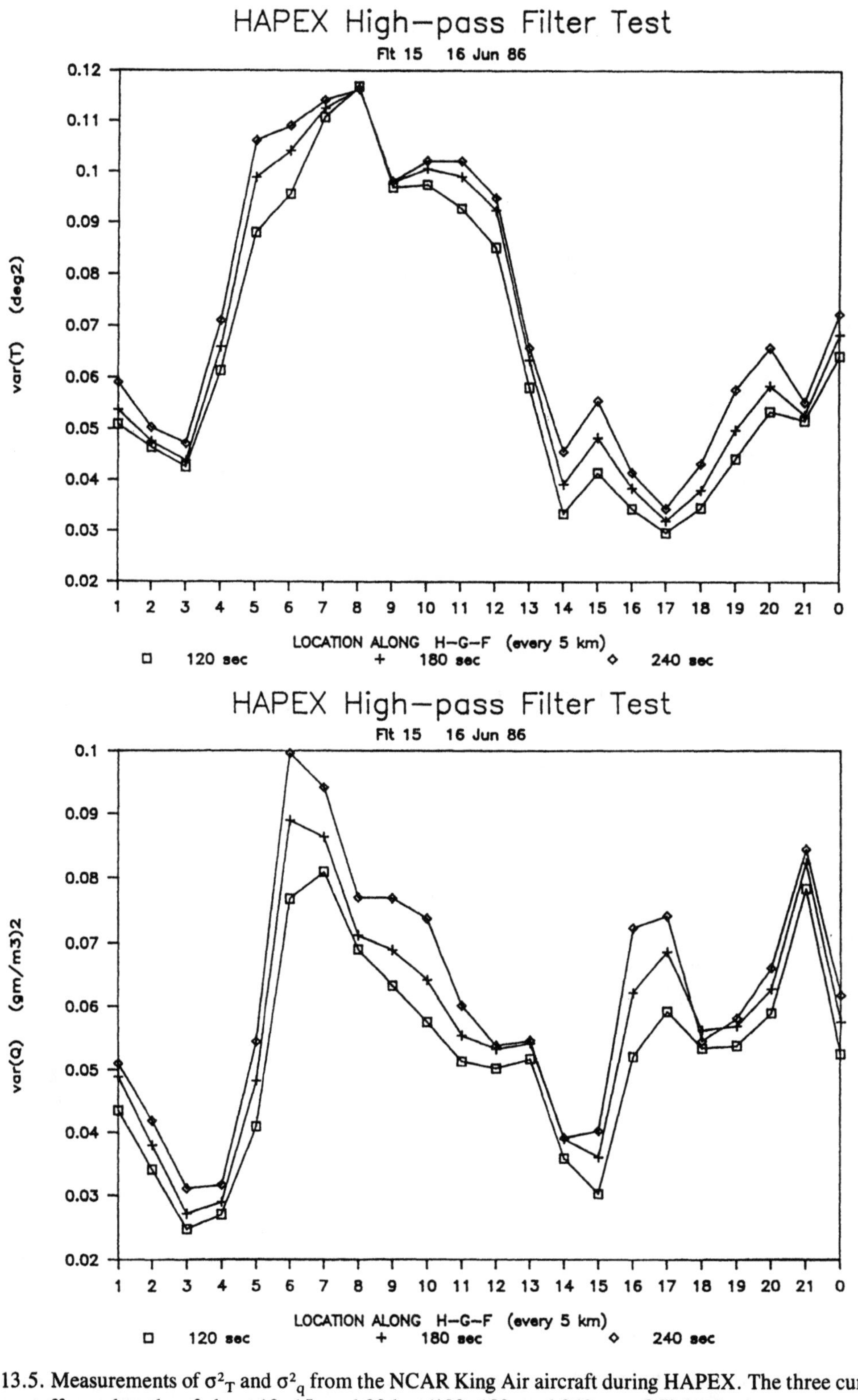

FIGURE 13.5. Measurements of σ^2_T and σ^2_q from the NCAR King Air aircraft during HAPEX. The three curves are for filter cutoff wavelengths of about 10, 15, and 20 km (120, 180, and 240 sec of flight time).

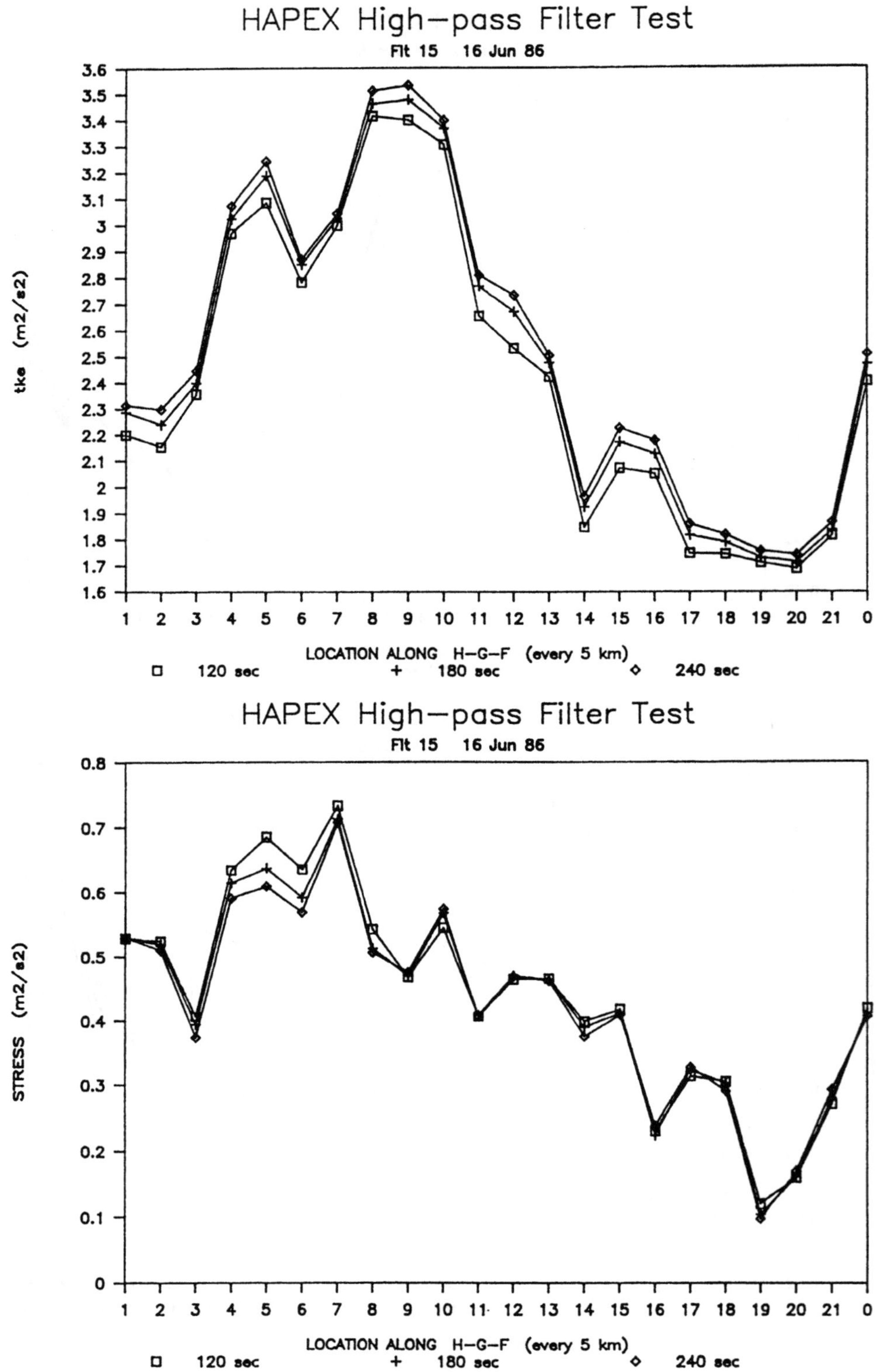

FIGURE 13.6. Measurements of turbulent kinetic energy and stress from the NCAR King Air aircraft during HAPEX. The three curves are for filter cutoff wavelengths of about 10, 15, and 20 km (120, 180, and 240 sec of flight time).

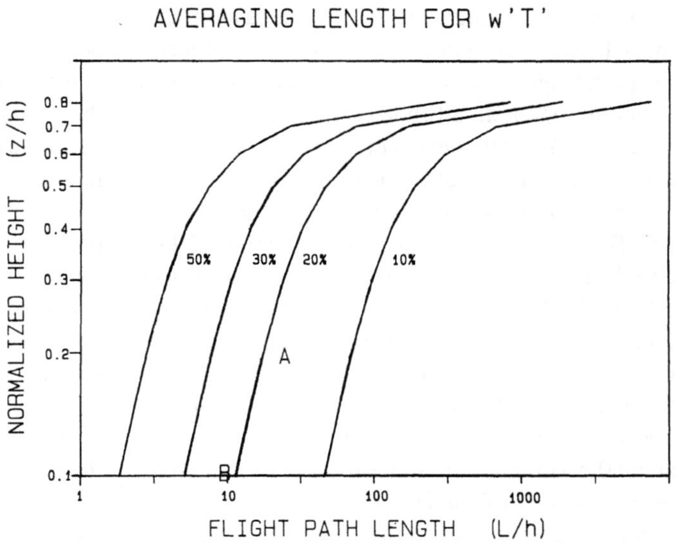

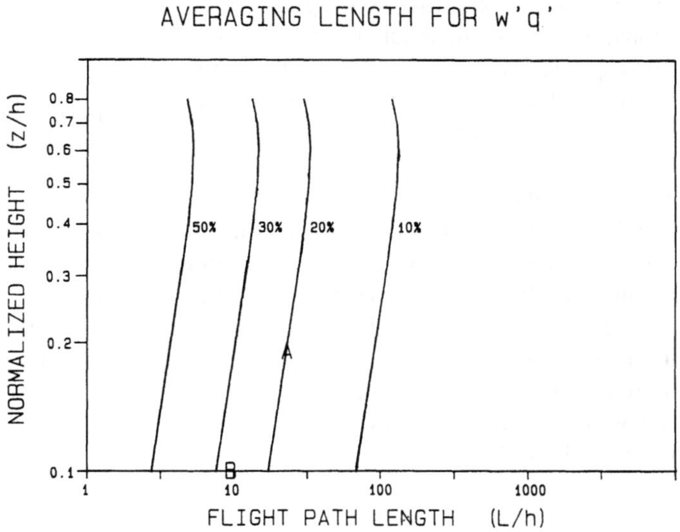

FIGURE 13.7. Vertical temperature and moisture flux accuracy as a function of flight path length and altitude in the boundary layer. The vertical scale is normalized boundary layer depth. The horizontal scale is flight path length, normalized by boundary layer depth. The curves of constant standard error of the temperature and moisture flux include 10, 20, 30, and 50% error curves, denoted by the numbers 1, 2, 3, and 5. Points A and B refer to examples from HAPEX measurements as explained in the text.

basis, and on the basis of the typical HAPEX terrain/land-use scales, the 15-km wavelength cutoff was selected for HAPEX. This 15-km wavelength corresponds to about 10 to 30 times the typical HAPEX boundary layer depths of 0.5 to 1.5 km.

5.3 Sampling Errors Relating to Flight Track Length

Lenschow and Stankov (1986) evaluate the accuracy of boundary layer turbulence estimates due to random sampling error, as a function of aircraft flight path length. For vertical temperature flux, they find

$$\frac{L_{wT}}{h} = \frac{0.32 z_*^{1/2}}{e^2}\left[3.2\left(\frac{1-0.8z_*}{1-1.2z_*}\right)^2 + 1\right]$$

where L_{wT} is the flight track length, $z_* = z/h$, h is the boundary layer depth, and e is the standard error of the estimate of $w'T'$. For the vertical humidity flux, they find

$$\frac{L_{qT}}{h} = \frac{0.32 z_*^{1/3}}{e^2}\left[3.2\left(\frac{1-0.8z_*}{1-0.5z_*}\right)^2 (1+2z_*)^3 + 1\right]$$

These relations are plotted in Figure 13.7 for $0.1 < z_* < .9$ and for values of e ranging from 0.1 through 0.5. These curves show that path lengths of about 10–100h will be needed to achieve a 20%

error in measurement of $w'T'$ for measurements made between altitudes of $0.1h$ and $0.7h$. Limiting the path length to about $30h$ will increase the error to about 30% at an altitude of $0.6h$. For measurements of $w'q'$, path lengths of about $18–30h$ will be needed to achieve a 20% measurement error.

For the HAPEX measurements, the decomposition filter was set to a fixed value of about 15 km, and the turbulence and fluxes were evaluated from 10 km block averages of the high-frequency filter output. The accuracy of these measurements can be estimated from the curves in Figure 13.7. When the measurements were made in the morning at 100 m agl, when the boundary layer height was about 0.5 km, the $w'T'$ and $w'q'$ values should have been accurate to about 20% (points A in Fig. 13.7). Later in the day when the boundary layer had grown to about 1 km, the 100 m agl measurements should be accurate to about 20% for $w'T'$ and 27% for $w'q'$ (points B in Fig. 13.7). The curves of Figure 13.7 also indicate that the 10 km averaged values of $w'T'$ should be considerably less accurate for the mid and high level flight levels that were at about $0.3h$ and $0.6–0.7h$. However, for the full length 100-km HAPEX flight track, the vertical flux measurements should have been accurate to within 10% plus any time variability due to diurnal changes. The variability of these turbulence measurements agrees well with those from previous, homogeneous boundary layer projects, e.g. (Lenschow, 1986, Lenschow and Stankov, 1986, Wyngaard et al., 1978).

6. Summary of Sampling Errors

The net effect of the different sampling errors is to limit the accuracy of aircraft turbulent flux measurements to about 10–20%. These errors are primarily due to uncertainties related to the formation of an ensemble average of the turbulence as measured along the aircraft flight track. The aircraft flight track provides spatial sampling only through its length, and unfortunately, a long flight track will produce problems due to mesoscale atmospheric variability and diurnal effects. Thus, it seems the best measurement approach is to plan for data collection that includes estimation of diurnal effects, while also providing a long flight path that includes multiple flights over the same surface.

Instrumental errors and errors resulting from the decomposition of the total measurement into mean and turbulent values are small, in comparison to the sampling problems relating to acquiring an adequate sample of the turbulence. It appears likely that other such errors can be held to 5–10%.

The major hope for improved measurement of turbulence intensities and transport should thus be directed toward use of remote probing techniques to extend the domain of the aircraft horizontally to the sides, and if possible also vertically. Multiple wavelength lidar and radar systems may possibly be developed that can extend the measurements out a few kilometers from the aircraft. Even extending the measurements out to $1h–1.5h$ on either side of the aircraft could produce a significant improvement in sampling capabilities.

Acknowledgment. The National Center for Atmospheric Research is supported by the National science Foundation.

References

Andre JC, Goutorbe JP, Perrier A (1986) HAPEX-MOBILHY: A hydrologic atmospheric experiment for the study of water budget and evaporation flux at the climatic scale. Bull Am Meteorol Soc 67:138–144.

Brown E, Friehe C, Lenschow D (1983) Thus use of pressure fluctuations on the nose of an aircraft for measuring air motion. J Climat Appl Meteorol 22: 171–180.

Desjardins RL, MacPherson JI (1990) Water vapor flux measurements from aircraft. In "Measurement and Parameterization of Land-Surface Evaporation Fluxes" (T. Schmugge, J-C. Andre, this volume eds.). Springer-Verlag, New York.

Friehe C, Grossman R, Pann Y (1986) Calibration of an airborne Lyman-alpha hygrometer and measurement of water vapor flux using a thermoelectric hygrometer. J. Atmos Ocean Tech 3:299–304.

Grossman R (1990) Temporal variation of heat and moisture flux within the atmospheric boundary layer over a grassland. In "Measurement and Parameterization of Land-Surface Evaporation Fluxes" T. Schmugge, J-C. Andre, this volume eds.). Springer-Verlag, New York.

Hildebrand P (1988) "Flux and Sounding Data from the NCAR King Air Aircraft during HAPEX" NCAR Tech. Note, NCAR/TN-319+STR.

Kaimal JC, Wyngaard JC, Haugen D, Cote O, Izumi Y, Caughey S, Readings C (1976) Turbulence structure in the convective boundary layer. J Atmos Sci 33:2152–2169.

Lanczos D (1956) "Applied Analysis." Prentice Hall, Englewood Cliffs, NJ.

Lenschow D (1986) Aircraft measurements in the boundary layer. In "Probing the Atmospheric Boundary Layer" (D. Lenschow, ed.), pp. 39–55. American Meteorological Society, Boston.

Lenschow D, Stankov B (1986) Length scales in the convective boundary layer. J Atmos Sci 43:1198–1209.

Resaerch Aviation Facility (1987a) Airborne Humidity Measurements. "Research Aviation Facility, Bulletin 22."NCAR, Boulder, CO.

Research Aviation Facility (1987b) Measurements Techniques: Air Motion Sensing. "Research Aviation Facility, Bulletin 23." NCAR, Boulder, CO.

Research Aviation Facility (1988a) The King Air: Overview and Summary of Capabilities. "Research Aviation Facility, Bulletin 2." NCAR, Boulder, CO.

Research Aviation Facility (1988) Flight Planning: King Air. "Research Aviation Facility, Bulletin 5." NCAR, Boulder, CO.

Research Aviation Facility (1988c) Standard Output Data Products from the NCAR Research Aviation Facility. "Research Aviation Facility, Bulletin 9." NCAR, Boulder, CO.

Spyers-Duran PA, Baumgardner D (1983) In-flight estimation of the time response of airborne temperature sensors. "PrePrints, Fifth Conference on Meteorological Observations and Instrumentation" Toronto, pp. 352–357. American Meteorological Society, Boston.

Stull R (1986) Observer's Logs from HAPEX. (Available from the author.)

Wyngaard J (1986) Measurement physics. In "Probing the Atmospheric Boundary Layer" (D. Lenschow, ed.), pp. 5–18. American Meteorological Society, Boston.

Wyngaard J (1990) On the maintenance and measurement of Scalar fluxes. In "Measurement and Parameterization of Land-Surface Evaporation Fluxes" (T Schmugge, J-C. Andre, eds.), this volume. Springer-Verlag, New York.

Wyngaard J, Pennell W, Lenschow D, LeMone M (1978) The temperature-humidity covariance budget in the convective boundary layer. J Atmos Sci 35:47–58.

14 Water Vapor Flux Measurements from Aircraft

R.L. Desjardins and J.I. MacPherson

1. Introduction

Large-scale measurements of land surface evaporation are required to develop techniques of integrating small-scale measurements into regional estimates of evaporation. These are necessary inputs into general circulation models that are used for weather forecasting (Mintz, 1984). They are also important for assessing and controlling water resources. Most of the techniques that have been developed over the years to estimate regional evaporation do not account for the heterogeneity in landform, soil, and vegetative cover (Brutsaert, 1982). Interpretation of satellite data in combination with meteorological data represents a realistic approach for taking this heterogeneity into account and providing representative regional estimates of evaporation (Andre et al., 1986; Sellers et al., 1988). Regional near-surface observations of evaporation fluxes may serve to evaluate the usefulness of such remote sensing models. A direct and promising approach of obtaining near-surface evaporation estimates is from turbulent flux measurements of water vapor by the eddy correlation technique. Such measurements have been obtained for extended periods using ground-based systems and for short periods using aircraft-mounted sensors.

Airborne flux measurements, which allow for spatial integration of data, provide a particularly promising link for relating small-scale ground-based measurements to large-scale estimates provided by satellites and meteorological data. However, there are many factors that affect airborne flux estimates of evaporation, such as type of sensor used, run length, altitude, environmental conditions, and averaging procedures. The purpose of this chapter is to provide information on the instrumentation, recorded parameters, flight scenarios, and analysis procedures to obtain water vapor flux measurements from aircraft. Examples of evaporation fluxes over land surfaces for various conditions and scales will be presented and discussed.

2. Literature Review

2.1 Eddy Correlation Technique

One of the most direct techniques to estimate water vapor flux is based on the Reynold's approximation; it computes the following mean products:

$$F_C = \frac{1}{N} \sum_{t=1}^{N} (C_t - \overline{C})(W_t - \overline{W}) \quad (14.1)$$

$$= \frac{1}{N} \sum_{t=1}^{N} (C_t' W_t')^{1} \quad (14.2)$$

where C_t, the mixing ratio of water vapor, equals the mean mixing ratio, $\overline{C}$, plus the deviation from the mean, C_t', and W_t, the vertical air velocity, equals the mean vertical wind, $\overline{W}$, plus the fluctuation from the mean, W'. These values are measured at equal intervals of time over a total time duration T. Equation 14.2 provides a fundamental method for measuring flux. Its simplicity is due to the following assumptions:

1. Horizontal inhomogeneities of variable C are neglected
2. Mean values can be defined unambiguously.

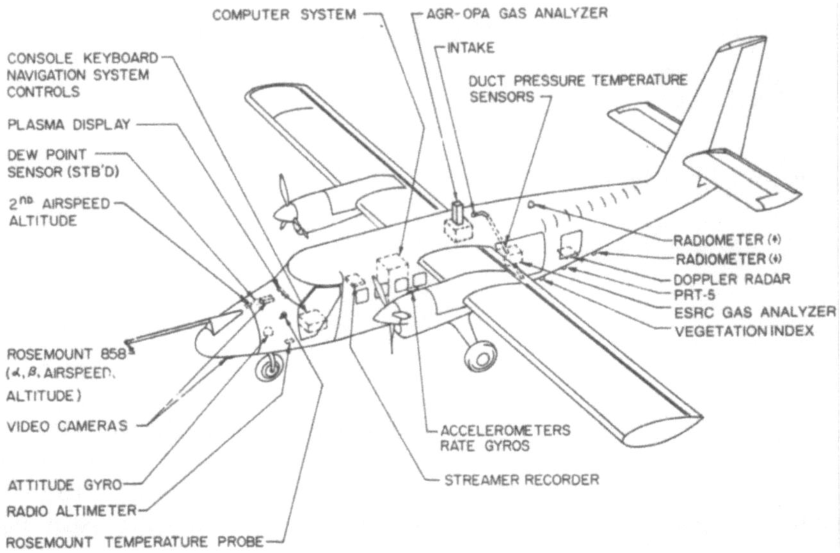

FIGURE 14.1. Schematic of atmospheric research aircraft to measure evaporation flux.

With these assumptions, we are dealing with a model that can be considered to describe stationary conditions, in a time average sense, over a rough plane of infinite extent, with a flow parallel to the X axis at all heights.

The use of aircraft for eddy flux measurements is justified by the assumption that turbulence structures can be sampled equally well from a fast moving platform as by recording in time at a fixed location. Aircraft have the advantage of being capable of probing the entire depth of the mixed layer rapidly and of obtaining statistically significant measurements of an order of magnitude faster than fixed point measurements. Another important advantage is the ability to sample both perpendicular and parallel to the wind. In addition, the advantage of measuring over inaccessible terrain is obvious. Development of fast response sensors for turbulence and water vapor measurements and the availability of high capacity data-gathering systems have opened the door to widespread use of aircraft as observation platforms for regional fluxes of gases.

2.2 Aircraft Measurements of Evaporation Flux

Aircraft measurements provide a good spatial picture of the mean and turbulent structure of the mixed boundary layer. During the past 15 years, fluxes of water vapor using aircraft-mounted sensors have been measured by many research groups (Grossman and Bean, 1973; McBean and Peterson, 1975; Bean et al., 1976). The NOAA WP-3D weather research aircraft equipped with gust probe systems and a microwave refractometer has been used to estimate evaporation over the ocean (Greenhut and Khalsa, 1982). Eddy flux measurements of water vapor recorded simultaneously with three powered gliders along the sides of a triangle at heights of 150, 300, and 450 m, although highly variable, showed a remarkable degree of vertical consistency with time (Hacker, 1982). Durand et al. (1987) measured sensible and latent heat fluxes over complex terrain and found no direct relationship between the relief and these fluxes at 100 m above the main ground level. The NCAR King Air measured H_2O flux in a variety of experiments. Recently, as part of HAPEX-MOBILHY, water vapor flux has been measured using two Lyman-alpha units over distances of 100 km (Hildebrand 1988). The Canadian Twin Otter has also been used for estimating regional evaporation in Manitoba to compare with ground-based measurements (Desjardins et al., 1988b) and in Québec to compare with estimates of evaporation from a regional water balance model (Schuepp et al., 1987, 1988). As part of FIFE

TABLE 14.1. Characteristics of variables recorded for water vapor flux measurements.

Variable	Reliability of absolute value	Resolution	Frequency response	Instrument type
Vertical air velocity	0.1 m s^{-1}	0.01 m s^{-1}	10 Hz	Gust probe; inertial velocity
Horizontal velocity components	0.5 m s^{-1}	0.01 m s^{-1}	10 Hz	Gust probe; inertial velocity
Humidity	0.5 g m^{-3}	0.1 g m^{-3}	2°C/s	Dew point hygrometer
Humidity fluctuations	1 g m^{-3}	0.01 g m^{-3}	> 15 Hz	Radiation absorption
Air temperature	0.5°C	0.01°C	10 Hz	Platinum resistance
Pressure	1 mb	0.1 mb	1 Hz	Quartz crystal transducer
Absolute altitude	2–3 m	0.2 m	1 Hz	Radar altimeter
Heading	2°	0.1°	1 Hz	Inertial navigation system; gyro compass
Position	0.5 km	0.1 km	1 Hz	INS, GNS, Loran-C
Surface temperature	1.0°C	0.01°C	1 Hz	Infrared radiometer
Solar radiation (upward and downward)	20 W m^{-2}	2 W m^{-2}	0.2 Hz	Thermopile

(Sellers et al., 1988), extensive H_2O flux measurements were recorded over a 15 by 15 km area of relatively uniform vegetation with enhanced meteorological and satellite coverage.

3. Instrumentation

Flux measuring aircraft can collect data on scales from a few meters to hundreds of kilometers (Desjardins et al., 1982; Lenschow, 1986). Results from various experiments indicate that at an aircraft speed of 100 m s^{-1} most of the turbulent transport can be attributed to wavelengths > 10–20 m (Lenschow et al., 1981; Nicholls, 1983). Data are usually low pass filtered at 5–10 Hz and sampled at 16–32 Hz to ensure adequate sampling of air motion. Figure 14.1 shows the relative position of the various sensors on the Twin Otter aircraft. A list of the variables used to measure aircraft position, air motion, water vapor concentration, air temperature, atmospheric pressure, and solar radiation conditions is given in Table 14.1 along with their reliability, resolution, and frequency response. Technical details of the main variables are given below.

3.1 Position

Various methods are available for calculating aircraft position. One of them is the Loran-C system. Although being fairly reliable, its use in some conditions can give rise to significant errors. Depending on the location of the area relative to the transmission stations, a fairly consistent offset can occur in the Loran measured position. For example in Kansas, during FIFE, the Loran-C recorded a position 0.9 km northeast of the actual position. This type of offset can be removed by adjusting the waypoint data entered to the Loran-C, in effect moving the earth into the Loran reference frame. A second source of error is a lag within the Loran system in computing aircraft position. For example, reciprocal east–west runs over a ground marker will produce longitudes that are too low on westbound runs and too high on eastbound runs. For both north–south and east–west tracks, the computed lag was 0.5 km. A flight track plot produced from the recorded Loran-C data is shown in Figure 14.2a for two grid flights. The positioning error is shown by the calculated position of the aircraft at 32 points during four complete grid runs identified as A, B, C, and D (Fig. 14.2b). A difference of approximately 0.01°, equivalent to 1 km, results for the location of the points when the aircraft is moving in opposite directions. In this case grids A and C were flown with reciprocal heading to grids B and D. Based on the speed of the aircraft, the Loran position data can be corrected by shifting the recorded data by the estimated time interval.

Geographic position is also recorded using other navigation systems. Monitoring of the Litton-90 inertial reference system data indicated that for the case just presented the recorded position drifted an average of about 2 km/h to the south. In general,

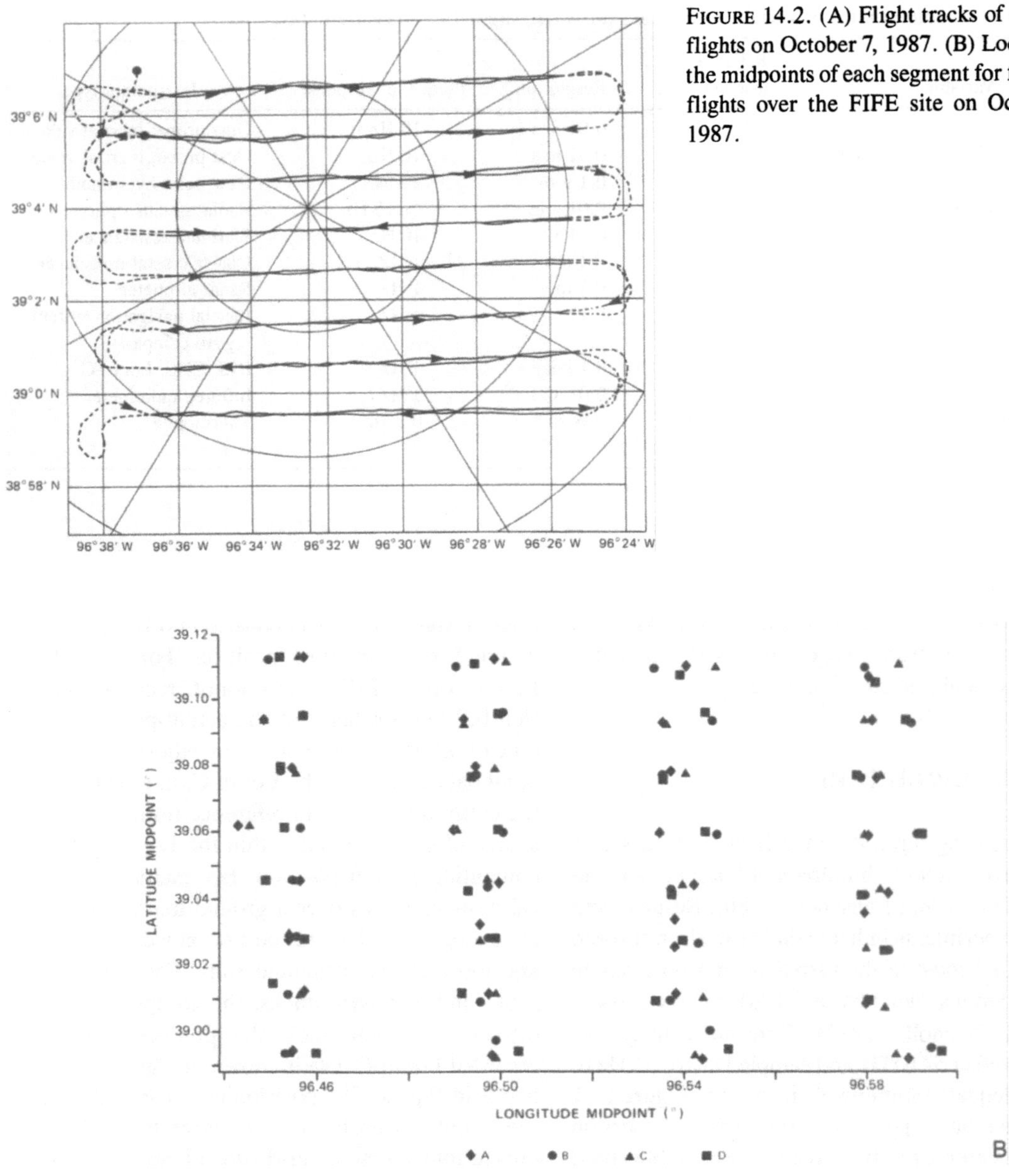

FIGURE 14.2. (A) Flight tracks of two grid flights on October 7, 1987. (B) Location of the midpoints of each segment for four grid flights over the FIFE site on October 7, 1987.

accuracy of ± 1/2 km can be expected with most navigation systems. With the use of video camera and satellite imagery, it is possible to locate the track of the aircraft within 10s of meters, which is much more accurate than our present estimate of the actual area being sampled (Schuepp et al., 1988). The satellite-based Global Positioning System (GPS) offers real promise of much improved accuracies over the next decade.

3.2 Air Motion

The mobility of aircraft is well recognized as an asset to quantify fluxes for the entire depth of the mixed layer but this mobility requires precise correction for the motion of the aircraft during computation of the true air motion. To calculate the atmospheric motion, both the velocity and angular orientation of the aircraft relative to the earth as

FIGURE 14.3. Wind sensors for turbulence measurements on aircraft. (A) Differential pressure probe; (B) fixed vane probe; (C) gust probe with balsa vanes.

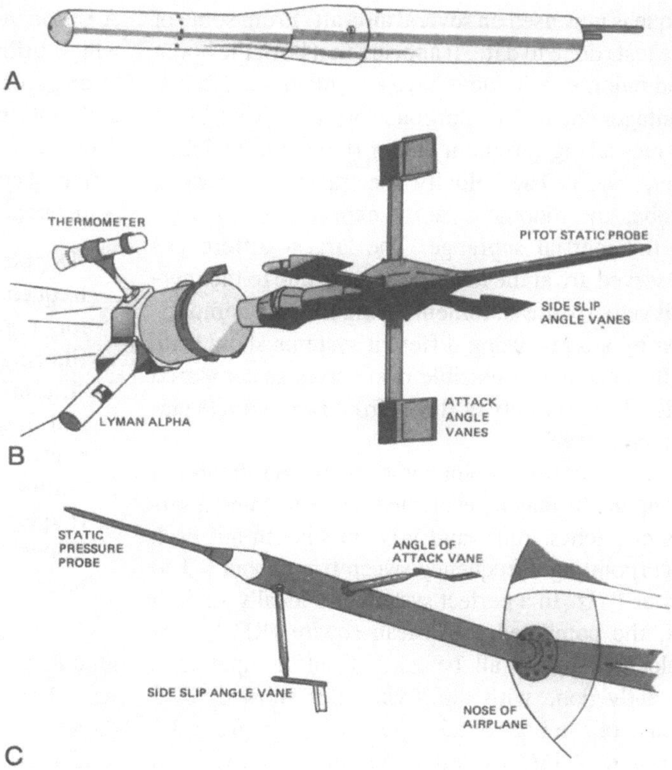

well as the velocity of the air relative to the aircraft need to be measured. Airplane velocity relative to the earth can be obtained from accelerometers on an inertial navigation system or from either a Doppler radar or a radio navigation technique. The air velocity relative to the aircraft is obtained from the difference between the dynamic and static pressure in a pitot tube usually installed on the nose of the aircraft (Fig. 14.3). A variety of techniques have been developed to measure the transverse air velocity components (Brown et al., 1983; Lenschow, 1986). Most gust boom installations use a Rosemount 858 5-hole probe (Fig. 14.3a) and the associated pressure transducers to measure static pressure (altitude), dynamic pressure (airspeed), and the angles of attack and sideslip. Fixed vane probes (Fig. 14.3b) have also been used to measure both attack and sideslip angles (Johnson et al., 1978). They have the advantage of fast frequency response, up to approximately 100 Hz. The limited reliability of strain gauge bridges and poor performance in clouds and rain have been a problem. Light and rugged balsa vanes, which align themselves with the airstream (Fig. 14.3c), have also been extensively used to measure air turbulence. With low mass/area vanes, accurate counterbalancing of the vane mass and an angle transducer with minimal inertia and resolution of 0.01°, accurate measurement of turbulent air motion is possible up to frequencies of 10 Hz (Nicholls, 1983). They are not suitable for flight in icing conditions, however.

All these sensors are installed on a noseboom, approximately 2–4 m in length, that places the sensors far enough ahead of the aircraft's nose to minimize airflow distortion effects on the measurements and yet maintain sufficient structural integrity to avoid undue vibrations in the range of interest. Noseboom structure is usually such that the natural frequency is at least 14 Hz, that is a frequency where there is very little contribution to the flux.

In the past 5 years differential pressure measurements have also sometimes been obtained directly on the airplane nose. This so-called "radome" technique is now used on several aircraft. From some of the tests done to date, it appears that both the boom and radome techniques have advantages and disadvantages and neither approach is the limiting factor in measuring turbulent fluxes (Greenhut and Gilmer, 1985). The velocity spectra from the vane, probe, and radome have the expected $-5/3$ slope in the inertial subrange. The largest differences observed are at the low frequencies due to the inertial velocity measurement. Wing to wing comparison by aircraft using different systems show little difference in the sensible heat fluxes as compared with the variability of fluxes from run to run (MacPherson 1988).

The adequacy of compensation for aircraft motion components may be evaluated by performing a series of pitches, rolls, and yaws in smooth air, each incorporating a frequency sweep from about 0.1 to about 1 Hz. In a perfect system and ideally smooth air, the computed root mean square (RMS) gust velocities would all be zero. Control input tests recently done with the Twin Otter showed RMS values of the horizontal wind component from 0.1 to 0.5 m s^{-1}. Pitches caused the worst disturbance in the vertical component, while yaws affected the horizontal components most.

To estimate potential effects such variations have on the RMS gust velocities, the mean square of the control-induced gust velocity should be subtracted from the mean square of typical measured gust velocities. For the horizontal component, a typical RMS value measured during evaporation flux measurements would be about 1.3 m s^{-1}. Using the worst case control-induced RMS of 0.5, the mean square of the true gust would be $(1.3 \times 1.3 - 0.5 \times 0.5) = 1.44$ giving a corrected RMS of 1.2 m s^{-1}. For a typical vertical gust RMS of 1.0 m s^{-1}, the corresponding corrected RMS would be 0.87 m s^{-1}. Thus the worst-case control-induced errors would add about 0.1 m s^{-1} to measured RMS gust velocities. In practice, straight and level project runs produce aircraft attitude disturbances very much less than those used in the control input tests. RMS values of 0.2 are representative of error levels on normal runs, which would translate to an error in a typical RMS vertical gust velocity of only 0.02 m s^{-1}.

Alternative air motion computations are frequently done on aircraft. For example, precipitation can cause degradation of the inertial velocity derived from the Doppler radar on the Twin Otter. A Litton 90 Inertial Reference System is also used which utilizes precision accelerometers and ring laser gyros to compute aircraft position, attitude, and motion in both aircraft- and earth-fixed axes. This provides the capability to estimate winds with different combinations of inertial velocity systems. An example of three is given here:

1. "Doppler" winds rely on Doppler radar for low frequencies combined with the accelerometers for high frequencies using complementary filtering (MacPherson et al., 1981).
2. "Litton" winds rely on Litton system for low- and high-frequency inertial velocity component.
3. "Litton/Doppler" winds rely on Doppler for low frequencies and Litton accelerometers for high frequencies, again using complementary filtering.

Comparing average winds computed for one minute flights along each cardinal heading in a wind box gave RMS variations of 2–4° in wind direction and 0.4–0.8 m s^{-1} in wind speed. For longer averaging times the variations would be slightly reduced.

Three test runs along a calibrated railway track have shown that the Doppler ground speed is accurate to within 1% (0.6 m s^{-1}). Results are presented in Table 14.2 using the above three wind calculation methods obtained for each of the three successive runs along the railway track. One can observe the magnitude of the differences using various systems. The agreement is especially good in the RMS values, which are most important for eddy flux measurements. One notes that the effect of high pass filtering on the sensible heat flux is more important than the choice of inertial data used. This limited data set is presented primarily as an indication of the magnitude of the variability in the flux estimates due to the wind measurements rather than an evaluation of various systems.

3.3 Water Vapor Measurements

The most commonly used device for measuring absolute humidity is the dew point hygrometer, which cools a mirror until a change of reflectance of the mirror surface is detected optically. The mirror is then heated until the water starts to disap-

TABLE 14.2. Wind data from three flux runs flown along a railway track.

Variable	Run	Doppler winds	Litton winds	Litton/ Doppler winds
Direction	01	349/7.5	351/9.3	352/7.5
(deg)/speed (m s^{-1})	02	349/7.6	353/6.7	346/7.4
	03	360/6.3	359/6.7	358/6.6
U RMS (f)[a]	01	1.48	1.48	1.49
(m s^{-1})	02	1.40	1.36	1.38
	03	1.45	1.45	1.46
V RMS (f)	01	1.25	1.19	1.18
(m s^{-1})	02	1.06	1.05	1.11
	03	1.24	1.19	1.22
W RMS (f)	01	0.87	0.87	0.89
(m s^{-1})	02	0.89	0.91	0.94
	03	0.85	0.86	0.89
Heat flux (f)	01	96	104	103
(W m^{-2})	02	74	75	77
	03	99	113	109
Heat flux	01	114	128	117
(unfilt) (W m^{-2})	02	95	83	75
	03	100	118	113
UW (f)	01	−0.67	−0.67	−0.71
(N m^{-2})	02	−0.78	−0.83	−0.81
	03	−0.70	−0.77	−0.71
		Track distances (km)		
	Run	Integrated Doppler	Integrated Litton	Actual (map)
	01	10.47	10.23	10.45
	02	10.48	10.47	10.45
	03	10.48	10.46	10.45

[a](f) means high-pass with a cutoff frequency of 0.012 Hz (5 km wavelength) was used.

pear. Although this device has a slow response (2° C/s), it provides a good reference for a faster responding sensor (Coantic and Friehe, 1980).

For flights near the surface, faster responding sensors are essential. As reviewed by Hay (1980), Lyman-alpha is most frequently used; it is based on the absorption of light by hydrogen in the open path at the 121.5 nm wavelength. The device consists of a source tube of a hydrogen neon mixture and a detector tube containing nitrous oxide (Buck, 1976). Magnesium fluoride windows are usually used for the source and detector tubes. The sensor is small and is sometimes placed directly on the noseboom of the aircraft. Problems associated with these sensors include the following: (1) transmission properties of the windows and source strength changes with time and temperature (Mestayer and Rebattet, 1985), (2) source tubes emit undesirable wavelengths in addition to the Lyman-alpha wavelength, and (3) the effective path length may be difficult to determine, particularly on an aircraft, where flow distortion and pressure fluctuations are large.

FIGURE 14.4. Infrared H_2O analyzer mounted in a duct within an aircraft.

Microwave refractometers are also sometimes used on aircraft (Bean and Emmanuel, 1980). At microwave frequencies of 9 GHz, the humidity-dependent refractive index of the air is measured in a cavity simultaneously with the air temperature and pressure. This permits the water vapor density to be determined with a resolution of 0.02 g m^{-3} (Dobson, 1980).

Infrared analyzers have also found applications in aircraft studies. They measure the transmittance through a path length of the order of 1 m of air at 2.6 μm where H_2O absorption is strong and at 3.9 μm where absorption of H_2O is minimal. In the instrument shown in Figure 14.4, the light undergoes 10 internal reflections from the mirrored top and bottom of the sampling section, which is mounted to form part of the duct through the aircraft. An internal reference compensates for variations in the operation of the instrument but does not completely eliminate the effect of dirt on the mirrors, which for the given instrument causes a slight drift in the absolute value and a reduction of a few percent in sensitivity after a couple of hours of flight. Its resolution is 0.01 g m^{-3}. The sensitivity of such instruments changes with altitude due to pressure broadening (Burch et al., 1962). This can be corrected for easily, since most flux runs are done at a constant pressure altitude. To compute the mixing ratio with respect to dry air for use in

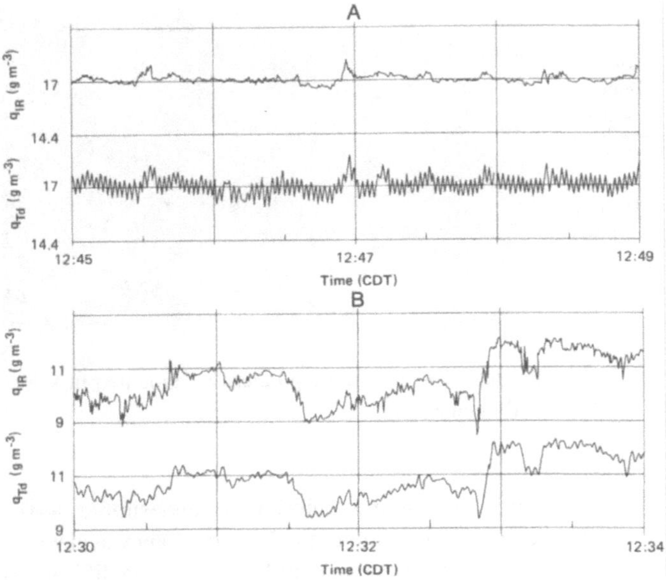

FIGURE 14.5. Time traces of water vapor density from a fast response infrared analyzer, q_{IR}, and slower responding dew point hygrometer, q_{Td}, recorded at 170 m above ground: (A) 15 August 1987; (B) 17 August, 1987.

the flux calculation, pressure and air temperature also need to be measured in or adjacent to the sensing volume at the same sampling frequency as the water vapor signal. The air intake should also be a sufficient distance from the fuselage such that flow distortion at the point of air intake is minimized.

Figure 14.5 shows analog traces of the water vapor density measured at 170 m above ground with a dew point sensor and a fast response infrared sensor. In case B, because large-scale eddies determine the transfer, the time traces of the water vapor concentration from both sensors are very similar while in case A the fluctuations are of a higher frequency and much more difficult to detect with the dew point hygrometer, which has a 2 sec oscillation period at an amplitude of 0.5 g m^{-3}.

3.4 Air Temperature

A variety of temperature sensors have been used in aircraft such as thermometers, thermocouples, and platinum resistance thermometers. Since airspeed has a significant effect on temperature measurement, such that at 100 m s^{-1} the total adiabatic heating (which is proportional to the square of the airspeed) is approximately 5°C, a device to stop the flow is required. The widely used Rosemount probe is designed to measure the total temperature and also provide protection to the 0.025-mm platinum wire sensing element. It has a time constant of approximately 0.03 sec, a resolution of 0.01°C, and absolute accuracy of 0.5°C at Twin Otter airspeeds (60–70 m s^{-1}). Additional heating resulting from flow distortion might cause erroneous temperature fluctuations.

3.5 Radiometers

Measurements of incident and reflected solar radiation, surface temperature, and vegetative greenness index are often recorded simultaneously with water vapor flux measurements to parameterize evaporation. Incident and reflected solar radiation have been measured by either Kipp and Zonen or Eppley pyranometers. These sensors have a spectral range of 305–2800 nm and a time constant of a few seconds. Incident radiation values need to be corrected for the aircraft attitude, which changes during flight.

Radiation thermometers, such as the Barnes PRT-5, can provide a continuous measurement of surface temperature. This instrument has a spectral bandpass of 950–1150 nm, and a time constant of 0.5 sec. With a field of view of 2° and a footprint of 3.5 m at a flight altitude of 100 m, it is useful to fine tune the position of the aircraft by detecting, for example, the crossing of highways.

A greenness index has been measured using silicon photocells exposed to filtered light reflected

TABLE 14.3. Flux of CO_2, H_2O, sensible heat, and momentum recorded on July 9, 1987 during FIFE for six sequential runs along the same flight track.[a]

Local time	Altitude (m)	Distance (km)	Sensible heat ($W\ m^{-2}$)	Latent heat ($W\ m^{-2}$)	CO_2 flux ($mg\ m^{-2}\ s^{-1}$)	Momentum flux ($N\ m^{-2}$)
1201-05	102	15.7	62	208	−1.03	−0.47
1206-10	98	15.6	54	229	−0.96	−0.34
1224-28	102	15.7	42	148	−0.63	−0.32
1230-34	99	15.5	64	209	−0.99	−0.43
1212-16	42	15.6	70	215	−1.03	−0.46
1218-22	38	15.3	70	215	−1.05	−0.43

[a] In this case flux data were high-pass filtered at 0.02 Hz, which is equivalent to a wavelength of 3 km.

from the earth's surface in wavelength bands centered at 660 and 730 nm. The ratio of the two signals is proportional to the vegetation greenness index, a similar ratio to that available from various NOAA satellites (Asrar et al., 1985).

4. Data Analysis

4.1 Flux Calculations of H_2O

To calculate the flux density of water vapor it can be measured either in terms of density, that is g $H_2O\ m^{-3}$ of air, or mixing ratio, g $H_2O\ g^{-1}$ of dry air. If measured in terms of density, one must account for air density fluctuations using, for instance, the correction suggested by Webb et al. (1980). If measured in terms of mixing ratio, there is no need to apply such a correction. Both approaches give very similar results, however the mixing ratio method is superior because it also corrects for any density variations resulting from altitude excursion during a run.

Flux values can be dominated by relatively rare events (Mahrt, 1987). The chance inclusion or exclusion of one of these events can significantly influence the computed flux. Table 14.3 presents sensible and latent heat fluxes as well as CO_2 and momentum fluxes obtained over a grassland ecosystem as part of FIFE. These data were obtained from six successive runs in east and west directions with a cross wind of 9.5 m s^{-1} from the south during relatively constant radiation conditions. To remove trend effects, the flux data were high pass filtered at 0.02 Hz. The repeatability of the flux estimates is very good but one can note that during the 1224–1228 run the flux values were substantially smaller. Such a discrepancy is most likely due to inadequate sampling of the few events. This is commonly observed during repeated measurements; there is general consistency with an occasional higher or lower flux value due to the exclusion or inclusion of a few singular events.

4.2 Longitudinal Displacement of Sensors

Turbulence sensors, which are more sensitive to the effects of the flow distortion, are usually mounted ahead of the aircraft, often separated from other sensors. The same turbulent eddies (neglecting possible distortion effect) are therefore sampled at different times by other sensors mounted further aft on the aircraft. Time lag adjustment needs to be made to account for this displacement. In the case of the Twin Otter, the distance from the tip of the noseboom to the duct inlet is 9.9 m, with the H_2O analyzer centered 2.4 m from the duct inlet. At 60 m s^{-1} aircraft speed, taking the slower air speed in the duct into account, the expected lag is 4.1 data samples at a 16 Hz sampling rate. Similarly, the distance from the tip of the noseboom to the temperature probe on the nose is 3.9 m, so at an airspeed of 60 m s^{-1} the lag is expected to be 1 data sample. To verify the above lags, data from two low-altitude runs have been analyzed with lags from 0–7 data samples, and correlation coefficients plotted in Figure 14.6. For the sensible heat flux calculation, the best lag is 1 data slice and for latent heat flux H_2O the curve peaks at 4–5 data slices. This corresponds closely to the theoretical expected values derived above. The maximum correlation occurred with the same lag data slice for two elevations. Over flat terrain the curves become less peaked with altitude, hence the correction is not as critical at higher

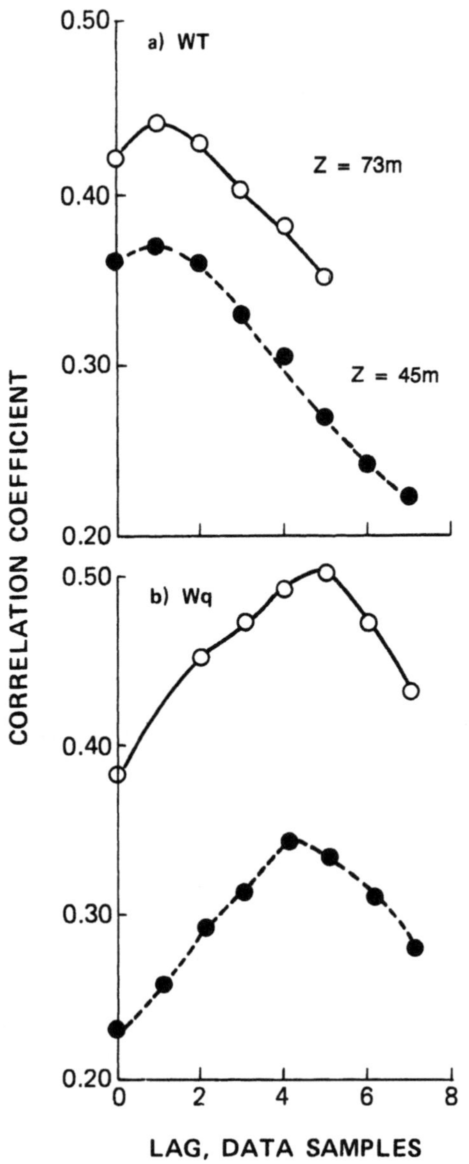

FIGURE 14.6. The correlation between the vertical gust velocity and air temperature, and water vapor density as a function of the number of data samples the gust velocity is lagged for runs at two altitudes over the FIFE site.

elevations (Druilhet et al., 1983). Over hilly terrain, as in the case of Figure 14.6, the smaller eddies extend to higher altitudes and the lag correlation curves are still quite peaked at 73 m. Not accounting for lag between vertical wind, *W*, and water vapor, *q*, at 73 and 45 m would have caused errors in the flux estimates of 24 and 33% respectively, in those situations.

4.3 Spectral and Cospectral Analyses

The analysis of power spectra and cospectra for time series of data is necessary for its inherent information as well as its role in the quality control of the data. Figure 14.7 shows typical spectra for vertical gust velocity, *W*, and H_2O density, *q*, as a function of wavenumber obtained over a grassland ecosystem during FIFE. Both spectral estimates demonstrate a well-behaved spectral shape with $-5/3$ slope down to a wavelength of 10 m. A steeper slope generally implies a poor frequency response while a flattening of the spectra denotes the presence of noise.

The cospectrum of *Wq*, which is a measure of the spectral contribution to the H_2O flux, is presented as a function of wavenumber *K* and as a function of a dimensionless wavenumber $K \times Z$, where *Z* is the sampling height, for a series of data collected over wheat fields in Manitoba (Fig. 14.8). These plots show mean cospectra obtained from doing 4 passes per day at each of three altitudes on 6 days between July 4 and 11, 1985 (Desjardins et al., 1988b). The 4-km runs were done in all four cardinal directions during mainly sunny conditions. Figure 14.8 shows the expected shift in flux contribution to lower wavenumber, i.e., longer wavelength, with increasing sampling height. With a dimensionless wavenumber scale $K \times Z$, the *Wq* cospectra at 25 and 50 m are practically identical while the *Wq* cospectra at 10 m indicates an underestimation of the water vapor flux of approximately 10% due to high frequency losses.

Because of flux divergence, vertical flux profiles are required to extrapolate water vapor flux estimates to the surface. Due to the variability of flux with time, a flight pattern should be designed in such a way that the means at all levels are collected at the same average time.

Cumulative contributions of the cospectra of *Wq* as a function of wavenumber are given in Figure 14.9 for the 25- and 50-m runs. Each is based on the integration of 24 cospectra with different limits. It provides information on the importance of various turbulence scales to regional flux estimates. As expected, it shows that more contribution to the flux is contained at lower wavenumbers or longer wavelengths at 50 than at 25 m. This information is helpful to estimate average loss in water vapor flux estimates made from short runs.

FIGURE 14.7. Spectral density of vertical wind, W, and water vapor, q, as a function of wavenumber, K.

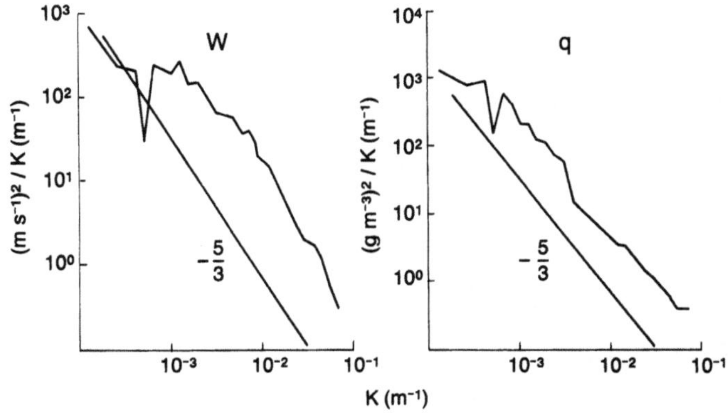

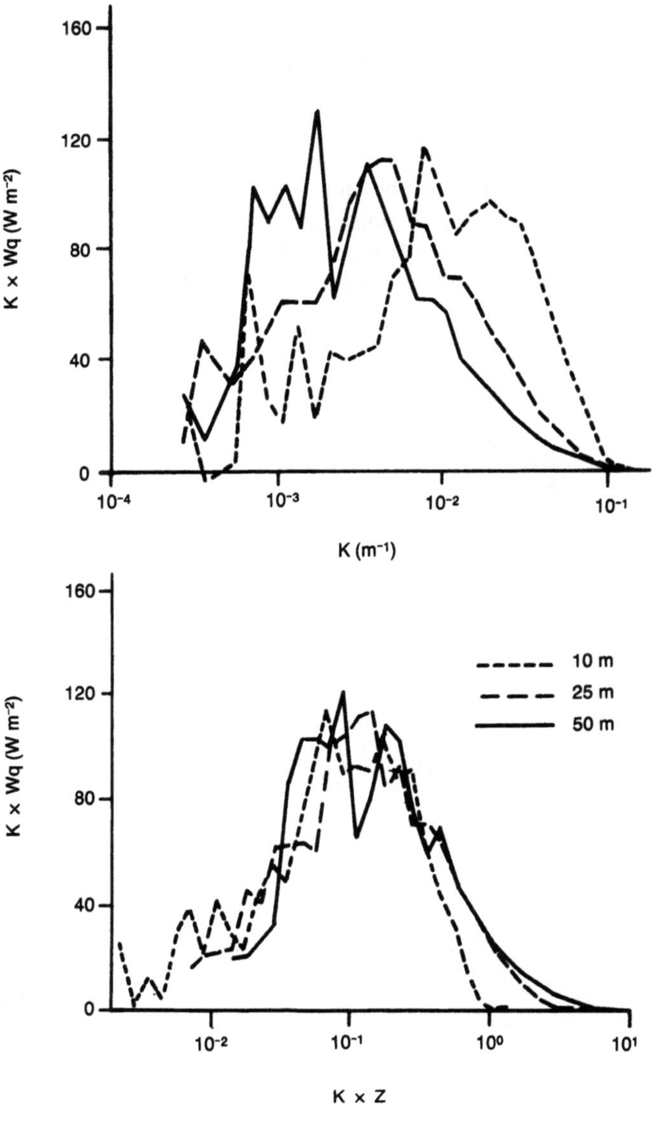

FIGURE 14.8. Average cospectra of latent heat (Wq) as a function of wavenumber, K, and dimensionless wavenumber $K \times Z$ for three elevations near Fannystelle, Manitoba from July 4 to 11, 1985.

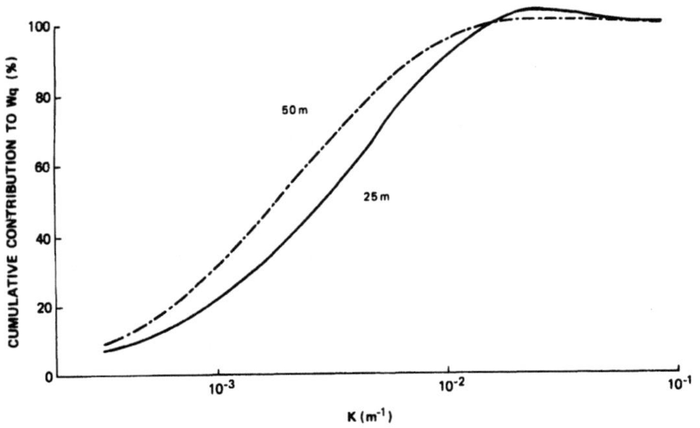

FIGURE 14.9. Average cummulative contribution to water vapor, *Wq*, versus wavenumber *K* measured at 25 and 50 m above a wheat crop in Manitoba between July 4 and July 11, 1985.

4.4 Data Filtering

All the raw data for flux calculation are low pass filtered at a cutoff frequency of 5–20 Hz. This usually corresponds to approximately half of the sampling frequency. It is a region that does not contribute significantly to the flux and filtering prevents problems due to noisy signals. High pass filtering is also frequently used to remove trends in the data with little or no detriment to the final estimates. In many cases, nonstationary effects arise from long wavelength components that are related to local terrain but are unimportant for vertical flux estimates. It is also used to permit comparison of data collected on runs of different lengths or by different aircraft. A filter frequently used is a third-order algorithm with the breakpoint set at 0.02 Hz that is a wavelength of approximately 3 km (Durand et al., 1987; Desjardins et al., 1988b). Flux estimates for water vapor from a series of 15-km runs during unstable conditions were reduced by 12% by such a filter (Desjardins et al., 1988b).

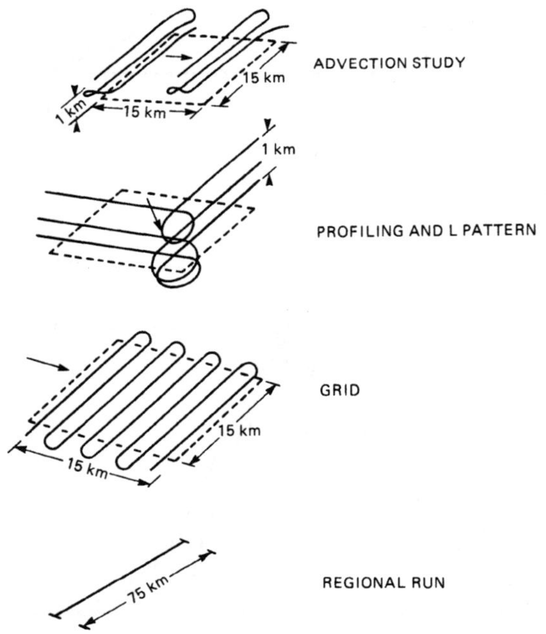

FIGURE 14.10. Flight patterns to determine areal flux estimates of water vapor.

4.5 Flight Scenarios

The choice of flight patterns is very dependent on the objectives. Cross-wind profiling runs done at four altitudes upwind and downwind of the site of interest can provide the required information to quantify advection and flux divergence (Fig. 14.10). It is often very important to extrapolate aircraft flux measurements to the surface. Mean and turbulent fluctuations of a scalar must then be measured with sufficient accuracy and resolution that spatial and temporal changes of its mean and its vertical flux divergence may be estimated. This has been done with limited success for temperature and water vapor (Warner and Telford, 1965; Lenschow et al., 1981; Desjardins et al., 1988a). The maximum altitude must be close to the top of the mixed layer. Such a flight pattern consisting of 16 runs, two sets of four runs at each end of a 15 by 15 km site, can be flown in such a way that all pairs of runs on the same track are flown at the same aver-

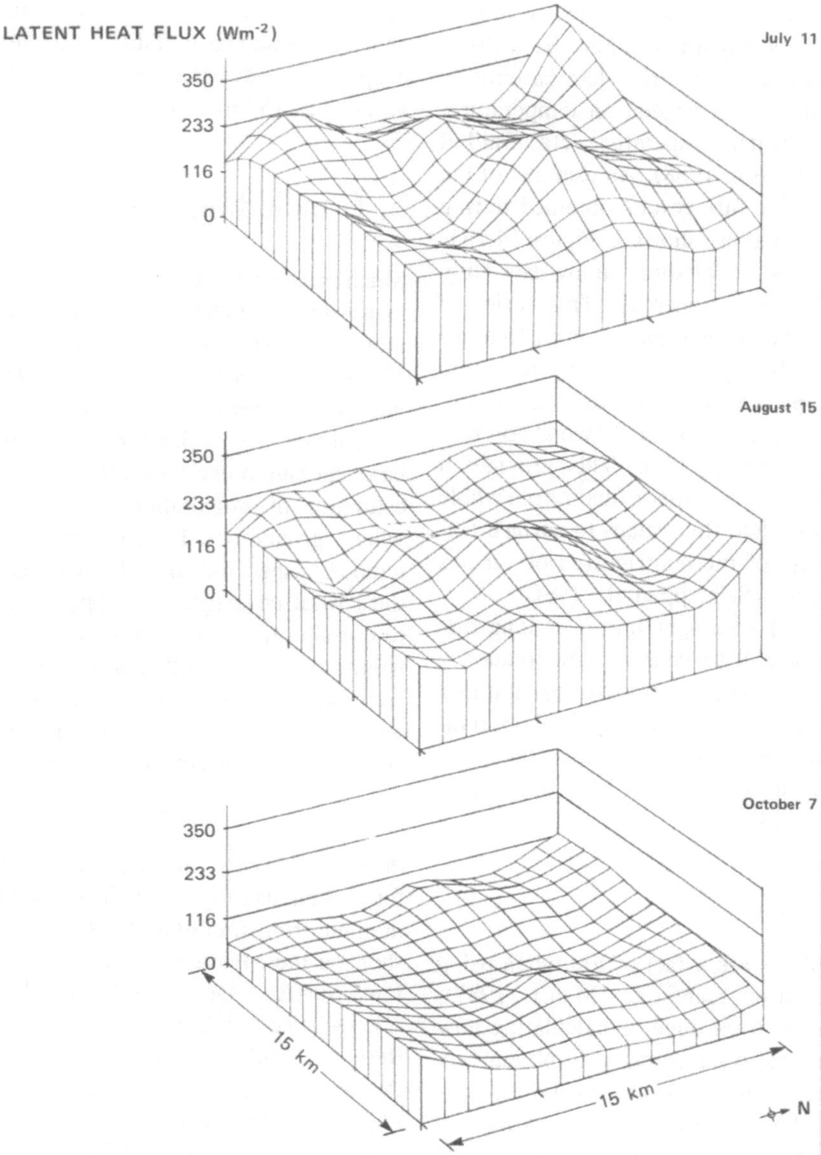

FIGURE 14.11. Three-dimensional plots of latent heat flux over a 15 by 15 km grassland area on July 11, August 15, and October 7, 1987.

age time. This minimizes the effect of linear drift on the measurements.

L-shaped flight tracks along the downwind edges of the area of interest are a good pattern to determine area average fluxes for comparison with a group of ground-based stations. To extrapolate flux profiles to the surface and account for the possibility of large temporal variability of water vapor fluxes at any given height, it is important to alternate the flight levels. The length of the runs should be approximately 15 km to ensure that most of the long wavelength contribution to the flux is accounted for. Even over this distance an underestimation due to an unmeasured long wave contribution cannot be ruled out. Its magnitude is thought to depend on the stability and has not yet been quantified experimentally.

An area average can also be obtained from grid flight patterns. Such flight tracks have been flown at constant pressure altitude to relate patterns in the fluxes to changes in vegetation temperature, greenness index, water temperature, etc. (Said et al., 1985; MacPherson, 1988). An example of the kind of water vapor flux results obtained during FIFE by flying the pattern shown in Figure 14.2a is shown in Figure 14.11 for three sunny days (July 11, August 15, and October 7). This series of three-dimensional plots was made using a 16 by 16 bivariate spline (Meinguet, 1979). By repeating the whole grid pattern in the opposite direction, each pair of runs on each track is flown at approximately the same average time. This provides a good estimate of the water vapor flux for the whole area. In order to relate flux values to surface features, an estimate of the contribution of various features is required (Schuepp et al., 1990).

Regional runs in a straight line over long distances are also useful to quantify low-frequency contributions to flux and to compare regional flux estimates with satellite data. Some initial observations on the variability of water vapor flux estimates over a distance of 75 km have been made (Schuepp et al., 1988).

4.6 Integral Scale

Averaging problems are a fact of life when using the eddy correlation technique (Mahrt, 1987). Much research has been done on calculating how much flux data has to be averaged to obtain flux results to a given accuracy (Wyngaard, 1983). Lenschow and Stankov (1986) used observations to estimate that an averaging length of 200 km is required to estimate fluxes with 10% accuracy, if a realistic level of complexity and scaling to long wavelength is incorporated. Schuepp et al. (1988) estimated that the sampling variability of water vapor flux would be reduced by approximately 37% by going from 4 to 10 km sampling runs, and approximately 50% by going from 10 to 40 km runs, if flux estimates were distributed normally about the mean. Although this is not necessarily correct, experimental evidence indicates that it might be an acceptable simplifying assumption in certain cases. These estimates are very much a function of the flight level. Close to the surface, the integral scales are smaller and the sampling distances required to obtain stable estimates are then shorter. As shown in Table 14.3, very consistent evaporation estimates are obtained over distances as short as 15 km under certain conditions.

5. Concluding Remarks

Much progress has been achieved in recent years in measuring regional flux of H_2O using aircraft mounted sensors. Direct measurements by aircraft of regional evaporation are still rare but are becoming more common with advances in experimental capabilities. A combination of aircraft and ground-based measurements is essential to arrive at regional estimates of evaporation for extended periods, where ground-based measurements provide the temporal aspect and aircraft measurements provide the spatial coverage. It is apparent that if we are to use aircraft-based measurements to verify algorithms to employ satellite observations, the limiting factor at this time is not instrumentation but inadequate insight into downwind displacement and averaging requirements of the turbulent transfer.

Acknowledgments. We are grateful to J.C. André for inviting us to speak on the subject of flux measurement from aircraft at the Banyuls, France workshop on the Measurement and Parameteriztion of Land-Surface Evaporation Fluxes. We would like to acknowledge the assistance of the Land Resource Research and the Engineering and Statistical Research Centers of Agriculture Canada and of the Flight Research Laboratory of the National Aeronautical Establishment of NRC for collecting and analyzing these data. Thanks are also given to P.H. Schuepp of McGill University for his helpful discussion in reviewing this manuscript. The financial support of NASA as part of the FIFE program is also gratefully acknowledged. LRRC Contribution No. 88-59.

References

Andre JC, Goutorbe JP, Perrier A (1986) Hapex-Mobilhy—A hydrologic atmospheric pilot experiment for the study of water budget and evaporation flux at the climatic scale. Bul Meteorol Soc 67:138–144.

Asrar G, Kanemasu ET, Jackson RD, Pinter PJ Jr (1985) Estimation of total above-ground phytomass production using remotely sensed data. Rem Sens Environ 17:211–220.

Bean BR, Emmanuel CB (1980) "Aircraft in Air-Sea Interaction: Instruments and Methods" (F Dobson, L Hasse, R Davis, eds.), pp. 571–588. Plenum Press, New York.

Bean BR, Gilmer RO, Hartmann RF, McGavin RE, Reinking RF (1976) Airborne measurement of vertical boundary layer fluxes of water vapor, sensible heat and momentum during GATE. NOAA Tech Memo ERL WMPO-36, NOAA Environmental Research Laboratories, Boulder, CO.

Brutsaert W (1982) "Evaporation into the Atmosphere: Theory, History and Applications." D. Reidel, New York.

Brown EW, Friehe CA, Lenschow DH (1983) The use of pressure fluctuations on the nose of an aircraft for measuring air motion. J Clim Appl Meteorol 22:171–180.

Buck AL (1976) The variable-path Lyman-alpha hygrometer and its operating characteristics. Bull Am Meteorol Soc 57:1113–1118.

Burch DE, Singleton EB, Williams D (1962) Absorption line broadening in the infrared. Appl Opt 1:359–363.

Coantic M, Friehe CA (1980) Slow-response humidity sensors. In "Air-Sea Interaction: Instruments and Methods" (F Dobson, L Hasse, R Davis, eds.), pp 399–411. Plenum Press, New York.

Desjardins RL, Brach EJ, Alvo P, Schuepp PH (1982) Aircraft monitoring of surface carbon dioxide exchange. Science 216:733–735.

Desjardins RL, MacPherson JI, Betts AK, Schuepp PH, Grossman R (1988a) Divergence of CO_2, latent and sensible heat fluxes: A case study. Proceedings on Lower Tropospheric Profiling: Needs and Technologies, Boulder, CO, pp. 71–72.

Desjardins RL, MacPherson JI, Schuepp PH, Karanja F (1988b) An evaluation of airborne eddy flux measurements of CO_2, water vapor and sensible heat. Bound Layer Meteorol 47:55–70.

Dobson FW (1980) Air pressure measurement techniques. In "Air-Sea Interaction: Instruments and Methods" (F Dobson, L Hasse, R Davis, eds.), pp. 231–254. Plenum Press, New York.

Druilhet A, Noilhan J, Bénech B, Dubosclard G, Guedalia D, Frangi JP (1983) Étude expérimentale de la couche limite au-dessus d'un relief modéré proche d'une chaîne de montagne. Part II: Caractéristiques moyennes et turbulentes de la couche limite en situation instable. Bound Layer Meteorol 25:3–16.

Durand P, Druilhet A, Hedde T, Bénech B (1987) Aircraft observations of the structure of the boundary layer over a rugged hilly region (MESOGERS 84 experiment) Ann Geophys 5B:441–450.

Greenhut K, Gilmer RO (1985) Calibration and accuracy of the NOAA/ERL gust probe system and intercomparison with other systems. NOAA Tech Memo ERL ESG-22.

Greenhut GK, Khalsa SJS (1982) Updraft and downdraft events in the atmospheric boundary layer over the equatorial Pacific Ocean. J Atmos Sci 39:1803–1818.

Grossman RL, Bean BR (1973) An aircraft investigation of turbulence in the lower layers of a marine boundary layer. NOAA Technical Report ERL 291-WMPO, No. 4.

Hacker JM (1982) First results of boundary layer research flights with three powered gliders during the field experiment PUKK. Beitr Phys Atmos 55:383–402.

Hay DR (1980) Fast-response humidity sensors. In "Air-Sea Interaction: Instruments and Methods" F Dobson, L Hasse, R Davis, eds.), pp. 413–432. Plenum Press, New York.

Hildebrand P (1988) Flux and sounding data from the NCAR King Air aircraft during HAPEX. NCAR/TN-319.

Johnson HD, Lenschow DH, Danninger K (1978) A new fixed vane for air motion sensing. "Preprints, Fourth Symposium on Meteorological Observations and Instrumentation, Denver, CO," pp. 467–470. American Meteorological Society, Boston.

Lenschow DH (1986) Aircraft measurements in the boundary layer. In "Probing the Atmospheric Boundary-Layer" (DH Lenschow, ed.), pp. 39–55. American Meteorological Society, Boston.

Lenschow DH, Stankov BB (1986) Length scales in the convective boundary layer. J Atmos Sci 43:1198–1209.

Lenschow DH, Pearson Jr R, Stankov BB (1981) Estimating the ozone budget in the boundary layer by use of aircraft measurements of ozone eddy flux and mean concentration. J Geophys Res 86:7291–7297.

MacPherson JI (1988) NAE Twin Otter Operations in FIFE. NAE Laboratory Technical Report LTR-FR-104, National Research Council of Canada, Ottawa.

MacPherson JI, Morgan JM, Lum K (1981) The NAE Twin Otter atmospheric research aircraft. National Research Council of Canada Report LTR-FR-80.

Mahrt L (1987) Boundary layer eddies and surface inhomogeneity and their impact in measurement strategy and data analysis for FIFE. Report from Oregon State University for NASA-Goddard Space Flight Center.

McBean GA, Peterson RD (1975) Variations of the turbulent fluxes of momentum, heat and moisture over Lake Ontario. J Phys Oceanogr 5:523–531.

Meinguet J (1979) Multivariate interpolation at arbitrary points made simple. J Appl Math Phys 30:292–304.

Mestayer P, Rebattet C (1985) Temperature sensitivity of Lyman-alpha hygrometers. J Atmos Ocean Tech 2:656–664.

Mintz Y (1984) The sensitivity of numerically simulated climates to land surface conditions. In "The Global Climate" (J Houghton, ed.), pp 79–105. Cambridge University Press, Cambridge.

Nicholls S (1983) An observational study of the mid-latitude, marine atmospheric boundary layer. Ph.D. dissertation, University of Southampton, U.K.

Said F, Durand P, Koehler B, Druilhet A (1985) Analyse des mesures aéroportées dans la couche limite marine au cours de l'expérience. TOSCANE. Proc 3rd Int Coll on Spectral Signatures of objects in Remote Sensing, Les Arcs, France, pp. 67–71.

Schuepp PH, Desjardins RL, MacPherson JI, Boisvert J, Austin LB (1987) Airborne determination of regional water use efficiency and evapotranspiration: Present capabilities and initial field tests. Agr Forest Meteorol J 41:1–9.

Schuepp PH, Desjardins RL, MacPherson JI, Boisvert J, Austin LB (1988) Interpretation of airborne estimates of evapotranspiration. Estimation of Areal Evapotranspiration, IAHS publication no. 177, pp. 185–196.

Schuepp PH, Leclerc MY, MacPherson JI, Desjardins RL (1990) Footprint prediction of scalar fluxes from analytical solutions of the diffusion equation. Bound Layer Meteorol 50:355–373.

Sellers PJ, Hall FG, Asrar G, Strebel DE, Murphy RE (1988) The first ISLSCP field experiment (FIFE). Bull Am Meteorol Soc 69:22–27.

Warner J, Telford JW (1965) A check of aircraft measurements of vertical heat flow. J Atmos Sci 22:463–465.

Webb EK, Pearman GI, Leuning R (1980) Correction of flux measurements for density effects due to heat and vapor transfer. Quart J Meteorol Soc 106:85–100.

Wyngaard JC (1983) "Mesoscale Meteorology–Theories, Observations and Models (DK Lilly, T Gal-Chen, eds.), pp. 216–224. D. Reidel, New York.

15
Heat and Moisture Fluxes over the Pine Forest in HAPEX

L. Mahrt

1. Introduction

This study analyzes heat and moisture fluxes computed from repeated flight legs 100 m above ground on a 120-km east–west flight track over the pine forest in HAPEX. Sampling problems are first studied in terms of structure functions in order to distinguish between small-scale intermittency of the fine-scale gradients and larger scale global intermittency. The error of flux estimates is then posed in terms of the filter cutoff wavelength, the length of the record or subrecord, and the scale of the surface inhomogeneity. Both the temporal and spatial variation of the flux and other quantities are examined. Finally, the estimated fluxes are compared with results of a simple one-dimensional model of the boundary layer, canopy, and soil.

Turbulent fluxes and other statistics computed from data depend on the scales included as dictated by the record length and filtering procedures. Filtering is used to remove larger scale motions, which are otherwise observed with inadequate sample size. The distinction between turbulent scales and small mesoscale motions is generally not very clear. The textbook *spectral gap* between turbulence and larger scale motions is mainly limited to vertical motions near the surface. Other variables, such as moisture, temperature, and horizontal velocity components, do not always exhibit this separation of scales, particularly in the middle and upper portions of the boundary layer. The scale of the largest turbulent eddies is not always distinct from forcing by overlying internal gravity waves and underlying surface inhomogeneity, both of which occur on a variety of scales. A plausible summary of different spatial scales for HAPEX flow over the forest is included in Figure 15.1.

The choice of scales for computation of fluxes, required for selection of the filter cutoff wavelength, is somewhat arbitrary without a spectral gap. For sampling reasons, the cutoff wavelength is often chosen to be roughly 1/10 of the record length regardless of the suspected size of the main transporting eddies.

Application to gridded numerical models is further complicated by the requirement that the formulation for subgrid scale transport include all motions not resolved by the grid. Therefore, the nature of the subgrid scale parameterization changes with the resolution of the model. Existing formulations of subgrid scale parameterizations are based on the expected behavior of turbulence over homogeneous terrain and ignore the influence of surface inhomogeneity and the contribution of various mesoscale motions not resolved by the grid.

We will briefly examine the dependence of fluxes and other statistical quantities on horizontal scale by analyzing six horizontal legs over 100 km long and approximately 100 m above the nearly flat pine forest on each of two fair weather days. The flight track is east–west and passes over the Lubbon central site. These two HAPEX days provide the best starting point of our analysis because:

1. These particular data contain one of the largest sample sizes ever accumulated by low-level aircraft flights over land with approximately stationary conditions.
2. The inhomogeneity is simpler compared to other flight locations and is limited mainly to pine forest and occasional clearings.

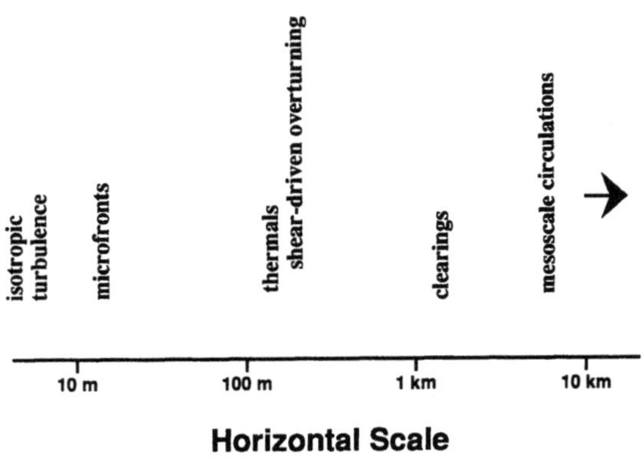

FIGURE 15.1. Idealized partition of scales thought to occur in the turbulence over the pine forest.

In Section 2, we examine intermittency statistics and their dependence on scale. In Section 3, we examine the dependence of fluxes on scale and note various sampling problems intrinsic to flux calculations. In Section 4, we examine the physical contributions to the fluxes in terms of differences between forested areas and clearings, and the impact on comparisons between local surface data and aircraft data. These comparisons are briefly interpreted in terms of a one-dimensional model of the boundary layer and surface energy budget. In Section 5, we study the importance of various scales to the horizontal transport and discuss the impact on the subgrid formulations.

2. Intermittency and the Microfront Scale

A useful introduction to the dependence of the turbulence on horizontal scale can be posed in terms of the intermittency of the turbulence. The degree and scale of the intermittency partly determine the severity of sampling problems and the ability to estimate fluxes on various scales. In this section, we study the nature of the intermittency and then study the variability of the fluxes in the following two sections.

Intermittency of turbulence includes organization of small-scale gradients by individual coherent eddies (Fig. 15.1). With such *small-scale or microscale intermittency*, dissipation of turbulence kinetic energy is confined primarily to small subregions of individual eddies. In addition, the turbulence may be patchy or *globally intermittent* due to organization on scales larger than the main coherent eddies. The ambiguous term "intermittency" may refer to either small-scale intermittency of dissipation, or larger scale global intermittency of the main eddies.

With global intermittency, the turbulence develops as patches of eddies with large intervening areas of weaker turbulence activity. Global intermittency can lead to sampling problems for computation of fluxes and other turbulence quantities (Shaw and Businger, 1985; Baker and Gibson, 1987). In the HAPEX data sets, global intermittency is expected to be forced by variations of surface conditions.

For example, over the forest, global intermittency can occur in terms of variation of thermal strength and location partially forced or modulated by variable tree height and occasional clearings. Preliminary analysis of data from early in the drying season indicates that over the pine forest, the sensible heat flux is larger and the moisture flux smaller compared to the clearings. In the afternoon, the pine trees are characterized by greater stomatal resistance (lower stomatal conductance) compared to the crops for this case leading to lower evapotranspiration over the pine forest compared to exposed moist soil or crop-covered clearings. Higher stands of pines lead to more effective heating of the atmosphere since the heat source occurs at a higher elevation above the ground and less energy is lost to soil heat flux.

The variation between the clearings and the pine canopy may reverse as the drying season advances. The surface moisture flux in nonirrigated clearings might become quite small as the upper layers of the soil dry out. However over the pine canopy, significant evapotranspiration continues since the trees extract moisture from a deeper layer of soil. Then the latent heat flux may no longer be greatest over the pine forest.

Small scale intermittency may result from overall modulation of the turbulence by the main eddies or may occur in connection with sharp edges of the main eddies. With overall modulation, the small-scale turbulence inside the eddies is sometimes larger than in the fluid outside the eddies, as often occurs with thermals (Khalsa and Businger, 1977 and others). In some cases, the strongest small-scale gradients are concentrated as sharp edges of the

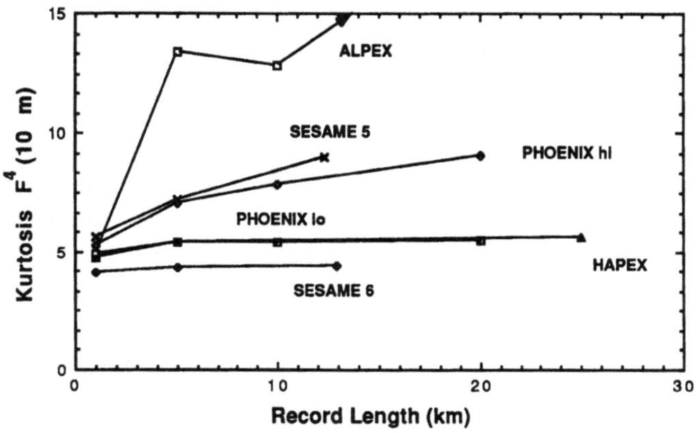

FIGURE 15.2. Averaged values of the kurtosis as a function of the subrecord length; subrecords are created by equal partitioning of the entire record. To facilitate comparison between cases, records were partitioned into 1, 5, 10, and when possible, 20 km segments. The HAPEX and low-level PHOENIX lines coincide. The significant difference between the SESAME 6 values compared to the PHOENIX low-level and HAPEX values is somewhat dwarfed in this particular figure by the large influence of global intermittency on the other cases.

main eddies. These shear zones are referred to as *microfronts*. In other observations of turbulence fields, the small-scale intermittency occurs as numerous narrow zones of shear with uncertain relationship to the main coherent structures. In the present analysis, concentrated shear zones seemed to occur as edges of thermals and low-level shear-driven overturning. These microfronts appear to contribute significantly to the overall turbulent transport as in the studies of Schols (1984), Chen and Blackwelder (1978), and Kikuchi and Chiba (1985).

In the present section, we will study both types of intermittency in terms of the dimensionless structure function without reference to specific surface features. In Section 4, differences between clearings and the pine forest will be examined explicitly. Although the structure functions do not allow direct comparison with existing models of turbulence based on Reynolds averaging, they avoid the somewhat arbitrary process of determining the local mean from geophysical data and are less vulnerable to sampling problems (Mahrt and Gamage, 1987).

The usual higher order structure function is defined as

$$D^n(r) = [\{\phi(X+r) - \phi(X)\}^n] \qquad (15.1)$$

where ϕ is a flow variable such as temperature, specific humidity, or one of the three velocity components, r is the separation distance between the observations, and the outer brackets indicate an ensemble average. The dimensionless structure function for studying intermittency can be defined as (e.g., Frisch et al., 1978)

$$F^n(r) = D^n(r)/\{D^2(r)\}^{n/2} \qquad (15.2)$$

For n=3, Equation 15.2 becomes the structure skewness while for n=4, Equation 15.2 becomes the structure kurtosis and so forth.

To study the influence of small-scale and global intermittencies, the kurtosis of horizontal velocity gradients were computed for the two HAPEX days and compared with aircraft data from several other field programs (Fig. 15.2). A description of the other field programs and the data analysis can be found in Mahrt (1989). The records with the largest values of the kurtosis occur with strong stability; however, the scatter in the relationship with stability is large, and the link between stratification and dimensionless structure functions is rather indirect. Other factors are, at least collectively, more important.

The kurtosis is smallest in turbulence that is more or less continuous (not globally intermittent) and relatively homogeneous. For example the kurtosis at 10 m separation distance averages only 4.4 for the windy SESAME class. The kurtosis increases modestly to about 5.5 for the low level legs in the heated boundary layers during HAPEX and PHOENIX where the sharp edges of updrafts are often characterized by large concentrated gradients of vertical velocity. The larger kurtosis and intermittency in these cases are also due to greater intensity of fluctuating gradients inside the thermals compared to the regions between thermals.

Most of the remaining records are characterized by global intermittency associated with strong stratification or characterized by other larger scale inhomogeneity of the turbulence. Both effects greatly enhance the kurtosis values. This is because the structure variance appearing in the denominator of the kurtosis is influenced by both global regions of weak turbulence and regions of strong turbulence. However, the fourth moment of the gradients (numerator of the kurtosis) is dominated by regions of especially strong turbulence. Therefore, regions of weaker turbulence decrease the denominator more than the numerator and, therefore, increase the kurtosis compared to the homogeneous case of no global intermittency.

For example, kurtosis values are typically on the order of 10 in the globally intermittent turbulence in the strongly stratified SESAME flow, the inhomogeneous turbulence in the upper part of the PHOENIX boundary layer, and the globally intermittent turbulence in the bora flow (ALPEX) far downstream from the mountain range and also upstream from the mountain range. Kurtosis values are extremely large, exceeding 20, in the developing globally intermittent turbulence above the mountain range. Here the developing patches of turbulence are intense leading to larger contrast with the intervening regions of little turbulence. This large contrast augments the kurtosis values. The sharp edges of the drafts in the turbulence patches lead to a smaller enhancement of the kurtosis.

We tentatively conclude that global intermittency significantly increases the kurtosis and other higher moments to the extent that values of the dimensionless structure function represent the variation of the turbulence more than the nature of the turbulence itself. Fortunately, the global intermittency for the two HAPEX days examined here is quite low. We expect the global intermittency to be greater on some of the other HAPEX flights where the surface inhomogeneity is greater. Global intermittency will be most likely on days with weak surface heating where local variations of surface evaporation may actually determine the local stability.

With significant global intermittency, the statistics are not well posed and depend on the record size as can be shown by breaking the records into smaller records. As the record length becomes long enough to include a sample of the main eddies, the kurtosis value becomes nearly independent of the record length in the cases where global intermittency is weak; this includes the HAPEX, low level PHOENIX turbulence, and windy-SESAME cases (Fig. 15.2). For these cases, the kurtosis increases only slightly as the record length increases beyond 5 km suggesting that the influence of inhomogeneity and global intermittency is minimal. In fact the kurtosis values for different record lengths in the HAPEX and low level PHOENIX cases reach the same value of approximately 5.6. *Apparently, the occurrence of clearings has not seriously affected the HAPEX statistics in terms of global intermittency and attendant sampling problems as long as the sample length is at least 5 km. The statistics for the two HAPEX days are considerably more robust compared to those other data sets with global intermittency.* In the turbulence with global intermittency or other significant inhomogeneity, the kurtosis continues to increase as the record length increases beyond 5 km and more of the inhomogeneity is captured (Fig. 15.3). This sensitivity to record length implies that the statistical characteristics of the turbulence itself cannot be quantitatively established in these cases without further analysis such as record partitioning.

Confident that the records for the two HAPEX days are not vulnerable to sampling problems associated with global intermittency, we proceed to examine the dimensionless structure functions for other variables as a function of separation distance. The structure kurtosis for all the variables shows a maximum at the smallest resolvable scale (Fig. 15.3) indicating the possible influence of microfronts. The kurtosis for temperature is much larger than for the velocity components. Horizontal gradients (in the x-direction) of the u-component are restricted by mass continuity while the horizontal gradients of the other two components are possibly restricted by the onset of small scale shear-instability.

Structure skewness responds to preferred sign of the sharpest gradients and serves as an index of the

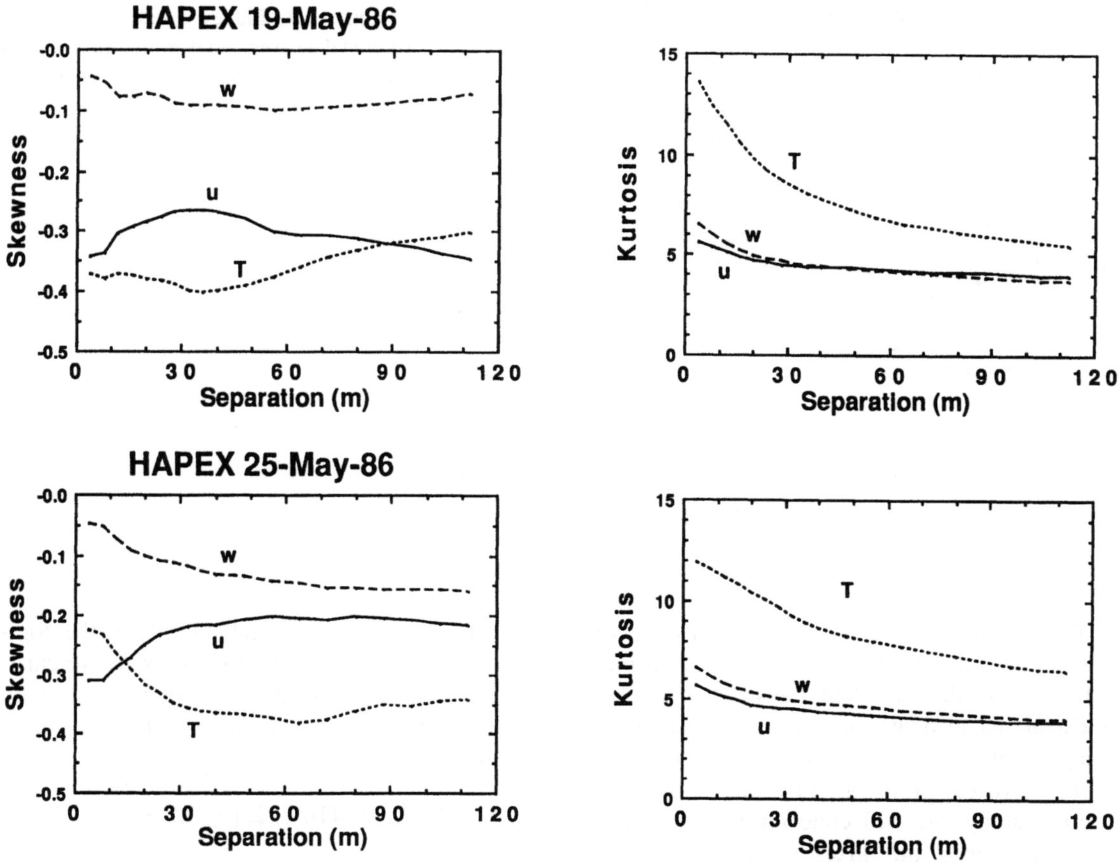

FIGURE 15.3. Structure skewness and kurtosis as a function of separation distance composited over the flight legs for each day.

relative importance of mean shear. For example, updrafts rising into faster moving ambient air would be expected to form the strongest or sharpest gradients at the upstream edge where horizontal convergence between the faster ambient motion and updraft motion intensifies horizontal gradients. At the downstream edge, horizontal gradients are expected to be more diffuse possibly due to a downstream wake effect.

On 19 May and 25 May, vertical motion and temperature are both negatively skewed (Fig. 15.3) indicating that the sharpest gradients occur with $\partial T/\partial x < 0$ and $\partial w/\partial x < 0$ probably corresponding to the upstream edges of updrafts associated with the increase of the mean easterly component with height. The structure skewness for the *u*-component of the motion (Fig. 15.3) shows strong negative values consistent with strong convergence of the easterly flow at the upstream edges of the updrafts.

The correlation between the strong gradients of u, w, and T were verified by computing eigenvectors of the lagged correlation matrix for these variables based on samples selected for concentrated horizontal gradients of vertical motion (see Mahrt and Frank, 1988, for methodology). This analysis (not shown) also indicates that the strongest gradients were edges of larger coherent structures consisting of warm rising air. *One might tentatively refer to these structures as thermals although shear-driven overturning and upward scooping of warmer air is also a possibility as discussed in Schols and Wartena (1986).* In fact, based on conditional sampling, the downdraft widths averaged only slightly larger than the updraft widths in contrast to the usual situation with pure thermal motions. The Monin–Obukhov length L is -30 to -50 m on 19 May suggesting the importance of shear. This length is near -150 m on 25 May indicating even greater influence of the mean shear.

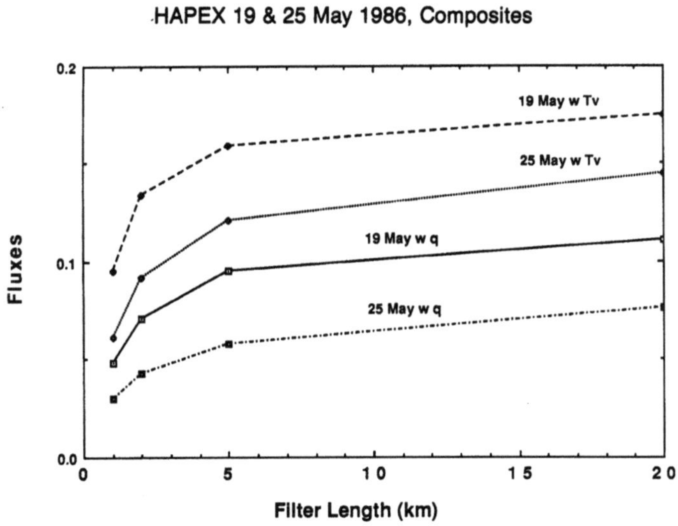

FIGURE 15.4. Fluxes based on high-pass filtered data averaged over the flight legs as a function of the filter cutoff wavelength.

The nature of the main boundary-layer eddies and associated microfront edges exerts a major influence on the turbulent transport of moisture and heat, and also influences the nature of the sampling problems associated with attempts to estimate the flux from the aircraft time series. In the present study, we will concentrate more on the determination of the fluxes and comparison with surface-based estimates and modeling results. Further study of the nature of the main boundary-layer eddies and their contributions to the fluxes will appear in a future manuscript.

3. Turbulent Fluxes

Although cospectra are somewhat noisy, the main dependence of the heat flux on scale can be simply illustrated using a few different filter cutoff wavelengths. Figure 15.4 shows that *most of the flux at the flight level of about 100 m occurs on scales smaller than 5 km.* However the flux continues to increase slowly with inclusion of larger scales up to the largest scale permitted by sampling problems (roughly 20 km). It may be that weak circulations associated with clearings and variations of soil moisture due to recent precipitation patterns or larger scale variability contribute to the flux at larger scale.

The 125 km record lengths appear to be of sufficient length to obtain an adequate estimate of the flux. To study the statistical stability of the computed fluxes as a function of record size, we partition the record into N_j smaller subrecords of width L_j and define the *estimated error of the flux* calculation as

$$\sigma_{\text{flux}}(L_j, \lambda) = \{(1/N_j)\, \Sigma\, [F_{ij}(\lambda) - F(\lambda)]^2\}^{1/2} \tag{15.3}$$

where $F_{ij}(\lambda)$ is the ith flux estimate for a subrecord of size L_j, $F(\lambda)$ is the flux computed from the entire record, and the sum in Equation (15.3) is over all of the subrecords of size L_j. Noting that for a given filter length and subrecord width, the sum of fluxes $F_{ij}(\lambda)$ over the subrecords is equal to the flux for the entire record, the above error estimate Equation (15.3) becomes simply the standard deviation of the subrecord flux. Nonzero values of Equation (15.3) are due to errors associated with problems of obtaining enough samples of the main transporting eddies, the influence of a few anomalous events, and also spatial and temporal variability of the mesoscale and larger scale motions. The standard deviation Equation (15.3) can be viewed as a typical error when estimating a "representative" flux from a given subrecord length.

As expected, this estimated error decreases with increasing subrecord width L_j (Fig. 15.5) due to increased sample size and increases with increasing wavelength of the filter cutoff since larger scales are observed with smaller sample size. The above results tentatively suggest that for moisture, the estimated error of the flux will become an order of magnitude smaller than the magnitude of the flux, here about 0.09 (g/kg) m/s, when the subrecord width is at least an order of magnitude larger than the cutoff filter length. The estimated error of the heat flux generally decreases with increasing subrecord length and decreasing cutoff filter length; however the estimated error of the heat flux does not appear to be well described by a simple linear ratio of these two lengths as is evident in Figure 15.5. In particular, the standard deviation of the heat flux has decreased to about 15% of the flux when the subrecord width has increased to 25 km regardless whether the filter length is 1, 5, or 20 km. *It may be that the subrecord length is now sufficiently long compared to the main scale of the spatial variability of the flux such as a typical width of the clearings.* This speculation suggests a third length scale is needed to describe the surface inhomogeneity. More specifically, moisture is found to vary more than temperature on scales greater than a few kilometers, a subject of present investigation.

The results in Figure 15.5 indicate that in most cases, the variation of the estimated flux increases rapidly beyond the 10–15% value as the filter length increases to values greater than 1/10 of the record length. The strategy is to select a filter length of this size and then estimate whether such a selection includes most of the turbulent flux. The problem can be reduced by averaging the flux values over a number of records. This was the sampling strategy in FIFE where repeated "short" legs were flown over the same track and can be applied in a more limited sense to the HAPEX fluxes reported by Hildebrand (1988). The above study of flux errors resulting from finite record lengths is indeed tentative and serves more as an example rather than conclusive evidence. A larger number of cases will be examined.

The variability of the subrecord flux is due partly to transient charges on various scales and partly due to spatial variation as is evident from the spatial distribution of the flux that has been averaged over the flight legs for different subrecords approximately fixed with respect to the ground (Fig. 15.6). Most notable is the east–west trend, which varies with the type of flux and varies between days. The smaller scale spatial variation based on

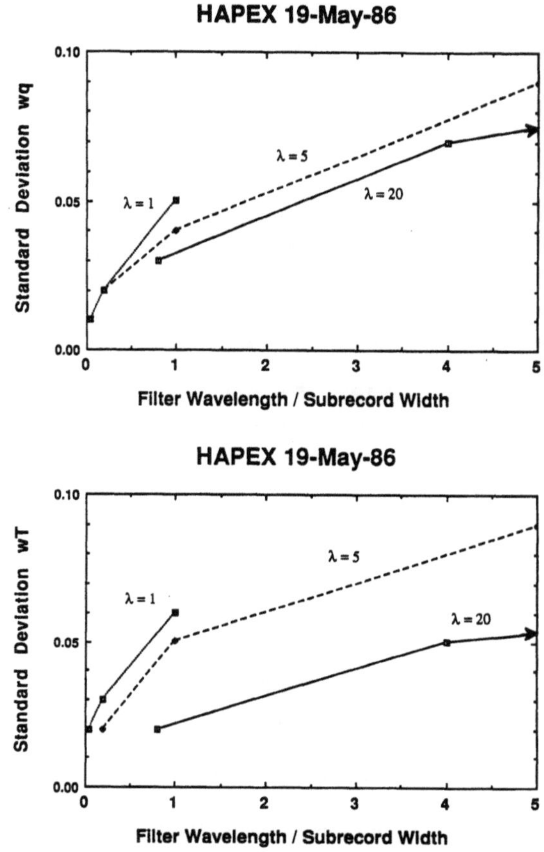

FIGURE 15.5. Estimated error [Equation (3)] of the fluxes computed from subrecords of width L_j for data that have been high-pass filtered with cutoff wavelength λ labeled in kilometers. Moisture fluxes are expressed in (g/kg)(m/s) while heat fluxes are expressed in (Km/s).

averages of 5 km subrecords fixed with respect to the ground is of unknown significance. There is little correspondence between this smaller scale spatial variation on 19 May compared to 25 May. Of course the above partitioning of the record did not recognize locations of specific clearings. The variability between clearings and the pine forest is briefly examined in the next section.

4. Fluxes and Surface Conditions

In this section, we initiate a study of the relation of the low level fluxes to the characteristics of the surface, more specifically the occurrence of the pine canopy versus clearings. The surface is sufficiently

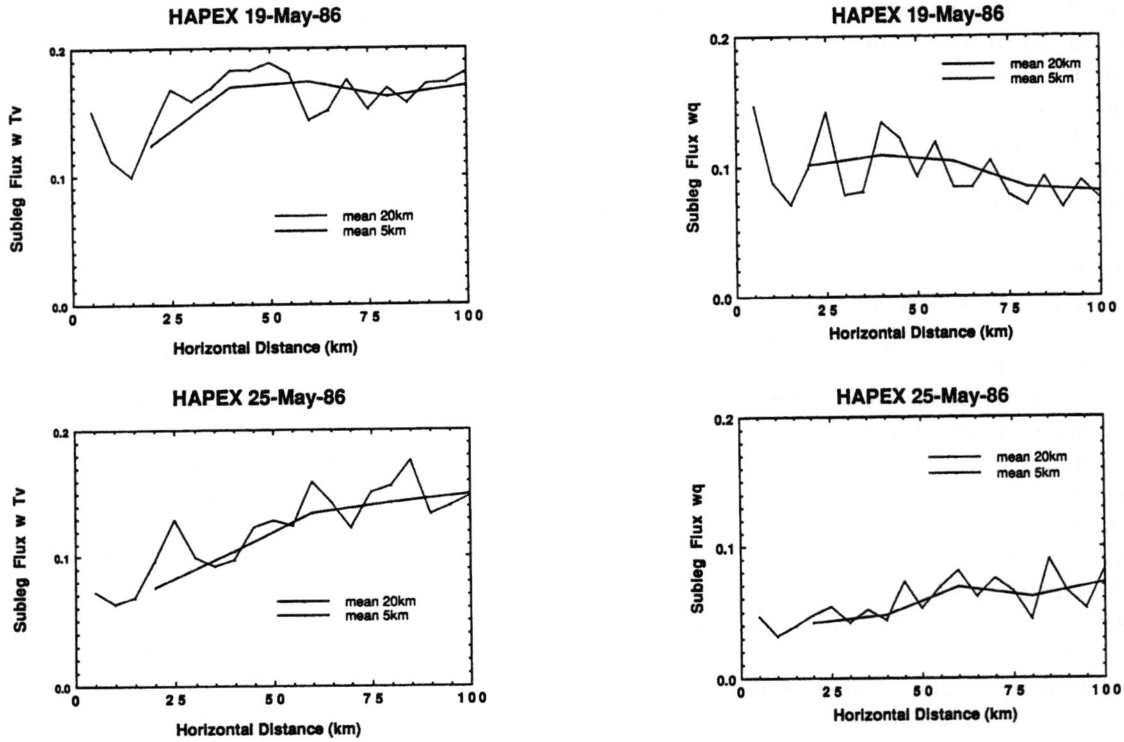

FIGURE 15.6. Fluxes averaged over the flight legs for subrecords fixed with respect to the ground. For this calculation, the legs were truncated to 100 km so that they could be evenly partitioned into 20-km subrecords and at the same time include legs smaller than 120 km. East is to the right. The Lubbon central site is near the right hand edge at 100 km.

flat that the surface elevation deduced from the radiometric altimeter and pressure altitude of the aircraft (*effective surface elevation*) can be used as a rough local indicator of tree height or the absence of trees. The spatial variation of the effective surface elevation due to the slow increase of ground elevation toward the east seems to have been removed by the usual linear detrending. Since the radioaltimeter over the forest "sees" some ground between the branches and trees, the actual estimated effective surface elevation will be lower than the average canopy height. Time series of the effective surface elevation indicates that the variation between clearings and the forest is typically on the order of 10 m and that local variations of the actual surface elevation also contribute to the effective surface elevation. Here, we are interested in only correlations between the atmospheric variables at flight level (approximately 100 m above ground) and the effective surface elevation. More specifically, we will examine the low pass variables (5 km cutoff wavelength) computed from the detrended data.

The correlations of air characteristics with the effective surface elevation seem insignificant on 19 May while some correlation is found on 25 May when the moisture flux is a little weaker (Fig. 15.7). *The air temperature at flight level is warmer ($r=0.46$) and dryer ($r=-0.54$) above the taller trees* where r is the correlation with the effective surface elevation. The correlations between the air at the aircraft level and at the surface are probably reduced by horizontal advection occurring between the flight level and the tree canopy and due to the approximate nature of effective surface elevation measured by the radioaltimeter. The air is more effectively heated over the higher trees because the effective heat source (level of absorbed radiation) is higher compared to the clearings. In addition, the trees are normally characterized by a lower albedo compared to clearings as is evident from comparison of reflected upward radiation with the location of clearings determined from the downward looking video camera. The dryer air above the pine canopy is probably due to the greater stomatal control of the pine needles compared to the vegetation in the clearings.

The surface radiation temperature of the trees is not necessarily larger than that of the clearings. The effect of lower albedo and reduced evaporation at the pine canopy is countered by better ventilation in the canopy compared to the surface of the clearing. Therefore, the main effect of the tall trees on the atmosphere as observed at 100 m appears to be the elevated height of the heating.

The difference between the surface energy budget of the clearing compared to the pine forest can be studied by comparing aircraft fluxes with those at the Lubbon central site (Table 15.1). Here we compare values at 1200 Universal Time close to the time of maximum solar radiation. The aircraft fluxes in Table 15.1 are "local" values representative of the Lubbon region and have been augmented by about 8% to crudely simulate a linear decrease of the fluxes from the surface to the boundary-layer top. The aircraft measurements are dominated primarily by the pine forest, which possesses significant stomatal control while the measurements at Lubbon occur over oats (SAMER 1) with weak stomatal control during the growing phase of the phenology (see, for example, Pinty and Mascart, 1988) and emerging corn (SAMER 5). *Therefore the moisture flux measured by the aircraft is less and the Bowen ratio greater compared to the Lubbon clearing although the SAMER 5 fluxes are closer to the aircraft values than the dramatically larger moisture flux at nearby SAMER 1.* On 25 May, the aircraft data indicated considerably less energy available for sensible and latent heat fluxes compared to the SAMER sites. Scattered boundary-layer cumulus reduced the spatially averaged solar radiation over the region. However the solar radiation and heat and moisture fluxes at the SAMER sites in the Lubbon clearing did not experience serious reduction of solar radiation until 1330 UT. This may partly explain why the spatially averaged fluxes measured by the aircraft are less than the local surface fluxes measured at Lubbon at 1200 UT. Early development of a few innocent scattered clouds in the western part of the

TABLE 15.1. Aircraft-measured, model, and SAMER estimates of the sensible and latent heat fluxes (W/m²) and sum of the two.[a]

	Aircraft	Model	SAMER 1	SAMER 5
19 May				
Q_H	220	238	−18	178
Q_{LE}	310	344	536	389
Total	530	582	518	567
Bowen ratio	0.71	0.69	0	0.46
25 May				
Q_H	216	287	0	233
Q_{LE}	268	288	615	341
Total	484	575	615	574
Bowen ratio	0.81	1.00	0	0.68

[a] Reported aircraft fluxes are local to the Lubbon region and have been crudely extrapolated to the surface by assuming linear variation with height (roughly 8% augmentation).

forest seems to have led to modest horizontal inhomogeneity of the heat and moisture fluxes on 25 May (Fig. 15.6).

Both the aircraft data and the SAMER 5 data indicate larger Bowen ratios (smaller latent heat fluxes) on 25 May (Table 15.1). *The increase of the Bowen ratio appears to be greatest over the forest, which is consistent with the observation of warmer dryer air above the pine forest compared to the clearings on 25 May.*

The near zero sensible heat flux at SAMER 1 may be partly due to advection of warm air from the adjacent pine forest, reducing or eliminating the unstable stratification at the surface of the clearing. Additionally, the turbulence may not be in equilibrium with surface conditions leading to errors in the estimation of the flux from vertical gradients although the SAMER 1 moisture fluxes compare well with the neutron probe data (Cuenca and Noilham, this volume). Later in the phenological cycle, the moisture flux over the oats is smaller leading to significant sensible heat flux (Bessemoulin et al., 1987).

Both 19 May and 25 May were tentatively interpreted in terms of a one-dimensional model of the soil, surface energy balance, and boundary layer (Troen and Mahrt, 1986; Mahrt and Pan, 1984). The initial soil moisture in the model is specified to be 0.20 based on the neutron probe data at Lubbon. The surface albedo is specified to be 0.13 to approximate spatially averaged conditions. The albedo of the clearings is probably about 0.2 while

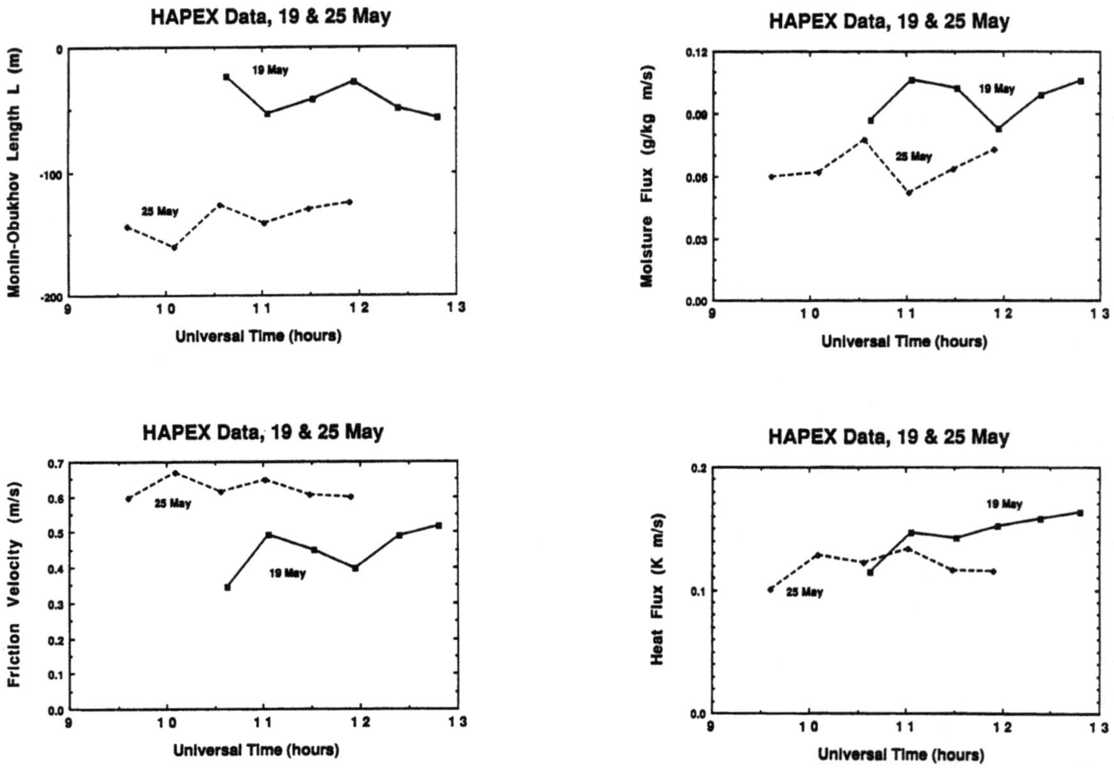

FIGURE 15.7. Heat and moisture fluxes, the Bowen ratio, and Monin–Obukhov length computed from the aircraft data (5 km filter cutoff) versus Universal Time (GMT).

the albedo of the mature pine forest is probably about 0.1 (Pinty and Mascart, 1988). To include the understory vegetation in the pine forest, the vegetation coverage (shading factor) is specified to be 0.85 (Pinty and Mascart, 1988). The plant coefficient representing the stomatal control is 0.7 (Pan and Mahrt, 1987) and the wilting point soil moisture content is 0.07. The diurnal variation of the stomatal control was neglected in these model studies. Such variation may be partly responsible for a mid-day dip in the latent heat fluxes observed in Pinty and Mascart (1988) and hinted by the aircraft data (Fig. 15.7). The values of the subsidence and horizontal advection are additional sources of uncertainty.

The model simulations, using initial radiosonde data from W. Brutsaert and aircraft soundings, indicate that the observed and modeled evolution of the boundary-layer temperature profile and boundary-layer depth are consistent with sensible heat fluxes estimated by the aircraft and SAMER 5 stations whereas the heat fluxes measured at the SAMER 1 station would lead to little boundary-layer development. The boundary layer as a whole probably responds mainly to the pine forest. On 19 May, the modeled surface fluxes are quite close to the aircraft-measured fluxes and somewhat similar to the SAMER 5 values but quite different from the SAMER 1 fluxes. On 25 May, the model predicts a Bowen ratio near unity whereas both the aircraft data and SAMER sites predict the latent heat flux to be significantly larger than the sensible heat flux. It may be that the stomatal control in the model is too strong for this day. The favorable comparison between the model and data probably indicates that the boundary-layer development on these days can be approximated by rather simple physics but does not serve as a formal validation of the model.

5. Horizontal Fluxes

Horizontal fluxes are normally thought to be unimportant because their variation occurs over length scales that are large compared to the boundary-layer depth leading to unimportant horizontal divergence of the horizontal flux. This assumption breaks down in regions of significant local variation of surface conditions. Usually surface inhomogeneity is studied in terms of sharp discontinuities between uniform surface areas where the chance of physical understanding is greatest. Many surfaces consist of continually changing surface conditions forced by variations of soil conditions even when the vegetation appears to be uniform. The boundary-layer flow may never achieve equilibrium with the local surface conditions and although flow perturbations may be induced on a variety of scales. These perturbations together with transient motions such as inertial gravity waves or cloud-induced circulations lead to motion variance and fluxes on a variety of scales larger than the main turbulent eddies. From a modeling point of view, the horizontal flux divergence may not be negligible and the vertical flux may include more than just turbulent contributions.

The horizontal flux divergence can be neglected if the ratio

$$[\delta(v'\phi')/\Delta y]/[\delta(w'\phi')/\Delta z]$$

is small where Δy and Δz are the grid spacings where δ indicates a change of the flux between grid points or grid levels. In the boundary layer, the vertical flux divergence is typically on the order of the surface flux divided by the boundary-layer depth and not sensitive to the vertical resolution. However the magnitude of the horizontal flux appears to be more sensitive to the choice of the horizontal resolution.

As a plausible estimation of subgrid scale fluxes due to motions larger than turbulent scales, consider a hypothetical grid box where the flight track coincides with either the north or south edge of the grid box, say at $y=y_0$ (fig. 15.8). In the analysis below, the width of the grid box will be imagined as either the entire flight leg or the flight leg will be broken into subrecords corresponding to hypothetically smaller grid boxes. For each grid box (subrecord), we divide the flow into the mean (estimated grid-averaged value), the linear trend ϕ_t, the low pass filtered part of the detrended data ϕ_ℓ, and finally the remaining high pass part ϕ_h so that

$$\phi' = \phi_t + \phi_\ell + \phi_h$$

The filter cutoff wavelength is 5 km. We then compute the fluxes due to the various subgrid components. The high pass flow is thought to be mainly turbulence, while the low pass flow is associated with subgrid scale circulations on scales larger than 5 km. The trend is associated with the largest scales not resolved by the grid, or may be partly identified with the resolved gradient between model grid points, depending on the interpretation of the grid-volume averaging. The calculation of fluxes will be only exploratory since fluxes on the largest scales do not fulfill sampling requirements.

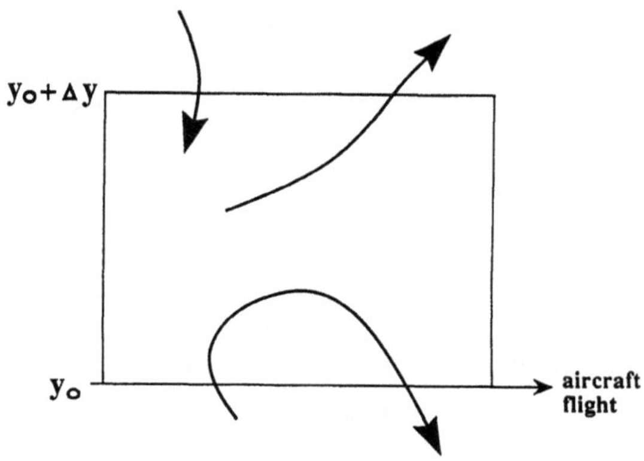

FIGURE 15.8. Sketch of the grid geometry and plausible subgrid scale motions.

Consider the horizontal flux due to $v'\phi'$ averaged over the flight leg and assume that it is associated with scales smaller than the grid scale and thus not highly correlated with the subgrid scale flux on the opposite side of the box at $y=y_0 + \Delta y$ (Fig. 15.8). Then the order of magnitude of the horizontal divergence of the flux can be estimated as

$$v'\phi'/\Delta y$$

Fluxes due to cross-products between the different components such as $v_\ell\phi_h$ do not average to zero as in rigorous Reynolds averaging without use of filtering but for the present data are always at least one order of magnitude smaller than the fluxes associated with a single component of the flow.

The various components of the horizontal fluxes were computed for hypothetical grid boxes of various sizes of 19 May. In general, the horizontal flux divergence becomes more important as the size of the grid box becomes smaller but does not become significant compared to the vertical flux divergence until the grid box size becomes at least smaller than 5 or 10 km. For example, with a 5 km grid size, the northward subgrid flux of moisture averages almost double the vertical flux. However, more than half of this flux is due to the 5 km high pass fluctuations, which for most hypothetical grid boxes, agree in sign with that expected from vertical motions acting on the mean vertical gradients. That is, turbulent updrafts lead to positive moisture fluctuations and weaker northward velocity, and vice versa, leading to a southward flux of moisture. However, since this horizontal flux is partly related to the larger scale vertical gradients, the divergence of this part of the flux is weak. In other terms, this part of the larger scale divergence of the flux is weak even though the flux is due to small scale motions.

Similar analysis of the heat flux for this case indicates that the horizontal flux divergence is less important than that for the moisture. Perhaps fluxes due to surface inhomogeneity are forced mainly by variations of evapotranspiration and variations of the heat flux follows as a secondary effect. *From Figures 15.5 and 15.7, one can determine that the 5 km standard deviation of the vertical flux scaled by the average magnitude is about twice as large for moisture as compared to the variability of the vertical heat flux.*

The "subgrid" horizontal fluxes of momentum are even larger exceeding the vertical flux by more than a factor of two. The relatively larger size of the horizontal momentum flux, compared to horizontal heat fluxes, is probably due to transport by pressure effects. However divergence of the horizontal momentum flux appears to be generally weak. Apparently much of the horizontal momentum flux is again induced by the vertical gradients associated with the larger scale flow.

However, larger horizontal flux divergences at a given location are not significant on the usual time scales of meso- and larger scale models unless they exhibit persistence in time as opposed to random variations of the flux which usually average to small values. This problem will be addressed by examining the horizontal fluxes with respect to specific surface features as viewed by each of the six flights. The above analysis must also be extended to more cases including flights over the rolling farmland part of the HAPEX region and days with weak surface heating. Then the vertical flux divergence is smaller and horizontal flux divergences associated with surface inhomogeneity are likely to be larger. In cases of near neutral stability, even the sign of the local stability may be determined by local soil moisture and evaporation leading to significant horizontal fluxes of moisture on the subgrid scale.

6. Conclusions

The above analyses indicate that the inhomogeneity due to clearings and variation of the height of the stands of pine trees does not lead to serious statistical sampling problems in the analysis of various structure functions as computed from flight legs 100 m over the pine forest. In more specific terms, the data are not characterized by significant global intermittency (Fig. 15.2). An analysis of flux sampling problems indicates that a filter cutoff wavelength of at least 5 km is needed to include most of the vertical transport at 100 m above the forest (Fig. 15.4). However, this cutoff wavelength cannot be arbitrarily increased to scales much larger than 10 km to avoid inclusion of larger scale motions which cannot be adequately sampled with the 120 km legs (Fig. 15.5).

The above analysis yields some support for the unwritten rule that the time series must be at least 10 times longer than the largest scales admitted by the filtering. However, the flux errors also depend on the relative length scales of the inhomogeneities and the factor of 10 cannot be categorically applied to all situations even as a rough rule.

Analysis of the structure functions and eigenvectors of the lagged correlation matrix indicates that the warm undrafts are influenced by shear leading to sharp microfronts at the "upstream" edges of the updrafts. It is not yet clear whether to consider the updrafts as shear-modified thermals or shear-driven overturning leading to updrafts of warm moist air from the surface layer.

Although the statistics for the flight legs are quite robust and determined mainly by the pine canopy, the variations associated with the occasional clearings are of scientific interest. The occurrence of clearings from the aircraft was detected by the downward video camera, infrared measurements, and radioaltimeter values. The air over the clearings is cooler and more moist on the 2 days examined. SAMER fluxes in the Lubbon clearing indicate greater latent heat flux and less sensible heat flux compared to over the pine canopy (Section 4 and Table 15.1). However, the actual flux values over the clearing vary substantially and depend on the vegetation and crop phenology. The application of a simple one-dimensional model of the soil, vegetation, and boundary layer indicates that the evolution of the boundary layer structure and depth is responding to the fluxes over the pine canopy as expected from the predominance of the areal coverage by the canopy.

Horizontal "turbulent fluxes" appear to exceed the corresponding vertical fluxes in the case of moisture and horizontal momentum (Section 5). However, the horizontal divergence of the horizontal flux appears to be generally smaller than the vertical divergence of the vertical flux depending on the choice of scales. This subject will be investigated in closer detail with respect to the surface inhomogeneity toward the goal of assessing the importance of horizontal fluxes between grid boxes in high-resolution meso-scale models.

Acknowledgments. This material is based on work supported by the Experimental Meteorology Program of the National Science Foundation under Grant ATM-8820090. The computational assistance of Wayne Gibson and Michael Ek is greatly appreciated. The King Air research aircraft was provided by the National Center for Atmospheric Research, USA.

References

Baker MA, Gibson CH (1987) Sampling turbulence in the stratified ocean: Statistical consequences of strong intermittency. J Phys Ocean 17:1817–1836.

Bessemoulin P, Desroziers G, Payen M, Tarrieu C (1987) Atlas des données SAMER. EERM/CNRM, Toulouse.

Chen CHP, Blackwelder R (1978) Large-scale motion in turbulent boundary layer: A study using temperature contamination. J Fluid Mech 85:693–704.

Cuenca R, Noilhan J (1990) Use of soil moisture measurements in hydrological balance studies. In TJ Schmugge (ed.). Land Surface "Evaporation: Measurement and Parameterization." Springer-Verlag, New York.

Frisch U, Sulem P, Nelkin M (1978) A simple dynamical model of intermittent fully developed turbulence. J Fluid Mech 87:719–736.

Hildebrand P (1988) Flux and sounding data from the NCAR King Air aircraft during HAPEX. NCAR Technical Note.

Khalsa SJS, Businger JA (1977) The drag coefficient as determined by the dissipation method and its relation to intermittent convection in the surface layer. Bound Layer Meteorol 12:273–297.

Kikuchi T, Chiba O (1985) Step-like temperature fluctuations associated with inverted ramps in a stable surface layer. Bound Layer Meteorol 31:51–63.

Mahrt L (1989) Intermittency of atmospheric turbulence. J Atmos Sci 46:79–95.

Mahrt L, Frank H (1988) Eigen structure of eddy microfronts. Tellus 40A:107–119.

Mahrt L, Gamage N (1987) Observations of turbulence in stratified flow. J Atmos Sci 44:1106–1121.

Mahrt L, Pan H-L (1984) A two-layer model of soil hydrology. Bound Layer Meteorol 29:1–20.

Pan H-L, Mahrt L (1987) Interaction between soil hydrology and boundary-layer development. Bound Layer Meteorol 38:185–202.

Pinty JP, Mascart P (1988) Evaluation d'un modèle d' interface sol/végétation/atmosphère à l'aide de données SAMER et HYDRA de la campagne HAPEX-MOBILHY 1986. Observatoire de physique du globe de Clermont-Feerrand, Note O. P. G. C. No. 96.

Schols JLJ (1984) The detection and measurement of turbulent structures in the atmospheric surface layer. Bound Layer Meteorol 29:39–58.

Schols JLJ, Wartena L (1986) A dynamical description of turbulent structures in the near neutral atmospheric surface layer: The role of static pressure fluctuations. Bound Layer Meteorol 34:1–15.

Shaw WJ, Businger JA (1985) Intermittency and the organization of turbulence in the near-neutral marine atmospheric boundary layer. J Atmos Sci 42:2563–2584.

Troen I, Mahrt L (1986) A simple model of the atmospheric boundary-layer: Sensitivity to surface evaporation. Bound Layer Meteorol 37:15–49.

16
Temporal Variation of Heat and Moisture Flux Within the Atmospheric Boundary Layer over a Grassland

Robert L. Grossman

1. Introduction

This chapter serves to introduce some aspects of the role the atmospheric boundary layer (ABL) plays in evaporation from the land surface since the ABL couples the surface to the free atmosphere (that region of the atmosphere not greatly affected by surface properties). Time variability of moisture is the focus, especially variability on the subdiurnal time scale. Several physical processes are presented that could account for subdiurnal time variability and an example of the effect of turbulent transport throughout the depth of the ABL is demonstrated.

1.1 The Atmospheric Boundary Layer (ABL)

The role of the atmospheric boundary layer is paramount in any investigation of land surface climatology, since the ABL is the medium through which momentum and radiative, sensible, and latent heat energy is transported to and from the surface. The fair-weather ABL can be divided into three layers (Fig. 16.1): a shallow layer, usually one-tenth the depth of the ABL, called the surface layer, which is in direct contact with the surface; a deep layer, called the mixed layer, which is often 70–80% of the total depth of the ABL; and the entrainment layer, which is the interface between the mixed layer and the free atmosphere above (see Stull, 1988, Ch. 1). In fair weather the free atmosphere above the ABL layer is subsiding and the resulting potentially warm and dry air forms an inversion over the ABL. The entrainment layer often contains shallow cumulus clouds. The presence of even shallow cumulus clouds makes the transport of latent and sensible heat through the ABL substantially different and more complex than that which occurs in clear conditions.

The energy transports associated with these three layers are coupled to the radiative properties of the underlying surface, its roughness, and a combination of vegetation and soil characteristics, the most important of which is the amount and vertical distribution of soil moisture. Soil moisture is an important component of the overall heat capacity of the soil, determines the amount of vegetation stress (and thus to some degree surface radiation characteristics and roughness), and is a controlling factor in latent heat exchange between the land surface and atmosphere. The nonlinear interaction between radiation, vegetation, soil moisture, and the ABL is important to our understanding of global and regional climate variation (Sellers et al., 1986; Avissar and Pielke, 1989; Shukla et al., 1990).

1.2 Fair-Weather Convection in the ABL

During the growing season, most of the energy exchange during fair weather occurs during daylight hours in the form of upward-moving, warm, moist convective cells that also transport carbon dioxide, trace chemical species, and particulate matter from the surface to the interface between the ABL and overlying free atmosphere. These convective cells are driven by density anomalies created near the surface when the surface is radiatively warmed to temperatures greater than the overlying air. These anomalies consist of a component associated with

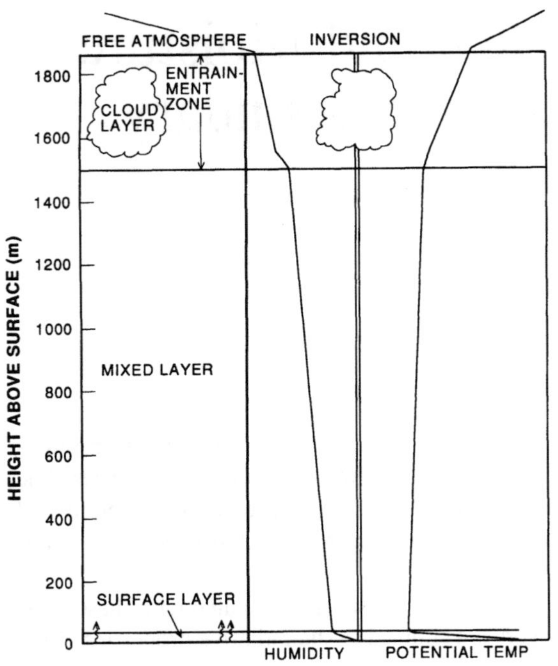

FIGURE 16.1. Schematic diagram of the atmospheric boundary layer under conditions of clear skies and scattered shallow cumulus cloud. Heights roughly correspond to those measured on 6 June 87 about mid-day.

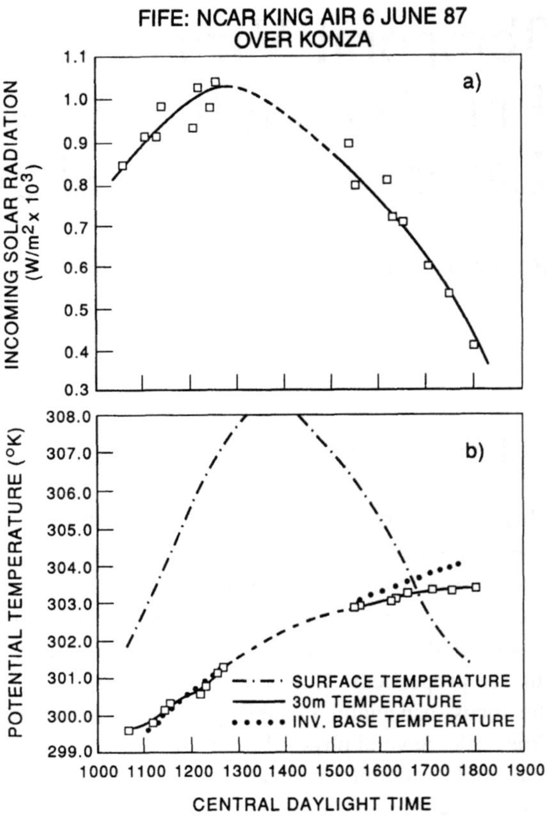

FIGURE 16.2. Time variation of (a) incoming solar radiation and (b) surface temperature (dash–dot) inferred from an infrared radiometer plotted with low-level (solid) and near entrainment level (dotted) potential temperature.

dry air temperature (sensible heat) and to a lesser extent moisture (latent heat). These upward-moving, warm, moist convective cells are primarily driven by the dry-air temperature component. Because of the latent heat of condensation, if soil moisture is high and plants are transpiring, latent heat energy transport is often greater than sensible heat transport. Sensible heat can exceed latent heat transport when soil moisture is depleted or when plants are stressed and transpiration is reduced.

However, it is important to appreciate that in the spring–summer convectively active ABL downward motions also play a role in energy and mass exchange. For the most part downward motions are part of the mass compensation associated with the energetic upward moving cells. These downward compensating motions are usually smaller in size than the upward moving convective cells. In some cases the downward moving air originates just above or within the entrainment zone (Grossman, 1984) so the air is dry and the potential temperatures associated with them are relatively warmer than mixed-layer potential temperatures (but not as warm as temperatures near the surface). This compensating downward moving air from the entrainment layer and free atmosphere above is lighter than its surroundings and will be positively buoyant. If not dynamically forced downward by strong upward moving cells, this air will rise. Some downward motions are relatively cool and moist; these were upward moving cells that have moved above their level of neutral buoyancy because of inertia, have become negatively buoyant, and are falling back. Both upward and downward motions mix with each other at their respective boundaries. This mixing tends to reduce the buoyancy within either an upward or downward moving cell.

Thus the vertical profiles of potential temperature and moisture in the ABL and their variation with time are the result of a complex interaction of upward and downward motions. During the day scalar transport from the surface exceeds that from the free atmosphere while mass balance is maintained.

1.3 Diurnal and Subdiurnal Time Variation in the Clear Sky ABL

Continental climate regimes are characterized by high variability both in time and space, mainly because of the low heat capacity of the underlying surface, which is usually not spatially homogenous. Thus throughout the course of a clear day, such as the example discussed in the coming pages, the incoming solar radiation slightly leads the surface temperature and significantly leads the air temperature (Fig. 16.2). The resulting superadiabatic lapse rate of temperature (potential temperature decreasing with height) initiates and maintains the upward-moving convective cells described in section 1.2. Moisture plays a relatively minor role. Once the superadiabatic layer is established, the depth of the ABL increases dramatically up to the time of maximum surface temperature then increases at a much slower rate (Fig 16.3). Explosive growth, shown in Figure 16.3, is often associated with the presence of an elevated well-mixed layer (EML) above a low-level inversion formed by radiative cooling of the surface during the night (Fig. 16.4). The temperature and moisture profiles in Figure 16.4a,b evolve into the schematic profiles shown in Figure 16.1 (see also Fig. 16.4c,d) during the early and late morning. As the surface warms and the radiation inversion is destroyed, warmer and warmer upward-moving convective cells are formed and begin to penetrate into the less stable layer above the nocturnal inversion. This layer is progressively eroded by sensible heat transport until the ABL reaches the potential temperature of the EML and the ABL depth appears to suddenly increase. Upward-moving convective cells can now go to much higher levels. This explosive growth is often associated with nearly simultaneous explosive formation of cumulus clouds, since the top of the EML is often very close to the condensation level.

1.4 Advection

Up to this point the time variation of temperature and moisture in the ABL has been described as a one-dimensional (1-D) process: only vertical motions have been considered. However, if the wind blows across a scalar gradient, that scalar quantity (temperature, moisture, chemical compound) will be advected (or moved) with the wind. This advection can also contribute to time variation and in some cases overwhelm the variation associated with vertical motions. Especially when describing mesoscale ($<$ 100 km length scales) situations, vertical as well as the more familiar horizontal advection should be considered. Vertical advection is by vertical motions, which are of much larger scale than the 1-D situation described above, acting on a vertical gradient of a scalar. Thus time variation of mean quantities in the ABL is a three-dimensional (3-D) problem over a variety of spatial scales.

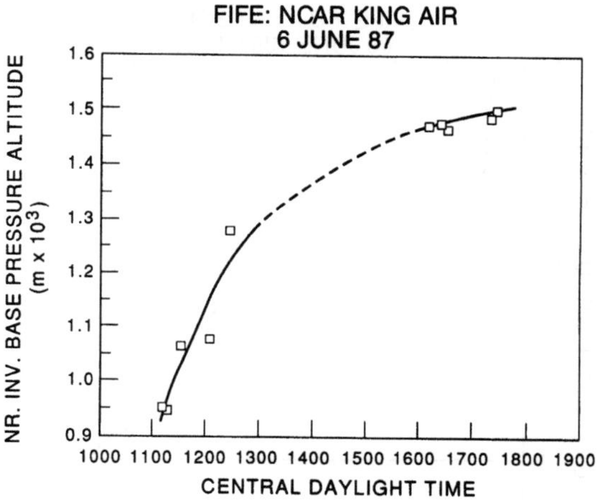

FIGURE 16.3. Time variation of aircraft altitude near entrainment layer, a measure of the depth of the ABL.

1.5 Sources and Sinks

A source or sink of a scalar can also contribute to time variation. For instance, rain falling into a very dry layer of the atmosphere will evaporate. This phenomenon acts as a source of water vapor (moisture) and a sink of heat (evaporative cooling). Cloud formation in the entrainment zone acts as a sink of water vapor (which changes to condensate) and a source of heat; this is the primary reason cloud formation in the ABL substantially complicates our understanding of time variation.

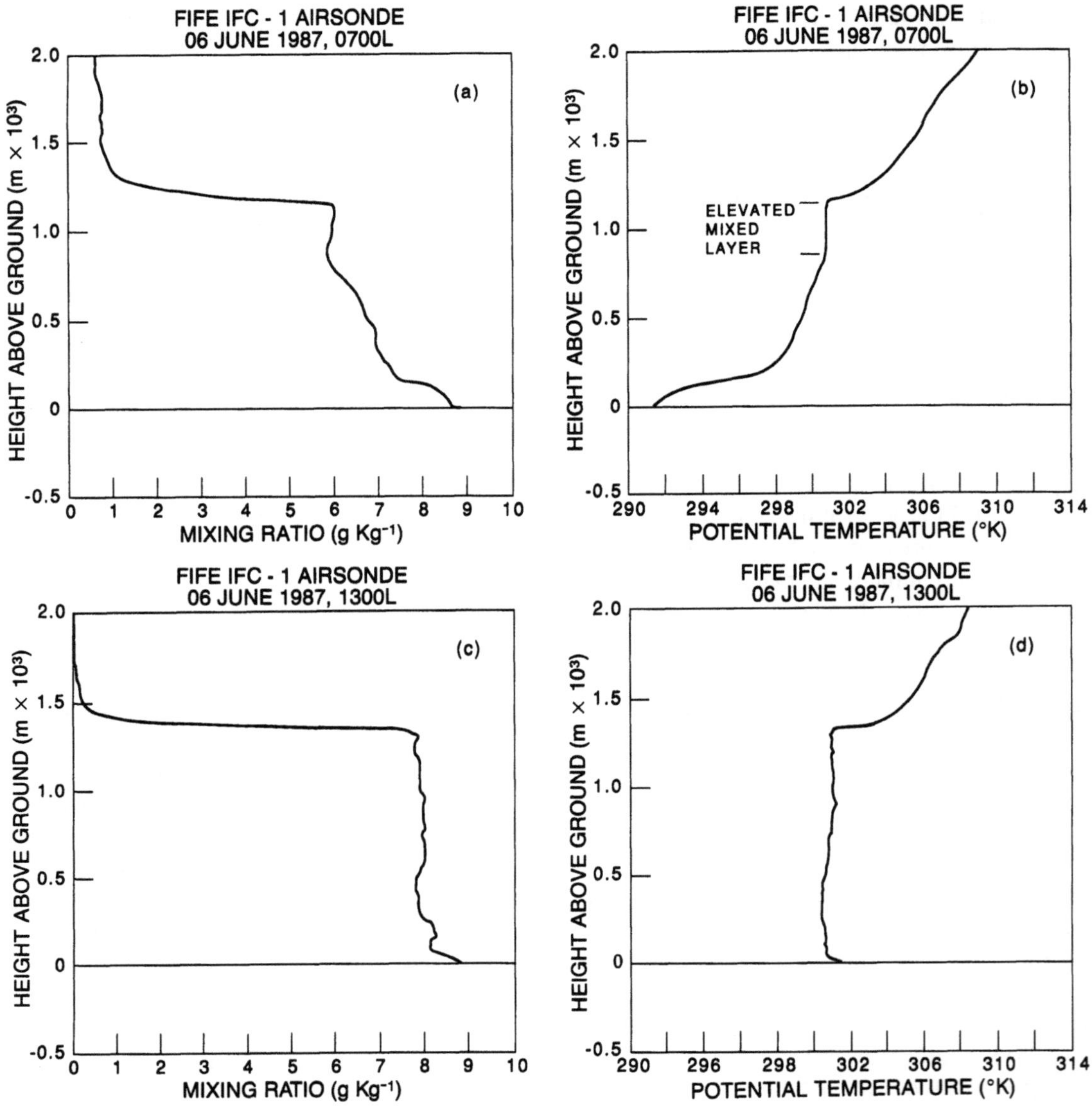

FIGURE 16.4. Vertical profiles of mixing ratio and potential temperature from balloon-sonde ascents on 6 June 87. From data obtained by W. Brutsaert, Cornell Univ.

1.6 Quantifying the Components of Time Variation: ABL Budgets

The ideas presented in the previous subsections are mathematically described in the Eulerian budget or balance equation for a scalar:

$$\frac{\partial \Psi}{\partial t} = -\bar{u}\frac{\partial \Psi}{\partial x} - \bar{v}\frac{\partial \Psi}{\partial y} - \bar{w}\frac{\partial \psi}{\partial z} - \frac{\partial \overline{w'\psi'}}{\partial z} + S_O - S_I \qquad (16.1)$$

From left to right the terms are local time variation, horizontal advection (u, x is positive to the east and v, y is positive to the north), vertical advection (w positive upward), turbulent flux divergence, and sources minus sinks.

Note that this equation is predictive and is part of a set of equations that describes the ABL (see Stull, 1988, Ch. 3). Thus it is an equation that can be used in computer models of the ABL and its interaction with the layers of the atmosphere above it. Modelers would like to know if this equation can be simplified or at least what the limits of certain types of simplification might be. The work presented in the following sections explores the hypothesis that for some cases (16.1) can be approximated by

$$\frac{\partial\overline{\psi}}{\partial t} = -\frac{\partial\overline{w'\psi'}}{\partial z} \qquad (16.2)$$

even under conditions of relatively high surface winds. This means that horizontal and vertical advection and sources and sinks can be neglected. Equation (16.2) says that the time variation is primarily dependent on the turbulent flux divergence (See Section 2.3 for definition of w' and ψ'. The turbulent flux divergence is a consequence of the relatively small scale (100 m to 5 km) motions associated with the upward and downward motions described in Sections 1.2 and 1.3.

If we consider the mixed layer as the layer associated with the ∂z in the turbulent flux divergence term, then when the turbulent flux divergence is negative, the turbulent flux at the higher level is transporting the scalar out of the mixed layer at a lower rate than it is being imported from the surface; this results in an increase of the scalar with time. Conversely, a positive turbulent flux divergence means that the scalar is being exported out of the mixed layer at a faster rate than it is being supplied from below and a decrease in the scalar with time occurs.

We will explore the applicability of Equation (16.2) using potential temperature and moisture data obtained during an aircraft investigation of the mixed layer. Though there have been many investigations of the continental ABL, few of the investigations have had the resources available to study the time variation of the turbulent flux divergence *throughout the depth of the mixed layer.* This is best done by use of aircraft, since, when properly instrumented with eddy correlation systems (gust probes), they can measure the turbulent fluxes near the top the ABL as well as near the surface. This study was carried out as part of the First ISLSCP (International Satellite Land Surface Climatology Project) Field Experiment (FIFE; Sellers et al., 1988), one of a series of experiments sponsored by the USA National Aeronautics and Space Administration to study the interaction between the living surface and the atmosphere. FIFE had five intensive field campaigns (IFC): four during the growing season of 1987 and one (IFC-5) during a period of decreasing soil moisture in 1989.

2. Flight Mission Plan, Meteorological Scenario, and Data Acquired

One of the main objectives of FIFE is to "obtain a rigorous understanding of the processes controlling energy/mass exchange at the surface and how these are manifested in satellite-resolution radiometric data." FIFE extensively investigated five periods during the May to October life cycle of the grasses covering one of the last natural prairie lands in the United States, the Konza Prairie near Manhattan, Kansas in the central United States. The Konza is owned by the Nature Conservancy and managed by the Kansas State University. A wide variety of surface-based, satellite, and aircraft observations were utilized in FIFE (Sellers et al., 1988). This study concentrates on 1 day (6 June 87) of the first of the five IFCs of FIFE. The aircraft used was a Beechcraft King Air from the USA National Center for Atmospheric Research, Research Aviation Facility. This King Air is a highly instrumented aircraft capable of making in situ measurements of atmospheric parameters (Miller and Friesen, 1985).

2.1 The ABL Aircraft Mission Plan

The aircraft mission plan for 6 June 1987 is shown in Figure 16.5. This "vertical race track" was flown between points A and C throughout the day except for a short break for refueling both aircraft and crew. Point B marked the boundary between the natural prairie and agricultural land to the east and each crossing from prairie to farmland was noted by the author, who acted as observer during the flight. The upper flight leg was about 50 to 75 m below the height of the convectively mixed layer. This height was sensed by making short excursions upward until the characteristic sharp drop in moisture content was observed and by information relayed to the aircraft from ground-based balloon soundings and an acoustic sounder, both of which were near the flight track. The lowest level was flown about 30–40 m above the highest point of the

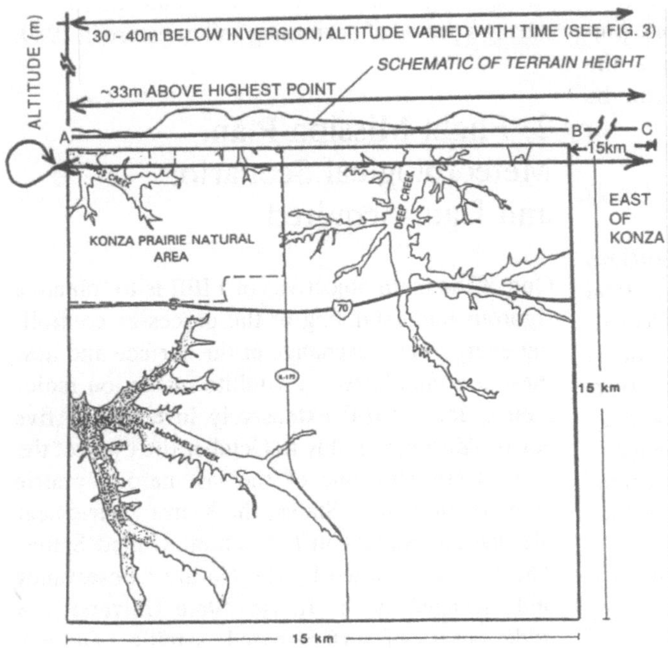

FIGURE 16.5. Schematic of NCAR King Air flight mission profile for 6 June 1987.

Konza Prairie so the turbulent flux divergence measured was characteristic of the mixed layer. The flight track was near the northern edge of the rolling hills of the Konza Prairie and was oriented nearly east–west.

2.2 Meteorological Scenario

The meterological scenario was dominated by a synoptic scale high-pressure system, southerly surface winds of approximately 10 m s^{-1}, and clear skies. However, widespread rainfall characterized the week previous to this day so that the ground was very wet and little stress was evident in the emerging grasses covering the prairie. Cloud cover was less than 10% throughout the flight mission.

2.3 Data Obtained

Data acquired on board the King Air mission reported here included atmospheric state variables (pressure, temperature, humidity), horizontal wind velocity, surface and sky temperature (from infrared radiometers), and upwelling and downwelling visible and infrared radiation. These measurements supplemented the turbulence measurements from a gust probe (Lenschow, 1984), which measured the horizontal and vertical gust components of the wind (e.g., deviations from a detrended level leg mean, $\psi = \overline{\psi} + \psi'$) as well as fast response (approx. 10 Hz) temperature and moisture fluctuations.

2.4 Turbulent Flux Measurement

Vertical fluxes of sensible and latent heat are obtained by estimating the covariance of vertical velocity deviations, w', and potential temperature deviations, θ', or absolute humidity variations, ρ_v',

$$\overline{w\psi} \cong \overline{w'\psi'} \qquad (16.3)$$

This is also known as the eddy correlation method. For this estimate to be correct, it is important to statistically account for all turbulent eddies that substantially contribute to the flux. This is done by choosing an averaging distance that is much longer than the average scale size of the eddies important to flux transport. Concern about the short flight legs (approx. 15 km) over the Konza Prairie caused me to carefully test the flux estimates. This was done by looking at cospectra of the fluxes and by averaging over successively longer and longer distances ("distance" in this case is measured by multiplying true air speed with time).

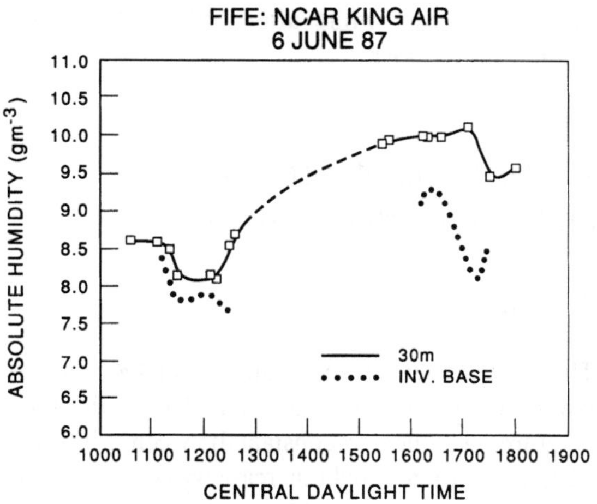

FIGURE 16.6. Time variation of low-level (solid) and entrainment layer level (dotted) absolute humidity.

From these tests, I determined that the 15 km average was just adequate for flux estimates. More confidence in the flux estimate would result from a weighted average of two runs. The consistency of the flux variation with time (Section 5) and the good agreement between the variation of flux divergence and mean quantities also indicate the adequacy of the 15 km averaging distance *for this particular day.*

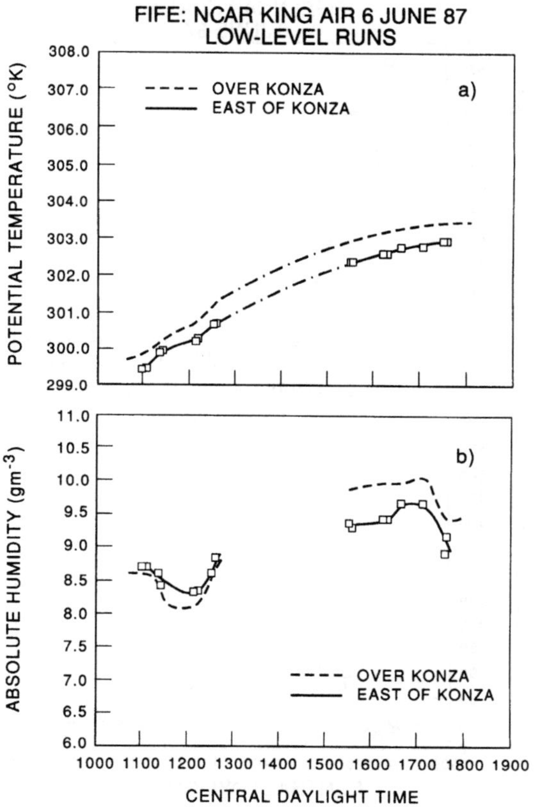

FIGURE 16.7. Time variation of (a) potential temperature over Konza prairie (dashed) and agricultural land to east (solid); (b) absolute humidity over Konza (dashed) and agricultural land (solid).

3. Mean Temperature and Moisture Variation

3.1 Time Variation

Since only two levels were flown, a linear variation of mean potential temperature and moisture with height between the two aircraft levels is assumed. This is supported by the aircraft profiles obtained when the aircraft changed levels. The difference between the low and high level potential temperature (Fig. 16.2b) was nearly zero until about 1600 CDT. This indicates that the ABL was nearly well mixed with respect to temperature and is consistent with the low sensible heat fluxes which were observed (see Section 5). After 1600 a slight increase with height is observed indicating a tendency to more stable conditions from the mid-afternoon instability. The onset of more stable conditions occurred about 40 min before the time at which the surface temperature dropped below the air temperature.

The time series of mean moisture shows a more complex behavior (Fig. 16.6). Between 1000 and 1200 CDT moisture in the ABL *decreased* then increased to about 1600 CDT when a decrease was again observed. There was a slight decrease in moisture with height around 1000 to 1100 CDT, which increased in magnitude throughout the rest of the day. This is consistent with the overall unstable conditions observed in the mixed layer, since a constant potential temperature with height and decreasing moisture with height (Fig. 16.4) combine to a decrease in mean density with height (e.g., a decrease in virtual potential temperature).

3.2 Spatial Variation

A comparison of mean potential temperature between the Konza Prairie and the adjacent agricultural land to the east (Fig. 16.7a) shows that at the

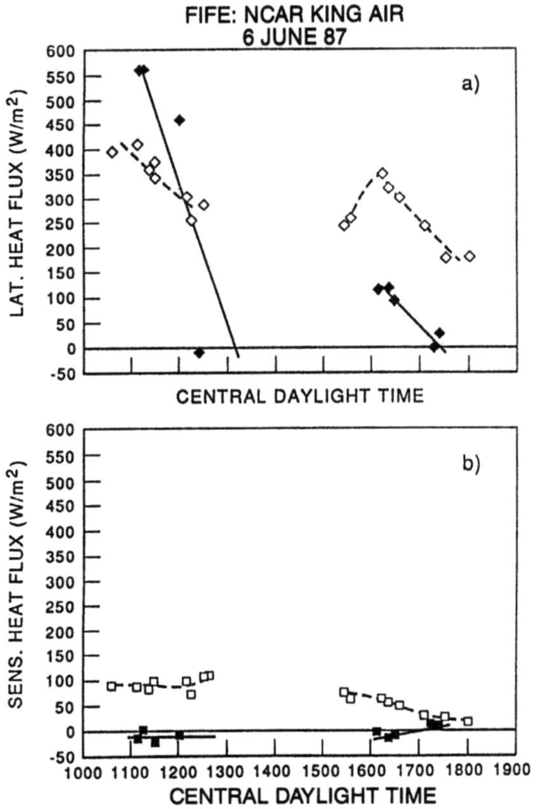

FIGURE 16.8. Time variation of (a) entrainment layer sensible (solid) and latent (dashed) heat flux; (b) same for low-level.

same pressure altitude the air over the Konza was about 0.5 K warmer than the adjacent (and lower) agricultural land even though the mean surface temperature over the Konza was about 0.5 K cooler. This variability may be due to the possibility that ABL characteristics well above the surface closely follow the large scale variation of the terrain, which in this case slopes downward to the east. One possibility is that over the Konza the aircraft was sensing the effect of the very warm, superadiabatic surface layer and was slightly above it east of the Konza. However, from Fig. 16.2b it appeared that the surface layer temperature gradient reversed after about 1630 CDT, which should have caused the Konza air temperature to be slightly cooler than the lower land to the east after that time; however, this was not observed.

A similar comparison of mean absolute humidity (Fig. 16.7b) revealed that before 1300 CDT the air over the Konza was slightly drier than the adjacent land to the east while after 1500 CDT it was slightly more moist.

4. Variation of Sensible and Latent Heat Fluxes

4.1 Time Variation

The low-level sensible heat flux, calculated from the covariance of vertical velocity and potential temperature, was nearly constant between 1000 and 1200 CDT, then a slight increase was observed; the flux decreased from 1500 to 1800 CDT (Fig. 16.8b). The sensible heat flux at the level near the entrainment layer was small, negative, and had no time variation. Other investigators have often found small, negative heat fluxes near the entrainment layer (Mahrt and Paumier, 1984). Thus the low level sensible heat flux exceeded the high level flux throughout the day.

The latent heat flux (calculated from the covariance of vertical velocity and absolute humidity) showed a complex behavior quite unlike the sensible heat flux. The low-level flux decreased from 1000 to 1230 CDT; after 1500 CDT an increase was observed until 1600 CDT when the flux decreased to values below those measured around 1230 CDT. At the upper level the flux appeared to decrease from very high values at 1000 CDT to nearly zero by 1230 CDT. The number of upper level measurements was smaller than the lower level measurements because some of the level runs penetrated the entrainment layer into the free atmosphere and had to be excluded. These "penetrations" indicated that the top of the mixed layer sloped downward to the east generally following the large-scale variation of terrain height. The flux values at the upper level exceeded those in the lower level until after 1200 CDT when they were always lower.

The ratio of low-level sensible to latent heat flux showed a sharp rise from 1000 CDT to 1200 CDT as the sensible heat flux increased slightly while the latent heat flux decreased. Form 1500 CDT to 1800 CDT the ratio decreased dramatically because the relative decrease in the sensible heat flux was almost twice that of the latent heat flux. The maximum ratio was 0.33 near noon and the minimum was 0.12 near 1800 CDT.

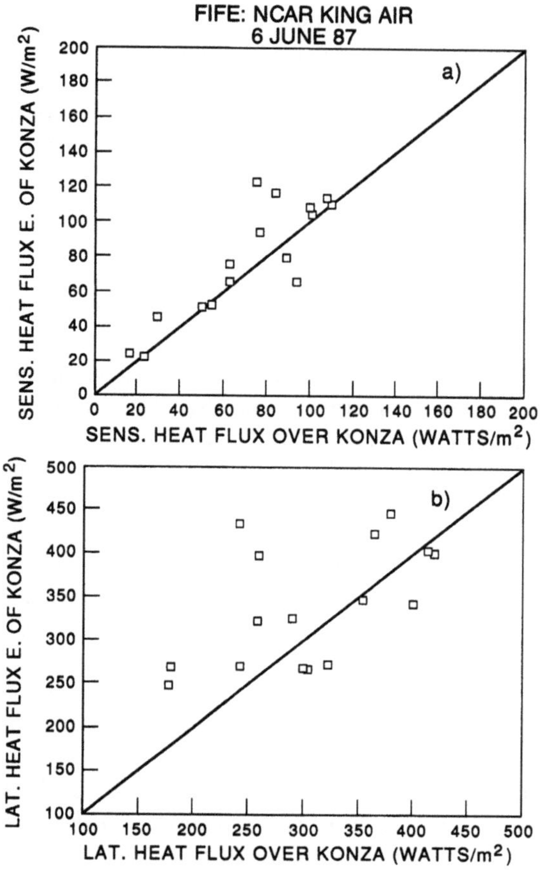

FIGURE 16.9. Comparison of (a) sensible heat flux over Konza and agricultural land to east; (b) latent heat flux over Konza and agricultural land.

4.2 Spatial Variation

The low-level sensible and latent heat fluxes were generally higher over the agricultural land to the east (Fig. 16.9a,b). The difference in sensible heat showed less scatter and was relatively larger for the smaller values of sensible heat flux. The same general rule held for the latent heat flux as well but the scatter was much larger.

Near the inversion base no difference was seen between the two areas for sensible heat flux. Recall that there was a change in the vertical gradient of latent heat flux from positive (before 1200 CDT) to negative (after 1200 CDT). A systematic difference appeared when this is taken into account: when the gradient was positive fluxes were higher over the Konza and when the gradient was negative fluxes were lower over the Konza. This can be explained by the fact that the aircraft, flying at constant altitude, was probably further from the entrainment layer over the Konza compared to the area east of the Konza, since the inversion appeared to be sloping downward to the east.

TABLE 16.1. Comparison of observed and calculated time rate of change of potential temperature and absolute humidity.

Time (CDT)	Calculated (kh^{-1})	Observed (kh^{-1})	Calculated ($gm^{-3}h^{-1}$)	Observed ($gm^{-3}h^{-1}$)
1115	0.7	0.8	−0.5	−0.4
1145	0.5	0.8	0.0	0.0
1220	0.4	0.8	0.4	0.5
1610	0.2	0.3	0.3	0.3
1700	0.1	0.2	0.3	0.3
1745	0.0	0.1	0.3	−1.4

5. Sensible and Latent Heat Budget Analysis

Equation (16.2) was volume integrated along the aircraft track (length) with a height corresponding to the difference between the upper and lower flight levels and a width 1 m wide (unit width). Estimates of flux divergence were centered on three times during the morning mission (1115, 1145, and 1220 CDT) and three during the afternoon mission (1610, 1700, and 1745 CDT). The estimates were made from curves drawn through the individual data and centered on the respective times. Table 16.1 shows the results of the computation of Equation (16.2) and comparison with observed values of layer averaged $\partial\overline{\psi}/\partial t$. Given a flux and height accuracy of about 10% of the absolute value, the error associated with flux divergence is about 14%. In general the results encourage the use of Equation (16.2) for this case for the agreement between $\partial\overline{\psi}/\partial t$ estimated from the flux divergence and observed values is quite good, given the error sources inherent in this type of computation. There was good agreement (within 14% of the calculated value) in five of six cases for water vapor and four of six cases for potential temperature.

The lack of agreement for the other cases can be explained by referring to Equation (16.1). For the sensible heat budget the source-sink term is radiation, which has been assumed negligible for the clear sky, daytime situation reported here. This

TABLE 16.2. Fraction of moisture flux associated with updrafts and downdrafts.

Low-level legs			Near inversion legs		
Leg start time	Flux up (%)	Flux down (%)	Leg start time	Flux up (%)	Flux down (%)
1057:23	64	36	1116:00	72	28
1105:36	66	37	1123:45	72	28
1133:30	67	33	1152:07	77	23
1141:59	62	38	1200:45	76	24
1211:30	69	31	1229:38	−04	104
1219:30	70	30	1238:00	−47	147
1248:04	74	26			
1256:13	64	36			

would not be true after sunset! For the moisture budget the source-sink term is evaporation–condensation, neither of which occurred in this clear sky example. After accounting for measurement accuracy, the difference between the observed local time change and flux divergence, called the residual in Equation (16.1), could be a combination of the horizontal and vertical advection terms. The flight mission plan described in Section 2 was not designed to evaluate these advection terms.

Other flight missions during IFC-4 of the 1987 FIFE were designed to evaluate those terms but could not investigate time variation (Betts et al., 1990). For the cases they studied, they concluded that horizontal advection was a term that could not be neglected while vertical advection was assumed negligible. Since temperature and moisture are often coupled variables in the boundary layer (positively correlated in the lower layers and negatively correlated in the upper layers), a question remains as to why the lack of agreement for potential temperature (1145 and 1220 CDT) was not also seen in the moisture nor was the one case for moisture (1745 CDT) reflected in the potential temperature. A similar lack of coupling between moisture and temperature advection is apparent in the results of Betts et al. (1990).

The morning agreement between the calculated and observed estimates of $\partial \bar{\rho}_v / \partial t$ is very encouraging. The variation of ρ_v was quite complex with drying, then moistening. Here flux divergence appeared to play a dominant role. Importantly, the drying of the ABL was due to the fact that the flux divergence was positive. Because of copious rainfall over the previous week, the ground was quite wet so potential evaporation could be expected. The data show an unexpected pattern in the morning mission with low-level evaporation decreasing and high-level evaporation exceeding the low-level values but decreasing at a faster rate so that by noon the flux divergence had switched from positive to negative.

It is likely that the large values of water vapor flux observed near the inversion base are connected to the explosive growth of the mixed layer depth through the EML. Grossman (1984) and Mahrt and Paumier (1984) used conditional sampling to explore eddy structure in the ABL. This phenomenological approach is well suited to determine whether or not the large positive water vapor flux measured near the inversion base was mainly due to predominantly downward moving dry air, which would produce a positive water vapor flux, being entrained into the ABL. Table 16.2 shows that at all times the moisture flux due to upward motions was generally twice that due to downward motions. Wyngaard and Moeng (1990, personal communication) report a similar ratio from results for a large eddy simulation of a convective boundary layer. From the conditional sampling analysis I conclude that *vigorous, upward moving convective cells were exhausting moisture from the BL in the layer explored by the aircraft* (approx. 30 m above the highest terrain to just below the inversion level) *and that the drying of the BL was not due to the entrainment of dry air from the inversion level.* Quite the contrary, the conditional sampling analysis suggests that the inversion layer was being moistened by detraining upward moving, moist convective cells. Thus BL growth was primarily the result of bottom-up diffusion (Wyngaard and Brost, 1984) of moisture with less than 30% contribution from top-down diffusion.

6. Discussion

Although this case study appears to argue for a simplified version of the sensible and latent heat budget for modeling purposes, it also shows that time variation of the ABL is complex even in the most simple conditions (clear skies, constant wind, potential evaporation). In particular, this study shows that the ABL reacts not only to surface conditions but to conditions at the entrainment layer (which may be coupled in some way to changing surface conditions) and upstream of the experimental site. The study has raised some interesting questions as well such as the reason for the difference in potential temperature and moisture between the prairie and adjacent agricultural land during a period of little vegetative stress. A further question is why the low-level moisture flux decreased with increasing surface temperature during the morning period. This is probably associated with the overall drying of the mixed layer during the late morning, a subject of continuing research.

Acknowledgments. This work was performed as part of NASA Grant NAG-5-904. The pilots of the NCAR King Air, Jerry Tejcek and Gil Summers, are gratefully acknowledged for precise and safe flying. Thanks also to the NCAR Research Aviation Facility, especially Vince Glover and Paul Spyers-Duran, for program management and the first phase of aircraft data reduction.

References

Avissar R, Pielke RA (1989) A parameterization of heterogeneous land surfaces for atmospheric numerical models and its impact on regional meteorology. Mon. Wea. Rev. 117:2113–2136.

Betts AK, Desjardins RL, MacPherson JI, Kelly RD (1990) Boundary Layer heat and moisture budgets from FIFE. Bound Layer Meteorol 50(1–4): 109–138.

Grossman RL (1984) Bivariate conditional sampling of moisture flux over a tropical ocean. J Atmos Sci 41: 3238–3254.

Lenschow DH (1984) Aircraft measurements in the boundary layer. In "Probing the Atmospheric Boundary Layer" (D.H. Lenschow, ed), pp. 39–56. American Meteorological Society, Boston.

Mahrt L, Paumier J (1984) Heat transport in the atmospheric boundary layer. J Atmos Sci 41:3061–3075.

Miller ER, Friesen RB (1985) Standard output data products from the NCAR Research Aviation Facility. NCAR, Research Aviation Facility Bulletin No. 9.

Sellers PJ, Mintz Y, Sud YC, Dalcher A (1986) A simple biosphere model (SiB) for use within general circulation models. J Atmos Sci 43:505–531.

Sellers PJ, Hall FG, Asrar G, Strebel DE, Murphy RE (1988) The First ISLCP Field Experiment (FIFE). Bull Am Meteorol Soc 69:22–27.

Shukla J, Norbe C, Sellers PJ (1990) Amazon deforestation and climate change. Science 247:1322–1325.

Stull B (1988) "An Introduction to Boundary Layer Meteorology." Kluwer Academic Publishers, Dordrecht.

Wyngaard J, Brost RA (1984) Top-down and bottom-up diffusion of a scalar in the convective boundary layer. J Atmos Sci 41:102–112.

17
Use of Soil Moisture Measurements in Hydrologic Balance Studies

Richard H. Cuenca and Joël Noilhan

1. Introduction

Experiments such as the HAPEX-MOBILHY regional hydrologic experiment conducted in southwest France during 1985–1986 require numerous measurement techniques for land surface processes such as evapotranspiration (André et al., 1986). For fluxes from the surface layer to be useful as lower boundary conditions in general circulation models (GCM), the net radiation, sensible heat, latent heat, and soil heat fluxes must be accurately determined. Through the experience gained in HAPEX-MOBILHY it has been recognized that soil moisture content and property data are key elements in initialization and verification of hydrologic and energy balance models as well as models of the plant canopy. Soil moisture content is also one of the most, if not the most, cost-effective data sets that can be collected regionally on a large scale hydrologic experiment. This article discusses application of soil moisture measurement techniques for estimation and verification of the latent heat flux, which includes evaporation from soil and plant surfaces and transpiration from plants. Use of soil moisture measurements made with a neutron probe in a hydrologic balance model to estimate evapotranspiration is described. The expected spatial variability in soil moisture measurements and comparisons with other evapotranspiration measurements from the HAPEX-MOBILHY data base are presented.

2. Soil Moisture Measurement Techniques

Numerous techniques have been applied to measure soil moisture content. Important features of soil moisture measurements in hydrologic balance studies are that measurements be made to an adequate depth, i.e., covering the full extent of the root zone, and that they can be made repetitively in the same area. The measurements also should not be seriously destructive to the field being observed during the course of the experiment. This section reviews a number of soil moisture measurement techniques.

2.1 Gravimetric

The gravimetric method is the original and most basic soil moisture measurement technique capable of quantifying soil moisture content. This method requires taking a soil sample and weighing it in the field or placing it in an air-tight container for weighing in a laboratory. The sample is dried at 105°C for 24 hr or until it comes to a constant weight. The water content on a mass basis is computed as,

$$\theta_m = \frac{M_w}{M_s} \tag{17.1}$$

where θ_m = water content on a mass basis, fraction
M_w = mass of water, g
M_s = mass of dry soil, g

If the soil sample is of known volume, which can be accomplished by careful field sampling techniques, the volumetric soil moisture content can be computed by,

$$\theta_v = \frac{V_w}{V_b} \tag{17.2}$$

where θ_v = volumetric moisture content, fraction
V_w = volume of water, cm^3
V_b = bulk soil volume, cm^3

The bulk soil volume V_b includes the volume of solids plus the volume of interstitial pores that are occupied by air and water at the time of sampling.

Advantages of gravimetric sampling are that the method is relatively simple, reliable, and does not require extensive equipment. Because of the accuracy of this method, it is used as the standard against which other methods are calibrated or verified. There are two main limitations of gravimetric methods. First, it is awkward to sample at depth using this method and therefore very difficult to monitor the complete root zone in most situations. Second, this method is destructive to the field being sampled and therefore cannot be applied extensively nor over a long measurement period without seriously altering the condition of the field.

2.2 Tensiometric

Tensiometric techniques require making measurements of soil moisture tension that are then related to soil water content through the soil water characteristic curve, sometimes called the soil moisture release curve. Soil moisture tension is measured using a porous ceramic material connected to a water-filled tube. The water in the tube comes into equilibrium with the tension with which water is held in the soil matrix. Various gradations of ceramic material have been used that are sensitive to different ranges of soil moisture tension. In any case, tensiometers cannot measure soil moisture tensions in excess of approximately 100 kPa since this tension causes vaporization of water thereby breaking hydraulic continuity within the water column.

Different output devices have been applied to measure soil moisture tension using tensiometers. Water, usually dyed for ease of reading, in a transparent tube aligned next to a graduated scale has been applied, but this method is limited by the range of tensions that can be read using tubes of reasonable length. Mercury in contact with the water column has also been used. Mercury-filled tubes have a larger range of reading than water-filled tubes. The specific weight of mercury is 13.6 allowing a mercury-filled tube of the same length to read 13.6 times the range of tension of a water-filled tube. Other systems incorporate pressure measurement devices. These include dial type pressure gauges calibrated for tension and battery-powered pressure transducers. The pressure transducer system uses a hypodermic needle that is inserted through a rubber stopper into the air pocket that forms above the water column in a sealed tensiometer tube. This system has the advantage of being precise, extremely compact, about the size of a small transistor radio, and has a fast response time. Care is required with this device to ensure that the rubber stopper does not become cracked due to exposure to sunlight.

There are two distinct disadvantages of tensiometers that limit their application for soil moisture balance studies. First, is that the range of tension that can be measured is so limited, realistically up to approximately 80 kPa tension, as to be impractical in experiments which monitor cropped fields or natural vegetation that have root zones capable of attaining much higher tensions. Second, the tensiometer does not measure soil moisture content that is the normal parameter of interest in a hydrologic balance model. The tensiometer measures soil moisture tension that must be correlated with moisture content through application of the soil water characteristic curve. The spatial variability of the soil water characteristic curve on a field scale is of such magnitude (Warrick and Nielsen, 1980) as to make the translation of tension to moisture content extremely questionable for applications in precise hydrologic balance studies.

2.3 Gypsum Block

Gypsum blocks are electrical resistance measuring devices that have been applied to monitor soil moisture content. They are made of small cylinders usually 2–3 cm in depth and of equal diameter composed of gypsum and other ceramic materials. The ceramic materials used are similar to those applied in tensiometers. Imbedded within the ceramic block are conductive wires through which a battery-supplied electrical current is passed. The conductivity reading taken between the two wires is a function of the conductivity of the block, which is a function of soil moisture tension.

An advantage of gypsum blocks over tensiometers is that they can be applied over a wider range of soil moisture tensions. The range for any particular block is dependent on the make-up of the ceramic material. Gypsum blocks can be applied to tension measurements from approximately 33 to

1500 kPa spanning the full range of tensions normally found within a root zone. The blocks are not equally responsive over the entire range of tension and usually are considered less reliable at tensions under 80 kPa. In other respects gypsum blocks suffer from some of the same disadvantages as tensiometers. They do not directly measure soil moisture content and in fact do not even directly measure soil moisture tension. The measurement made is one of electrical resistance, which is related to soil moisture tension. This factor adds another degree of uncertainty to soil moisture estimation compared to the tensiometer. The electrical resistance is also dependent on the salinity level within the soil matrix and not only the water content. For these reasons, gypsum blocks have been found useful to measure relative soil moisture content but are not usually considered satisfactory for quantifying components of the hydrologic balance.

2.4 Neutron Probe

The neutron probe employs a low level radioactive source that constantly emits neutrons as it decays. The shielded source is placed at the depth required for measurement of soil moisture content and the probe detector is activated. The emitted neutrons have the same mass as hydrogen ions and are deflected on striking a hydrogen ion. In the course of the collision, energy is absorbed and the neutrons are slowed down. The probe detector is sensitive to slow neutrons and so the number of neutrons counted by the detector is proportional to the hydrogen ion content in the soil, most of which is bound up in the water molecule. By simultaneously making neutron probe readings and taking gravimetric samples for moisture content and bulk density, a regression equation can be developed correlating detector count with moisture content. This equation is typically given as

$$\theta_v = A\left(\frac{NMC}{STD}\right) + B \tag{17.3}$$

where θ_v = volumetric moisture content, percent
NMC = neutron meter count
STD = standard count for the particular neutron probe
A and B = statistical regression coefficients

Neutron probe readings are made by inserting the part of the probe apparatus with the radioactive source into a vertical access tube at the depth at which soil moisture is to be measured. The combination source and detector is connected by an electrical cable to the unit housing electronic components, which sits atop the access tube. The total system weight is about 10 kg. Access tubes are typically made of aluminum but polyvinylchloride (PVC) materials have also been used. Probe measurements made in aluminum access tubes are more sensitive than those made with PVC since the chloride ion absorbs some of the deflected neutrons. The access tubes are installed to a depth corresponding to the maximum required depth of measurement. Since the neutrons are scattered in the vicinity of the radioactive source, the probe samples a sphere of soil volume unlike the measurement devices previously discussed, which make point measurements. The volume sampled is dependent on the soil moisture content. An expression to estimate the radius of influence of the neutron probe reading is given as (Burman et al., 1983)

$$R = 15.0\left[\sqrt[3]{\frac{100.0}{\theta_v}}\right] \tag{17.4}$$

where R = radius of influence, cm
θ_v = volumetric soil moisture content, percent

A typical radius of influence is in the range of 20 cm.

The neutron probe can be applied to repetitively monitor the same site using one or more access tubes. With proper calibration, as described in the following section, the probe yields accurate estimates of soil moisture content and is sufficiently sensitive to changes in soil moisture content to be applied in hydrologic balance studies. The sensitivity to change in soil moisture content can be observed in Figure 17.1. The top section of the figure demonstrates the change in soil moisture content with time for individual depth layers in a grass field. The bottom section indicates depth of applied water as a function of time, in this case due to irrigation. It can be noted that the upper depth layers, in which the roots are most active, clearly show declining soil moisture content until the morning of day 175. (All measurements for soil moisture content were made within 120 min of

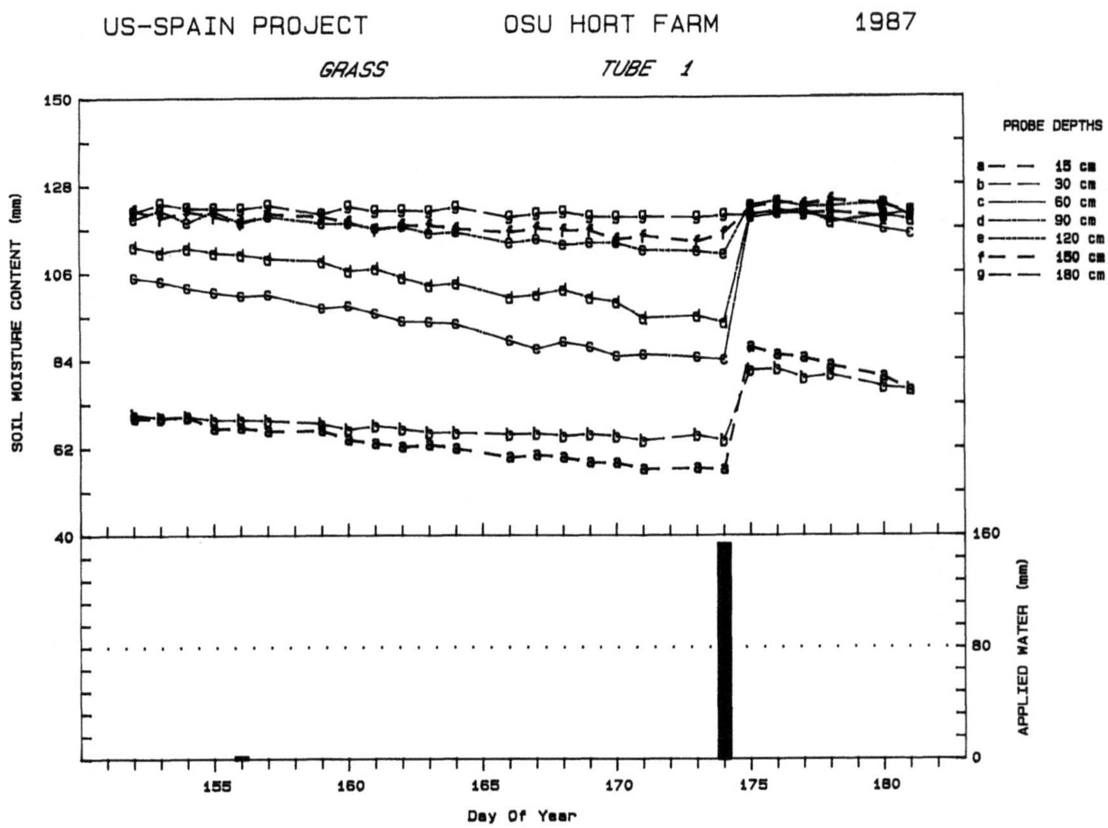

FIGURE 17.1. Variation of soil moisture content with time in specific depth layers as monitored by the neutron probe in a grass plot. The four layers closest to the soil surface (the four bottom-most layers in the plot) are the most active showing a gradual decline in soil moisture content up to day 174 and an immediate response on day 175 to the irrigation applied during the night of day 174.

sunrise for the experiment shown.) On day 175 the four upper depth layers show a distinct response to the irrigation made during the night of day 174. Due to the advantages of application of the neutron probe in hydrologic balance studies as well as its application in a regional network within the HAPEX-MOBILHY experiment, we will more closely examine its application in the following section.

3. Application of the Neutron Probe

3.1 Calibration

As indicated in the previous section, calibration of the neutron probe is generally made by performing regression analysis between gravimetrically determined soil moisture content and the ratio of the neutron detector count to the probe standard count. In the HAPEX-MOBILHY project, the calibration was performed by the French Commissariat de L'Enérgie Atomique (CEA) at Cadarache according to the procedure prescribed by Couchat (1974). This procedure requires the bombardment of soil samples by neutrons accompanied by measurement of neutron absorption and diffusion. Functions are given by Couchat (1974) to describe the soil moisture content as a function of four parameters determined in the calibration procedure. Nevertheless, for HAPEX-MOBILHY the results of the CEA calibration were modified in the intercept term of Equation (17.3) to agree with a gravimetric calibration made as previously described. The result is therefore a quasigravimetric calibration.

The first step in the calibration process is to divide the soil profile into layers of distinct bulk densities and to separate out the surface layer down to a depth of approximately 15 cm. The surface layer must be separated since the radius of influence extends beyond the soil surface for shallow readings and some of the neutrons are lost to the atmosphere. These neutrons do not see the same background conditions as neutrons emitted deeper in the soil profile and therefore a separate calibration is required for measurements made in this surface layer. A separate calibration equation is required for each layer of bulk density since the bulk density controls the relation between soil moisture content on a mass and volume basis as indicated by the following equation:

$$\theta_v = \theta_m \left[\frac{\rho_b}{\rho_w} \right] \quad (17.5)$$

where ρ_b = soil bulk density, g/cm³
ρ_w = density of water, g/cm³

Table 17.1 indicates the distribution of calibration and bulk density layers used for the Lubbon #1 site of HAPEX-MOBILHY.

The emission of neutrons from the radioactive source follows a binomial distribution, which, due to the large number of emissions in a sampling interval, can be approximated by a normal distribution (Devore and Peck, 1986). To correctly calibrate the probe, the sampling procedure should take into account this distribution. An adequate number of samples should be taken to verify the normal distribution of the soil samples for soil moisture content. This test requires 30 samples for each bulk density layer and 30 samples for the surface layer (Devore and Peck, 1986). This type of intensive sampling program is not generally carried out due to time required and the impacts on the field being calibrated. Nevertheless, a sampling program of this type is necessary to fully determine the statistical significance of the calibration error.

3.2 Frequency of Measurement in Time and Space

Let us now investigate the question of the recommended frequency in time and space of neutron probe measurements for soil moisture content. There is a relationship between the standard error of the estimate of soil moisture content and the strength of the radioactive source, number of measurements taken at each depth layer, spacing between successive depth layers monitored, and total time of counting of reflected neutrons (Carrijo, 1988; Havercamp et al., 1984; Vauclin et al., 1984). The total time of counting varies with the neutron probe and can generally be programmed by the probe operator. The acceptable value of the standard error depends on the application of the soil moisture data. If the data are to be used to estimate rates of evapotranspiration, the standard error must be reduced to a low enough level that the *signal* of the amount of change in soil moisture content is clearly distinguishable from the *noise* of the standard error over the time period of interest.

In the HAPEX-MOBILHY regional evapotranspiration experiment, 12 main sites were monitored with the neutron probe. (Site descriptions are given in a later section.) For various reasons, including the distance between the sites covering the 100 by 100 km square grid, readings were made on a weekly basis during the Special Observation Period (SOP) of the experiment which lasted from 07 May to 15 July, 1986. Standard probe counts taken in a tube immersed in a barrel of water were made before and after measurements in access tubes in the field. The standard count provides for a check on the probe electronics at each site. Four tubes monitoring essentially the same field conditions were installed at each site. In the case of fields

TABLE 17.1. Soil bulk density and calibration layers for the Lubbon #1 site of HAPEX-MOBILHY.

Neutron probe depth layer	Depth of neutron probe reading (cm)	Soil bulk density (g/cm³)	Calibration layer
1	5	1.38	A
2	15	1.40	B
3	25		
4	35	1.53	C
5	45		
6	55		
7	65	1.60	D
8	75	1.69	E
9	85		
10	95		
11	115		
12	125		
13	135	1.77	F
14	145		
15	155		

having row crops, two tubes were placed in the crop row and two tubes placed in the adjacent space between crop rows. In accordance with procedures to reduce the standard error, measurements were made at depth intervals of 10 cm starting at 5 cm from the soil surface. Two measurements were made at each depth and the neutron probe count was averaged. Previous experience with the neutron probe measurements in hydrologic balance studies indicates this procedure should produce sufficiently precise results to allow calculation of weekly rates of evapotranspiration, which corresponds to the HAPEX-MOBILHY data set (Cuenca, 1988; Havercamp et al., 1984). Current experiments at Oregon State University indicate that the neutron probe, with proper procedures and levels of atmospheric forcing, can probably be used to determine evaporation rates over a period of approximately 3 days.

4. Hydrologic Balance Model Using Neutron Probe Input

4.1 Application Under Ideal Conditions

Ideal conditions for the hydrologic balance model employed in this study are made up of evapotranspiration (ET) periods that start with water application events and end with a neutron probe reading of the soil profile. The water application event may be an irrigation or precipitation. An ideal ET period need not have a water application at the start, but any water application event starts an ideal period. An ET period is referred to as ideal because the soil moisture status is monitored by the neutron probe at the critical times immediately prior to a water application and at the end of the period. The soil moisture status is known at the time of water application because the previous ideal period ends with neutron probe readings. These readings are assumed to indicate the status of the soil profile immediately preceding the water application event.

The hydrologic balance model applied to the HAPEX-MOBILHY neutron probe data base separates the surface soil layer from the rest of the soil profile. This is due to the difference in calibration for the surface layer and because the surface layer may be of particular importance in studies of other components of the hydrologic balance such as infiltration and surface runoff. The hydrologic balance model distributes any applied water to individual depth layers up to the limit of field capacity. The field capacity may be taken as the maximum soil moisture content for each depth layer monitored based on the data set being analyzed, or a data set for field capacity may be input at the option of the user. If the amount of applied water distributed exceeds the field capacity for the total profile, the excess amount is assumed to percolate through the soil profile, pass the bottom of the root zone as deep percolation, and not be available for evapotranspiration. Under special conditions, the calculated percolation may be applied as evapotranspiration over short limited periods of time. This recognizes the fact that soil water is available for use by plants in the evapotranspiration process as it passes through the soil profile.

The evapotranspiration at the end of an ideal period is calculated by the following:

$$ET_{act} = DR_s + DR_p + \Delta\theta \tag{17.6}$$

where ET_{act} = actual evapotranspiration over the ET period, mm

DR_s = depth of applied water stored in the surface layer, mm

DR_p = depth of applied water stored in the remainder of the soil profile, mm

$\Delta\theta$ = change in soil moisture content over the ET period, mm

Neutron probe readings are used to evaluate the change in soil moisture content over the ET period. If the depth of evapotranspiration calculated using Equation (17.6) proves to be greater than a physically based upper limit, the excess may be indicated as slow drainage from the bottom of the soil profile over the ET period. The upper limit is generally computed using a meteorological-based model (Burman et al., 1983). Activation of the maximum evapotranspiration option to compute slow drainage is an option of the user. Full details of the governing set of equations for the model applied to the HAPEX-MOBILHY data in both the ideal and general field conditions (described in the next section) are given by Cuenca (1988).

4.2 Application Under General Field Conditions

In a regional hydrologic balance experiment such as HAPEX-MOBILHY, which includes many different soil and plant types spread over a wide geographic area, numerous situations arise that do not conform to ideal ET periods. One example is when a precipitation event occurs without a set of neutron probe measurements to determine the soil moisture status preceding the precipitation event. In this case, some other scheme for evaluation of soil moisture status before a water application event must be used in the hydrologic balance model.

The method applied to the HAPEX-MOBILHY data base was unique in that a coefficient given as the ratio of actual to computed maximum evapotranspiration for the previous ideal period was applied for an ET period that was not ideal, i.e., which did not end with a set of neutron probe readings. This coefficient requires that the maximum evapotranspiration be calculated for each ET period, normally using meteorological data (Burman et al., 1983). The actual evapotranspiration from the previous ideal ET period is used to calculate a relative crop coefficient giving the ratio of actual to maximum crop evapotranspiration. This relative crop coefficient reflects the actual stage of plant growth, type of soil, and nutrient levels because it is based on soil moisture measurements made for the same site (Cuenca, 1988).

The same value of the computed relative crop coefficient for the previous ideal period is assumed valid for the following nonideal ET period. This is a reasonable assumption because similar plant growth and soil fertility conditions will be maintained during the nonideal ET period. The relative crop coefficient is applied with a meteorological-based estimate of maximum evapotranspiration for the nonideal period to estimate actual evapotranspiration over the period. This estimate of actual evapotranspiration is used to update the status of the soil moisture profile at the end of the nonideal period, i.e., prior to the water application event. Any errors using this procedure are corrected when the subsequent set of neutron probe readings are made to verify the status of the soil moisture profile (Cuenca, 1988). Adjustments made to the relative crop coefficient in the hydrologic balance model include functions to account for the effects of soil moisture depletion in the root zone on evapotranspiration and wetness of the soil surface following irrigation or precipitation. The corrections applied follow Wright's procedure (Wright, 1982). These same corrections are inversely applied to adjust for the effects of soil moisture depletion and surface wetness for the nonideal ET period. The full set of governing equations is given in Cuenca (1988).

4.3 Testing Estimating Methods for Components of the Hydrologic Balance

With proper calibration and operating procedures, measurements of soil moisture content made using the neutron probe can be used to test parameterizations of various components of the hydrologic balance. A sample procedure is indicated in Figure 17.2.

In the case demonstrated, different meteorological-based estimating methods for evapotranspiration are to be evaluated. Neutron probe readings of the status of the soil moisture profile are made at the beginning of the time period to be evaluated. Meteorological data are then applied to estimate effective precipitation and the maximum evapotranspiration required for the hydrologic balance model. (Note that for an unbiased test of the meteorological models, only those evapotranspiration periods with no precipitation and no slow drainage calculated based on the maximum periodic evapotranspiration can be used in the final analysis.) The meteorological data are also applied in the three models to estimate evapotranspiration in the example shown in Figure 17.2.

At the end of the time period to be evaluated, a second measure is taken of the soil moisture status using the neutron probe. The actual value of soil moisture content can be compared to that calculated using the meteorological-based ET estimate applied to the initial status of the soil moisture profile. Alternatively, the estimated ET using the three methods being tested can be compared to the ET calculated based on the hydrologic balance model described. A statistical analysis can be made of the accuracy of the estimating methods over varying time periods.

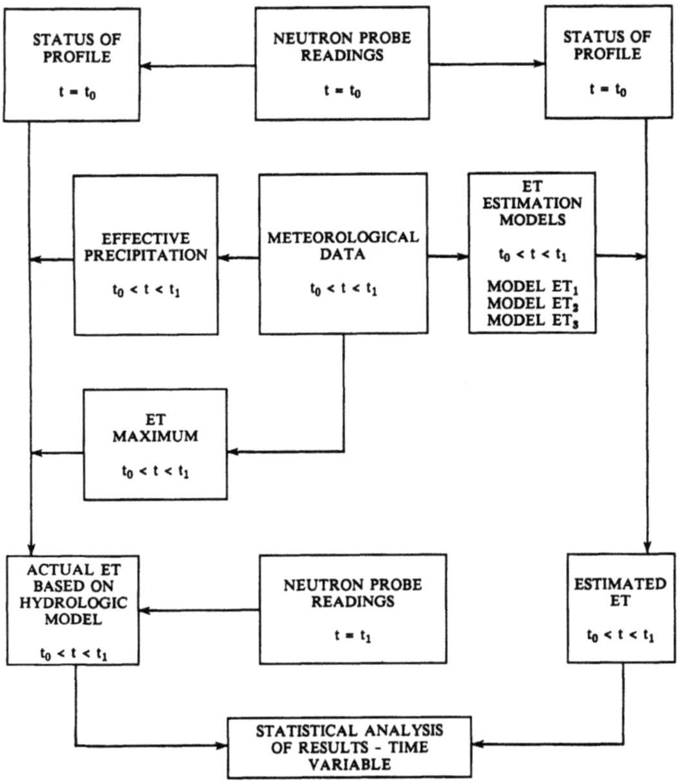

FIGURE 17.2. Flow chart for application of hydrologic balance model to test estimates of components of the hydrologic cycle.

5. Results from HAPEX-MOBILHY

5.1 Sites Monitored

The main sites monitored using the neutron probe in the HAPEX-MOBILHY experiment are indicated in Table 17.2. This table indicates the type of soil at each site, total depth monitored, type of crop, and whether it was irrigated or unirrigated. Some of the sites had a limited root zone monitored, which made it impractical to apply the hydrologic balance model (i.e., Lubbon #2 and Tieste). Difficulties such as this and interference from a water table that intersected the root zone have been described in other publications (Goutorbe et al., 1989; André et al., 1988). Unless specifically noted, the sites discussed in this article were monitored over the total depth of root zone and were not subject to an intersecting water table.

5.2 Variation of Individual Access Tubes

Soil physicists have spent considerable effort to describe the spatial variability of soil water properties at the field scale (Warrick and Nielsen, 1980). It is natural that this spatial variation would cause a local variation in evapotranspiration as measured by the neutron probe. It should be noted that other estimations of evapotranspiration made using the neutron probe data base from HAPEX-MOBILHY (Goutorbe et al., 1989; André et al., 1988) and other experiments (Cuenca, 1989) generally apply the average of data from individual access tubes. This is equivalent to assuming that the measured soil moisture content and subsequent calculated evapotranspiration are normally distributed in the field.

In fact, if results from monitoring individual access tubes are analyzed, the considerable spatial variation expected from variation in soil water properties becomes evident. This result is demonstrated in Figure 17.3, which displays the total evapotranspiration for individual access tubes from the Lubbon #1 site of HAPEX-MOBILHY over the SOP. The data shown are typically for weekly ET periods. There is clearly a difference in the evapotranspiration calculated using data from individual access tubes, although this difference is

TABLE 17.2. Soil, crop, depth monitored, and irrigation information for neutron probe access tube sites corresponding to SAMER station locations in HAPEX-MOBILHY.

SAMER station number	Location	Soil texture	Crop	Depth monitored (cm)	Irrigation
1	Lubbon #1	Sand	Oats	160	No
2	Casteljaloux	Silty clay	Maize	160	No
3	Caumont	Sandy clay loam	Soybean	160	Yes
4	Courrensan	Sandy clay	Wheat	150	No
5	Lubbon #2	Sand	Maize	70	Yes
6	Sabres	Sand	Maize	160	No
7	Bats	Clay loam	Maize	160	No
8	Vicq d'Auribat	Loamy sand	Maize	200	Yes
9	Tieste	Sandy clay loam	Maize	90	Yes
10	Castelnau	Sandy clay loam	Maize	170	No
11	Fusterouau	Sandy clay loam	Maize	170	Yes
12	Lagrange	Sandy clay loam	Maize	180	No

generally less in the case of the Vicq d'Auribat data compared to Lubbon #1. Similar variation was observed from the other sites in HAPEX-MOBILHY and in other experiments applying the neutron probe to evaluate evapotranspiration over shorter time periods. This variation could be further analyzed to give an idea of the confidence of the evapotranspiration estimate made at the field scale with the neutron probe. It should be noted that soil moisture measurements made over the soil profile using the neutron probe spatially integrate a much smaller portion of the field than measurements made in the atmosphere at some height above the soil and crop surface. The resulting evaporative flux measurement made with the probe is therefore much more localized.

5.3 Comparison with SAMER Data

Comparisons were made between evapotranspiration rates calculated using neutron probe data and that measured by the SAMER energy balance system (Itier, 1980, 1982). Although data from the neutron probe network was collected on a weekly basis, the frequency for energy balance data from the SAMER stations was 15 min. Daily totals from the SAMER stations were applied in the comparison demonstrated in Figure 17.4. The differences in time scale were managed by distributing the evapotranspiration calculated over the ET period using the neutron probe as a daily average. The normal length of an ET period was 7 days. The cumulative daily evapotranspiration computed using both methods was then plotted versus time in days. Results for Lubbon #1 are indicated in Figure 17.4. These results are comparable to those from other sites.

If the slopes of the two cumulative evapotranspiration lines in Figure 17.4 are equal, the two methods produced equal estimates of daily evapotranspiration. The figure shows that in general there is surprisingly good agreement between the two estimating methods for the majority of the SOP. Toward the beginning and at the end of the SOP, the disagreement between the slopes of the two lines is largest. There are a number of possible physical explanations for these differences. Early during the SOP, a high water table condition existed at Lubbon #1. This situation is observed in the traces of soil moisture content with depth for different dates shown in Figure 17.5. The high moisture content below 120 cm depth for the first four dates corresponds to the position of the water table that is seen to rapidly decline by the final date of measurement. Analysis of piezometer records at this site also indicates probable intersection of the water table with the root zone. This is a case that can cause errors in application of neutron probe data because moisture that may move directly from the water table to the roots is not measured by the probe. The neutron probe can measure changes only in soil moisture content. The neutron probe

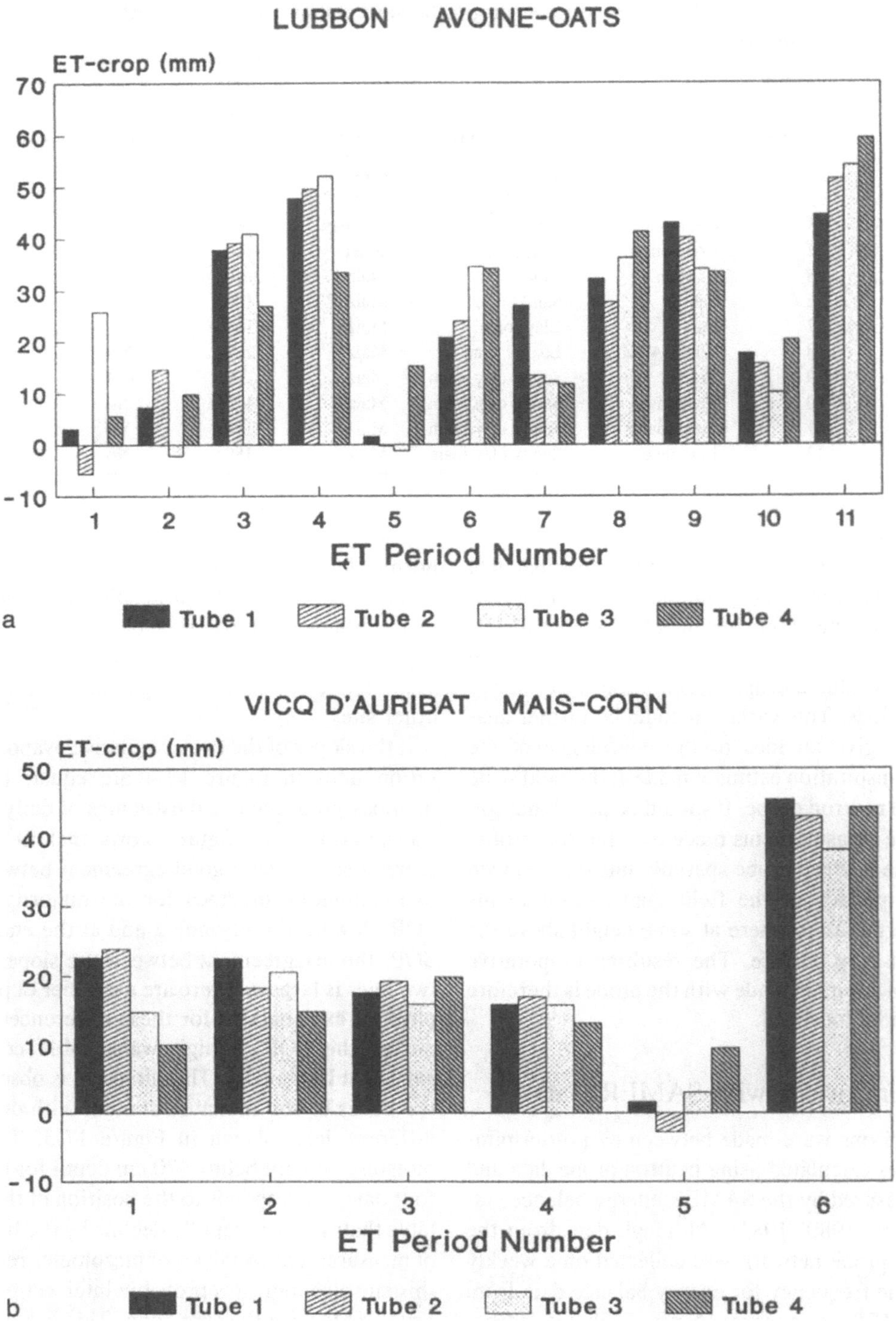

FIGURE 17.3. Variation in computed evapotranspiration rates for individual access tubes at (a) Lubbon #1 and (b) Vicq d'Auribat over the special observation period.

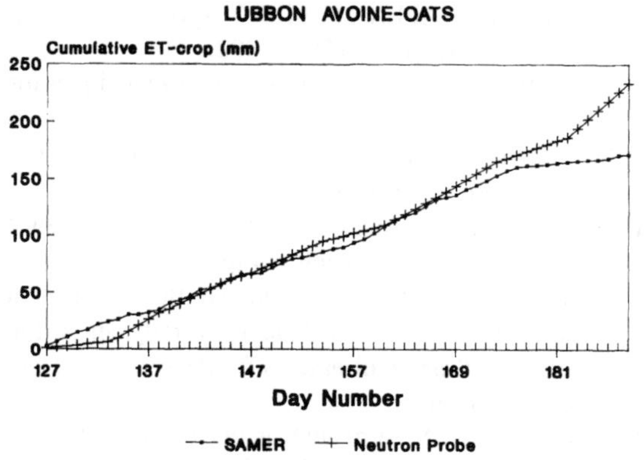

FIGURE 17.4. Comparison of equivalent daily evapotranspiration as measured by SAMER and using neutron probe data in the hydrologic balance model for the Lubbon #1 site over the special observation period. (Days with missing SAMER data were removed from the comparison data set).

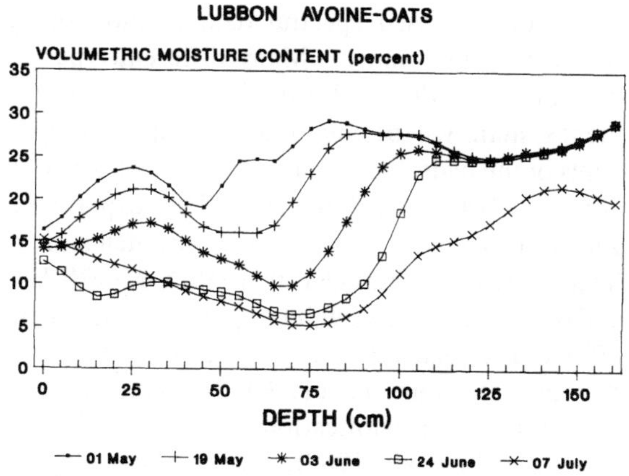

FIGURE 17.5. Variation of soil moisture content with depth for different dates of the special observation period at Lubbon #1.

analysis would therefore tend to underestimate evapotranspiration under these conditions. This may be the case shown in comparison with the SAMER data early during the SOP although the differences shown are not remarkable considering the two very different types of measurement systems used. Nevertheless, the intersecting water table is a difficult situation to analyze using the neutron probe.

If there is a decline of the water table, the neutron probe would show the subsequent change in soil moisture content as evapotranspiration when in fact it would only be drainage of the soil profile. This would lead to an overestimation of evapotranspiration during the period of drainage using neutron probe data with the hydrologic balance model. The fact that rapid drainage occurred at Lubbon #1 late in the SOP is indicated in Figure 17.5, which verifies other reports of this condition (Goutorbe et al., 1989; André et al., 1988). The high evapotranspiration rates indicated by the neutron probe in comparison to the SAMER station in Figure 17.4 toward the end of the SOP is probably caused by the rapid decline of the water table.

6. Conclusions

The neutron probe can be an effective tool for determination of soil moisture content if proper calibration procedures are followed. Using carefully developed operating procedures, neutron probe data can be applied in a hydrologic balance model to estimate evapotranspiration (Cuenca, 1988). Field data obtained with the neutron probe have been used in experiments to determine differential evapotranspiration rates caused by various irrigation treatments in the same crop over the horizontal space of a few meters (Cuenca, 1989).

A network of soil moisture sites to be monitored by a portable neutron probe was established as part of the HAPEX-MOBILHY regional experiment due to the practicality and cost-effectiveness of this method. The neutron probe sites were normally placed adjacent to SAMER energy balance stations for comparison of the two methods in estimation of evapotranspiration. The results shown in this article indicate that the spatial variability of soil water properties definitely affects the estimated evapotranspiration from individual access tubes. However, when data from individual tubes are averaged and compared to SAMER data on an equivalent daily basis, there is generally good agreement between the two methods. Differences in the estimates may be explained by physical situations not correctly accounted for by the neutron probe.

Future HAPEX scale experiments are currently being considered for arid regions in Niger, Africa, and Spain. These experiments will be aimed at improving formulations of energy and hydrologic balances at the land surface for application in GCMs, improving estimates of carbon dioxide fluxes, and developing algorithms for application of remotely sensed spectral data in monitoring the earth's environment. Many different types of ground-, aircraft-, and satellite-based measurement systems will be employed to compute components of the hydrologic and energy balances as was done in HAPEX-MOBILHY. The experience gained using soil moisture measurements in a hydrologic balance model in HAPEX-MOBILHY and other experiments will be applied in these new efforts. The relatively recent development of time domain reflectometry (TDR), a technique discussed in other publications (Topp and Davis, 1985; Zegelin et al., 1989), may also be applied in these upcoming experiments. This technique may be especially well-suited to application in the surface layer, which requires special calibration in the case of the neutron probe but which is very important in evaluation of remote microwave radiometer data (Schmugge, 1985; Schmugge et al., 1980). In these experiments, monitored status of the surface and profile soil moisture content will be a key element in the development and verification of land surface and plant canopy models (Feddes et al., 1988; Noilhan and Planton, 1989). The efforts involved in these large scale, multinational experiments will give new information to aid in evaluating changes in and ultimately protection of the Earth's environment.

Acknowledgments. The work presented in this article was supported by the Centre National de Recherches Météorologiques in Toulouse and the Agricultural Experiment Station at Oregon State University. Assistance with data collection and analysis was given by Charles Valencogne of the Institut National de la Recherche Agronomique (INRA), Bordeaux and Osmar Carrijo, William Nichols, and Gabriel Katul of the Department of Agricultural Engineering at Oregon State University.

References

André JC, Goutorbe JP, Perrier A (1986) HAPEX-MOBILHY: A hydrologic atmospheric experiment for the study of water budget and evaporation flux at the climatic scale. Bull Am Meteorol Soc 67(2):138–144.

André JC, Goutorbe JP, Perrier A, Cuenca RH, Noilhan J, others (1988) Evaporation over land-surfaces: First results from HAPEX-MOBILHY special observing period. Ann Geophys 6(5)477–492.

Burman RD, Cuenca RH, Weiss A (1983) Techniques for estimating irrigation water requirements. In "Advances in Irrigation," Vol. 2 (D. Hillel, ed.), pp. 335–394. Academic Press, Orlando, Fl.

Carrijo OA (1988) Analysis of a hydrologic balance model and Penman-Monteith evapotranspiration estimating methods. Thesis presented to the Department of Soil Science, Oregon State University, in partial fulfillment of the requirements for the degree of Doctor of Philosophy.

Couchat P (1974) Mesure neutronique de l'humidité des sols. Thesis presented to the L'Université Paul Sabatier, Toulouse, in partial fulfillment of the requirements for the degree of Doctor of Philosophy.

Cuenca RH (1988) Hydrologic balance model using neutron probe data. J Irrigation Drainage Eng, American Society of Civil Engineers, IR4(5):644–663.

Cuenca RH (1989) "Irrigation System Design—An Engineering Approach." Prentice-Hall, Englewood Cliffs, N.J.

Devore J, Peck R (1986) "Statistics, The Exploration and Analysis of Data." West Publishing, St. Paul, MN.

Feddes RA, Kabat P, Van Bakel PJT, Bronswijk JJB, Halbertsma J (1988) Modelling soil water dynamics in the unsaturated zone—state of the art. J Hydrol 100: 69–111.

Goutorbe JP, Noilhan J, Valencogne C, Cuenca RH (1989) Soil moisture variations during HAPEX-MOBILHY. Ann Geophys 7, No. 4:415–426.

Haverkamp R, Vauclin M, Vachaud G (1984) Error analysis in estimating soil water content from neutron probe measurements: 1. Local standpoint. Soil Sci 137(2):78–90.

Itier B (1980) Une méthode simplifiée pour la mesure du flux de chaleur sensible. J Recherch Atmos 14:17–34.

Itier B (1982) Révision d'une méthode simplifée pour la mesure du flux de chaleur sensible. J Recherch Atmos 16:85–90.

Noilhan J, Planton S (1989) A simple parameterization of land surface processes for meteorological models. Mon Weather Rev 117:536–549.

Schmugge TJ (1985) Remote sensing of soil moisture, In "Hydrological Forecasting" (M.G. Anderson, T.P. Burt, eds.), pp. 101–124. Wiley, Chichester.

Schmugge TJ, Jackson TJ, McKim HL (1980) Survey of methods for soil moisture determination. Water Res 16:961–979.

Topp GC, Davis JL (1985) Time-domain reflectometry and its application to irrigation scheduling. In "Advances In Irrigation," Vol. 3 (D. Hillel, ed.), pp. 107–127. Academic Press, Orlando, Fl.

Vauclin M, Haverkamp R, Vachaud G (1984) Error analysis in estimating soil water content from neutron probe measurements: 2. Spatial standpoint. Soil sci 137(3): 141–148.

Warrick AW, Nielsen DR (1980) Spatial variability of soil physical properties in the field. In "Applications of Soil Physics" (D. Hillel, ed.), pp. 319–344. Academic Press, New York.

Wright JL (1982) New evapotranspiration crop coefficients. J Irrigation Drainage Div, American Society of Civil Engineers, IRI(108):67–74.

Zegelin SJ, White I, Jenkins DR (1989) Improved field probes for soil water content and electrical conductivity measurement using time domain reflectometry. Water Res Res 25:2367–2376.

18
Indirect Measurements of Fluxes Using Doppler Sodar

A. Weill

1. Introduction

Since 1970, Sodar (Sound Detection and Ranging) has been developed as an economical and simple instrument giving wind profiles in the atmospheric boundary layer. The effort was initiated with a research point of view; however, rapidly considering the potential of the results, commercial systems have been built. Remote sensing with Sodar is now recognized as the most simple remote sensing technique to estimate wind and turbulence in the lower part of the atmospheric boundary layer (ABL). Doppler Sodar data joined with advances in turbulence theory and ABL modeling (Brown and Hall, 1978; Neff and Coulter, 1986) have led to a better understanding of the different parts of the boundary layer. For more information about these results and a historical point of view of acoustic sounding, see Little (1969), Weill (1981), Underwood (1984), Singal (1988), and Kallistratova (1959).

It also has to be considered that acoustic remote sensing and remote sensing in general have been stimulated by microturbulence theory (Kraichnan, 1953) and that understanding acoustic backscattering intensity in a turbulent medium is due to the work of Kallistratova (1959) and Monin (1962). This explains that since the first Doppler Sodar was developed by McAllister (1968), new observations and a better understanding of the boundary layer have been possible.

My intention is not to give a comprehensive course on acoustic remote sensing, which is now so large a subject that an extensive course would cover too much (Neff and Coulter, 1986). My purpose is to discuss flux measurements as estimated by Sodar in the general context of flux measurements, their representativeness, and the problem of large-scale flux extension as already discussed (Weill et al., 1988; André and Bougeault, 1988). I will try to discuss this topic without bias, taking not only the point of view of the Sodar user, but also considering Doppler Sodar as a complementary technique to conventional techniques to measure fluxes, having both advantages and limitations.

First, I will describe Sodar techniques, and then discuss how to obtain fluxes, virtual temperature flux, and momentum flux using Sodar. This discussion will be followed by comments about knowledge of the uncertainties in the different estimates: I will try to show that whatever instrument is used to obtain fluxes in the well-mixed layer, indirect methods with parameterization may be most "accurate." To conclude, I shall describe several methodologies that could be decisive in research about large-scale flux measurements.

2. Principle of Acoustic Sounding

2.1 How a Sodar Works

The acoustic backscattering cross section (improperly termed "reflectivity") is the echo received $\sigma(\theta)$ along a scattering angle θ from an acoustic transmitter in a turbulent atmosphere (Monin, 1962):

$$\sigma(\theta) \approx 0.03k^{1/3} \cos^2\theta \, (C_v^2/C^2 \cos^2\theta/2 + C^2_{T/T^2}) \sin\theta/2^{-1/3} \quad (18.1)$$

where k is the acoustic wave number, C_v and C_T are the structure function of velocity and tempera-

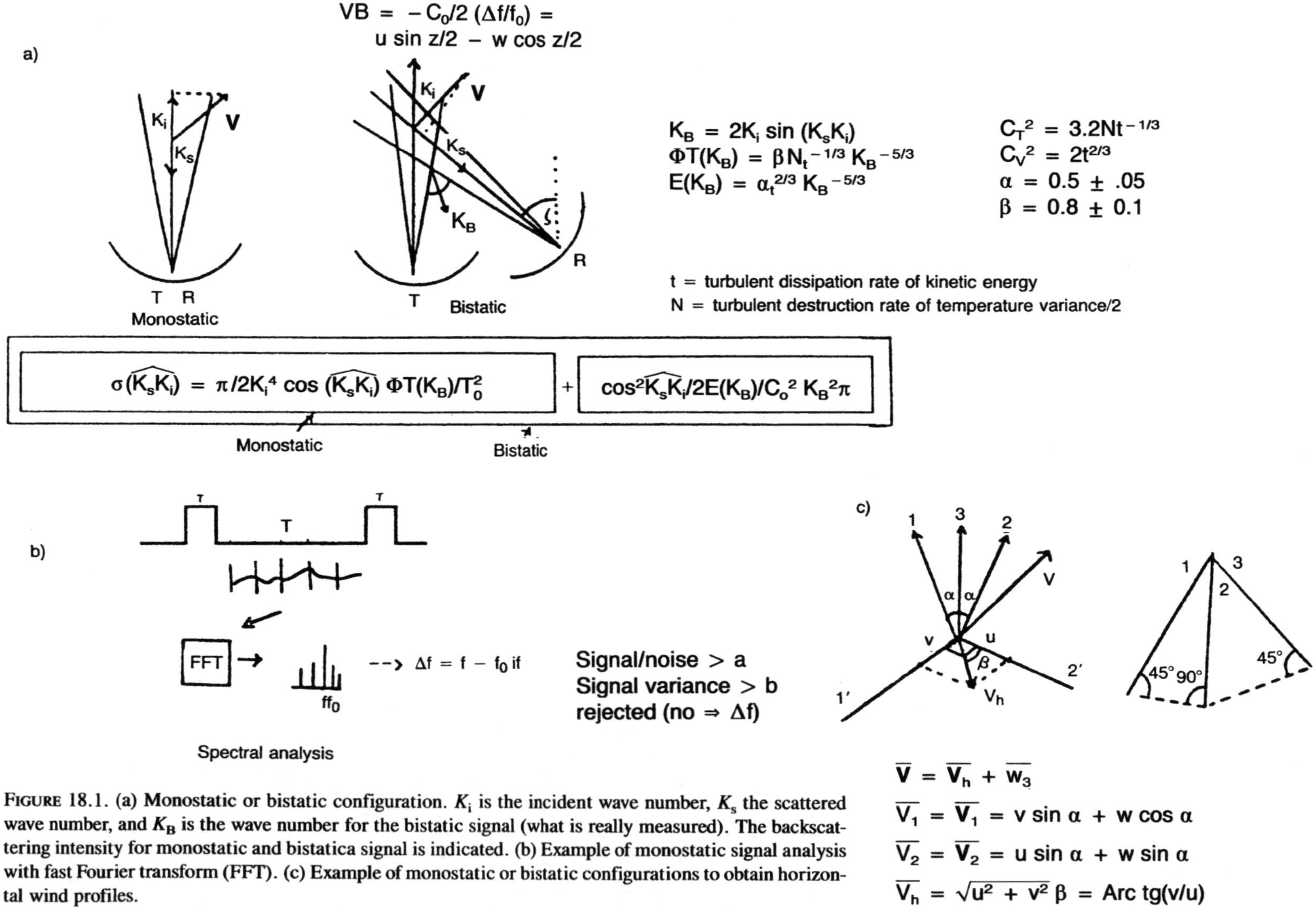

FIGURE 18.1. (a) Monostatic or bistatic configuration. K_i is the incident wave number, K_s the scattered wave number, and K_B is the wave number for the bistatic signal (what is really measured). The backscattering intensity for monostatic and bistatica signal is indicated. (b) Example of monostatic signal analysis with fast Fourier transform (FFT). (c) Example of monostatic or bistatic configurations to obtain horizontal wind profiles.

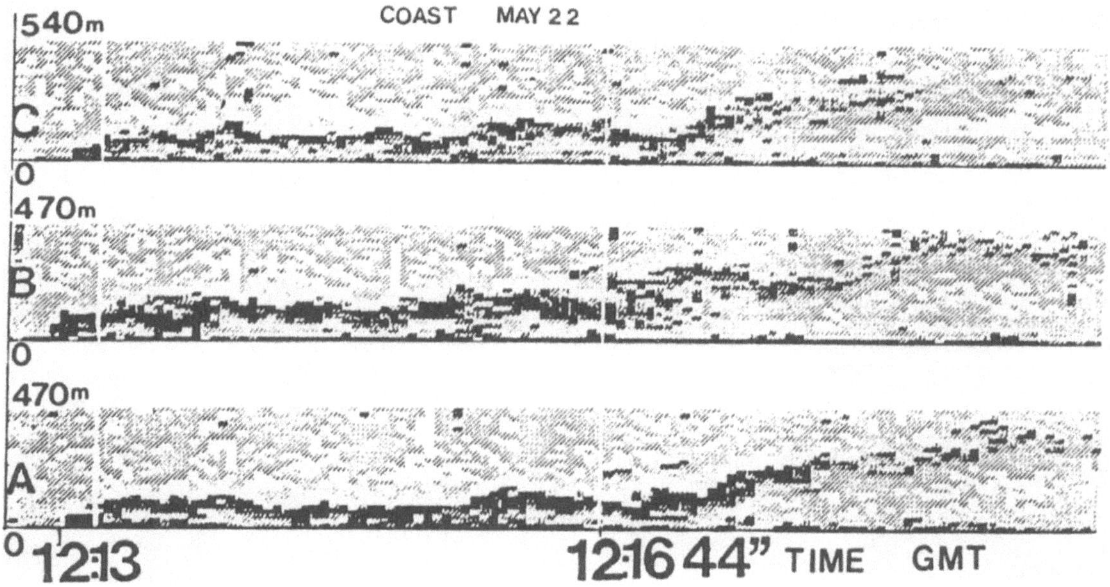

FIGURE 18.2. Example of facsimile record on the three antennas of a monostatic Sodar, giving reflectivity as a function of time and height: the interpretation of the facsimile record in terms of convective structures related to convective colds could not be done without complementary information.

ture, respectively, C is the sound velocity, and T is the mean temperature. In general, an acoustic pulse of duration τ (50–200 msec) is transmitted with a repetition period T (1–16 sec). After a time τ' of a few milliseconds to protect the electronics of the system or to have a full damping inside the antennas, the transmitted signal is analyzed inside distance gates of the same duration as the pulses and corresponding to the spatial dimension of $c\tau/2$. Fast fourier transform or other techniques are employed from which Doppler shift and reflectivity inside gates are obtained (Underwood, 1984; Baudin, 1977).

As indicated in Figure 18.1 two Sodar techniques are generally used:

1. The monostatic technique, in which transmitter and receiver are collocated, the reflectivity being proportional to C_T^2 as expressed in Equation (18.1). The Doppler shift is colinear to the acoustic antenna axis.
2. The bistatic technique, in which transmitter and receiver are separated, the reflectivity being proportionnal to C_T^2 and C_V^2. The Doppler shift is colinear to the bisector between the transmitting and the receiving antennas. Though the bistatic signal can be considered a priori as stronger than the monostatic signal, the problem of refraction and the difficulty of common volume knowledge have resulted in monostatic systems being most commonly used.

For several years acoustic sounding has been used in environmental studies (pollution monitoring and survey) from which reflectivity has been mainly used (Singal et al., 1985), giving a picture of turbulent intensity as a function of time and height (Fig. 18.2). However, this kind of information is not sufficient, and in many cases leads to misleading interpretations if not joined to complementary meteorological information. Only a Sodar "expert" is really able to use reflectivity facsimile records as the only information. Figure 18.1 indicates how a Doppler Sodar works.

It is important to note here that acoustic reflectivity makes a Sodar more efficient in the lower boundary layer than electromagnetic reflectivity for a radar in clear air (Little, 1969): "The sensitivity of the interaction in normal conditions of temperature and pressure shows that 1°C in tempera-

TABLE 18.1. Synoptic view of possible turbulent parameters measurement.

	Direct measurement	Indirect measurement	References	Precision
ε	Yes with bistatic Sodar	With structure functions	Gaynor (1977)	40%
C_v^2		With spectral density of W	Weill et al. (1978)	
C_T^2	Yes	With parameterization of σ_w	Coulter and Wesely (1980); Weill et al. (1980)	30%
N	Yes with bistatic Sodar		Neff (1975)	
U_*	Yes with direct correlation between antennas	With parameterization of dV/dz	Eymard and Weill (1979); Strauss (1980); Klapisz and Weill (1982)	30%
Q_0	Yes but needs corrections (from Sodar equation)	With σ_w^3/z and/or TKE budget	Coulter and Wesely (1980); Weill et al. (1980); Desbraux and Weill (1986)	30%
σ_w	Yes		Spizzichino (1974)	Function of sample duration
σ_u	Yes + FFT	With $\sigma_i \neq$ antennas + correlations		
σ_v	Yes			
σ_w^3	Yes	With distribution function fitting	Taconet and Weill (1983)	Function of sample duration
Z_i	Yes if $z_i < 10^3_w$	With $Q(z)$ if $Z_i > 10^3$ m		20 m
W		Combination		20%
Z/L		Z_i, Q_0		90%
Z_i/L		U_*, Q_0		
		Z_i, Q_0, U_*		100%

ture fluctuations is equivalent to a 1700 N change in refraction units ($N = 10^{-6}$) for the sonic refractive index, whereas for radio wavelengths the resultant change is of the order of 1 N unit. For wind, the situation is more striking and 1 m/s corresponds to a change of 300 N units for acoustic refractivity and 2×10^{-6} N for electromagnetic waves."

In summary, basically with a monostatic Doppler Sodar using three antennas, one obtains reflectivity and Doppler shift along the three antennas from which vertical velocity (vertical antenna) and mean horizontal wind speed are computed. In Table 18.1 a synoptic point of view demonstrates the variable that can be estimated. In every configuration, secondary beams and problems of reverberation related to the transmitter are such that the first levels are never close to the surface. For the Sodar developed at CRPE the first levels are 10 m for the 6-kHz minisodar and 20 m for the 2-kHz Sodar.

Signal processing from Sodar is rather different from Radar processing due to the slow sound velocity. We can really speak of specific Sodar processing and in the literature numerous investigations have been made to find the best signal processing scheme (Underwood, 1984; Baudin, 1977; Neff and Coulter, 1986). Using numerical methods, it has been found that fast Fourier transform (FFT), pulse pair processing (PPP), and vector phase change (VPC) are the more precise methods. It must be mentioned that signal processing from the mathematical point of view is not sufficient to ensure relevant data. Some other criteria have to be added, for example, data based on the maximum admissible shear between adjacent gates. About precision of the measurement, as discussed by Spizzichino (1974), we assume that signal and noise are gaussian, and in averaging the spectrum over N pulses, we find the uncertainty s_u and s_w on horizontal and vertical velocity, respectively:

SCHEME 18.1.

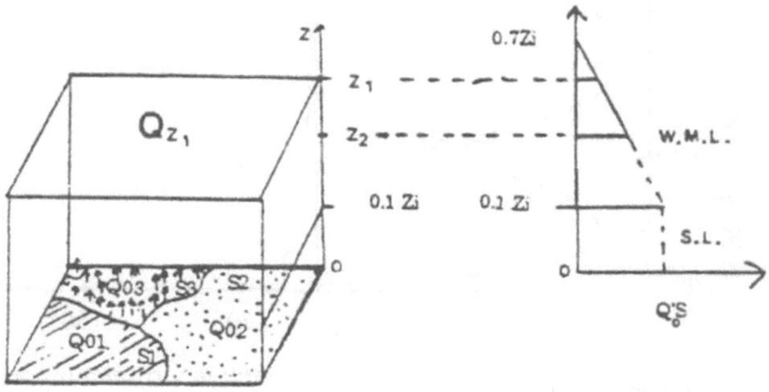

$$\sigma_u \approx 1.3b^{0.5}/2N^{0.5}\,\tau^{0.5}\cos\theta$$
$$\sigma_w \approx 1.3b^{0.5}/2N^{0.5}\,\tau^{0.5} \qquad (18.2)$$

where τ is the pulse duration, b the frequency band on which the signal is found, and θ is the antenna inclination. Hence a 2000 Hz frequency, 4 sec pulse duration, and average approximation of 20 min gives

$$\sigma_u \approx 0.11\cos\theta \text{ m/s} \quad \text{and} \quad \sigma_w \approx 0.11 \text{ m/s}$$

One consideration that we observe is that the Doppler or reflectivity information is relative to a volume of radial section equal to $C\tau/2$ and corresponds to ensemble averages inside this volume (Chong, 1976).

3. Flux Measurements

Flux measurement (humidity, temperature, and momentum fluxes) at scales comparable to meteorological models is one of the most important and difficult problems. Indeed, on complex terrain, which corresponds to the most general case, surface measurements are local measurements and the problem of extension of local measurements up to a large surface is not obvious and would at least need a surface network not compatible with realities and possibilities.

Experiments such as Mesogers 84 (Weill et al., 1988) or HAPEX-MOBILHY 86 (André et al., 1988) have led to the possibility of testing different methods to estimate fluxes. Acoustic remote sensing is considered as a means to extend flux measurements considering the mixing properties of the well-mixed layer.

3.1 Methods of Measurement

3.1.1 Virtual Temperature Flux

The Sodar estimation is based on the following assumptions:

1. On inhomogeneous terrain, during convective situations, due to convective mixing in the well-mixed layer (WML), fluxes have lost "memory" of the local surface (Taconet and Weill, 1983).
2. Thermals (updrafts and downdrafts) observed in the WML are transporting more than 85% of the sensible heat flux (Greenhut and Bean, 1981), and have horizontal dimensions larger or equal to 1.5 z_i (where z_i is the inversion height) (Taconet and Weill, 1983; Lenschow and Stephens, 1980); therefore, fluxes estimated in the WML are representative of flux output of the surface layer after turbulent mixing.

These very simple considerations can explain the method (Weill et al., 1980) used in the WML:

$$(\overline{w_2'})^{3/2}/z \approx Ag/T\,\overline{w'\theta'v} \qquad (18.3)$$

where w' is the vertical velocity fluctuation, $\theta'v$ is the virtual potential temperature fluctuation, and A is a constant close to 1.4. A linear extrapolation of the flux down to 0.1 Z_i, the surface layer top, gives the flux that we would have in the surface layer for a homogeneous surface (see Scheme 18.1).

The flux estimated with our method is Q's and can be considered as a priori different from Q_os, the arithmetic average of the fluxes on the three surfaces (S_1,S_2,S_3):

$$Q\text{'s} \neq Q_O\text{s} = (Q_{01}\cdot S_1 + Q_{02}\cdot S_2 + Q_{03}\cdot S_3)/(S_1 + S_2 + S_3) \qquad (18.4)$$

It should be mentioned that during different experiments in the Beauce region, several vertical Doppler Sodar were used on three different canopies close to each other. It was also observed that the "Sodar fluxes" were not very different from each other and were very close to the flux related to the preponderant canopy in the region (Weill 1983).

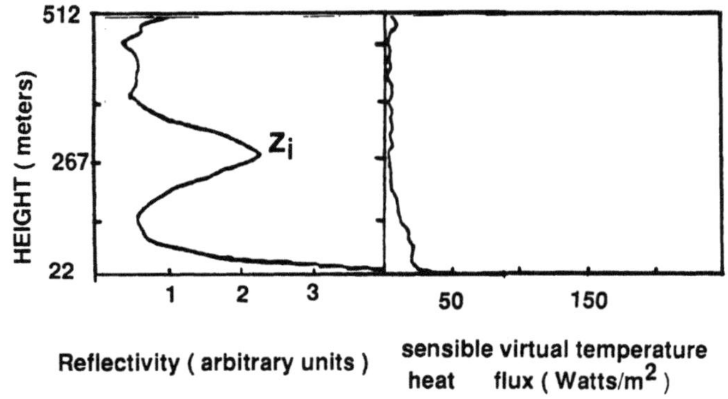

FIGURE 18.3. Example of output showing reflectivity as function of height and virtual temperature flux as function of height. It can be observed that the inversion height is characterized by a maximum of reflectivity and that the heat flux profile is rather linear.

Another method is to use calibrated reflectivity to measure C_T^2, from which virtual temperature flux is estimated. This method has been used by Melling and List (1977) and Dubosclard, (1980) and has been found to give good estimates. However, reflectivity is difficult to calibrate and transducer efficiency is not well known enough to be able to estimate C_T^2; therefore an uncertainty of 3 dB on C_T^2 can be expected.

3.1.2 Momentum Flux

The same concept holds for momentum fluxes, but as the linear height behavior is not obvious, extrapolation down to the surface is not applied. Different methods can be used.

A direct method, using the Fourier components of the radial velocity on the three antennas, computes the variance by integrating the spectrum in a frequency band. This method, which has been applied at C.R.P.E., has the great advantage of eliminating the problems of separation between antennas and uses only one hypothesis of homogeneity on the three antennas. However, a careful analysis of the fourier coefficients variation is necessary to observe that they change only a few percent if the temporal series for the vertical velocity is delayed $\pm$ 8 sec (4 sec is the pulse period repetition).

If V_{r1} is the radial velocity on the slanting antenna 1 and V_{r3} is the vertical velocity on antenna 3, we have $V_{r1} = (U_1 + w_1\sqrt{3})/2$, where U_1 and w_1 are, respectively, the horizontal and vertical velocity components along antenna one. Using the Fourier transform (TF) we get

$$\text{TF } V_{r1} = \text{TF } (U_1 + w_1\sqrt{3})/2 \qquad (18.5)$$

$$\text{TF } w_1 = \text{TF } w_1 \quad \text{(homogeneity)} \qquad (18.6)$$

hence we obtain TF (U_1). Integrating in the frequency interval (f_1,f_2) we obtain the U_1 variance. Then with

$$\overline{V_{R1}^2} = (\overline{U_1^2} + \overline{w_1^2}\cdot 3 + 2\sqrt{3}\,\overline{U_1 w_1})/4 \qquad (18.7)$$

we obtain the momentum flux component $\overline{U_1 w_1}$.

To obtain the other momentum component it is necessary to use the second slanting antenna. This "direct" method seems to be easier to use than the very sophisticated method proposed by Kristensen et al. (1986) but may be less precise. As indirect methods we suggest the methods proposed by Clarke (1970) and Klapisz and Weill (1978), which can also be found in Strauss (1980). These methods are based on WML parameterization and extension of surface layer classical parameterization (Businger et al., 1971). A method suggested by Clarke (1970) and applied by Mazaudier and Weill (1989) has been used in HAPEX experiments. It supposes Sodar use very close to each other at two locations on different canopies and assumes that the geostrophic wind does not differ too much in the two locations. Hence, if the fetch effects are not important, via the integrating motion equation, we obtain differential stress.

If $(\overline{u_1'w_1'}, \overline{v_1'w_1'})$ and $\overline{u'_2w'_2}, \overline{v_2'w_2'})$ are the momentum components at the two locations, with $\partial/\partial t = 0$ (t is the time)

$$-f(V_1 - V_2) = d/dz(\overline{u_1'w_1'} - \overline{u_2'w_2'}) \qquad (18.8)$$

$$f(U_1 - U_2) = d/dz(\overline{v_1'w_1'} - \overline{v_2'w_2'}) \qquad (18.9)$$

Integration of these two equations up to the level h from which the stress differences vanish (wind velocity and shear differences negligible) gives the momentum flux differences as the function of wind speed differences:

FIGURE 18.4. Examples of fluxes measurements with Sodar at four stations of measurement and fluxes integration with aircraft (A) during Mesogers experiment. Aircraft measurements have been performed too late to be really compared to Sodar measurements (From Weill et al., 1988.)

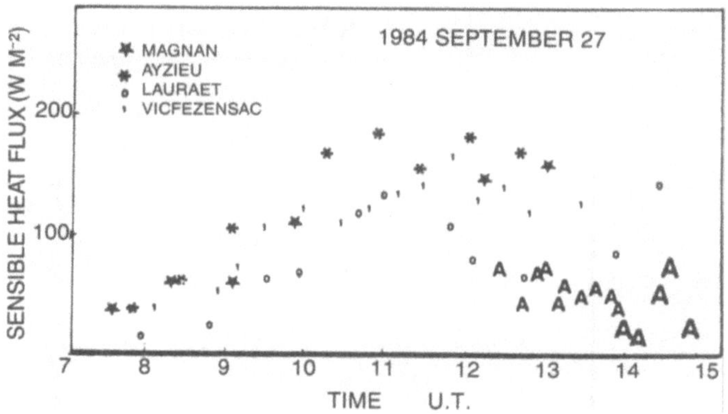

$$\overline{u'_1w'_1}(z_1) - \overline{u'_2w'_2}(z_2) \approx fB\,(h/2 - z_1 - z_1^2/2h) \tag{18.10}$$

where f is the Coriolis parameter, z_1 is the reference height (lowest level) from which the integration is performed, and B is an experimental constant.

3.1.3 Examples of Applications

Figure 18.3 shows an example of virtual heat flux obtained during the Mesogers experiment on complex terrain; one observes the characteristic linear height behavior, the level of zero virtual heat flux, and the surface heat flux. Figure 18.4 presents a comparison among different methods for flux integration using surface heat flux from Sodar, or using the temperature evolution equation and zero heat flux height h obtained by Sodar during the increasing temperature morning stage with

$$\partial/\partial t(\theta'v) = Q'_0v/\rho C_p h \tag{18.11}$$

following (Gupta et al., 1986), but using several Sodar. This method can be used to study homogenization effects when the level h between different Sodar becomes comparable, and when the temperature time evolution measured from the different surface stations becomes similar.

Figure 18.5, taken from Soarès et al. (1988), is an example of comparison of Sodar fluxes to fluxes obtained using a helicopter-borne microwave radar and a physical model.

For the momentum fluxes, we have chosen an example obtained during the HAPEX-MOBILHY experiment (Fig. 18.6) where wind speed and shear differences are close to zero and height is almost 130 m. The differences of momentum are found to be close to 10^{-2} m^2 s^{-2}.

Figure 18.7, taken from Masmoudi (1988), shows differences between momentum computed by direct and indirect methods using Sodar data. Of course the data validity from the previous results is difficult to obtain, therefore a numerical model (Pielke, 1974) is used to try to validate the different estimates. The discrepancies between the different methods and the numerical model suggest the validity of the different estimates and the necessity to increase future numbers of comparisons.

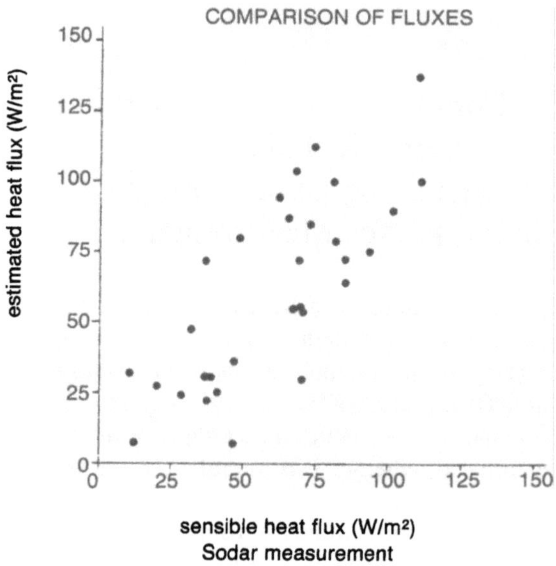

FIGURE 18.5. Comparison between sensible heat flux estimated from airborne measurement and measured Sodar fluxes. (From Soarès et al., 1988.)

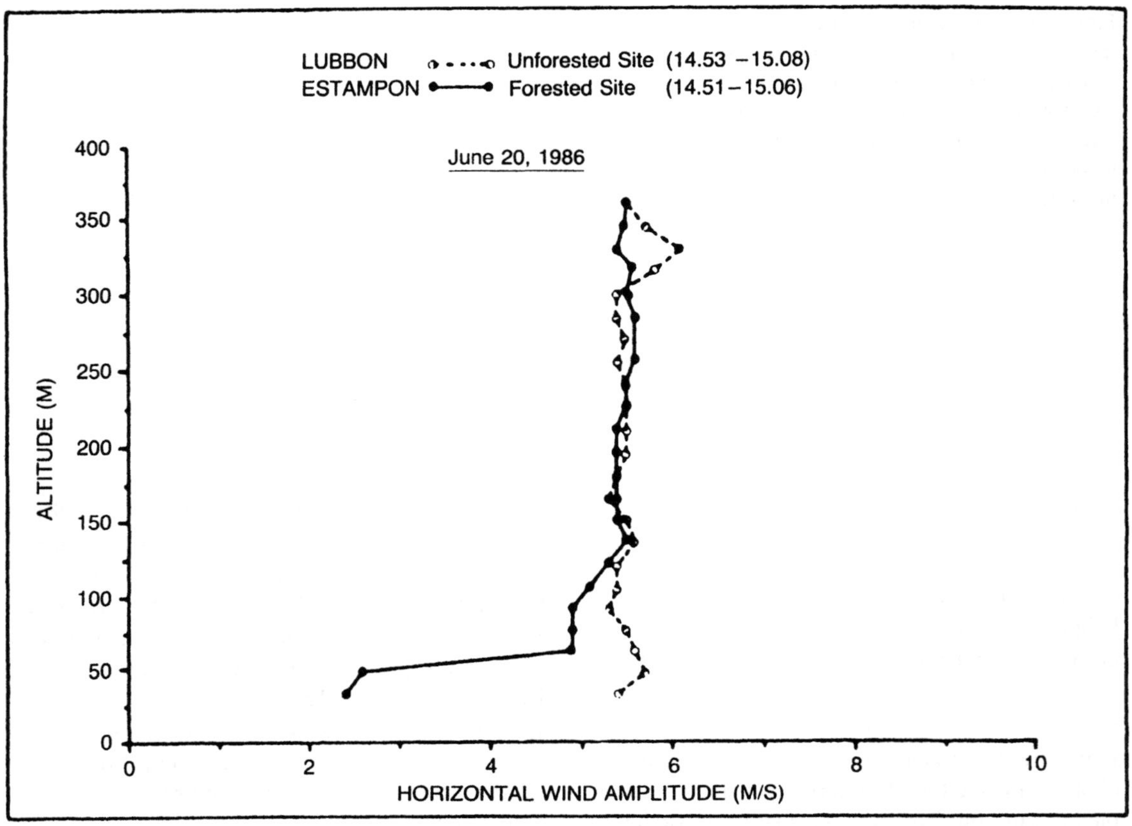

FIGURE 18.6. Height profiles of horizontal wind speed differences measured on forested and unforested terrain during HAPEX-MOBILHY experiment. (From Mazaudier and Weill, 1989, Journal of Applied Meteorology, Vol. 28, No. 8, Copyright American Meteorological Society.)

4. Comments on the Difficulties in Utilizing Techniques for Managing Uncertainties on the Fluxes Measurements

All the estimations of fluxes and generally those that use the direct method of flux measurement are derived from ensemble averages in a stationary flow (Wyngaard, 1973), however, nature does not respond to this condition regardless of the instrument used, whether it is aircraft, remote sensing techniques, or in situ measurements. Following uncertainty measurements using averaging time expression presented by Lumley and Panofsky (1964) and discussed by Wyngaard (1973), we observe the following:

1. For the same time period of averaging, the variance of a turbulent variable is better estimated than the correlation between two variables: for the unstable and neutral case the ratio of uncertainties seems to be around 2, which justifies indirect measurements, if similarity conditions are fulfilled.
2. The hypothesis of gaussian behavior does not always seem to be realized, inasmuch as some skewness that appears as soon as morning convective activity is developing makes a good estimate of the uncertainties difficult (Taconet and Weill, 1983); therefore averaging-time expression presented by Lumley and Panofsky (1964) is just a basis of discussion and comparison.

Considering indirect techniques based on parameterization and the problem of flux extension, we

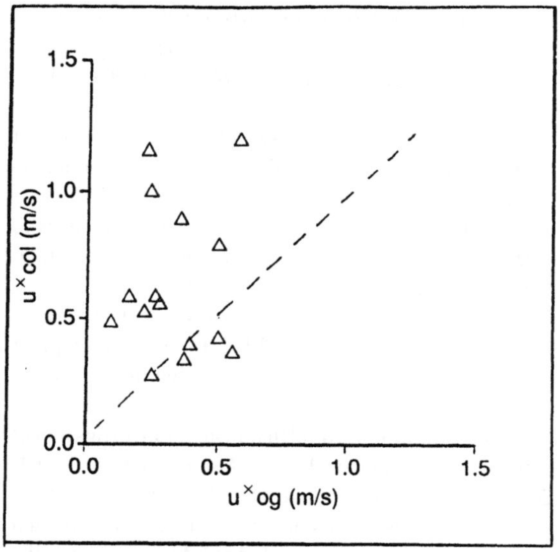

FIGURE 18.7. Scatterplot of friction velocity indirectly measured using parameterization versus directly estimated friction velocity. The poor correlation between the two estimates cannot be explained without independent in situ measurement. (From Masmoudi, 1988.)

observe that the limitations are due to the instrument:

1. Sodar data quality (a Sodar can give a good estimate only if the thermal turbulent level is large and if the acoustic noise of the environment is small);
2. the signal processing methods and the rate of signal repetition; and
3. the antenna beam, which limits the response of the signal (Weill et al., 1980).

But there are some advantages:

1. The possibility of using methods to estimate the fluxes with different variables given several degrees of freedom;
2. some physical stresses that validate the estimates giving consistency of variance profiles with similarity concepts such as

2a. good correspondence between some critical height (inversion height, zero heat flux height, relationships between mean plumes dimension and inversion height) and

2b. the possibility of having a picture of present phenomena using reflectivity facsimile records.

Particularly in the WML where quasiorganized phenomena and intermittent motions are coexistent (Haugen, 1973), the indirect method may be appropriate. Until recently we have used Sodar data and occasionally Doppler Radar data (not during optimal conditions) (Eymard and Weill, 1982; Kropfli and Hildebrand, 1980). We shall try to consider now what kind of instrument or type of experiment could give relatively large-scale flux estimates.

5. Conclusions

Acoustic sounding technique is an efficient technique from which virtual temperature flux and momentum fluxes can be obtained using mainly indirect methods based on similarity. The most drastic limitation is the signal limitation in a noisy or not very turbulent atmosphere. Though this technique gives estimates of fluxes on scales comparable or larger than 1 km², this is too small compared to mesoscale or regional scale estimates and justifies the use of other instruments such as aircraft or Doppler Radar. Aircraft gives the opportunity to integrate fluxes on a large horizontal path, but a nonstationary state limits the precision of the estimates and some correction has to be made based on surface variability, which requires the knowledge of the variability at the flight level (Druilhet et al., 1983). Therefore, identical sensors such as those at SAMER stations (Itier, 1981), deployed on homogenous terrain could solve the problem of surface extension, when trying to perform some conditional averaging, looking at the correlation function between the measurements. This last suggestion concerns only homogeneous terrain. Doppler Radar as used by Eymard and Weill (1982, 1988), and as suggested by Lenschow and Stankov (1986) would probably give the best estimate of "integrated" temperature or momentum flux. The condition, which has to be realized using the coplan methodology (Chong and Testud, 1983) or the velocity-azimuth-display (VAD), is to have a good spatial resolution not compatible with practical use of Doppler Radar. Typically it would be necessary to have a spatial resolution better than 100 m. Considering spatial averaging and the concept of mixing due to turbulence, an observation as a function of time and height of the structure of turbulence using a small towers network, and a minisodar network, would be an extraordinary means with which to understand what really happens and how the atmospheric flow can adjust to the mean or the most unstable atmospheric properties rather than local properties.

References

André JC, Bougeault P (1988) On the use of the Hapex Mobilhy data for the validation and development of parameterization schemes of surface fluxes. Report of the 2nd session of the SSG on Land-Surface Processes and Climate, WMO, Geneva. WCP publication Series No. 126.

André JC Goutorbe JP, Perrier A, Becker F, Bessemoulin P, Bougeault P, Brunet Y, Brutsaert W, Carlson T, Cuenca R, Gash J, Gelpe J, Hildebrand P, Lagouarde JP, Lyod C, Mahrt L, Mascard P, Mazaudier C, Noilhan J, Ottle C, Payen M, Phulpin T, Stull R, Shuttleworth J, Schmugge T, Taconet O, Tarrieu C, Thepenier RM, Valencogne C, Vidal-Madjar D, Weill A (1988) HAPEX-MOBILHY: First results from the special observing period. Ann Geophys 6:477–492.

Baudin F (1977) Estimateurs de fréquence pour mesurer la dérive Doppler Sodar. Rep. CRPE NT/88, Orléans Cédex, France.

Brown EH, Hall FF Jr (1978) Advances in atmospheric acoustics. Rev Geophys Space Phys 16:47–110.

Businger JA, Wyngaard JC, Yzumi Y, Bradley EF (1971) Flux profile relationships in the atmospheric surface layer. J Atmos Sci 25:1021–1025.

Chong M (1976) Measure des profils de vent par Sodar-Doppler. Note technique CRPE/22, CRPE, Orléan Cédex, France.

Chong M, Testud J (1983) Three dimensional wind field analysis from dual Doppler radar data. Part 3. The boundary conditions: An optimum determination based on a variational concept. J Clim Appl Meteorol 22:1227–1241.

Clarke RH (1970) Observational studies in the atmospheric boundary layer. Quart JR Meteorol Soc 96: 91–114.

Coulter RL, Wesely ML (1980) Estimates of surface heat flux from Sodar and Laser scintillations measurements in the unstable boundary layer. J Appl Meteorol 2(19):1209–1222.

Desbraux G, Weill A (1986) Mean turbulent properties of the stable boundary layer during the COAST experiment. Atmos Res 4(20):151–164.

Druilhet A, Frangi JP, Guedalia D, Fontan J (1983) Experimental studies of the turbulent structure parameters of the convective boundary layer. J Climate Appl Meteorol 22:593–608.

Dubosclard G (1980) A comparison between observed and predicted values for the entrainment in planetary boundary layer. Bound Layer Meteorol 18(11):473–483.

Dubosclard G (1986) Thèse d'Etat, IOPG (Institut Observatoire de Physique du Globe) de Clermont Ferrand.

Eymard L, Weill A (1979) A study of gravity waves in the planetary boundary layer by acoustic sounding. Bound Layer Meteorol 17:231–245.

Eymard L, Weill A (1982) Investigation of clear air convective structures in the P.B.L., using a dual Doppler radar and a Doppler sodar. J Appl Meteorol 21:1891–1906.

Eymard L, Weill A (1988) Dual Doppler investigation of the tropical boundary layer. J Atmos Sci 43:853–864.

Gaynor JE (1977) Acoustic Doppler measurements of atmoshperic boundary layer structure functions and energy dissipation rates. J Appl Meteorol 16:148–155.

Greenhut GK, Bean BR (1981) Aircraft measurements of boundary layer turbulence over the central Pacific Ocean. Bound Layer Meteorol 14:513–523.

Gupta SK, Kunhikrihnan V, Radhikha V, Narayanan N (1986) Estimating surface sensible heat flux and surface measurements in the evolving boundary layer. Atmos Res 20:119–123.

Haugen DA (Editor) (1973) "Workshop in Micrometeorology." American Meteorological Society, Boston.

Itier B (1981) Une méthode simple pour la measure de l'évapotranspiration réelle à l'échelle de la parcelle, Agronomie 1(10):869–876.

Kallistratova MA (1959) Procedure for investigating sound scattering in the atmoshpere. Akus Zh 5(5): 496–498.

Klapisz C, Weill A (1978) Modèle semi-empirique d'évolution matinale du profil de vent entre le sol et le sommet de l'inversion. J Recher Atmos 12:113–117.

Klapisz C, and Weill A (1982) Mean horizontal wind in an inversion capped boundary layer, J. Appl. Meteor. 24:648–655.

Kraichnan RH (1953) The scattering of sound in a turbulent medium. J Acoust Soc Am 25(6):1096–1104.

Kristensen L, Gaynor JE, Neff WD (1986) Can monostatic Doppler Sodar be used to determine fluxes? (invited paper). Ground Based Remote Sensing Conference, August 25/28, pp. 199–211. Max Plank Institute.

Kropfli RA, Hildebrand P (1980) Doppler radar measurement in the P.B.L. during Phoenix. preprints, 19th Conference on Radar Meteorology, Miami. pp. 637- 644. American Meteorological Society, Boston.

Lenschow DH, Stephens PL (1980) The role of the thermals in the convective boundary layer. Bound Layer Meteorol 19:509–532.

Lenschow DH, Stankov B (1986) Length scales in the convective boundary layer. J Atmos Sci 42:1198-1209.

Little CG (1969) Acoustic methods for the remote probing of the lower atmoshpere. Proc IEEE 53:571-578.

Lumley JL, Panofsky HA (1964) "The Structure of Atmospheric Turbulence." Wiley-Interscience New York.

Masmoudi M (1988) Etude de l'écoulement à moyenne échelle au cours de l'expérience Mésogers 84. Thèse Université Paris 6.

Mazaudier C, Weill A (1989) A method of determination of dynamic influence of the forest on the boundary layer using two Doppler Sodar. J Climate Appl Meteorol 28(8):705-710.

McAllister LG (1968) Acoustic sounding of the lower atmosphere. J Atmos Terr Phys 30:1439-1440.

Melling H, List R (1977) Characteristics of vertical fluctuations in a convective urban boundary layer. 19: 1184-1195.

Monin AS (1962) Characteristics of the scattering of sound in a turbulent atmosphere. Sov Phys Acoust, Engl Trans 7:370-373.

Neff WD (1975) Quantitative evaluation of acoustic echoes from the planetary boundary layer. Technical Report, ERL 322-WPL, NOAA, Boulder, CO, pp. 38-44.

Neff WD, Coulter RL (1986) Acoustic remote sensing. In "Probing the Atmospheric Boundary Layer" (D. Lenschow (ed.), pp. 201-266. American Meteorological Society, Boston.

Pielke RA (1974) A three dimensional numerical model of the sea breezes over south Florida. Mon Weather Rev 102:115-139.

Singal SP (1988) Acoustic sounding studies of the atmospheric boundary layer. Scientific Report 30, New Zealand Meteorological Service.

Singal SP, Aggarwal SK, Pahwa DR, Gera BS (1985) Stability studies with the help of acoustic sounding. Atmos Environ 19:221-228.

Soarès JV, Bernard R, Taconet O, Vidal Madjar D, Weill A (1988) Estimation of bare soil evaporation from airborne measurements. J Hydrol 99:281-296.

Spizzichino A (1974) Discussion of operating conditions of a Doppler Sodar. J Geophys Res 79(36):5585-5591.

Strauss B (1980) Estimation du bilan d'énergie cinétique turbulente par sondage acoustique. Document de Travail C.R.P.E./1063, Orléans Cédex, France.

Taconet O, Weill A (1983) Convective plumes in the atmospheric boundary layer as observed with an acoustic Doppler Sodar. Bound Layer Meteorol 25: 143-158.

Underwood K (1984) Signal processing methods and the Risöe 78 experiment. Pennstate University.

Weill, A, Klapisz C, Strauss B, Baudin F, Goutorbe JP (1980) Measuring heat flux and structure functions of temperature fluctuations. J Appl Meteorol 19(2):199-205.

Weill A (1981) Sodar micrometeorology. In "Proceedings of the First International Symposium on Acoustic Remote Sensing of the Atmosphere and Oceans" (T. Mathews, ed.), pp. 1-60. Calgary, Alberta, Canada.

Weill A (1983) Atmospheric applications of Sodar. In "Proceedings of the Second International Symposium on Acoustic Remote Sensing of the Atmosphere and Oceans" (G. Fiocco, ed.), pp. 1-41. Rome Italy.

Weill A, Baudin F, Goutorbe JP, Van Grunderbeeck P, Leberre P (1978) Turbulence structure in temperature inversion and in convection fields as observed by Doppler Sodar. Bound Layer Meteorol 15:375-380.

Weill A, Mazaudier C, Baudin F, Klapisz C, Leca F, Masmoudi M, Vidal Madjar D, Bernard R, Taconet O, Gera BS, Sauvaget A, Druilhet A, Durand P, Caneil JY, Mery P, Dubosclard G, Beljaars ACM, Monna WAA, Van Der Vliet JG, Crochet M, Thomson D, Carlson T (1988) The Mesogers 84 experiment: A report. Bound Layer Meteorol 42(3):251-264.

Wyngaard JC (1973) On surface-layer turbulence. In "Workshop on Micrometeorology." (D.A. Haugen, ed.), pp. 101-149. American Meteorological Society, Boston.

19
In Situ Water Vapor Measurements in the Lyman-alpha and Infrared Spectrum: Theory and Components

James E. Tillman

1. Introduction

The measurement of humidity and the direct measurement of fluxes of moisture, or their indirect parameterization, present one of the most important as well as most difficult problems in atmospheric–surface interaction. The primary reason is that there is no universal humidity instrument that will function accurately under all atmospheric conditions. Without protection, it is quite difficult to obtain good measurements from most sensors soon after they have been exposed to saturation conditions. Schemes to protect sensors from water, dust, organics, ice, etc. generally bias the results by some unknown amount, produce hysteresis, and slow down their response. The problems are less severe in the range of what might be called comfortable humidity–temperature conditions, and become more severe at lower and higher temperatures, especially at the lowest and highest relative humidities. Absorption by and adsorption on the components of the system leading to the transducer element can readily produce misleading results especially when the humidity rapidly decreases by an order of magnitude.

The pervasive nature of the humidity accuracy problem even with the most familiar techniques, is illustrated by the report of Wylie and Lalas (1985) that even the long used psychometric constants are incorrect. They state that the "World Meteorological Organization has adopted a type of psychrometer as the international standard for meteorological humidity measurements" and that the universally accepted psychrometer constant is up to 8% too high, due to organic monomolecular films: the 8% error is only for a limited humidity range. Support for their claims is provided by Fan (1987), at one temperature and relative humidity. This result uses a gravimetric hygrometer as the reference, which samples the input air to the test hygrometer. The main criticism of these results is that the source is ambient air in a basement and not a standard humidity generator and that no range of humidities is used. A careful verification of this discrepancy in the "universal psychrometer constant" and Wylie and Lalas's (1985) attribution of the problem to "organic monomolecular films" should be made before accepting these new constants as the ultimate basis of the international standards. If their conclusions are correct, we must address the question of how long, under what conditions, and with what accuracy the simple wet bulb temperatures can be determined. To resolve this fundamental problem, calibrations should be carried out at NIST, formerly, the National Bureau of Standards (or some other international standards facility having equally primary standards), which can cover the complete humidity and temperature range with high accuracy.[1] Along with this, we may need an unprecendentedly accurate and extensive set of field comparisons to restore confidence in, or correct past and future psychrometric measurements, and to determine their limitations in an atmosphere that has a generous supply of organics even without the presence of dirty fingers.

Assuming that we can satisfactorily solve the mean or slow humidity measurement problems, we

[1] NIST maintains its standards and calibration capability, and currently limits its range of research activity to the development of primary standards. National test facility capabilities are discussed in Moisture and Humidity (1985).

are then faced with the determination of the fluxes of moisture. Direct techniques require the simultaneous measurement of the components of the vector wind and their correlation with humidity. To produce an accurate spectrum of the humidity flux, the frequency and phase response of the transducers must match up to the cutoff frequency. Mismatches in response of interrelated sensors can produce up to 100% errors in the variance (Shaw and Tillman, 1985), and thus in flux estimates. Such problems have stimulated the development of alternative flux–gradient relations and even of "roving probe" systems whereby the same set of sensors is varied in height to reduce absolute accuracy sensor problems.

This chapter will concentrate on in situ, spectral absorption techniques of humidity measurements over short path lengths on the order of millimeters to at most 1 m and systems with response times of from 1 sec at most, to as short as milliseconds. The instruments will generally be of one form.

Source ⇒ *path* ⇒ Detector

The first section will cover Lyman-alpha technology in the 120nm spectral range, commonly referred to as the "vacuum ultraviolet." The Lyman-alpha emission line of atomic hydrogen results from the quantized decay of the electron of a hydrogen atom to its ground state, i.e., lowest energy orbit, from its next higher energy orbit. This radiation is often used to measure water vapor since, at its 121.56 nm wavelength, water vapor absorption is high, O_2 absorption is uniquely low, and most other common gases are relatively transparent, e.g., N_2, or low in concentration, e.g., O_3, in the lower atmosphere. The visible portion of the spectrum will be ignored due to its few lines and low absorption, even though these lines can be valuable for longer path sensing. The infrared portion of the spectrum between 800 and 8700 nm is covered in the second section. Primary attention will be focused on infrared wavelengths below 2500 nm, where the sources and detectors are not as affected by thermal radiation, and do not require significant cooling.

Infrared absorption will begin with theoretical calculations first using band models and then line-by-line calculations. These calculations assume a 0.5m path length, about the longest path length useful in fast boundary layer measurements at the surface. Theoretical calculations are discussed that indicate the possibility of relatively simple systems that might provide reasonable accuracy in the presence of clouds. Line-by-line calculations will illustrate the theoretical potential of systems with very high spectral resolution. The band and line calculations utilize the software and line parameters developed by the Air Force Geophysical Research Laboratories over the past 20 years. In both the Lyman-alpha and infrared spectral regions, key features of and problems with implementing systems will be discussed. A comprehensive summary of the state of the art in the early 1960s can be found in the proceedings of the 1963 International Symposium on Humidity and Moisture in Washington DC (Wexler, 1965), while an update is provided by the 1985 International Symposium on Moisture and Humidity.

2. Vacuum Ultraviolet, 100–150 nm

2.1 Absorption by H_2O and O_2

The measurement of humidity by the absorption of ultraviolet radiation in the vacuum ultraviolet portion of the spectrum in the region of the hydrogen Lyman-alpha emission line has been used as the basis for very fast and sensitive humidity systems for more than 30 years. The systems generally consist of a glow discharge source and a detector separated by a distance of millimeters to a few centimeters (Tillman, 1961, 1965; Randall et al., 1965; Buck 1985), Mestayer (1985). The advantages of an ideal Lyman-alpha system are

1. response time of $\ll 0.1$ sec,
2. simple, room temperature, low power sources (hydrogen Lyman-alpha emission line),
3. ion chamber detectors,
4. specific H_2O response at high and moderate humidities,
5. small and light weight, and
6. short, few centimeter, path length.

Disadvantages are

1. sensitive to O_2 at low humidities,
2. increased O_2 sensitivity with typical, multi-line sources,
3. magnesium fluoride windows degrade with use,
4. relatively expensive compared to polymer sensors, and
5. requires complex filtering to correct for O_2 at low humidities, or dual beam techniques.

The theory of absorption by water vapor and other gases in the vacuum ultraviolet, on which Lyman-alpha humidity instruments are based, is presented next while the following sections cover various practical aspects of some implementations and their limitations. For a system consisting of a collimated source of radiation and a detector separated by a distance x and operating at a single wavelength, the energy remaining is given by

$$I = I_0 e^{-\left(\frac{k_1\rho_1}{\rho_{1_0}} + \frac{k_2\rho_2}{\rho_{2_0}} + \frac{k_3\rho_3}{\rho_{3_0}} + \frac{k_4\rho_4}{\rho_{4_0}} + \cdots\right) x_1} \qquad (19.1)$$

where

I = energy incident on surface at $x_j = x_1$
I_0 = energy emitted from surface $x_j = 0$
k_i = absorption coefficient if ith gas in cm^{-1} referred to (STP)
ρ_i = density of ith gas
ρ_{i_0} = density of ith gas at STP[2]
i = subscript identifying gases
 1 → H_20, $k_1 = 387$
 2 → O_2, $k_2 = 0.34$ (slightly pressure sensitive)
 3 → CO_2, $k_3 = 1.97$
 4 → O_3, $k_4 = 640$
 etc. for other gases
x_1 = path length in cm
x_j = distance along path between source and detector

This equation assumes that the absorption depends only on the amount of absorber in the path and that the individual lines in the absorbing spectrum are not "pressure broadened" by a foreign gas or "self-broadened," by the same gas, a good first approximation for the Lyman-alpha region. Details of the rotational, vibrational, and electronic transitions responsible for water vapor absorption from the microwave to the vacuum ultraviolet are carefully presented by Johns (1965). Pressure and thermal broadening are covered in the following section on infrared absorption where they are more significant, since several transition levels above the ground state are populated by and accessible to thermally activated transitions. For absorption covering wavelengths between λ_1 and λ_2, the governing equation becomes

$$I = \int_{\lambda_1}^{\lambda_2} I_0(\lambda)\left\{e^{\left(\frac{k_1(\lambda)\rho_1}{\rho_{1_0}} + \frac{k_2(\lambda)\rho_2}{\rho_{2_0}} + \frac{k_3(\lambda)\rho_3}{\rho_{3_0}} + \frac{k_4(\lambda)\rho_4}{\rho_{4_0}} + \cdots\right)x_1}\right\}d\lambda \qquad (19.2)$$

where the spectral response of the detector and any intervening filters, absorbers, etc. is lumped with the source intensity, $I_0(\lambda)$. In the lower troposphere, for many typical Lyman-alpha systems, the gaseous absorption can be approximated by two gases and the absorption can be written at a series of discrete wavelengths as

$$I = I_{1_0} e^{-\frac{k_{1_1}\rho_1}{\rho_{1_0}}x} e^{-\frac{k_{2_1}\rho_{2}x}{\rho_{2_0}}} + I_{2_0} e^{-\frac{k_{1_2}\rho_1}{\rho_{1_0}}x} e^{-\frac{k_{2_2}\rho_{2}x}{\rho_{2_0}}} + \cdots +$$

where

I_{no} = intensity received at distance x at wavelength n with no absorber
k_{ij} = absorption coefficient (cm^{-1}) for gas i at wavelength j

The above equations cover two general cases and the specific case of the two dominant atmospheric gases absorbing at a finite number of wavelengths. The general equations are presented to illustrate the detail required for accurate calculations with nonideal spectral sources. The use of these equations for simple systems without spectral selectivity would require the knowledge of the source spectral distribution, the density of the other absorbers, the spectral response of all other elements, and the knowledge that none of them changes significantly with time, which is sometimes not the case. The absorption of both H_2O and O_2 are complex functions of wavelength, since Lyman-alpha sources often contain a suite of emission lines from excited electronic states of the hydrogen molecule as well as the desired atomic hydrogen Lyman lines. Figure 19.1 (Tillman, 1965) illustrates the absorption coefficients of H_2O and O_2 in the region of interest using data of Watanabe et al. (1953). The Lyman-alpha emission line at 121.56 nm fortuitously falls in the absolute minimum O_2 absorption feature in this spectral range (Fig. 19.2; Watanabe et al., 1958), where the water vapor absorption is simultaneously quite large.[3] The overall superiority of the Lyman-alpha

[2]Which may be fictitious in some cases. H_2O vapor pressure is far less than standard pressure at 25°C.

[3]The pressure dependence of absorption by O_2 at Lyman-alpha as determined by Watanabe et al. (1958) is $k_{O2}(121.56 \text{ nm}) = 0.27 + 5.265 \times 10^{-5} P$, where P is the pressure in Pascals.

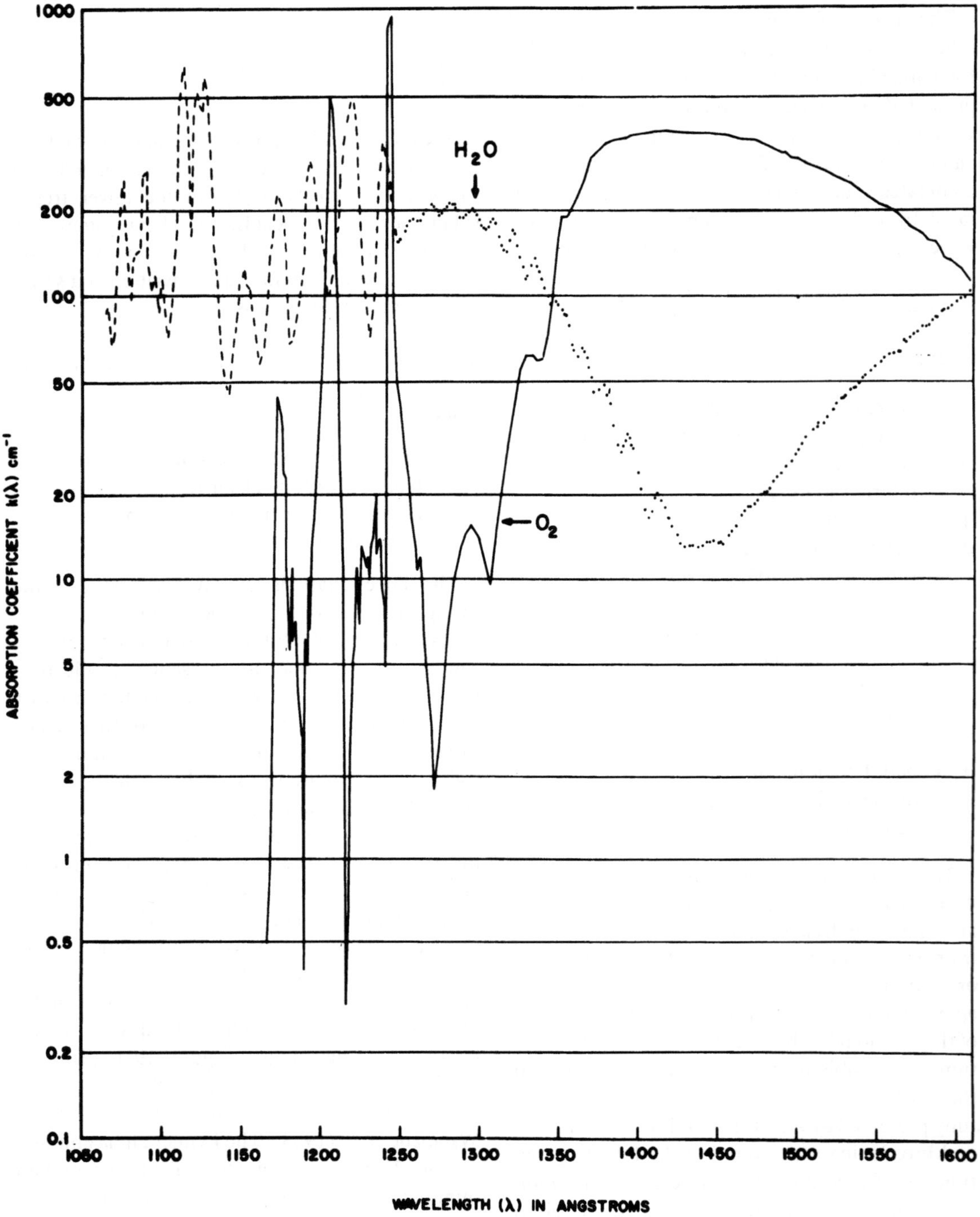

FIGURE 19.1. Absorption coefficients of O_2 and H_2O in cm^{-1} vs wavelength referred to STP. Abscissa, wavelength (λ) in angstroms; ordinate, absorption coefficient, $k(\lambda)$, cm^{-1}. [From Tillman (1965). Reproduced courtesy of van Nostrand-Reinhold Publishing Co.]

FIGURE 19.2. Absorption window of O_2 in the 1215.6-Å region (Lyman-alpha). (From Watanabe et al., 1958.)

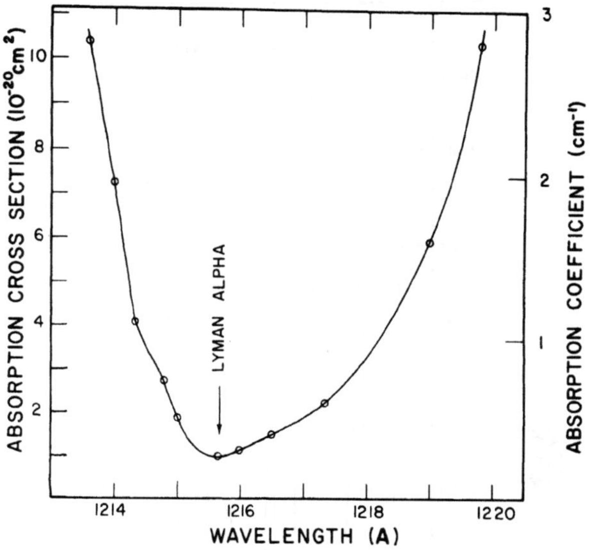

wavelength throughout the vacuum ultraviolet is better illustrated in Figure 19.3 (Tillman, 1965), where the ratio of the two absorption coefficients, defined as the "relative discrimination ratio" by Tillman, is plotted to show the relative sensitivity to each of the two gases.

In the simplest case of a monochromatic source and only one absorber, a normalized water vapor density is given by

$$\frac{\rho_1}{\rho_{1_0}} = \frac{1}{k_{1_1}x}\ln\left(\frac{I_0}{I}\right) \qquad (19.4)$$

In the more typical case of both water vapor and molecular oxygen absorbers at one wavelength for a solution we have

$$\frac{\rho_1}{\rho_{1_0}} = \frac{1}{k_{1_1}x}\left\{\ln\left(\frac{I_0}{I}\right)\right\} - \frac{k_{2_1}}{k_{1_1}}\frac{\rho_2}{\rho_{2_0}} \qquad (19.5)$$

For this case, we require a correction based on the two absorption coefficients and the density of the interfering absorber, oxygen. The theoretical optimum performance of a monochromatic system operating at Lyman-alpha is given by Table 19.1, where the relative absorptions by water vapor and oxygen are shown for saturation over water and ice at a number of temperatures. The path lengths for e^{-1} transmission and the relative contributions by water vapor and by oxygen, tabulated in the last two columns, become equal at about −38°C. If the absorption by O_2 is ignored, the absolute error is equivalent to approximately 0.15 g/m³ and the path length for e^{-1} transmission is on the order of 10 cm.

Estimating the density of oxygen to ± 10% would provide a water vapor density accuracy of ± 0.02g/m³, which is equivalent to the saturation density at −55°C. At high water vapor densities of 10 g/m³, an optimized Lyman-alpha system should have a path length on the order of 2 mm. If one could rely on monochromatic sources at Lyman-alpha, it would be relatively straightforward to provide theoretical calibrations for the effect of oxygen and other lesser interferring gases such as ozone. Unfortunately, this is not the case for many Lyman-alpha systems.

The basic problem with the source spectrum is that a pure hydrogen discharge source operating at low temperatures, such as the glow discharge shown in the top panel of Figure 19.4 (Buck, 1977),[4] emits far more radiation at other wavelengths within the range of simple detectors such as nitric oxide ionization chambers.[5] The spectrum can be made more nearly a pure atomic hydrogen emission source in the Lyman-alpha spectral region by using a far larger pressure of buffer gas in the discharge, which does not have any lines in the range of interest. The inert gases neon and argon are both useful, especially since they do not readily form

[4]See Tillman (1965) for other examples of the variation of spectral purity with different gases and operating conditions.

[5]These detectors have a long wavelength photoionization response limit of 135 nm and a short wavelength limit controlled by the magnesium fluoride windows below Lyman-alpha line.

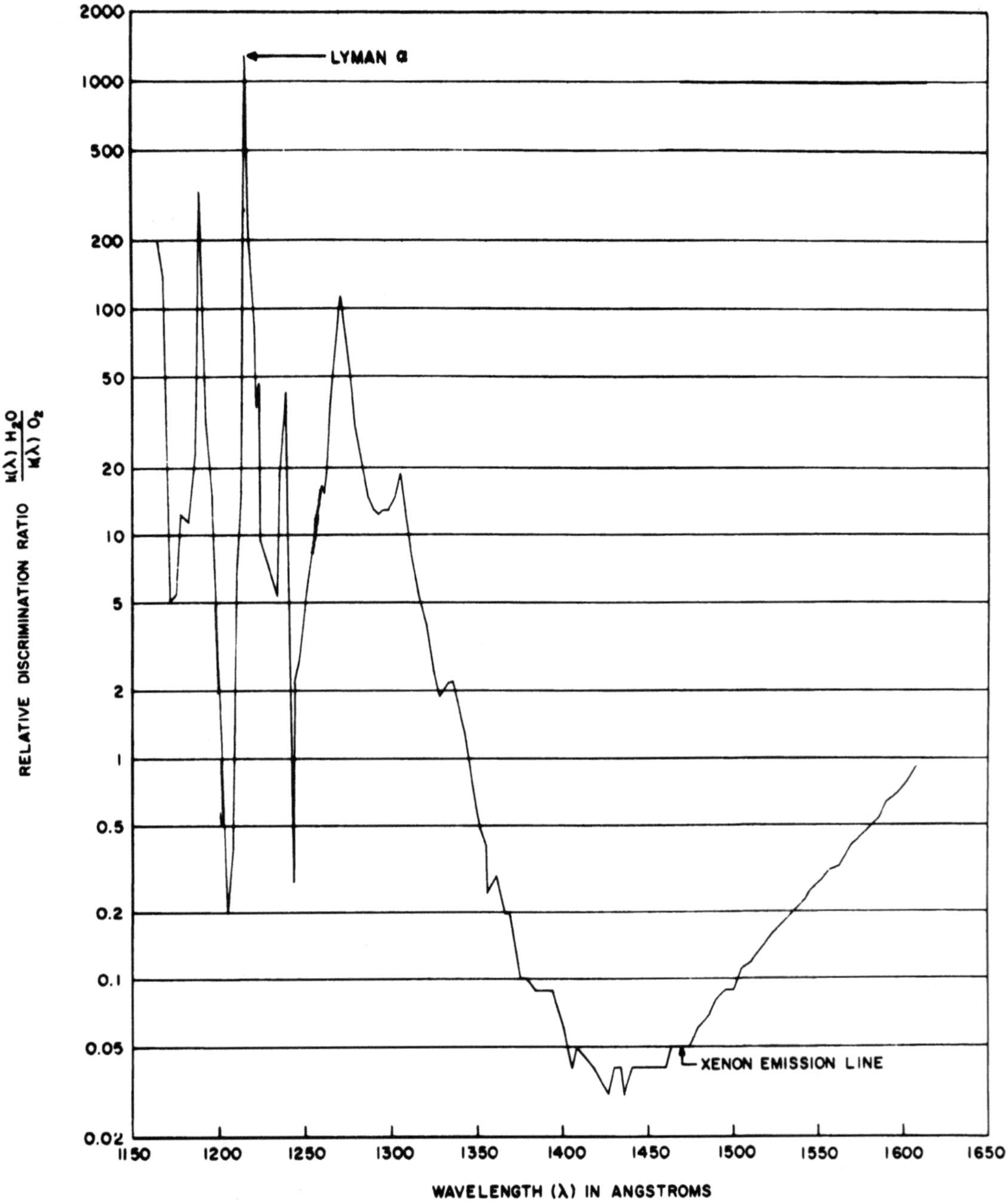

FIGURE 19.3. Relative discrimination ratio (RDR), $k(\lambda)H_2O/k(\lambda)O_2$, vs wavelength. Abscissa, wavelength, λ; ordinate, RDR. [From Tillman (1965). Reproduced courtesy of van Nostrand-Reinhold Publishing Co.]

compounds with the surfaces, windows, seals, and electrodes of the source. An advantage of argon is that it has metastable states, which are excited in some discharge conditions, that serve to disassociate molecular hydrogen during collisions. Spectra of buffered sources are illustrated in Tillman (1961, 1965), Mattix (1971), and Buck (1977), and the best spectra generally are obtained with very low hydrogen concentrations, such as the small hydrogen contaminant concentration found in a tank of commercial grade argon (Tillman, 1965).

While providing a pure Lyman-alpha spectrum, low concentrations of hydrogen result in a short useful life, often measured in minutes, due to the ready reaction of hydrogen with the contaminants and some materials often used in source construction. Such problems can be alleviated by the inclusion of a U–UH_3 mixture at a controlled temperature, which serves as a constant source of hydrogen as long as its temperature is controlled (Tillman, 1965; Buck, 1977). Operational temperatures in the 130°C range provide stable, spectrally satisfactory sources while the hydrogen pressure remains constant over a wide range of the U/UH_3 ratio as long as the temperature remains constant. Among the metal hydrides, uranium is especially constant in its stable pressure over a 10–90% composition range as described by Katz and Rabinowitch (1951). In addition to providing a long life hydrogen source, the free uranium in the compound is an excellent scavenger of any foreign gas, forming compounds with essentially all elements except the noble gases at these low temperatures.

Typical sources used for meteorological applications have from a litle non-Lyman-alpha radiation to a lot, especially where the U–UH_3 system is not used. In most cases, the lack of a pure Lyman-alpha spectrum is a compromise dictated by cost or the lack of the appropriate technology. An exception is the system developed by Mestayer et al. (1985), where additional radiation at other than Lyman-alpha wavelengths is used to expand the water vapor density measurement range easily attained with a single path length. Without the multiple source lines at different wavelengths having different absorption coefficients, the exponential absorption would produce signal levels covering many orders of magnitude for only modest water vapor density ranges. As the system is more sensitive to oxygen away from the Lyman-alpha line and the source emission wavelengths cannot readily be

TABLE 19.1. Absorption by water vapor and oxygen over water and ice.

T (°C)	P (mb)	ρ (g/m³)	x (cm)	$\frac{k_1\rho_1}{\rho_{1_0}}$	$\frac{k_2\rho_2}{\rho_{2_0}}$
		Saturated over Water			
+40	73.78	51.19	0.0407	24.60	0.0619
+30	42.43	30.38	0.0690	14.60	0.0640
+20	23.37	17.30	0.1210	8.32	0.0662
+10	12.27	9.40	0.221	4.52	0.0686
0	6.11	4.85	0.429	2.33	0.0711
−10	2.86	2.36	0.885	1.33	0.0738
−20	1.25	1.07	1.94	0.515	0.0767
		Saturated over Ice			
−10	2.60	2.14	0.971	1.03	0.0738
−20	1.03	0.884	2.35	0.425	0.0767
−30	0.38	0.338	6.21	0.162	0.0798
−40	0.128	0.119	18.20	0.057	0.083
−50	0.039	0.038	55.5	0.018	0.087

[a]From Tillman (1965), reproduced courtesy of Van Nostrand/Reinhold Publishing.

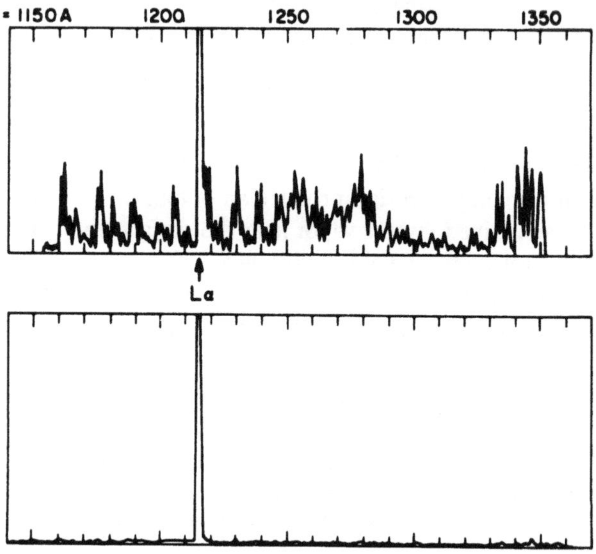

FIGURE 19.4. Relative spectral purity of conventional sealed dc discharge lamp with 8-Torr hydrogen pressure (top) vs UH_3 source (bottom). In both cases the Lyman-alpha line extends offscale by 10 times full scale. Resolution: approx 0.1 nm. [From Buck (1976). Reproduced courtesy of The American Meteorological Society.]

controlled, they remove the energy at the wavelengths where oxygen is a strong absorber by filtering with a flowing oxygen filter. A sealed filter cannot be used as O_3 is readily formed, which will cause the calibration to change with time. Another important characteristic of their implementation of the Lyman-alpha technology is that their system permitted the measurement of fluctuations to 100 Hz with a path length of only 3 mm! If there are more than two wavelengths or more than two gases, the calibrations of a system with a broad-band detector, as opposed to a spectrometer system, become more complex and require a constancy in the spectrum of the source and spectral transmission of the optics.

Any attempt to obtain an absolute measurement of humidity at a fixed height, or reasonable results over a range of altitudes requires that O_2 correction techniques be utilized. Since the absorbers in the lower atmosphere are mainly H_2O and O_2, Buck (1985) produced correction techniques for the series of Lyman-alpha systems developed at the National Center for Atmospheric Research since 1971. If stable and sensitive humidity measurements are to be obtained at Lyman-alpha without source problems, it is imperative that the additional effort of using the U–UH_3, or similar technology, be accepted, since it allows very low hydrogen pressure in a buffer gas and produces a fairly pure Lyman-alpha spectrum, thereby simplifying the O_2 correction. As the main constituent of these sources is an inert gas, the U–UH_3 produces a constant, low H_2 pressure, the free uranium removes other contaminating gases, and these sources are fairly stable in intensity and spectrum.[6] Without the spectral stability, no correction technique can work well. In some cases where raw Lyman-alpha results have been used in airborne applications, the range of aircraft altitudes combined with the lack of an oxygen correction produce an output that is most sensitive to water vapor near the surface and the pressure altitude at higher levels.

During the previous discussions, it has been assumed that the absorption by water vapor in the vacuum ultraviolet is independent of pressure and temperature.[7] Johns (1965) reviews the theoretical basis for absorption by water vapor throughout the spectrum. In the vacuum ultraviolet, he shows examples of rotational fine structure near Lyman-alpha that is generally beyond the resolution of many spectrometers and reports that there are conflicting accounts of any residual broadening effects. However, the band encompassing Lyman-alpha is diffuse and therefore should not exhibit changes in absorption coefficient with different resolution.

2.2 Lyman-alpha System and Components

2.2.1 Windows

One of the most severe practical problems limiting accurate Lyman-alpha measurements is the contamination of the windows by the environment or by handling. The very large absorption coefficient of liquids and solids means that any oils, greases, and even multimonolayers of water may cause serious problems. Even differential methods will fail unless both paths are exposed to the same environmental absorber history, such as atmospheric organics, and an invisible fingerprint will cause absolute errors of many tens of percent. A number of experiments with contaminants were carried out by Tillman (1961, 1965) on lithium fluoride crystals, and later magnesium fluoride, measuring the absolute and spectral absorption by such agents and the ability to clean them from the windows.[8] Current systems now use magnesium fluoride windows, which are less soluble, harder, and generally easier to handle.[9]

Various schemes have been developed to compensate for window degradation and other instability in the sources. These generally depend on either changing the path length, in single beam systems

[6]The U–UH_3 system was implemented by Buck at NCAR with the assistance of Tillman. Subsequently, the technology was passed on to commercial vendors of sources.

[7]In the infrared and microwave portions of the spectrum, this is clearly not the case, as the width of individual lines varies over the normal range of atmospheric temperatures and pressures. The quantized energy level transitions of the lower few vibrational and especially the rotational levels are small enough, compared to the thermal energy available, that transitions are continually taking place at room temperatures.

[8]A window cleaning procedure to restore transmission by polishing with 0.05 μm alumina optical polishing compound was developed by Tillman and Buck.

[9]Unfortunately, these two materials are the only practical bulk crystals suitable for use in this wavelength range.

(Buck, 1976), or a two beam approach using different path lengths (Randall et al., 1965). The latter was used in a laboratory demonstration and they clearly pointed out the advantages of such systems in compensating for window and source tube deterioration. Tillman (1965) also discusses a two beam technique that would use dry air in one path to compensate for the oxygen absorption, but to prevent O_3 buildup, flowing air would have to be used. He refers to the extensive, but unpublished, work by H.E. Hinteregger of Air Force Cambridge Research Laboratories, using dual beam sources as well as dual photodiodes as detectors.

2.2.2 Detectors

Both ionization chambers and photomultipliers can be used as detectors in this region of the spectrum. Nitric oxide ionization chambers provide some wavelength discrimination, having their lower wavelength cutoff determined by the window material and the upper wavelength cutoff determined by the ionization potential of the gas. They are fairly simple devices but they require careful construction and pure gases to preclude short lifetimes and sensitivity outside the desired wavelength region. Kreplin et al. (1962), of the Naval Research Laboratory, developed and used these detectors and provided notes on their fabrication. Careful machining of the special copper bodies was required. One disadvantage of ionization chambers is the very low currents produced and the associated difficulty in measuring due to the very high resistances required in the circuitry to minimize leakage. Currents on the order of 10^{-9} to 10^{-12} A are typical and require circuit design techniques unknown to the typical analog and especially digital system designer. All components from the ion chamber to the input lead of the low current amplifier must be insulated by teflon standoffs[10] and the components should not be handled with fingers, should not be cleaned with normal techniques, etc. so as not to leave organics and salts. All feedthrough insulators should be of the highest insulation resistance materials. To minimize response time, stray capacitances should be minimized in all associated components and "guarded" electrometer circuitry should be used. If any cable is used between the ion chamber and the electronics, it must be of the "low noise" coaxial or triaxial type, which minimizes capacitance change due to vibration or flexure. Electrons from the detector flow into the amplifier and charge the stray capacitance of the cable and amplifier input. Flexing the input cable changes its capacitance, causing current to flow back and forth between the cable and the equivalent input capacitance of the electrometer. Due to the very high impedance of the detector input, time constants are long enough for the noise to appear in the frequency range of interest rather than at many MHz or GHz. The special low noise coaxial cables, whose dielectric is coated with a conductive film, typically have more than two orders of magnitude less noise than regular coaxial cables. If at all possible, the detector system should either be hermetically sealed or should be ventilated through a drying element for all airborne and other applications susceptible to condensation.

Photomultipliers can be used in this wavelength range to provide much higher output currents and noise levels equivalent to a few photons per second. The main problems are size, expense, a generally higher long wavelength cutoff than ion detectors, and the requirement for a very highly regulated high voltage power supply (Tillman, 1965). High regulation is required as photomultipliers consist of a number of stages of electron avalanche multiplication, each of which amplifies by a factor greater than one. The amplification factor is dependent on the dynode material and the applied voltage to a power. Since typical photomultiplier contain up to 10 stages, the sensitivity of amplification to the power supply voltage for a 10 stage system would be proportional to 10^x where x will be on the order of 2 to 4.

2.2.3 Hollow Cathode Sources

There are many aspects of source tube design that can severely degrade lifetime, the spectrum, or produce changes in the spectrum with time. Poor seals of the windows may produce leaks or outgassing of the sealing material, especially if it is an epoxy. As previously mentioned, a low hydrogen

[10]None of the low current components or points should be mounted directly on printed circuit boards, as was found in one airborne system. This could explain noisy behavior, and the sudden calibration shifts, if as the systems pass through large temperature and dewpoint ranges they are subjected to moisture condensation and evaporation or sublimation.

pressure in a buffer gas is necessary for a good spectrum but without the U–UH_3 source, the hydrogen can quickly react with contaminants, diffuse into surfaces, or combine with the cathode or anode materials. Ion bombardment of the cathode causes sputtering and, depending on the material, can more or less easily form hydrides thereby scavenging hydrogen from the discharge. Sputtering of the cathode can also bury inert gases as well as hydrogen, this being the principle of the ion pump used for ultrahigh vacuum systems. An experimental source designed by Tillman and fabricated by NCAR to explore (1) small size, (2) different gasses, (3) higher power, and (4) potentially longer lifetime has several other features that may be of interest: a special hollow cathode, small size, and simple fabrication. Glow discharges generally operate at a constant current density, increasing the area of the discharge as the current increases. Once the cathode is covered, increases in current change the discharge to an abnormal glow and finally an arc discharge. However, with certain materials, specific pressure–current ranges and a spherical hollow cathode, current densities far greater than normal can be attained, thereby increasing the signal, while retaining the plasma characteristics of a glow discharge.

The initial hollow cathode can be made by drilling a 1-mm hole about 3 mm deep in a 3-mm-diameter niobium cathode, which is subsequently shaped to lead the discharge into the hole. After 500–1000 hr of operation, the cavity becomes spherical and the sputtering may be up to 100 times less than a planar cathode. The higher current possible with this cathode will provide more energy at Lyman-alpha than a similar size regular cathode, while the lower sputtering will decrease the scavenging after the initial break in period: This technique requires specific current–pressure combinations for the discharge to enter the cavity. This particular source was not intended to be a prototype, but merely easy to fabricate and to use for testing: the window, attached with epoxy, lasted far longer than expected. Other useful features are the sources of 7-mm-diameter-thick wall capillary tubing, with the cathode at the rear of the tubing, and the window at the front. The anode is attached to a branch in the tubing near the window. It is possible to design sources with smaller diameters that would be more suitable for boundary layer turbulence measurements requiring less flow blockage and disturbance.

2.2.4 Variable Path Length

The only conceptually clear ways to quantify and correct for slow transmission changes due to window surface contamination are (1) to insert a series of calibration gases without introducing more windows, (2) to provide independent measurements to monitor and correct slow variations in system response, or (3) to vary the path length during operation to calculate the absorption due to window contamination.[11] Buck developed a small plastic bellows with purging tubes to place between the source and detector allowing the introduction of specific gas mixtures during calibration. This can be useful for checking for the effect of O_2 absorption by varying a N_2–O_2 mixture. Jensen et al. (1988) calibrated a Lyman-alpha against a dew-point system during clear air soundings, in a program studying cumulus band clouds off the coast of Hawaii; Raga et al. (1990) studied reducing the error due to component and window problems. Varying the path length may be a troublesome engineering problem but is a conceptually viable method of correction. The basic idea is to vary the path from centimeters to as close to zero as possible, plotting the output vs distance to obtain an equivalent transmission loss at zero path length, and use this as the I_0 in the data reduction. The lack of collimation of the source must be taken into account, accurate control and measurement of the distances must be provided, and the window transmission must remain constant during the calibration.

2.3 Fluorescent Lyman-alpha

The normal Lyman-alpha techniques become less accurate at low humidities due to absorption by O_2 and O_3 as described in earlier sections. Wada et al. (1985) present a system and results using the photodissociation of H_2O into OH, which subse-

[11]None of these can compensate for fast changes such as would take place with liquid water on the windows and that may slowly be removed as the air becomes unsaturated.

quently emits a photon at 306.5 nm. This emission is detected by a photomultiplier operating in the photon counting mode, to measure the weak fluorescence. Between 0 and 2000 ppmv water vapor mixing ratios, the output signal is proportional to the mixing ratio. As the ratio continues to increase, the fluorescence decreases due to absorption of the Lyman-alpha radiation. Their system also contains an NO ion chamber detector, operating in a normal absorption mode with the source. Using the water vapor density measured by this combination, they can correct for the Lyman-alpha absorption by the water vapor, thereby correcting the fluorescence estimates. With this correction, their relation between corrected counting rate and water vapor mixing ratio is linear to at least 6000 ppmv. A 3-min aircraft comparison of the two different modes is shown covering the range of 1000 to 4000 ppmv along with a balloon comparison with a carbon element. The carbon element and the fluorescence compare well from the surface to 4 km, where the carbon element ceases to function. However, the fluorescence sensor continues to operate, indicating mixing ratios of 4000 ppmv at 4 km and decreasing to 0.6 ppmv at 11 km, then staying relatively constant to 14 km followed by an increase of 8 ppmv around 14 km.

2.4 Non-Lyman-alpha Wavelengths

Campbell and Tanner (1985) present an alternative to the Lyman-alpha instrument operating in the vacuum ultraviolet part of the spectrum using a krypton source with emissions at 116.49 and 123.58 nm as contrasted with 121.56 for Lyman-alpha. The significant advantage of their system is that it avoids the troublesome spectral properties of hydrogen as well as its reaction with the materials found in the sources. The disadvantage is a far greater sensitivity to oxygen than an ideal Lyman-alpha system. At the shorter wavelength krypton line, it is 8.3 times more sensitive to O_2 than Lyman-alpha, while at the longer wavelength krypton line, which is four times as intense, it is 62.5 times worse than at Lyman-alpha. Comparison of their data with the data in Table 19.1 shows that for saturation at 0°C the stronger krypton line has more absorption due to oxygen than water vapor. This can be corrected to some extent by a pressure measurement but, as the authors point out, a 30°C diurnal variation produces an error greater than 1 g^{-3}

2.5 Fast Wet and Dry Bulb Thermocouple Psychrometers Interdependent Sensor Problems and Comparison with Lyman-alpha

Numerous instrumental intercomparisons of fast humidity sensors have been carried out over the past two decades of different humidity sensors. Shaw and Tillman (1985) compare the Lyman-alpha with a fast wet bulb of a thermocouple psychrometer system, which was digitally compensated to enhance its response time by an order of magnitude. They show the error in water vapor mixing ratio caused by the slower wet bulb response, illustrating the importance of matching sensor responses for calculations using interdependent variables. At frequencies below the cutoff frequency of the slowest sensor, the error in the power spectrum is small while at frequencies on the order of 1 Hz, the overestimate of the variance of humidity reaches 500% while the error in the total variance before correction is 100%. Matching the time constants of interdependent sensors is a general problem that must be adequately addressed if one is to obtain accurate estimates with any confidence. If slow response of one sensor is the only problem, and the signal/noise ratio and the sampling frequency (or system bandwidth) are adequate, the response time of the slower sensor can often be improved by an order of magnitude or even more by analog or, preferably, digital filtering, and produce faster measurements from slower sensors.

In the following section, the theoretical band and line absorption by water vapor in the infrared will be treated and some examples of carbon dioxide absorption will be presented.

3. Infrared

The infrared portion of the spectrum has been used to measure water vapor, with good discrimination against other gases, for many decades. Path lengths as short as 50 cm, many kilometers along the earth's surface, and through the complete atmosphere have been routinely used to estimate water vapor concentration. In many applications, water vapor detracts from the desired observation, which has lead to much of the research on water vapor measurement techniques. Examples are radio propagation through the atmosphere, aircraft and missile detection, and the monitoring of surface and near surface properties of earth. As previously mentioned, a review of many techniques can be found in Humidity and Moisture, Measurement and Control in Science and Industry (1965) edited by Wexler, while a partial update in 1985 is provided by the International Moisture and Humidity conference proceedings, Instrument Society of America (1985). A serious limitation of these conferences is that they contain very little of the mainstream research of the spectroscopic and remote sensing community. In many of these areas, the techniques are more advanced than those reported here, although generally far more complex. A comprehensive reference for all aspects of Infrared research, development, design and applications is to be found in "The Infrared Handbook" prepared for the Office of Naval Research by the Environmental Research Institute of Michigan, edited by Wolfe and Zissis (1989). It contains 25 chapters on theory, materials, atmospheric processes, including band absorption, the AFGRL LOWTRAN programs (Kneizys et al., 1983), and separate chapters for each aspect of an infrared system design.

Quantitative measurement of gaseous species in the infrared by absorption generally can be separated into band or line absorption. In the current context, band denotes any technique that includes multiple groups of lines, while broad band could be construed to lump all absorption within a signal rotational or vibrational–rotational band into one absorption measurement. Line techniques utilize the absorption by a line or several lines. With non-overlapping lines, the exponential absorption in the form of Beer's law can provide a simple estimate of the absorber concentration based on physical constants related to each line. If the resolution is adequate to separate the individual lines, line techniques are superior to band techniques in that the effects of undesired, contaminating absorbers can be observed, and generally removed, while this is not the case with broad band techniques without spectral information or some method of modulating the observation by water vapor alone. Another advantage of line techniques is the ability to calculate the expected absorption and compare it with that measured. Although this can also be done for broad band systems, it requires either a calibration of the complete system, or of the individual components, to determine the response without and with gas combinations in the path, and associated measurements of the interfering gases. Narrow band absorption by H_2O and CO_2 will be presented for various water vapor concentrations and standard atmospheric carbon dioxide within the spectral range of 700–8700 nm. The effects of pressure and thermal broadening are illustrated along with the transmission typical of an instrumental design with a 0.5 path. Finally, the line-by-line absorption is illustrated for small portions of the bands within the narrow band figures.

3.1 Absorption of Water Vapor Between 700 and 8700 nm over a 0.5-m Path

An overview of the infrared absorption bands between 700 and 8700 nm will be first presented, followed by details of each band. The absorption is generally calculated for a 0.5-m path where the water vapor density is equivalent to saturation at −25, 0, and +25°C. An atmospheric pressure of 1013 mb will generally be used along with ambient temperature of +25°C. Due to broadening of the individual lines (by pressure-dependent collisions and doppler-dependent temperature effects) and their overlapping at surface pressures in much of the infrared spectrum, the transmission departs significantly from Beer's law. McCartney (1983) provides an excellent theoretical treatment of the physical and quantum mechanical foundations of absorption by gases.

The theoretical band calculations are produced by the Atmospheric Transmittance/Radiance program LOWTRAN 6 developed by Kneizys et al. (1983). Lowtran is a comprehensive set of software to compute atmospheric transmittance and thermal radiation emitted by the earth and atmosphere from 250 nm (40,000 cm^{-1}) to 28,500 nm (350 cm^{-1}) with a spectral resolution of 20 cm^{-1}. It was developed by the Air Force Geophysics Laboratory over the last 20 years. Included in the software are solar/lunar scattered radiation, spherical refractive geometry, improved water vapor continuum, wind-

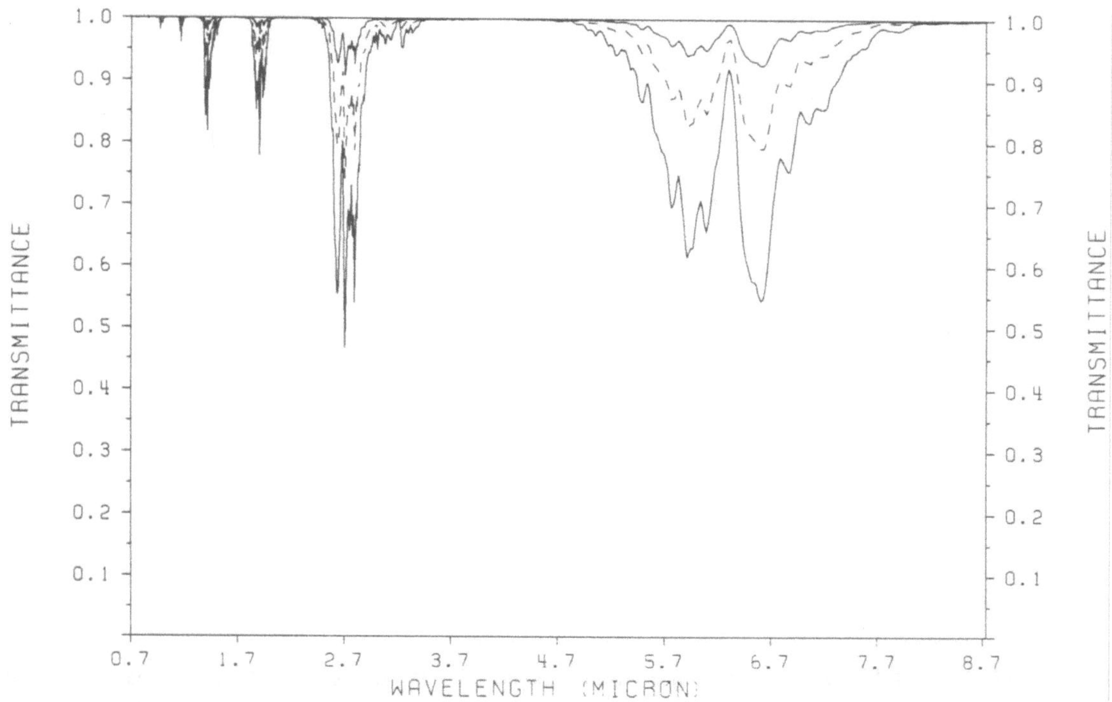

FIGURE 19.5. LOWTRAN: H_2O transmittance with dewpoints of −25° (top curve), 0° (center, dashed), +25° (bottom); temperature = +25°C; pressure = 1013 mb; path = 0.5 m. Spectral range of 700–8700 nm.

dependent maritime aerosol model, vertical structure aerosol model, cirrus cloud model, and a rain model. Representative (geographic and seasonal) atmospheric and aerosol models are included with options to replace them by user specified models.

The program can compute (1) atmospheric transmittance only, (2) atmospheric transmittance and radiance, or (3) atmospheric transmittance, atmospheric radiance, and scattered solar/lunar background radiance for a given path geometry, with fairly simple to use and well-documented options. It has 70 built-in scattering phase functions that approximate, within 20%, any function from the full set of 702 phase functions (26 separate models with 27 distinct wavelengths) that can be used for building aerosol models. In this application, no aerosol models are used and options have been selected to compute absorption over a 0.5-m path with fixed pressure and gas concentrations rather than those of the typical model atmospheres. An enhanced version LOWTRAN 7, is now available.

3.1.1 Band Absorption by H_2O and CO_2

Figure 19.5 illustrates the absorption by water vapor over a 0.5-m path with a gas temperature $T = 25°C$, $P = 1013$ mb, and with water vapor concentrations corresponding to dewpoints of −25, 0, and +25°C. The 2700-nm band is the strongest followed by the 6700, and then in decreasing intensity the 1900, 1375, 1190, 25,000 (not shown), and 935. For the moderately moist +25°C dewpoint, the peak absorption is 55% in the 2700-nm band and decreases to about 10% at −25°C. Figures 19.6, 19.7, and 19.8 individually illustrate the 935-, 1375-, and 6700-nm bands in detail while the 2700-nm band has been omitted.

Staats et al. (1965) describe a short path instrument in the 1375-nm band that operates on the principle of varying the angle of an interference filter, thereby changing its center frequency slightly, to capitalize on the sharp transmission peak in the spectrum. Although simple, such systems have

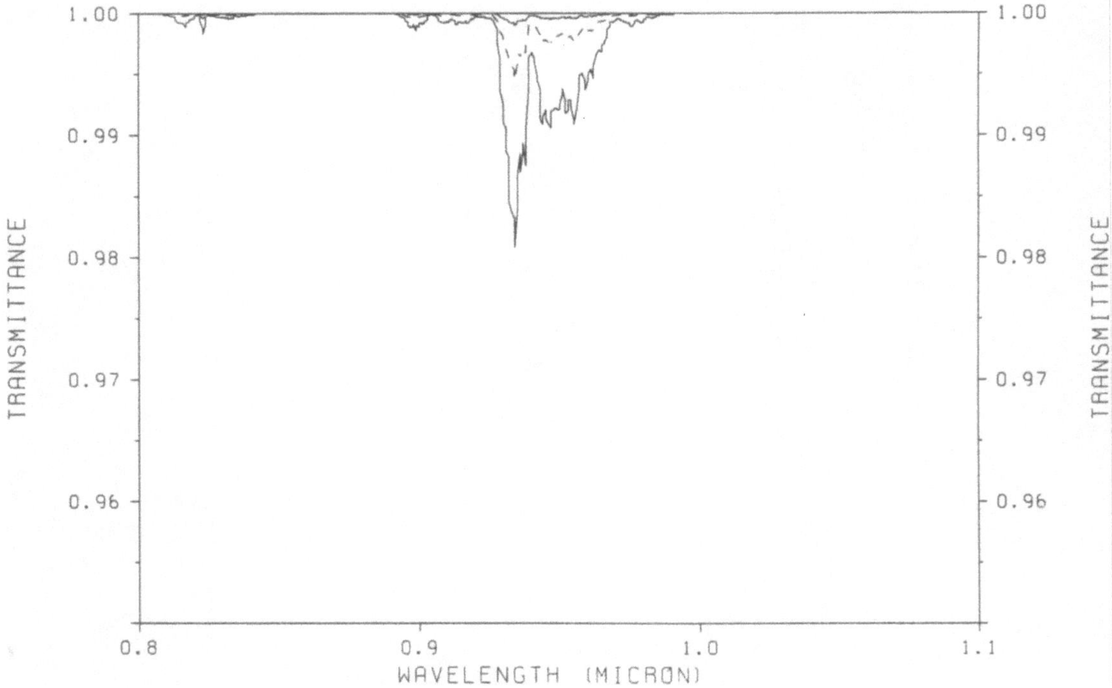

FIGURE 19.6. LOWTRAN: H_2O transmittance with dewpoints of −25° (top curve), 0° (center, dashed), +25° (bottom); temperature = +25°C; pressure = 1013 mb; path = 0.5 m. Spectral range of 800–1100 nm showing 935-nm water vapor band.

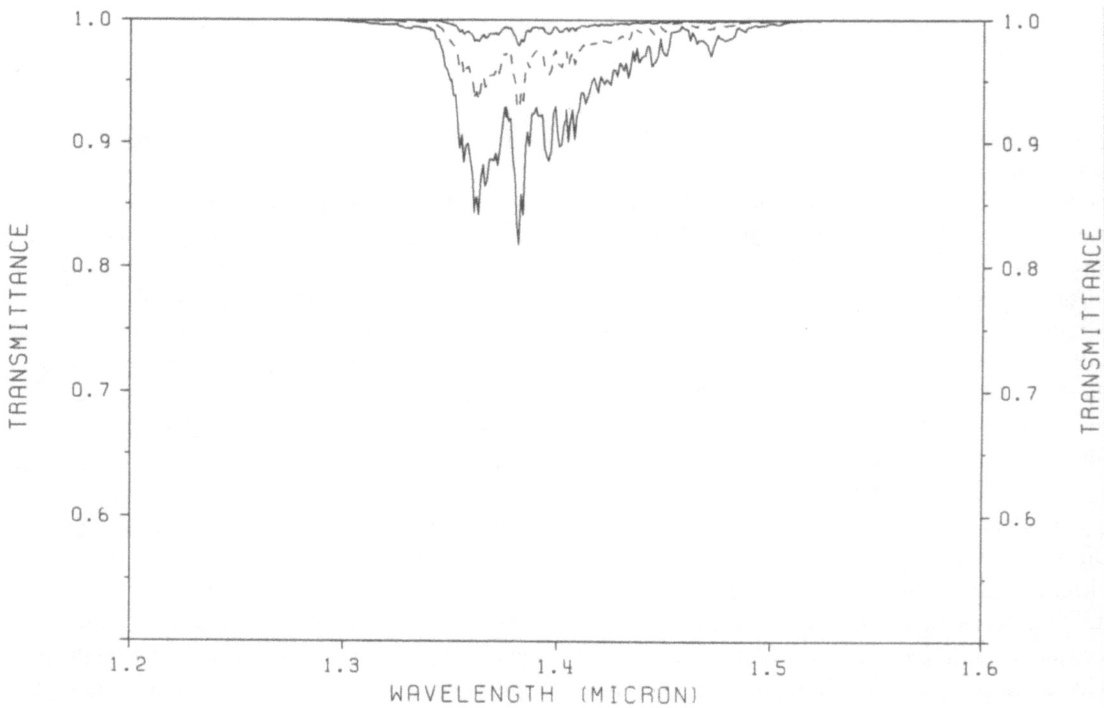

FIGURE 19.7. LOWTRAN: H_2O transmittance with dewpoints of −25° (top curve), 0° (center, dashed), +25° (bottom); temperature = +25°C; pressure = 1013 mb; path = 0.5 m. Spectral range of 1200–1600 nm showing 1375-nm water vapor band.

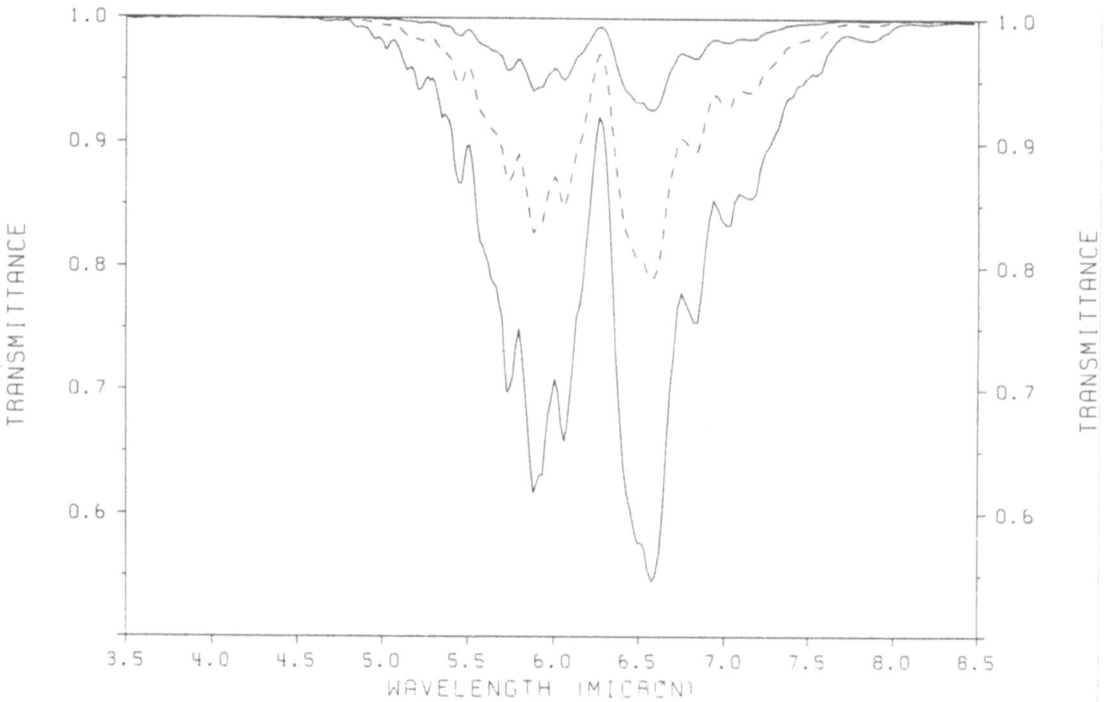

FIGURE 19.8. LOWTRAN: H_2O transmittance with dewpoints of −25° (top curve), 0° (center, dashed), +25° (bottom); temperature = +25°C; pressure = 1013 mb; path = 0.5 m. Spectral range of 3500–8500 nm showing 6700-nm water vapor band.

significant problems in establishing a constant sensitivity. In the same volume Wood (1965) describes a system operating in the 2700-nm band that was produced commercially. Tillman (1965) first proposed the use of semiconductor, solid-state diode sources with narrow spectral widths that could be produced at several wavelengths as the basis for very fast, spectrally selective systems. Since the diodes are spectrally much brighter than a black body at similar temperatures, their radiation is confined to narrow wavelength ranges, and they can be switched off and on at very high speeds; they are ideal for very fast humidity instruments provided the appropriate wavelengths can be produced. At the time of that publication, the Physics Division at MIT Lincoln Laboratory had just produced the first diodes composed of the three component semiconductor system GaInAs. By varying the composition, it was suggested that a wide range of wavelengths could be produced and the composition tuned to water vapor bands and nearby windows. Major advantages of such bright sources are that they can be made much smaller than a thermal light bulb of reasonable life and they are easier to collimate with small optics due to their small size. The small bandwidth of the source allows use of a narrow filter, thereby excluding more of the reflected and scattered solar radiation in open path applications and decreasing the noise. The main problem with the diodes developed at Lincoln Laboratory at that time was that they were readily made only near the weak 935-nm band.

Subsequent to these developments, Henry Lahore of the University of Washington Department of Atmospheric Sciences designed an instrument based on the above principles described by Tillman (1965) and Brooks (1971).[12] This system consisted of semiconductor light, emitting diodes at 880 nm

[12]The conceptual and electronic design and fabrication were completed by Lahore, the optical system was sketched by Tillman and fabricated by Lahore, while subsequent development and testing was carried out by Brooks (1971).

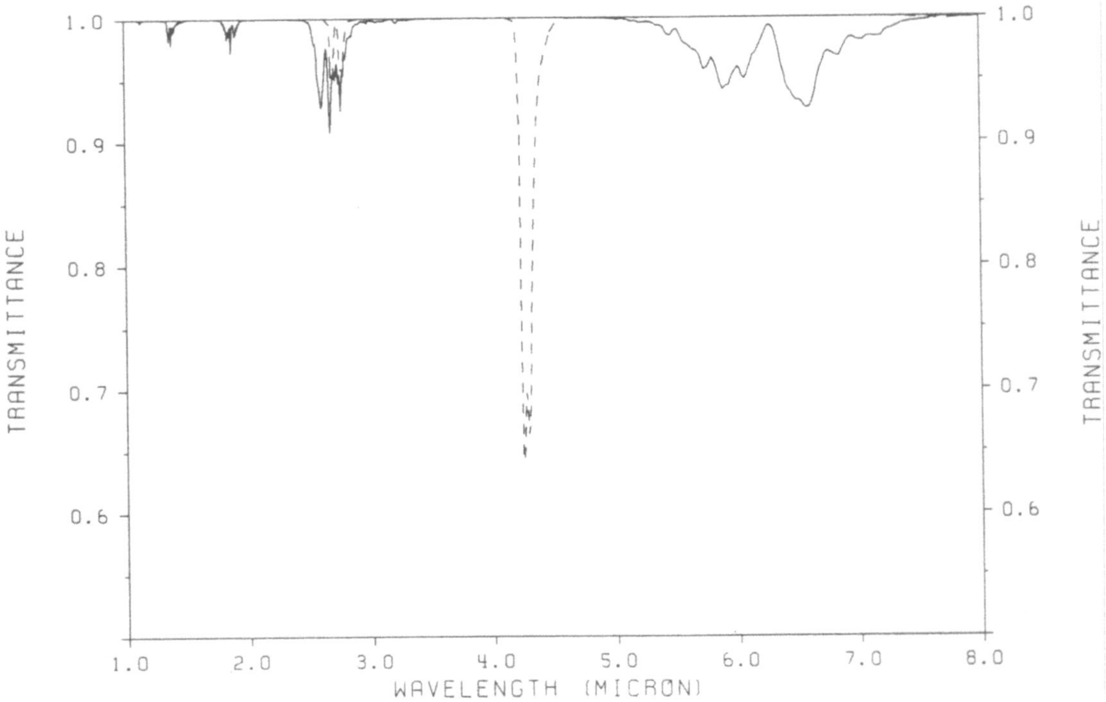

FIGURE 19.9. LOWTRAN: H_2O and CO_2 transmittance. Dewpoint = −25°C, temperature = +25°C, pressure = 1013 mb, path = 0.5 m. Spectral range of 1000–8000 nm. Two CO_2 bands are shown by dotted and H_2O by solid lines; one CO_2 band is encompassed by the 2700 nm H_2O band.

(Brooks, 1971), a non absorbing region, and at 935 nm, a very weak band (Fig. 19.6). The absorption is only about 2% maximum at the +25°C dewpoint in the center of the strongest part of the bank. Despite this weak absorption, Lahore's excellent design enabled the breadboard system to provide 100/1 signal/noise ratios with a 30-Hz bandwidth at water vapor densities around 5 g/m³ and > 1000/1 signal/noise with an 0.2-Hz bandwidth. It was also operated in an aircraft, circling a radiosonde to 15,000 feet, with a comparison agreement as good as could be expected with the slow radiosonde response.

A laboratory experiment by Lahore to determine the effect of absorption by liquid water produced such low absorption that Tillman (1985) calculated the potential for designing a system at this wavelength that might be able to operate in the presence of clouds or fog, or with liquid water on the optics, with little error. Simplified calculations with a C2 cloud model of Deirmendjian (1969) indicate that an instrument geometry might be possible that would reduce the scattering and absorption effects of typical clouds to less than 1% of the absorption due to a +25°C saturated atmosphere. Liquid water on the optics would also have a very small effect as the glossy part of the refractive index is quite small at this wavelength.

Figure 19.9 shows the absorption by water vapor and carbon dioxide between 1000 and 8000 nm. Several research and commercial instruments have been built in the 1375-nm band at various times. Sometimes the 2700-nm band is chosen since it is at a short enough wavelength that thermal emission is not a serious problem and the absorption is quite strong. A potential problem for low resolution systems is that in this band both H_2O and CO_2 absorb as can be more readily seen in Figure 19.10. However the CO_2 absorption is fairly weak, being somewhat less than the absorption due to a saturated atmosphere at −25°C and the short wavelength portion of the H_2O band is also relatively free of

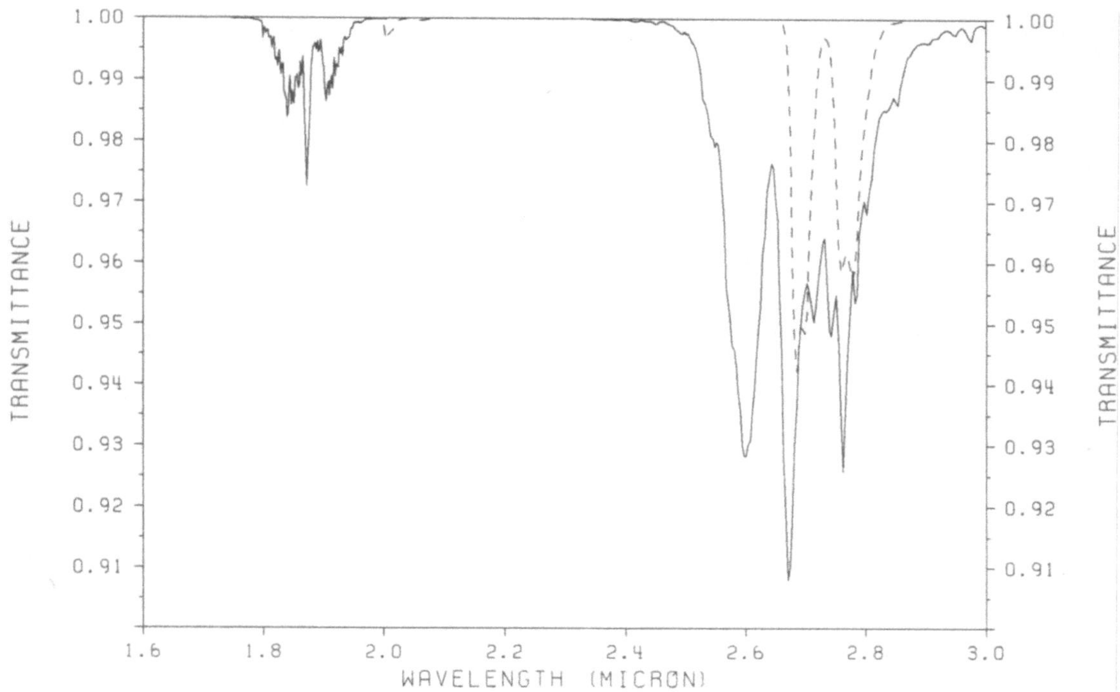

FIGURE 19.10. LOWTRAN: H_2O and CO_2 transmittance. Dewpoint = −50°C; temperature = +25°C; pressure = 1013 mb; path = 0.5 m. Spectral range of 800–2200 nm showing the overlapping of H_2O and CO_2 in the long wavelength portion of the band around 2700 nm.

CO_2 absorption. Ophir Corporation developed a humidity instrument in this wavelength range that operates at at least two wavelengths to provide compensation for contaminants on the optics. Advanced Systems of Okayama, Japan developed an instrument measuring H_2O at 2600 nm, CO_2 at 4300 nm, and using a reference measurement at 3900 nm. Systems without reference wavelengths are subject to serious error if undesired spectral components, whose absorption or scattering is wavelength dependent, are in the measuring path or collect on the optical surfaces.

With all infrared spectroscopic techniques, it should be remembered that the line broadening produce pressure and temperature dependent absorption. Figure 19.11 illustrates the effect of pressure broadening where the absorption at 910, 1013, and 1110 mb is presented for the 1375-nm band. With the LOWTRAN code, these effects can be quantitatively modeled to determine the errors inherent in ignoring pressure broadening or, alternatively, the pressure accuracy required for a given humidity accuracy. Figure 19.12 presents similar results for a temperature variation of ±25°C around 0°C.

3.1.2 Line-by-Line Absorption

Specific measurements of an absorbing gas in the context of other absorbers can be obtained if high enough resolution can be achieved to separate individual lines, due to quantized rotational and vibrational–rotational transitions. Typically, such measurements are made with grating spectrometers, although their resolution may not be adequate in some portions of the infrared with certain combinations of gases. Higher resolution is available with various types of interferometers, although they may give ambiguous results at times. The following theoretical calculations of absorption for portions of the 1375-nm band illustrate the complexity of the water vapor vibrational–rotational

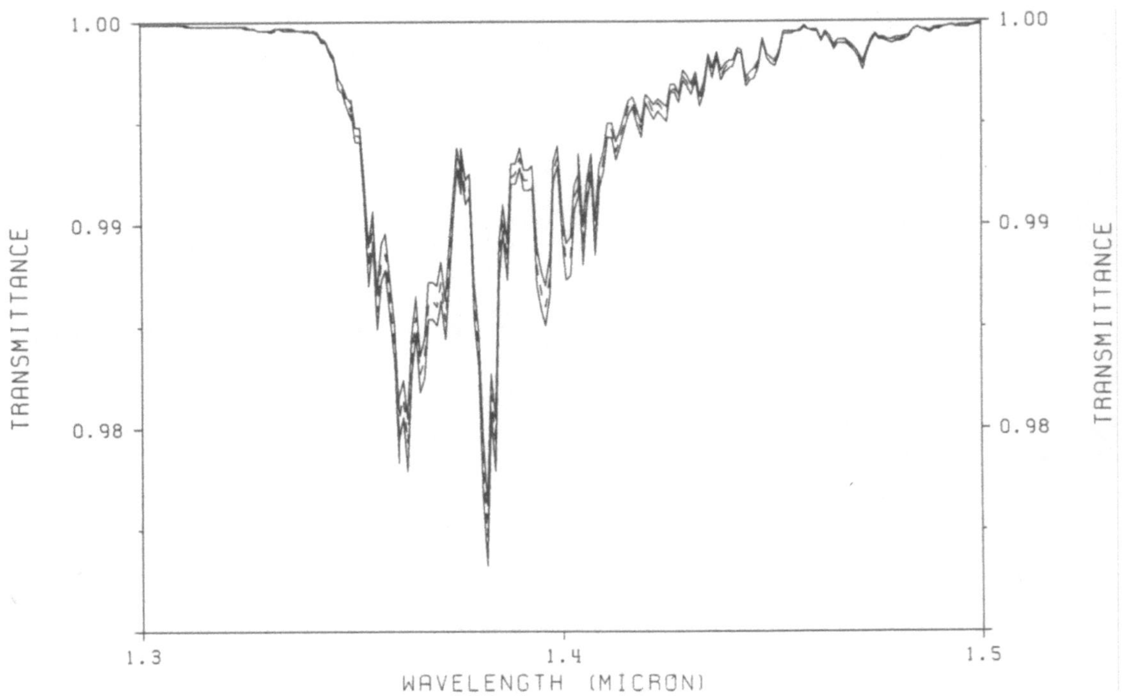

FIGURE 19.11. LOWTRAN: H_2O transmittance with pressure broadening. Dewpoint = −25°C; temperature = −25°C; pressures = 913, 1013, 1113 mb; path = 0.5 m.

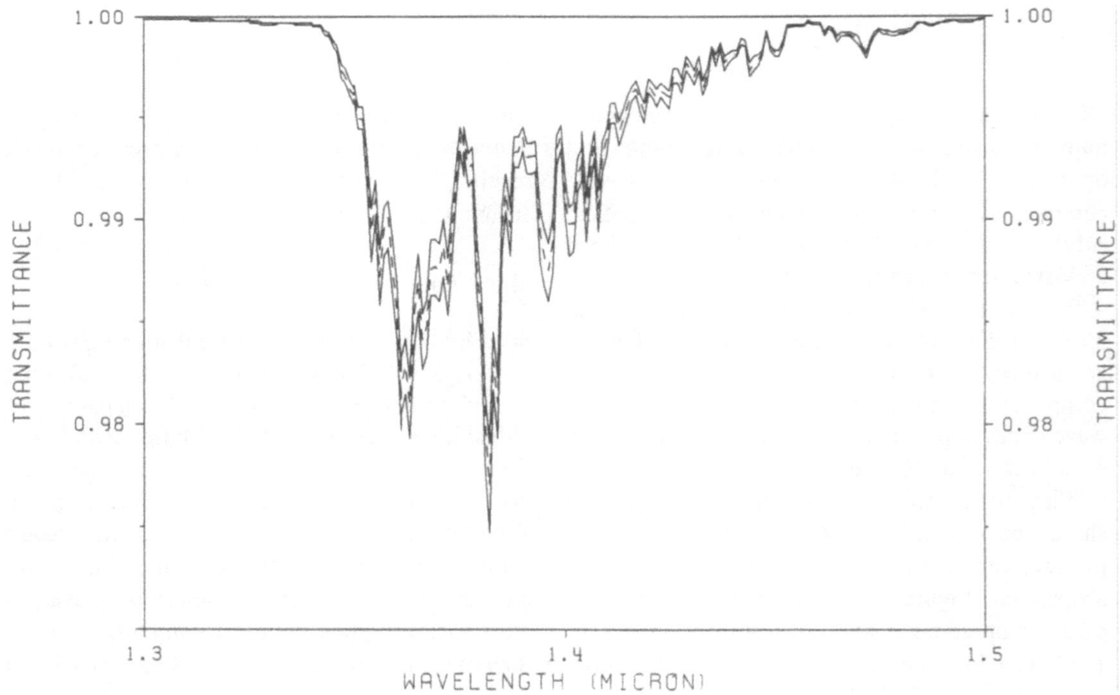

FIGURE 19.12. LOWTRAN: H_2O transmittance with thermal, doppler, broadening. Dewpoint = −25°C; temperatures = −25, 0, +25°C; pressure = 1013 mb; path = 0.5 m.

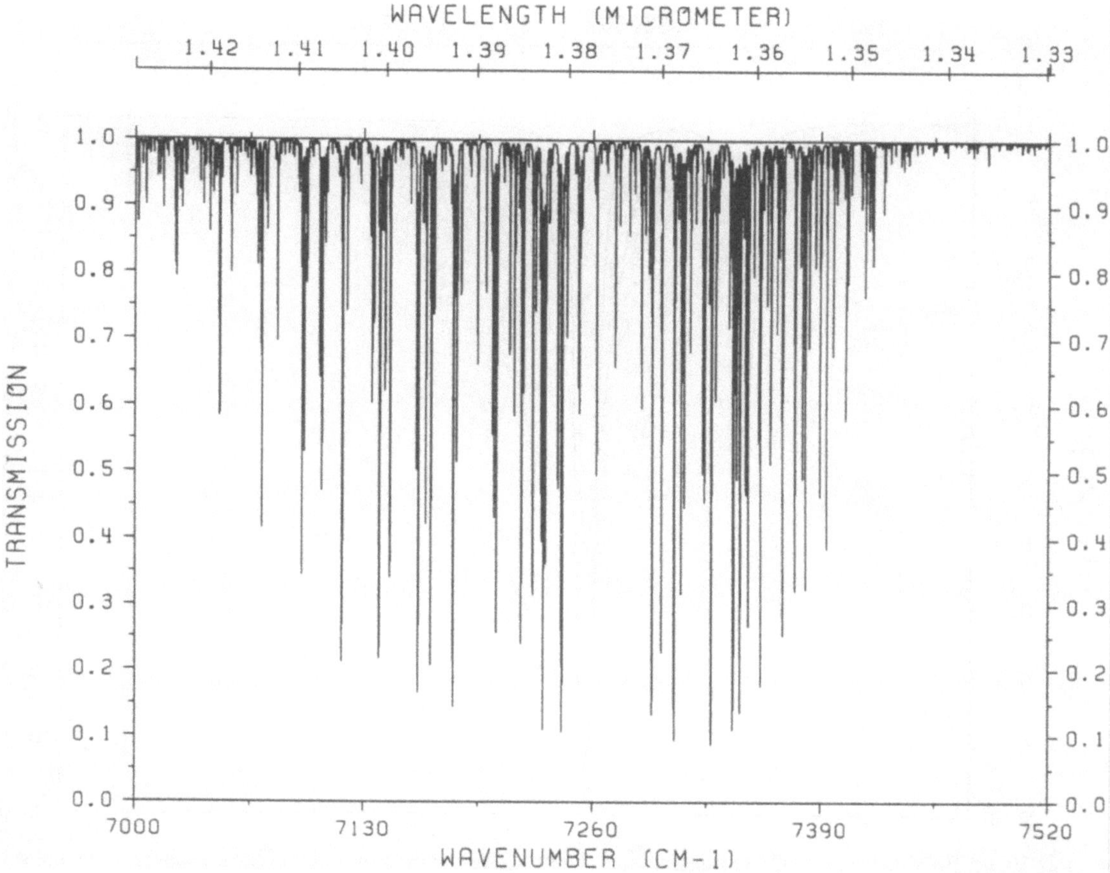

FIGURE 19.13. FASCODE: H_2O line-by-line transmittance. Dewpoint = +25°C; temperature = +25°C; pressure = 1010 mb; path = 0.5 m. Spectral range of ≈ 1430–1330 nm. (Note reversed order of abscissa from LOWTRAN plots.)

spectrum and the potential increase in sensitivity and specificity obtainable with high resolution. The calculations are accomplished with the AFGRL software FASCOD2 (Clough et al., 1986; Smith et al., 1978) and associated line parameter data base (Rothman et al., 1987). Figure 19.13 shows the absorption from 1330 to 1430 nm at a dew point of +25°C, ambient temperature of +25°C, path length of 0.5 m, and pressure of 1013 mb. Note that the absorption in the strongest line is on the order of 90% while the strongest narrow band absorption for the same conditions (Fig. 19.8) is less than 20%. Figure 19.14 illustrates the absorption by two fairly strong lines in the 1375-nm band, as calculated by FASCOD2 for a pressure of 100 mb, 0.5-m path length, ambient temperature of +25°C, and dew point of +25°C. For comparison, the same conditions except for a pressure of 1010 mb are plotted as a dashed line. Although this plot does not have high enough resolution to accurately represent the line profile in all of its detail, the broadening effect of the surface atmospheric pressure is very evident. High-resolution techniques that exploit absorption by individual lines are being developed by the author and colleagues that will be able to provide accurate measurements under the widest range of difficult environmental conditions, such as organic contaminants, water, and salts, on the optics.

4. Summary

Along with the spectroscopic techniques, there have been advances in the small polymer sensors, and an increased awareness of the problems of the interaction of water vapor with materials of the

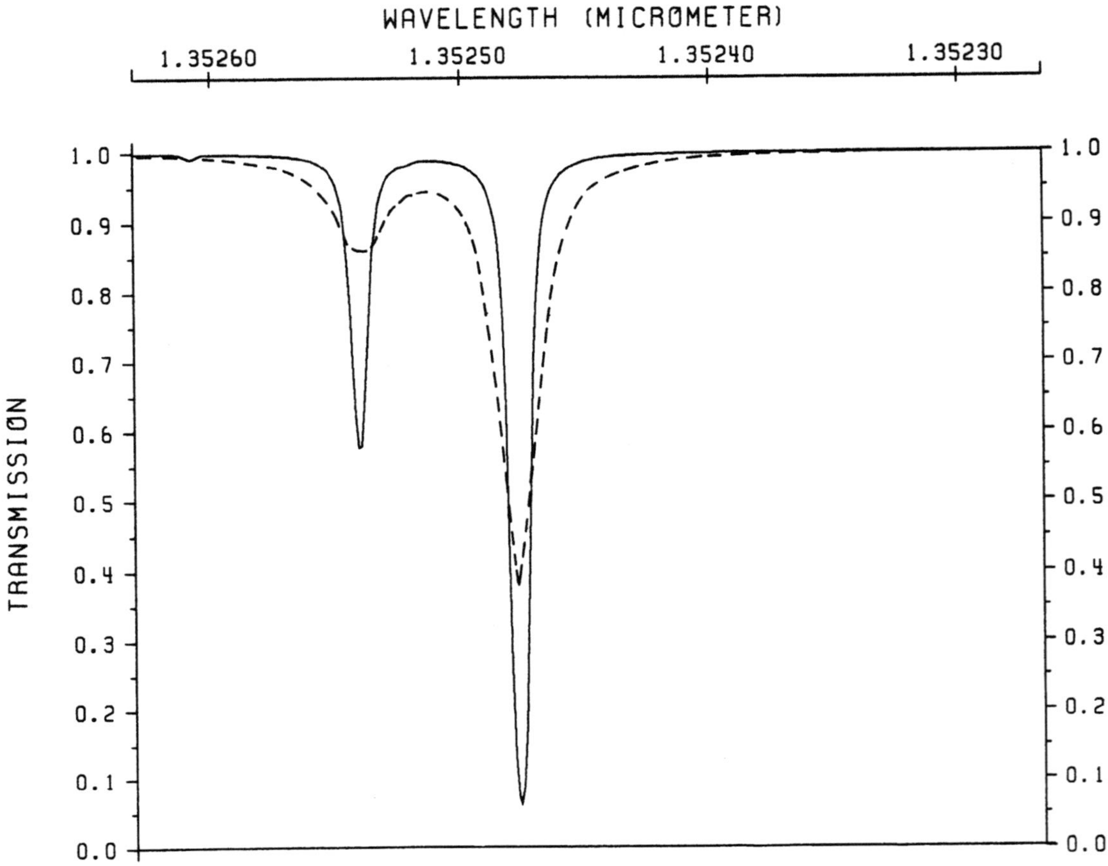

FIGURE 19.14. FASCODE plot of H_2O transmittance for two different pressures at the same dewpoint. Dewpoint = +25°C; temperature = +25°C; path = 0.5 m; pressure = 100 mb (solid line) and 1010 mb (dotted line).

system. Cortina (1985) quantifies the relative merits of different materials that may come in contact with the atmosphere to be measured, with regard to their lack of absorption of H_2O from a moist stream in terms of the time required to reach equilibrium after exposure to the ambient atmosphere. Stainless steel, teflon, and nickel tubing will dry to dewpoints of −70°C in 20–40 min, copper and polyethylene in 140 min, while nylon only reaches −30°C after 170 min. The fastest and best, stainless and nickel, ultimately reach dewpoints of −120°C after 120 min.

Advances in humidity measurements using the vacuum ultraviolet portion of the spectrum over the last decade have improved and refined the simple Lyman-alpha systems, developed corrections for O_2 and O_3, and developed fluorescence systems for very low dewpoints. The latter is quite important since at low dewpoints, the amount of water vapor available to produce a sensor output from an absorption/adsorption process is so small that time constants are on the order tens of minutes to an hour at dewpoints of −80°C for the more sensitive transducers. Under the same conditions, in the infrared, the absorption is weak enough that long or folded paths are required, making complicated, expensive systems. When deciding on instruments or developments for very low humidities, the work of King (1965, 1969a,b) can be ignored at the possible expense of reinventing the wheel, and possibly not doing a very good job. King invented the quartz sorption microbalance, which coats the surface of radio frequency crystals with a very hygroscopic material. Although the material has large hysteresis, he developed methods of avoiding the problems. Commercial units have been manufactured by either DuPont or Gilbarco for the past 25 years that have sensitivities of 2 ppm full scale. Although not used on the mission, he and others developed breadboard sensors for the Viking Mission to Mars and tested them at the National Bureau of Standards down to dewpoints of lower than −95 °C.

Dual beam Lyman-alpha systems are under development and may improve their reliability with respect to calibration stability and compensating for source or optics changes. The effectiveness of the improvement will depend on the exact nature of the design, and on informing the users of the basic limitations of the technology. The latter is quite important as in one major instance, users were provided (maybe they insisted on) raw Lyman-alpha results, rather than the corrected water vapor estimates, possibly without being carefully warned that the instrument is quite sensitive to O_2: the output is essentially a density measurement at higher altitudes in airborne applications. In that instance, there were excellent correction techniques, software and outputs available for the specific platforms, and it was mainly "customer training" the was lacking.

There have been modest advances in broadband infrared systems during the past 10 years in providing reference wavelengths, along with measuring ones, to correct for source changes and changes in the optical path. Although not a new idea, digital processing has greatly assisted the implementation and automation of these corrections. Not covered in this review, are high-altitude measurements of water vapor in the 6700-nm band, and of minor constituents by systems using folded paths and cryogenically cooled detectors. These developments have been excluded from this presentation for technological reasons, and not because they are unimportant. Unfortunately, there is still no universal humidity measuring system, regardless of the importance of water substance to the planet Earth and its inhabitants. Development of a semiconductor laser solid-state sensor system can provide an excellent solution to the humidity measurement problems and rapid progress can be possible with the proper combination of knowledge and adequate resources.

Acknowledgments. The author would like to acknowledge the efforts of Neal C. Johnson in converting LOWTRAN and FASCOD2 and in assisting in the preparation of the manuscript and to Conway Leovy for providing partial support of the conversion. The support of NATO under Grant 909/86 for foreign travel related to research and the presentation of this work during a workshop is gratefully acknowledged.

References

Brooks DL (1971) Development of an infrared absorption hygrometer using solid state energy sources. M.S. Thesis, Dept. of Atmospheric Sciences, University of Washington, Seattle, WA.

Buck AL (1976) The variable-path Lyman-alpha hygrometer and its operating characteristics. Bull Am Meteorol Soc 57:1113–1118.

Buck A (1977) Lyman-alpha radiation source with high spectral purity. Appl Optics 16:2364–2366.

Buck A (1985) The Lyman-alpha absorption hygrometer. In "Moisture and Humidity 1985, Measurement and Control in Science and Industry," pp.411–436. Proceedings of the 1985 International Symposium on Moisture and Humidity, Washington, D.C., 1985. Instrument Society of America, Research Triangle Park, NC.

Campbell GS, Tanner BD (1985) A krypton hygrometer for measurement of atmospheric water vapor concentration. In "Moisture and Humidity 1985, Measurement and Control in Science and Industry," pp. 609–614. Proceedings of the 1985 International Symposium on Moisture and Humidity, Washington, D.C., 1985. Instrument Society of America, Research Triangle Park, NC.

Clough SA, Kneizys FX, Shettle EP, Anderson GP (1986) Atmospheric Radiance and Transmittance: FASCOD2, Sixth Conference on Atmospheric Radiation, Williamsburg, VA.

Cortina VB (1985) Sampling systems for chilled mirror dewpoint hygrometers. In "Moisture and Humidity 1985, Measurement and Control in Science and Industry," pp. 849–854. Proceedings of the 1985 International Symposium on Moisture and Humidity, Washington, D.C., 1985. Instrument Society of America, Research Triangle Park, NC.

Deirmendjian D (1969) "Electromagnetic Scattering on Spherical Polydispersions." American Elsevier, New York.

Fan J (1987) Determination of the psychrometer coefficient *A* of the WMO reference psychrometer by comparison with a standard gravimetric hygrometer. J Atmos Ocean Tech 4:239–244.

Jensen JB, Baumgartner D, Raga GB (1988). On the

observation of cloud temperature and liquid water content using a Lyman α and an FSSP sensor. Proceedings of the 10th International Cloud Physics Conference, Bad Hamburg, FRG 1:40–42.

Johns JWC (1965) The absorption of radiation by water vapor. In "Humidity and Moisture, Measurement and Control in Science and Industry" (A. Wexler, Ed.), Vol. 1 (Ruskin, Ed.), pp. 417–427. Reinhold, New York.

Katz JJ, Rabinowitch E (1951) "The Chemistry of Uranium The Element, Its Binary and Related Compounds." Dover Press, New York.

King WH (1965) The piezoelectric sorption hygrometer. In "Humidity and Moisture, Measurement and Control in Science and Industry" (A. Wexler, ed.), Vol. 1, Ruskin, ed.), pp. 578–583. Reinhold, New York.

King WH Jr (1969a) "Using Quartz Crystals as Sorption Detectors . . . Part 1: Research/Development," Volume 20, No. 4, pp. 28–34. F.D. Chicago Ill. Thompson Publications.

King, H Jr (1969b) "Using Quartz Crystals as Sorption Detectors . . . Part 2: Research/Development," Volume 20, No. 5, pp. 28–33. F.D. Thompson Publications.

Kneizys FX, Shettle EP, Gallery WO, Chetwynd JH, Abreu LW, Selby JEA, Clough SA, Fenn RW (1983) Atmospheric transmittance/radiance computer code LOWTRAN 6. Air Force Geophysics Laboratory Report No. AFGL-TR-83-0187, Hanscom AFB, Massachusetts 01731.

Kreplin RW, Chubb TA, Friedman H (1962) X-Ray and Lyman-alpha emission from the sun as measured from the NRL-1 satellite. J Geophys Res 67. pp. 2231–2253.

Mattix SC (1971) Development and analysis of a Lyman-alpha absorption hygrometer. M.S. Thesis, Electrical Engineering Dept., University of Colorado.

McCartney EJ (1983) Absorption and Emission by Atmospheric Gases; the Physical Processes." Wiley, New York.

Mestayer P, Rebattet C, Goutail F (1985) Minaturized Lyman-alpha hygrometer to measure humidity turbulent fluxes in tunnel and atmosphere. In "Moisture and Humidity 1985, Measurement and Control in Science and Industry," pp. 437–452. Proceedings of the 1985 International Symposium on Moisture and Humidity, Washington, D.C., 1985. Instrument Society of America, Research Triangle Park, NC.

Moisture and Humidity (1985) "Measurement and Control in Science and Industry." Proceedings of the 1895 International Symposium on Moisture and Humidity, Washington, D.C., 1985, Instrument Society of America, Research Triangle Park, NC.

Raga GB, Jensen JB, Baker MB (1990) Characteristics of cumulus band clouds off the coast of Hawaii. J Atmos Sci 47:338–355.

Randall DL, Hanley TE, Larison OK (1965) The NRL Lyman-alpha humidiometer. In "Humidity and Moisture, Measurement and Control in Science and Industry" (A. Wexler, ed.), Vol. 1 (Ruskin, ed.), pp. 444–454. Reinhold, New York.

Rothman LS, et al. (1987) The HITRAN database: 1986 edition. Appl Optics 26(19), 4058–4097.

Shaw WJ, Tillman E (1985) A correction for different wet-bulb and dry-bulb response in thermocouple psychrometry. In "Moisture and Humidity 1985, Measurement and Control in Science and Industry," pp. 779–790. Proceedings of the 1985 International Symposium on Moisture and Humidity, Washington, D.C., 1985. Instrument Society of America, Research Triangle Park, NC.

Smith HJP, et al (1978) FASCODE–Fast Atmospheric Signature Code (Spectral Transmittance and Radiance). AFGL-TR-78-0081 Visidyne, Inc., 19 Third Avenue, Burlington MA 01803, January 16.

Staats WF, et al (1965) Infrared absorption hygrometer. In "Humidity and Moisture, Measurement and Control in Science and Industry" A. Wexler, ed.), Vol. 1 (Ruskin, ed.), pp. 465–480. Reinhold, New York.

Tillman JE (1961) Atmospheric humidity measurements by ultraviolet techniques. M.S. Thesis, Dept. of Meteorology, Massachusetts Institute of Technology, Cambridge, MA.

Tillman JE (1965) Water vapor density measurements utilizing the absorption of vacuum ultraviolet and infrared radiation. In "Humidity and Moisture, Measurement and control in Science and Industry" (A. Wexler, ed.), Vol. 1 (Ruskin, ed.), pp. 428–443. Reinhold, New York.

Tillman JE (1985) Near infrared humidity techniques using semiconductor sources: Incoherent sources and theoretical calculations in the presence of cloud and fog. In "Moisture and Humidity 1985, Measurement and Control in Science and Industry," pp. 791–795. Proceedings of the 1985 International Symposium on Moisture and Humidity, Washington, D.C., 1985. Instrument Society of America, Research Triangle Park, NC.

Wada M, Iwasaka Y, Murabayshi S, Yamashita Y (1985) Feasible study of monitoring the content of water vapor under a very dry atmospheric condition by means of Lyman α/*OH* hygrometer. In "Moisture and Humidity 1985, Measurement and Control in Science and Industry," pp. 451–461. Proceedings of the 1985 International Symposium on Moisture and Humidity, Washington, D.C., 1985. Instrument Society of America, Research Triangle Park, NC.

Watanabe K, Zelikoff M, Inn EC (1953) Absorption Coefficients of Several Atmospheric Gases. Air Force Cambridge Research Center Technical Report No. 52–23, Geophysical Res. Papers, No. 21, June.

Watanabe K, Saki H, Motti J, Nakayama T (1958) Absorption Cross Section of O_2, NO, and NO_2 with an Improved Photoelectric Method. Air Force Cambridge Research Center, Technical Note 58-658, Scientific Report No. 4, December.

Wexler A. (ed.) (1965) "Humidity and Moisture, Measurement and Control in Science and Industry," Vols. 1-4. Reinhold, New York.

Wolfe WL, Zissis GJ (eds.) (1989) "The Infrared Handbook." Environmental Research Institute of Michigan and SPIE, P.O. Box 10, Bellingham, WA 98227-0010.

Wood RC (1965) The infrared hygrometer—its application to difficult humidity measurement problems. In "Humidity and Moisture, Measurement and Control in Science and Industry" (A Wexler, ed.), Vol. 1 (Ruskin, ed.), pp. 492–504. Reinhold, New York.

Wylie RG, Lalas T (1985) Accurate psychrometer coefficients for wet and ice-covered cylinders in laminar transverse airstreams. In "Moisture and Humidity 1985, Measurement and Control in Science and Industry," pp. 37–56. Proceedings of the 1985 International Symposium on Moisture and Humidity, Washington, D.C., 1985. Instrument Society of America, Research Triangle Park, NC.

20
Remote Sensing Observations for the Monitoring of Land-Surface Fluxes and Water Budgets

Thomas J. Schmugge and F. Becker

1. Introduction

The monitoring at regional spatial scales and climatic temporal scales (from 1 month to 10 years) of the fluxes between the soil and the atmosphere and water budget of the root zone is recognized as important for applications such as the modeling of atmospheric behavior and the monitoring of water resources. This chapter will describe the contribution remote sensing techniques can make toward these applications.

To estimate the land surface fluxes it is necessary to determine

1. the energy driving forces, i.e., the incident solar energy (insolation), surface albedo, and resulting net radiation;
2. the moisture availability or status in the soil and the vegetation/soil interaction; and
3. the capacity of the atmosphere to absorb the flux, which depends on surface air temperature, vapor pressure gradients, and surface winds.

There has been considerable work recently on methods for estimating factors relating to the first two items from remotely sensed data. It is possible to estimate from remotely sensed data surface parameters related to the soil/vegetation system (leaf area indices and surface soil moisture), radiation forcing components (essentially incident solar radiation and albedo), and indicators of the surface response to them (surface temperature). However, there is no remote sensing method for estimating the surface atmospheric parameters. Therefore conventional surface measurements will be required for this factor.

To understand better how remote sensing can contribute, let us consider the basic energy and moisture balance equations. In the absence of advection or precipitation the energy balance is given by

$$R_n + G + H + LE = 0 \qquad (20.1)$$

where R_n is the net radiation, G the soil heat flux, H the sensible heat flux, and LE the latent heat or moisture flux into the atmosphere. The net radiation flux is the sum of the incoming and outgoing short and long wave radiation fluxes:

$$R_n = (1 - \alpha)R_s + \varepsilon(R_L - \sigma T^4) \qquad (20.2)$$

where α is the surface albedo, R_s is the incoming solar radiation, R_L is the incoming longwave radiation, ε the surface emissivity, and T the surface temperature in Kelvins.

There has been considerable progress in estimating R_s and α from geostationary satellite data. There are two approaches. One uses empirical regressions between simultaneous and collocated satellite radiances and ground-based pyranometer results (Tarpley, 1979). In the second, the information contained in the satellite radiances is interpreted in terms of scattering, reflection, and absorption parameters that are subsequently used in radiative transfer model calculations (Gautier et al., 1980; Diak and Gautier, 1983). The basis of both methods is that the major modulator of surface insolation is cloudiness. In recent paper Dedieu et al. (1987) describe the second method applied to Meteosat data. They also estimate

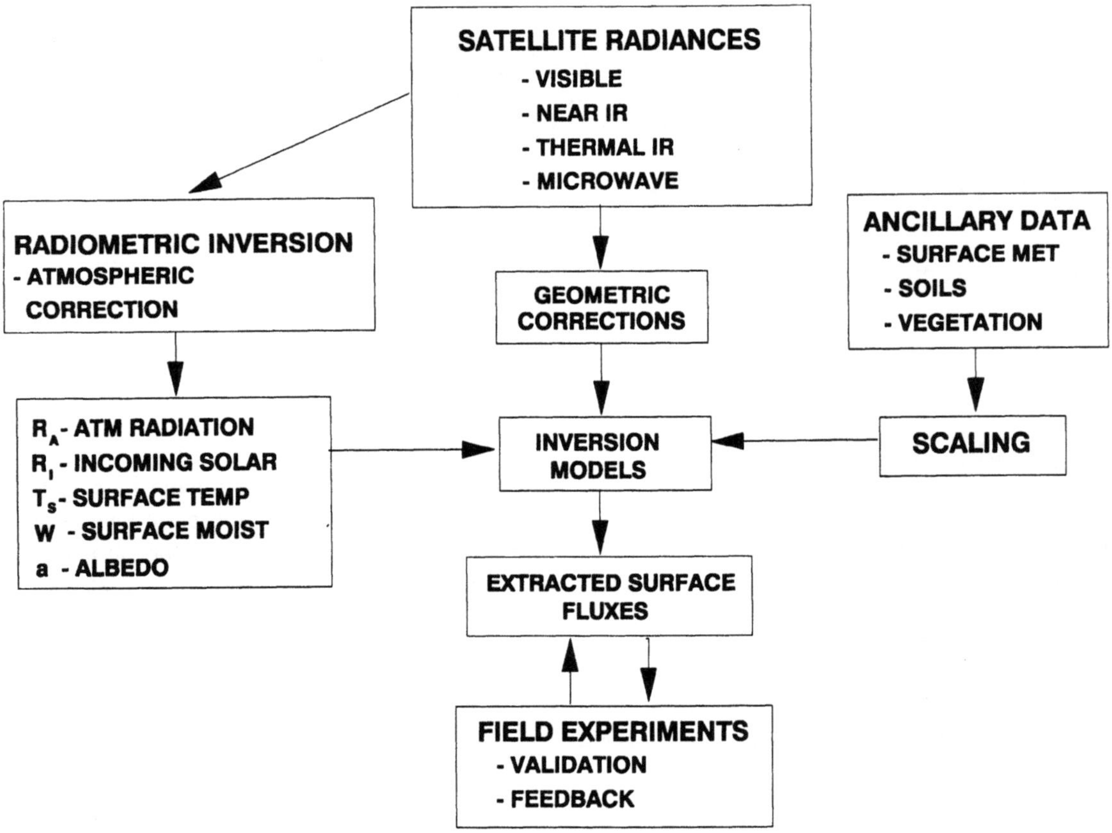

FIGURE 20.1. Schematic diagram of the process necessary to obtain surface fluxes from measured radiances.

albedo α from these data. R_L can be estimated from the atmospheric sounders. As for the land surface temperature T, it can be estimated, at least theoretically, if the emissivity is known from a number of different sensors, such as the AVHRR (Price, 1983, 1984). So it would appear that the components of the radiation flux in Equation (20.2) can be estimated reasonably well using remotely sensed data (Suttles and Ohring, 1986).

The difficulty arises then in determining the surface fluxes in Equation (20.1); both sensible and ground flux will involve temperature gradients, one in the soil and the other in the atmosphere. Thus

$$G = \lambda \, dT_{soil}/dz \qquad (20.3)$$

where λ is thermal conductivity of the soil. Although the soil temperature gradient cannot be determined from remotely sensed data the temperature profile can be modeled with sufficient accuracy to estimate G. The sensible heat flux is

$$H = \delta \, c_p \, (T_a - T) \, / \, r_a \qquad (20.4)$$

where δ is the air density, c_p is the specific heat of air at constant pressure, and r_a is the aerodynamic resistance. The latent heat flux is

$$LE = \delta c_p \, (e_a - e_s) / \, \gamma \, (r_a + r_s) \qquad (20.5)$$

where γ is the psychrometric constant, e_a is the atmospheric vapor pressure in the boundary layer, e_s is the saturation vapor pressure at the temperature T, and r_s is the stomatal resistance to water vapor transport. To get around the absence of these temperature and vapor pressure gradients in the soil and lower atmosphere one-dimensional models have been developed of the heat transfer in the soil and from the land surface into the atmosphere (e.g, Camillo et al., 1983 and the chapters by Shuttleworth and Camillo, this volume). These models use the remotely sensed surface temperature as a boundary condition and model param-

TABLE 20.1. Display of some of the models used to infer LE from satellite data (adapted from Becker et al., 1988).

Model type	Input parameters[a]	Output parameters	Type of model	Method of inversion	References
1	$T_s(t) R_N(t)$ $T_a(t)$ $e_a(t)$ $u(t)$	$LE(t)$ $H(t)$ $I,G(t)$ Resistance	Deterministic 2 layers (soil with or without cover, atmosphere)	Least squares of diurnal cycle Fourier analytic	Raffy and Becker (1985, 1986); Abdellaoui et al. (1986); Ho (1986)
2	T_{sm}	$\overline{LE}$	Statistical linear regression	Use the linear regression	Jackson et al. (1977); Hatfield et al. (1983); Seguin and Itier (1983); Carlson and Buffum (1989)
3	T_{sm} T_a	r_c M	Deterministic 3 layers (soil, vegetation, atmosphere)	Look-up table or multiple regression	Taconet et al. (1986a,b); Carlson (1986)
4	T_{sm}, T_{sn} T_a	I LE	Deterministic 2 layers	Fourier analytic	Price (1980, 1982)
5	$T_a(t)$, $e_a(t)$, $u(t)$ $R_N(t)$ (optional)	$LE(t)$, $H(t)$, $G(t)$ $R_N(t)$, $\theta(z,t)$ $T_{soil}(z,t)$ resistances LE,I	Deterministic multilayer soil, optional canopy layer atmosphere	Surface temperature and soil moisture	Camillo et al. (1983); Gurney and Camillo (1984); van de Griend et al. (1985)
6	$\Delta T_s(9\text{–}12) u_g$ NDVI,A_d	M	Statistical	Multiple regression	Wetzel et al. (1984)
	T_{sm},T_{sn}	LE,H,M	Deterministic		Soer (1980)
	T_{sm},T_{sn}	LE,M,I	Deterministic	Look-up table	Rosema et al. (1978); Rosema (1986)
	T_{sm},T_{sn}	M,I	Deterministic	Look-up table	Carlson et al. (1981, 1984)
	$T_{sm} \alpha_s$	LE	Statistical	Linear regression	Menenti (1984)

[a] T_{sm} = maximum surface temperature
T_{sn} = minimum surface temperature
$\Delta T_s(9\text{–}12)$ = difference of T_s at 12 pm and 9 am
u_g = geostrophic wind
α_s = albedo
A_d = advection
M = moisture availability.

eters are varied to obtain the best fit between the predicted and observed surface temperatures (Taconet et al., 1986(b); Carlson, 1986).

From this discussion it is clear that it is not possible to obtain the surface fluxes directly from remotely sensed radiances and inversion models of some type will be required. The process is represented schematically in Figure 20.1. A listing of the various models using surface temperature and other remotely sensed variables to infer the latent heat flux is given in Table 20.1. All the models incorporate three types of parameters: the measured input parameters (the data), the output parameters, typically the surface fluxes, and the model parameters, e.g., canopy resistance. They differ in these respects:

1. their parameterization, mainly the specification of the resistances and the number of layers in the soil and canopy;
2. the choice and the relative number of input, output, and model parameters, particularly the

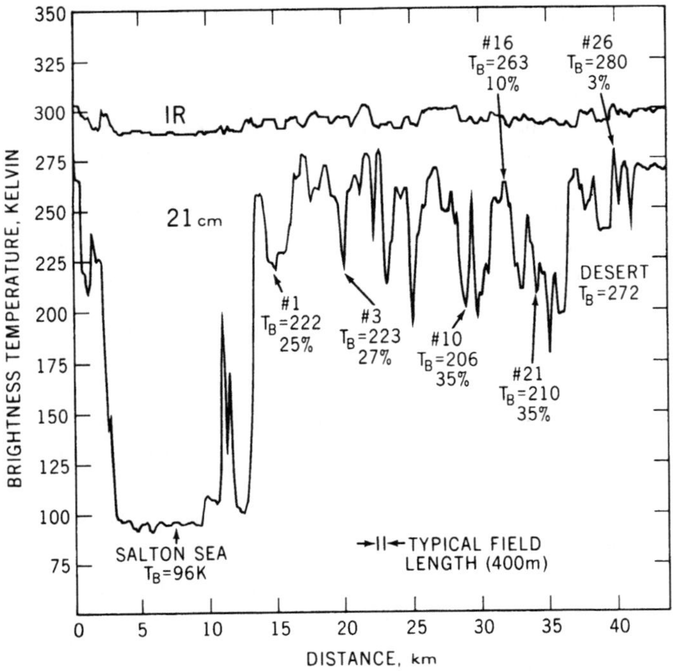

FIGURE 20.2. Thermal infrared (10- to 12-μm) and microwave (21-cm) brightness temperatures versus distance for a 600-m altitude flight path at the north end of the Imperial Valley, California. The 21-cm brightness temperatures and measured soil moisture are indicated for several of the irrigated fields.

number of parameters determined from satellites;

3. the inversion method (look-up table, various minimizations methods, analytic solution); and
4. the treatment of the unknown extra parameters (semiempirical relationships, average values from literature, determination by regression analysis on a set of calibration data).

Several of these methods have been used over the HAPEX test site. For example, Lagouarde (1988) using AVHRR data and an improved version of the Seguin and Itier (1983) model have been able to produce evaporation maps of the entire HAPEX square. To do this, they computed (1) an air temperature, map using an interpolation scheme based on the French Meteorological service mesoscale Peridot model, T_a, (2) a map of net radiation R_n, (3) a surface temperature image from AVHRR, T_s, and (4) a map of the roughness length, z_o, based upon a relation between roughness and the type of land use obtained from a classification scheme. With these data they computed the daily *ET* flux from

$$ET = R_n + a(z_o)\,(T_s - T_a) + b(z_o) \qquad (20.6)$$

Where $a(z_o)$ and $b(z_o)$ are semiempirical formulas. With an improved version of the inversion scheme described by Raffy and Becker (1985, 1986), and using the surface temperature, net radiation, and the wind, temperature, and moisture of the air at the 2 m height, it has been possible to retrieve the fluxes measured in situ with accuracy of the order of the experimental error from the remotely sensed data. Further examples are given in the next two chapters.

To use these models it is necessary to obtain surface temperature and moisture from space with as good an accuracy as possible. Methods to obtain these parameters are briefly described in the next section.

2. Measurement of Surface Temperature and Soil Moisture

Measurement of the thermally emitted radiation at various wavelengths from the earth's surface can yield much useful information about parameters such as surface soil moisture and temperature.

FIGURE 20.3. Calculated black body spectral radiance at three temperatures ad atmospheric transmission for the US Standard mid-latitude summer atmosphere.

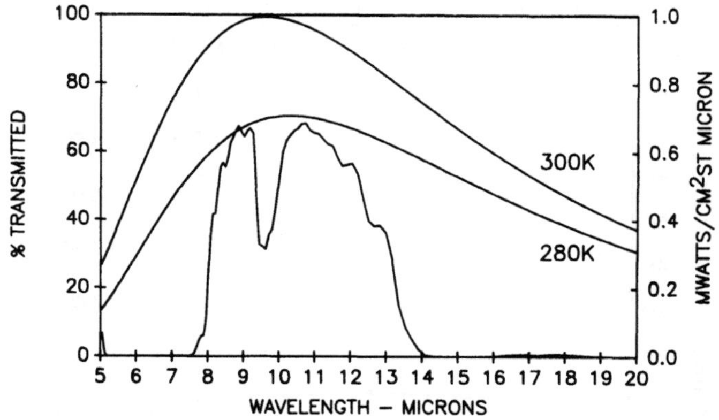

This radiation, which is emitted by any surface with a temperature above absolute zero, is described by the Planck black body equation:

$$E = \varepsilon C_1\, k^3/[\exp(C_2 k/\text{-}T) - 1] \qquad (20.7)$$

where C_1 is 1.19006×10^{-8} W/(m^2 sr cm^{-4}), C_2 is 1.4388 cm K, k is the wave number (cm^{-1}), and ε is the emissivity. For a perfect emitter or black body $\varepsilon = 1$, and for real surfaces $\varepsilon < 1$. To estimate surface temperatures, radiation at wavelengths around 10 μm is used because the peak intensity of Equation (20.7) occurs in this region for terrestrial temperatures ($\approx$300 K) and the atmosphere is relatively transparent in this region. Therefore the variations in the intensity of radiation are mainly related to surface temperature variations. However, it must be kept in mind that variations in emissivity of 1% yield brightness temperature variations of 0.3 K to 0.6 K depending on sky radiance. To estimate surface soil moisture, radiation at much longer wavelengths (10s of centimeters) is used. Changes in the intensity of the emitted radiation at these wavelengths are primarily due to the variation of the emissivity with the moisture content of the soil. This effect results from the large dielectric contrast between water and dry soils. This difference is demonstrated in Figure 20.2 where infrared (10–12 μm) and microwave (21 cm) brightness temperatures are presented. These data are from a NASA aircraft flight over the Imperial valley in southern California at an altitude of 600 m (Schmugge et al., 1976). They show about a 10 K variation in the infrared due to surface temperature and about an 80 K variation in the microwave due to soil moisture variations. The percentages indicate the ground measurements of soil moisture in the top 5 cm of the soil. The wide range of moisture values results from irrigation.

3. Thermal Infrared

In Figure 20.3 we plotted Equation (20.7) from 5 to 20 μm for temperatures 280 and 300 K, i.e., the low range of terrestrial temperatures. At these temperatures the peak of the emission occurs in the 8–10 μm range of wavelength. In this figure we have also plotted the atmospheric transmission calculated with the LOWTRAN-6 path radiance model (Kniezys et al., 1983) for the US standard mid-latitude summer atmosphere assuming the radiometer is at satellite altitude. The atmosphere is relatively transparent in the 8–12 μm range also, but there is still a significant effect, i.e., there is only 60–70% transmission with a major dip at about 9.5 μm due to ozone absorption. With the exception of this dip, water vapor is the dominant absorber in the 8–12 μm window and is due to what is called the "water vapor continuum" and not any individual absorption lines. Thus the magnitude of the atmospheric effect will depend on the water vapor content of the intervening atmosphere. And this unknown or uncertain atmospheric contribution is one of the problems for the remote sensing of surface temperature.

The radiation values given in Figure 20.3 are those that would be observed right at the surface; the relation between the radiation, R_i, seen by a

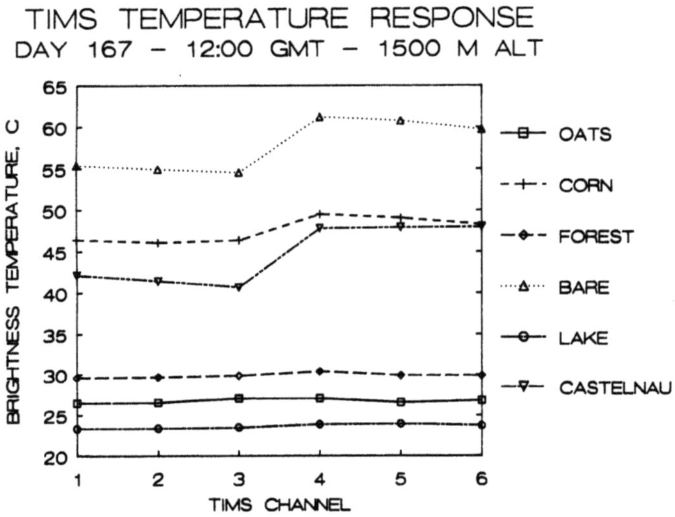

FIGURE 20.4. Spectral variation of the TIMS brightness temperature for several surfaces. The field at Castelnau was also planted in corn. Both corn fields had only partial cover. The data were obtained during HAPEX from an aircraft altitude of 1.5 km.

radiometer on a satellite or aircraft and the surface temperature is

$$R_i = \tau\,[\varepsilon_i B_i(T) + (1 - \varepsilon_i)R_{\mathrm{atm}}(dn)]_{\,i} + R_{\mathrm{atm}}\,(\mathrm{up}) \qquad (20.8)$$

where i indicates the integral of these quantities over the band width for channel i of the radiometer, B_i is the Planck function given by Equation (20.7), the R_{atm}s are the upward and downward components of the atmospheric radiation, and τ is the atmospheric transmission. The values of R_{atm} and τ can be calculated using a model for atmospheric radiance such as LOWTRAN-6, however, it would be desirable to eliminate the atmospheric effects using the multispectral information.

Several approaches have been developed for eliminating atmospheric effects in the estimation of sea surface temperature from space using multichannel thermal data. [Here the problem is simpler in that the temperature does not change rapidly with time and a week's worth of data can be used to estimate the surface temperature.] The technique used with the AVHRR data from the NOAA series of satellites involves the differential water vapor absorption in the 10–13 μm window, the so-called split window technique (McClain et al., 1983; Grassl, 1989). However, this assumes that the surface emissivity is constant over this spectral band, which is not the case for land surfaces (Price, 1984; Becker 1987; Grassl, 1989). This is demonstrated in Figure 20.4 where examples of multispectral observations in the 8–12 μm region from an aircraft platform are given. These data were obtained with NASA's Thermal Infrared Multispectral Scanner (TIMS) during the HAPEX-MOBILHY experiment in 1986 (Schmugge and Janssen, 1988; André et al., 1988). They show the spectral variation in brightness temperature (T_B) for a range of surfaces. For fully vegetated surfaces, the forest or oat field there is no variation in T_B, while for the bare soil there is a large difference. For a partially vegetated corn field there is a smaller variation. Thus if the surface is vegetated it may be possible to use differential absorption technique to eliminate the atmospheric contribution because the emissivity is close to 1 with little spectral variation.

4. Microwave Sensing of Soil Moisture

The microwave remote sensing of soil moisture relies on the large contrast between the dielectric constant of water (≈ 80) and that of dry soil (3.5). This arises from the ability of the electric dipole of the water molecule to align itself in response to the electric field at microwave frequencies. For ice the dielectric constant is large at low frequencies but because of the binding of the water molecule in the solid the motion is inhibited at about 10^4 Hz. For liquid water this reduction in the molecule's ability to rotate does not occur until about 10^{10} Hz, i.e., in the microwave range. The dielectric constant is of

FIGURE 20.5. Laboratory measurements of the dielectric constant for three soils ranging from a light sand to a heavy clay at a frequency of 1.4 GHz.

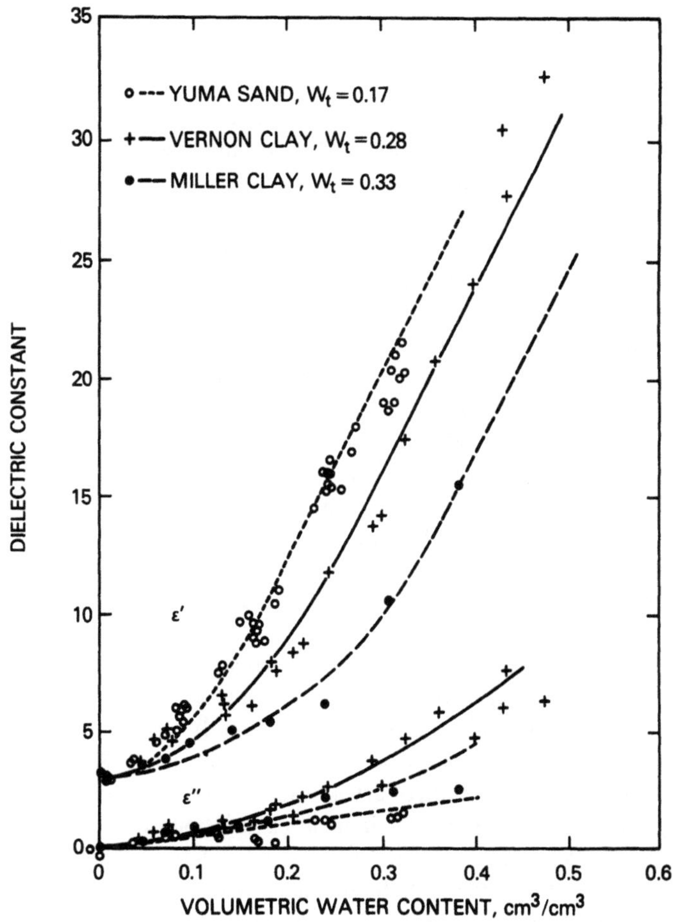

importance here because it describes the propagation characteristics of an electromagnetic wave in the medium. These characteristics include the velocity of propagation, the wavelength in the medium, and the absorption of energy in the medium. The square root of the dielectric constant is the index of refraction (n) for the material and it is the contrast in n at the boundary between two media that determines the reflection and transmission coefficients of the electromagnetic wave at such a boundary.

For the purposes of this discussion we will consider the frequency range from 1 to 100 GHz (1 GHz = 10^9 Hz; wavelengths between 30 and 0.3 cm) and for soil moisture sensing frequencies below 5 or 6 GHz are most effective. In this section we will describe how this variation of the soil's dielectric constant with moisture can produce a variation in the soil's emissivity from 0.95 for dry soils to 0.6 or less for wet soils with changes of a corresponding magnitude for soil's reflectivity. These variations have been observed by both passive and active microwave sensors. The former are radiometers that observe the variations in the thermal emission from the soil due to emissivity changes (see Schmugge et al., 1986; Jackson and Schmugge, 1986). The latter are radars that transmit a pulse of electromagnetic energy and then measure the backscattered return, which will be a function of the soil's reflectivity (see Dobson and Ulaby, 1986; Bernard et al., 1982). This capability to remotely sense soil moisture is limited to a surface layer about 5 cm thick and is affected by surface properties such as roughness and vegetation cover.

4.1 Dielectric Properties of Soils

When water is mixed with soil the dielectric constant of the mixture increases from about 3 or 4 for the dry soil to almost 30 for wet soils. Figure 20.5 presents measurements for three soils ranging from a sandy loam to a heavy clay. These measurements were at a frequency of 1.42 GHz or a wavelength of 21 cm. Qualitatively the behavior for the three soils is similar in that the real part of the dielectric varies from about 3 to greater than 20. However, it is clear that there are differences among the soils, namely with the rate of increase at the lower soil moistures. The heavier soils have a region of slow increase and then after a transition value there is a region of sharp increase. This flat region is a function of soil type being wider for the heavier soils (Wang and Schmugge, 1980). The simple explanation is that the first water added to a soil is tightly bound to the soil surface and has dielectric properties somewhere between those of bound molecules in ice and those of the freely rotating molecules in the liquid. It is only after there are several layers of water on the particle surface that the water begins to behave like a liquid in terms of its dielectric properties.

Dobson et al. (1985) found that the separation of the soil water into bound and bulk components depends on the soil texture or more directly the specific surface area of the soil. Their model treats the soil as a host medium of dry soil solids containing randomly distributed inclusions of bound water, bulk water, and air. With models such as these it is possible to study the sensitivity of a soil's dielectric properties on density, texture, salinity, etc. (Jackson and O'Neill, 1987).

4.2 Microwave Radiometry

A microwave radiometer measures the thermal emission from the surface and at these wavelengths the intensity of the observed radiation is proportional to the product of the thermodynamic temperature of the soil and the surface emissivity (Rayleigh–Jeans approximation to the Planck radiation law). This product is commonly called the brightness temperature (T_B) and is given by the linearized form or equation (20.7):

$$T_B = \tau\,[RT_{sky} + (1 - R)T_{soil}] + T_{atm} \qquad (20.9)$$

where R is the surface reflectivity and τ is the atmospheric transmission. The first term is the reflected sky brightness, which depends on the atmospheric conditions and frequency. For the frequencies of interest to us T_{sky} and T_{atm} are small (< 10 K) and will be neglected. The atmospheric transmission will typically be about 99% so are left with the emission from the soil, i.e., the second term as the main contributor to T_B. Thus Equation (20.9) reduces to

$$T_B = (1 - R)T_{soil} = \varepsilon T_{soil} \qquad (20.10)$$

where T_{soil} is a measure of the intensity of the upwelling thermal radiation from the soil and is a weighted average of the soil temperature over the electromagnetic skin depth of the soil. The factor ε, the emissivity, gives the fraction of this upwelling radiation, which is transmitted into the air, and it is this factor that shows the primary sensitivity to the soil moisture content. The critical factor is, what is the thickness of the layer at the surface whose dielectric properties determine ε. For it is this thickness that determines the soil moisture sampling depth for microwave sensors and has been determined to be on the order of a few tenths of a wavelength thick or about 2–5 cm at the 21-cm wavelength.

4.3 Experimental Results

As part of a study of salinity effects on the soil's emissivity Jackson and O'Neill (1987) made a careful series of field measurements over loamy sand having a smooth surface and compared the results with those expected from the dielectric constant models described above. The results are shown in Figure 20.6. The emissivity ranges from 0.6 for the wet soil ($\approx$30% volumetric soil moisture) to 0.9 for the dry soil ($\approx$8%). The calculations from the two models agree reasonably well with each other and with the data, which indicate that the basic sensitivity of microwave emissivity to soil moisture variations is well understood and the basic theory is verified. The complications arise when real factors such as surface roughness and vegetative cover are added to the problem. Vegetation is the most significant because an amount > 2 kg/m^2 can completely obscure the soil surface from observation. For a discussion of these effects the reader is referred to the reviews cited earlier.

5. Conclusion

We have presented an overview of the application of remote sensing techniques to the determination of surface parameters required for the estimation

FIGURE 20.6. Observed and predicted relationships between emissivity and volumetric soil moisture for a bare, smooth loamy sand at 1.4 GHz (L-band). (From Jackson and O'Neill, 1987.)

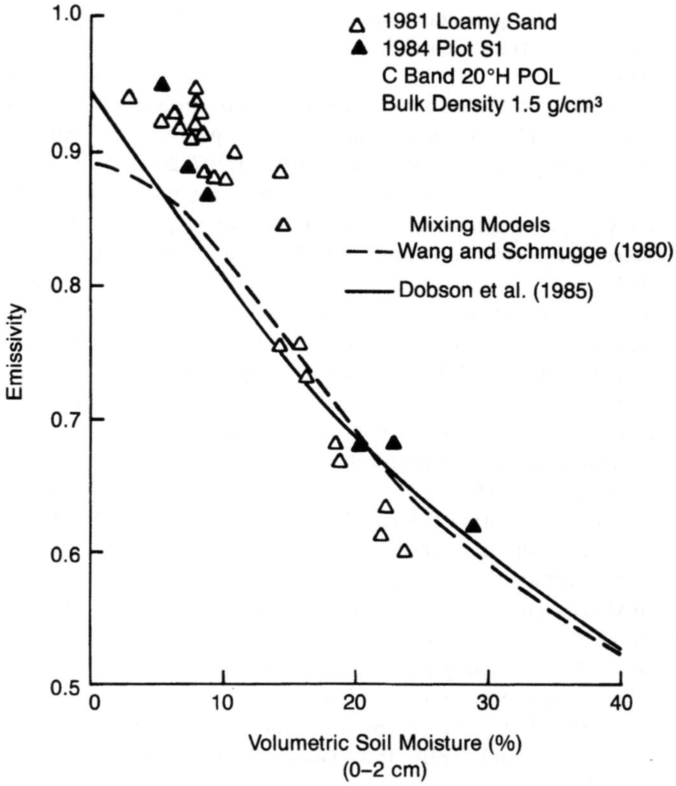

of the land surface fluxes. These include components of radiative fluxes, surface temperature, and surface soil moisture, with a concentration on the latter two quantities.

The need now is for the development of approaches for integrating these remotely sensed parameters with suitable algorithms to make estimates of the land surface fluxes. Examples of possible approaches are given in the papers by Prevot et al. (1984), Camillo et al. (1986), Bernard et al. (1986), and Soares et al. (1987). In the latter, remotely sensed measurements of surface temperature and soil moisture were used to derive regional estimates of the ET flux.

References

Abdellaoui A, Becker F, Olory-Hechinger E (1986) Use of METEOSAT for mapping thermal inertia and evapotranspiration over a limited region of Mali. J Climate Appl Meteorol 25:1489–1506.

André JC, Goutorbe JP, Perrier A Becker F, Bessemoulin P, Bougeault P, Brunet Y, Brutsaert W, Carlson T, Cuenca R, Gash J, Gelpe J, Hildebrand P, Lagourade JP, Lloyd C, Mahrt L, Mascart P, Mazaudier C, Noilhan J, Ottlé C, Payen M, Phulpin T, Stull R, Shuttleworth J, Schmugge T, Taconet O, Tarrieu C, Thepenier RM, Valencogne C, Vidal-Madjar D, Weill A, (1988) Evaporation over land surfaces: First results from HAPEX-MOBILHY special observing period. Ann Geophys 6:477–492.

Becker F (1987) The impact of spectral emissivity on the measurement of land surface temperature from a satellite. In J Remote Sens 8:1509–1522.

Becker F, Camillo PJ, Choudhury B (1988) Surface heat fluxes in Satellite Data Algorithms for Studies of the Land Surface. Proc of ISLSCP Workshop, Jet Propulsion Laboratory Pasadena Ca. January, 1987, pp. 4–14.

Bernard R, Martin P, Thony JL, Vauclin M, Vidal-Madjar D (1982) C-Band radar for determining surface soil moisture. Rem Sens Environ 13:329–344.

Bernard R, Soares JV, Vidal-Madjar D (1986) Differential bare field drainage properties from airborne microwave observations. Water Res Res 22:869–875.

Camillo PJ, Gurney RJ, Schmugge TJ (1983) A soil and atmospheric boundary layer model for evapotranspiration and soil moisture studies. Water Res Res 19:371–380.

Camillo PJ, O'Neill PE, Gurney, RJ (1986) Estimating

soil hydraulic parameters using passive microwave data. IEEE Trans Geosci Remote Sens GE-24:930–936.

Carlson TN (1986) Regional scale estimates of surface moisture availability and thermal inertia using remote thermal measurements. Remote Sens Rev 1:197–247.

Carlson TN, Buffum MJ (1989) On estimating total daily evapotranspiration from remote surface temperature measurements. Remote Sens Environ 29:197–207.

Carlson TN, Dodd JK, Benjamin SG, Cooper JN (1981) Remote estimation of surface energy balance, moisture availability and thermal inertia. J Appl Meteorol 20:67–87.

Carlson TN, Rose FG, Perry EM (1984) Regional scale estimates of moisture availability from GOES satellites. Agron J 76:972–979.

Dedieu G, Deschamps PY, Kerr YH (1987) Satellite estimation of solar irradiance at the surface of the earth and of surface albedo using a physical model applied to Meteosat data. J Climate Appl Meteorol 26:79–87.

Diak G, Gautier C (1983) Improvements to a simple model for estimating insolation from GOES data. J Climate Appl Meteorol 22:505–508.

Dobson MC, Ulaby FT (1986) Active microwave soil moisture research. IEEE Trans Geosci Remote Sens GE-24:23–36.

Dobson MC, Ulaby FT, Hallikainen MT, Reyes M (1985) Microwave dielectric behavior of wet soil-Part II: Dielectric mixing models. IEEE Trans Geosci Remote Sens GE-23:35–46.

Gautier C, Diak G, Masse S (1980) A simple physical model to estimate incident solar radiation at the surface from GOES satellite data. J Appl Meteorol 19: 1005–1012.

Grassl H (1989) Extraction of surface temperature from satellite data. In "Applications of Remote Sensing to Agrometeorology" (F. Toselli, ed.), pp. 100–200, Kluwer Academic Publishers, Dordrecht.

Gurney RJ, Camillo PJ (1984) Modeling daily evapotranspiration using remotely sensed data. J Hydrol 69: 305–324.

Hatfield JL, Perrier A, Jackson RD (1983) Estimation of evaporation at one time-of-day using remotely sensed surface temperature. Agric Water Manage 7:341–350.

Ho D (1986) Thermal inertia and soil fluxes by remote sensing. Proc. IGARSS 86 ESA-SP254:1215–1220.

Jackson RD, Reginato RJ, Idso SB (1977) Wheat canopy temperature: A practical tool for evaluation water requirements. Water Res Res 13:651–656.

Jackson TJ, O'Neill PE (1987) Salinity effects on the microwave emission of soil. IEEE Trans Geosci Remote Sens GE-25:214–220.

Jackson TJ, Schmugge TJ (1986) Passive microwave remote sensing of soil moisture. Adv Hydrosci 14: 123–159.

Kniezys FX, Shettle EP, Gallery WO, Chetwynd JH, Abreu LW, Selby JEA, Clough SA, Fenn RW (1983) Atmospheric Transmittance/Radiance Computer Code LOWTRAN 6, Air Force Geophysics Laboratory Report No. AFGL-TR-83-0187, Hanscom AFB, MA 01731.

Lagouarde, J-P (1988) Suivi des bilans energetique et hydrique a partir mesures spatiales dans l'infrarouge thermique. Proc of workshop Remote Sensing from Space: Physical Aspects & Modeling, pp. 949–965. Toulouse France August 1988.

McClain EP, Pichel WG, Walton CC, Ahamd Z, Sutton J (1983) Multichannel improvements to satellite derived global sea surface temperatures. Adv Space Res 2:43–47.

Menenti M (1984) Physical aspects and determination of evaporation in deserts applying remote sensing techniques. Thesis, Institute voor Cultuurtechnieck en Waterhuishouding, Wageningen, The Netherlands.

Prevot L, Bernard R, Taconet O, Vidal-Madjar D, Thony JL (1984) Evaporation from a bare soil evaluated using a soil water transfer model and remotely sensed surface soil moisture data. Water Res Res 20:311–316.

Price JC (1980) The potential of remotely sensed thermal infrared data to infer surface soil moisture and evaporation. Water Res Res 16:787–795.

Price JC (1982) On the use of satellite data to infer surface fluxes at meteorological scales. J Appl Meteorol 21:1111–1122.

Price JC (1983) Estimation of surface temperature from satellite thermal infrared data. Remote Sens Environ 13:353–361.

Price JC (1984) Land surface temperature measurements from the split window channels of the NOAA-7 advanced very high resolution radiometer. J Geophys Res 89:7231–7237.

Raffy M, Becker F (1985) An inverse problem for remote sensing in the thermal infrared bands and its solutions. J Geophys Res 90:5809–5819.

Raffy M, Becker F (1986) A stable iterative procedure to obtain soil surface parameters and fluxes from satellite data. IEEE Trans Geosci Remote Sens GE-24: 327–333.

Rosema A (1986) GAMP methodology for integrated mapping of rainfall, evapotranspiration, germination biomass development and thermal inertia based on conventional meteorological data. Proceedings of the ISLSCP Conference, Rome, Dec. 1985, ESA-SP248:549–557.

Rosema A, Bijleveld JH, Reiniger P, Tassone G, Blyth K, Gurney RJ (1978) TELL-US, a combined surface temperature, soil moisture and evaporation mapping approach. Proceedings of the 12th International Symposium Remote Sensing Environment, ERIM.

Schmugge T, Janssen L (1988) Aircraft remote sensing in HAPEX. Proceedings of the 4th International Colloquium of Spectral Signatures of Object in Remote Sensing, Aussois, France 18–22 January 1988, ESA SP-287:463–467.

Schmugge T, Wilheit T, Webster W, Gloersen P (1976) Remote sensing of soil moisture with microwave radiometers-II. NASA Tech. Note TN D-8321.

Schmugge TJ, O'Neill PE, Wang JR (1986) Passive microwave soil moisture research. IEEE Trans Geosci Remote Sens GE-24:12–22.

Seguin B, Itier B (1983) Using midday surface temperature to estimate daily evaporation from satellite thermal IR data. Int J Remote Sens 4:371–383.

Soares JV, Bernard R, and Vidal-Madjar D (1987) Spatial and temporal behaviour of a large agricultural region as observed from airborne C-band scatterometer and thermal infrared radiometer. Int J Remote Sens 8:981–996.

Soer GJR (1980) Estimation of regional evapotranspiration and soil moisture conditions using remotely sensed crop surface temperatures. Remote Sens Environ 9:27–45.

Suttles JT, Ohring G, Editors (1986) Surface radiation budget for climate applications. NASA Ref. Pub. 1169.

Taconet O, Bernard R, Vidal-Madjar D (1986a) Evapotranspiration over an agricultural region using a surface flux/temperature model based on NOAA AVHRR data. J Climate Appl Meteorol 25:284–307.

Taconet O, Carlson T, Bernard R, Vidal-Madjar D (1986b) Evaluation of a surface vegetation parameterization using satellite measurements of surface temperature. J Climate Appl Meteorol 25:1752–1767.

Tarpley JD (1979) Estimating incident solar radiation at the surface from geostationary satellite data. J Appl Meteorol 18:1172–1181.

van de Griend AA, Camillo PJ, Gurney RJ (1985) Discrimination of soil physical parameters, thermal inertia and soil moisture from diurnal surface temperature fluctuation. Water Res Res 21:997–1009.

Wang JR, Schmugge TJ (1980) An empirical model for the complex dielectric permittivity of soils as a function of water content. IEEE Trans Geosci Remote sens GE-18:288–295.

Wetzel P, Atlas D, Woodward R (1984) Determining soil moisture form geosynchronour satellite infrared data: A feasibility study. J Climate Appl Meteorol 23:375–391.

21 Recent Advances in Modeling the Infrared Temperature of Vegetation Canopies

Toby N. Carlson

1. Introduction

Remote sensing of substrate moisture and the surface energy balance using surface infrared temperature measurements is much more difficult and complex over vegetation than over bare soil. Unlike those of bare soil, the radiometric temperature of a vegetation canopy is usually close to the air temperature just above the canopy. Leaf temperature does not rise very far above air temperature because the plant makes use of available water from a relatively deep substrate layer (the root zone). Leaf temperature, however, depends not only on water uptake from the root zone but on plant constraints and atmospheric demand. Moreover, radiometric canopy temperature depends on the radiance reaching the detector from bare soil around and beneath the plants.

This chapter discusses three aspects of modeling plant radiometric temperatures: (1) the case of sparse vegetation cover, (2) plant constraints on transpiration, and (3) the effect of plant capacitance.

2. Sparse Vegetation Cover

Much horizontal variation in temperature over vegetation canopies can be attributed to variations in vegetation density and the amount of bare soil visible to the radiometer and/or exposed to the direct rays of the sun. Thus the surface radiometric temperatures over sparse vegetation can differ greatly from the pure sunlit leaf temperature. The image in Figure 21.1 exhibits a large range of radiometric surface temperature (roughly 30°C in the black areas and 50–55°C in the white areas).

These infrared surface temperatures can be devolved to determine bare soil and vegetation temperatures by considering an additional parameter, the *normalized difference vegetation index (NDVI)*. The NDVI is defined as

$$\mathrm{NDVI} = (R_{\mathrm{nir}} - R_{\mathrm{vis}})/(R_{\mathrm{nir}} + R_{\mathrm{vis}}) \quad (21.1)$$

where R_{nir} is the reflected radiance in the near infrared and R_{vis} is the reflected radiance in the visible. Measurements of surface radiance were made in HAPEX by the NASA NS001 multispectral radiometer aboard the NASA C-130 aircraft during HAPEX. Wavelength intervals were 0.633–0.697 µm for the visible radiance (channel 4) and 0.767–0.910 µm for the near infrared radiance (channel 5).

NDVI is known to be highly correlated with vegetation amount, which is sometimes expressed in terms of the leaf area index (LAI) (Tucker, 1979; Holben et al., 1980; Curran, 1983; Asrar et al., 1984; Best and Harlan, 1985; Hansen and Soegaard, 1987; Peterson et al., 1987). Almost all the measurements show a rapid increase in LAI with increasing NDVI until the former reaches a value of about 2.0–3.0, beyond which the latter increases very gradually with increasing LAI. Asrar et al. (1984) suggest that fractional vegetation cover approaches 1.0 as LAI approaches a value of about 2.0–3.0, at least for crops.

NDVI measurements were made during the HAPEX experiment aboard the NASA C-130 aircraft using the Thematic Mapper Simulator (NS001) and surface temperature using the Thermal

FIGURE 21.1. TIMS thermal infrared image made from a NASA C-130 aircraft flying at 1500 m on 16 June 1986 near Lubbon, France. Dark shading is cool and light shading is warm. Field numbers are indicated and field averages of temperature (°C) are listed (pixel size about 5 m).

Infrared Multispectral Scanner (TIMS). All aircraft measurements referred to in this chapter pertain to the 1500 m elevation over the region near Lubbon, France. Temperatures were corrected for atmospheric attenuation using a model developed by Price (1983). Surface pixels were resampled from the NS001 and TIMS and colocated as new pixels, which were about 5 m in diameter. No corrections were made to NDVI for solar or viewing angles. Further details pertaining to the aircraft missions and instrumentation can be found in Schmugge and Janssen (1988) and Carlson et al. (1990).

Consider the distribution of radiometric canopy temperature (T_c) as a function of the NDVI for the June 16 image shown in Figure 21.1 (Fig. 21.2). This type of figure was used by Nemani and Running (1989) to study canopy resistance of forests. Although various types of crop vegetation are combined, the figure shows a remarkably systematic relationship between T_c and NDVI. Gravimetric measurements made by NASA and Penn State show that the soil was mostly dry near the surface (less than 0.05 by volume) and wet in the root zone (greater than 0.18 by volume). We conclude that the spatial variation in surface temperatures shown in Figure 21.2 was due almost entirely to the fraction of dry bare soil illuminated by the sun and visible to the radiometer. The exception is the cluster of pixels at constant NDVI that represents a spatial variation in soil moisture over one field of bare soil.

The distribution of T_c can be viewed with the aid of another figure, the so-called "arch diagram," originally devised by Coakley and Bretherton (1982) to study cloud cover over oceans. The arch diagram shows the standard deviation of radiometric surface temperature for clusters of pixels versus the mean temperature for the clusters (Fig. 21.3). The two "feet" of the arch correspond to temperatures with a small scatter and therefore to relative uniform surfaces, which are sunlit bare soil and vegetation.

Both the arch diagram and the NDVI versus T_c distribution enable one to estimate two temperature extrema: that for sunlit, bare soil and that for sunlit leaves. We invert the boundary layer model for the bare soil and dense vegetation cases to determine the surface and root zone water contents that force agreement between measured and simulated radiometric surface temperatures. These two substrate water contents are used in the boundary layer model to calculate a range of pixel temperatures for a range of vegetation fractions. The only artificial constraint is that LAI has a fixed value (e.g., 3.0) for the part of the pixel covered by vegetation. Higher values of LAI are permitted where the vegetation fraction is 100%. The bare soil component is assumed to have an LAI of zero. Given a simulated distribution of vegetation fraction and LAI versus radiometric surface temperature and an observed distribution of the latter versus NDVI, we combine the two products to obtain a distribution of NDVI versus vegetation fraction. From this relationship we calculate the surface energy fluxes for each pixel. Simply stated, we convert a range of surface temperatures to a range of vegetation frac-

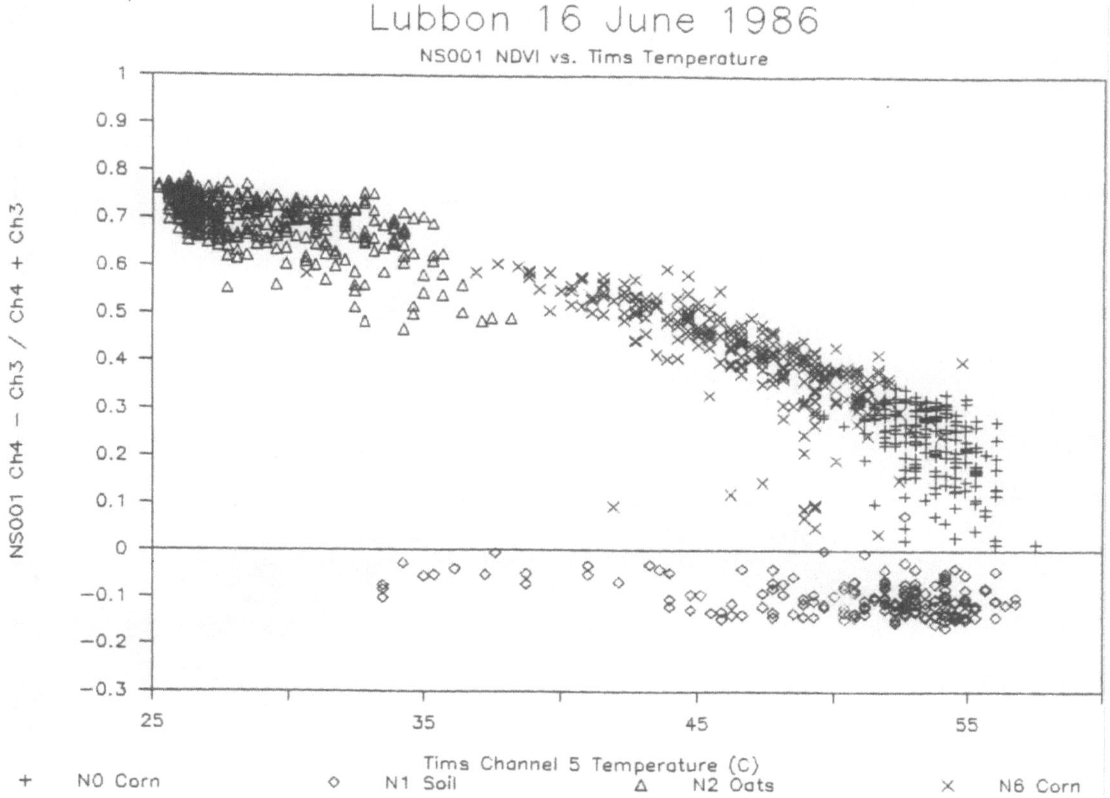

FIGURE 21.2. Distribution of NDVI versus radiometric surface temperature (°C; corrected for atmospheric attenuation) for individual pixels over the scene in Figure 21.1. The type of vegetation planted in each field is indicated by a symbol defined below the figure.

tion using the NDVI measurements and a simulated distribution of T_c versus vegetation fraction and LAI. Price (1989) also developed an analytic method for using NDVI and radiometric canopy temperature to infer vegetation fraction.*

We were able to make some limited verifications of the method for 2 days during HAPEX, 16 June and 27 June, 1986. Comparisons between the measured and modeled LAI versus NDVI and modeled vegetation fraction are shown in Figure 21.4. Agreement for June 27 is somewhat less exact than for June 16 because the distribution of T_c versus NDVI was not as sharply defined as on June 16. This case would have shown close agreement with ground measurements had we set the LAI for the vegetated part of the pixel at 2.5, instead of 3.0. Differences between these two curves may simply be due to the scatter in measurements.

*It has come to our attention since completing the analysis that the measured infrared surface temperatures were somewhat too warm due to a problem in the internal blackbody reference. This error is approximately 1°C for every 5°C in the measured infrared temperature above 25°C.

3. The Effects of Transient Water Stress

An important parameter governing leaf temperature (and therefore T_c) is the stomatal resistance. Taconet et al. (1986) discuss a simple mathematical parameterization of a stomatal resistance that is based on a model of Deardorff (1978). The latter relates soil moisture and solar flux directly to stomatal resistance. The threshold effect involves a sudden and rapid increase of stomatal resistance

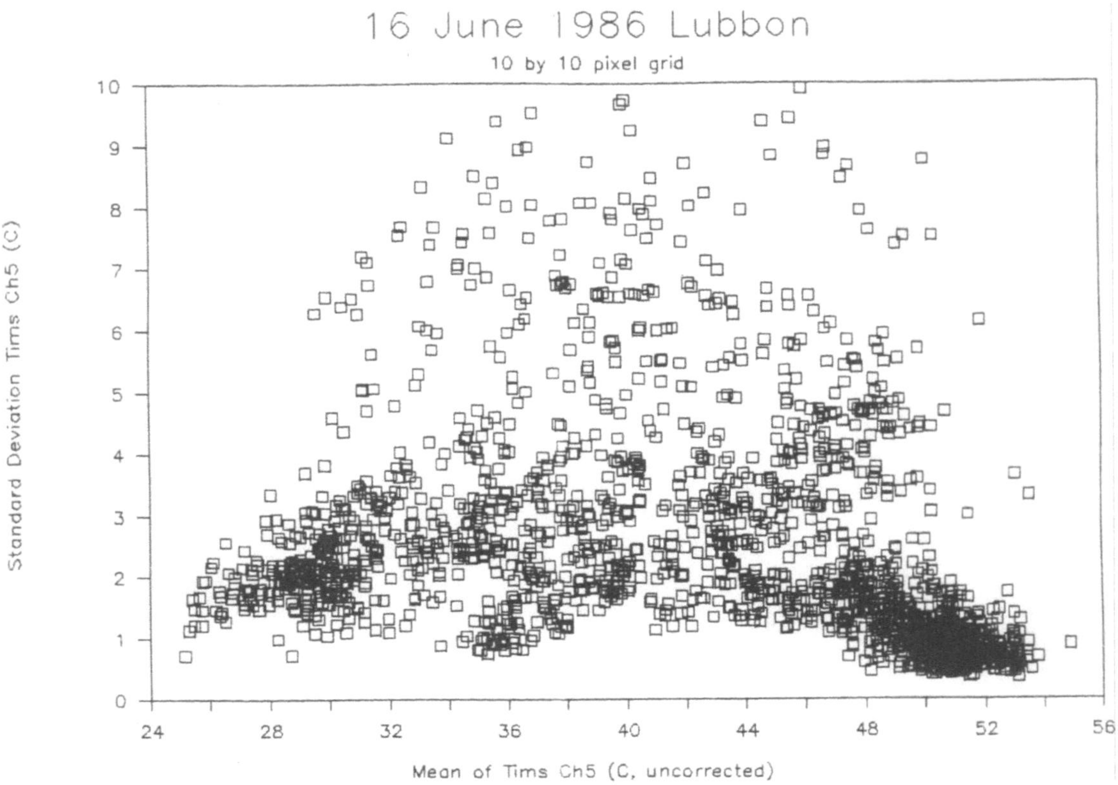

FIGURE 21.3. The arch diagram: standard deviation of radiometric surface temperature (uncorrected for atmospheric attenuation) versus uncorrected radiometric surface temperature (°C) for the scene in Figure 21.1. Each data point represents a subarea consisting of about 10 × 10 resampled pixels.

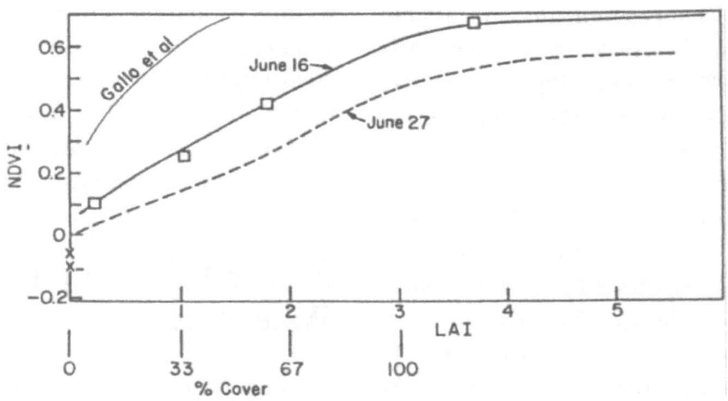

FIGURE 21.4. NDVI versus percent cover and LAI for model simulations for 16 June and 27 June 1986 at Lubbon, France. Squares represent simultaneous aircraft measurements of NDVI and ground measurements of leaf area index (LAI) made on 4 days over the Lubbon field site (field N6 in Fig. 21.1) during June 1986. Crosses represent measured NDVI and LAI for almost bare soil. Heavy solid line and dashed line represent simulated LAI and fraction vegetation cover for 16 and 27 June. The thin solid line obeys a formula proposed by Gallo et al. (1985) for NDVI versus LAI over corn.

with decreasing soil water content. In response to the decrease in soil water content, transpiration decreases and there is a rapid increase of the leaf temperature with decreasing soil moisture.

The Deardorff formulation, however, is limited in two respects: it does not account for leaf water status or vapor pressure deficit. It is thus incapable of describing the effect of transient water stress. We account for the threshold in a more mechanistic manner (Lynn and Carlson, 1990), which includes atmospheric demand and plant physiology. Soil moisture is still an important variable but its influence on stomatal resistance is indirect.

One key element in our plant model is the assumption that stomatal resistance is governed by leaf water potential. Turner (1970, 1974) and many others make a very strong case that the leaf water potential (rather than the soil water potential) governs the stomatal resistance threshold. This subject is still contentious, but most plant scientists agree that there is strong evidence of an empirical relationship between leaf water potential and stomatal resistance.

Our observations support the idea of a leaf water potential control. Accordingly, we postulated that stomatal resistance depends on leaf water potential and that there exists a critical leaf water potential (ψ_c) (Cowan, 1965). Until the critical leaf potential is reached (subthreshold), changes in stomatal resistance with changing leaf potential are small. Beyond the critical point, a further decrease in leaf water potential produces a rapid increase in the stomatal resistance. A typical value of critical potential for raw crops is about -15 bar.

Figure 21.5a and b illustrates the sensitivity of transpiration radiometric canopy temperature transpiration to critical leaf water potential. Reducing the magnitude of the critical leaf water potential (ψ_c) flattens the transpiration curve during the middle part of the day. For convenience, we refer to such a flattening of the transpiration curve as the "hat." The hat occurs as the result of the leaf epidermal water potential decreasing to the threshold value.

The threshold can be reached because of a combination of soil, plant, and atmospheric influences, e.g., the critical leaf water potential (ψ_c), the value of the vapor pressure deficit, the root–stem resistance, the ambient relative humidity, and the incident solar flux. Soil water content determines the hat indirectly via the soil water potential ψ_g and the soil resistance, Z_g, both of which depend on soil type, which we specify in the model.

This stomatal resistance threshold (and therefore the hat) occurs because atmospheric demand exceeds the ability of the soil to supply water to the plant in the face of physiological constraints. With increasing soil water deficit the hat becomes larger (it lasts longer), resulting in a lower transpiration during the middle of the day. Lower relative humidity (increase in the vapor pressure deficit) surrounding the plant can cause the hat to be reached earlier (and therefore remain for a longer time) following the onset of the hat. A sign that the vapor pressure deficit is important is a transient decrease in transpiration (a depression in the hat) near midday. Lynn and Carlson (1990) show that a seasonal decrease in minimum stomatal resistance of the corn may have occurred because of a seasonal decrease in the root–stem resistance Z_p.

Eventually, decreasing soil moisture completely flattens the transpiration curve. This collapsing phase is shown in Figure 21.6a and b for a soil moisture availability (M) $= 0.35$; we define soil moisture availability as the ratio of root soil water content to that at field capacity, which was 0.34 by volume for HAPEX simulations. Note that the changes are small between the two moist cases, $M = 0.65$ and $M = 0.50$, because the leaf water potential does not reach its threshold value. Intermediate values of M between 0.5 and 0.35 produce a hat. Figure 21.5b shows that a large soil moisture availability ($M = 0.65$) can produce a hat for some value of the critical leaf water threshold. Thus, *transpiration and radiometric canopy temperature depend on soil water content and plant physiology.*

A significant feature of Figures 21.5 and 21.6 is that the transpiration and radiometric canopy temperatures are identical (or nearly so) before and after the period of the hat. Water stress does not show in the radiometric temperatures unless and until the epidermal leaf water potential reaches the critical potential. Thus to measure remotely the effects of water stress, one must be able to model the hat.

We must emphasize that the hat is not an artifact of the model. There is considerable observational evidence showing the same behavior in the field. We have calculated surface evapotranspiration from the balance between measurements of net

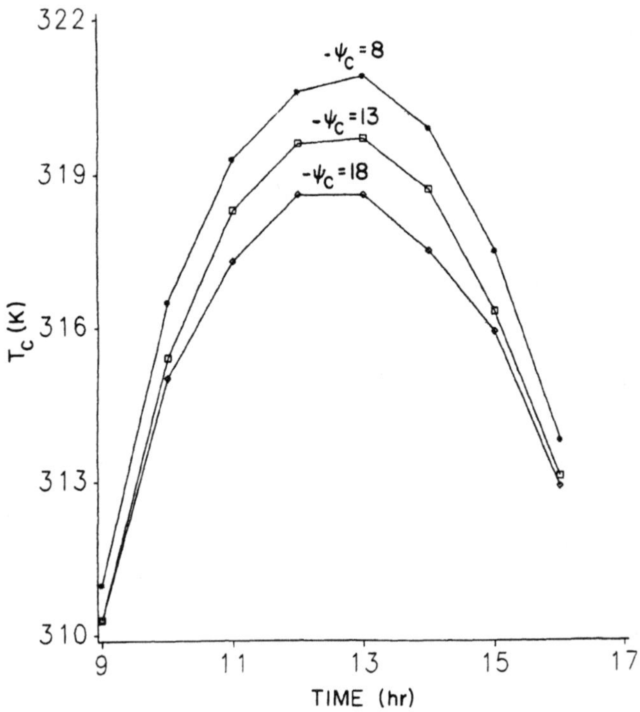

A

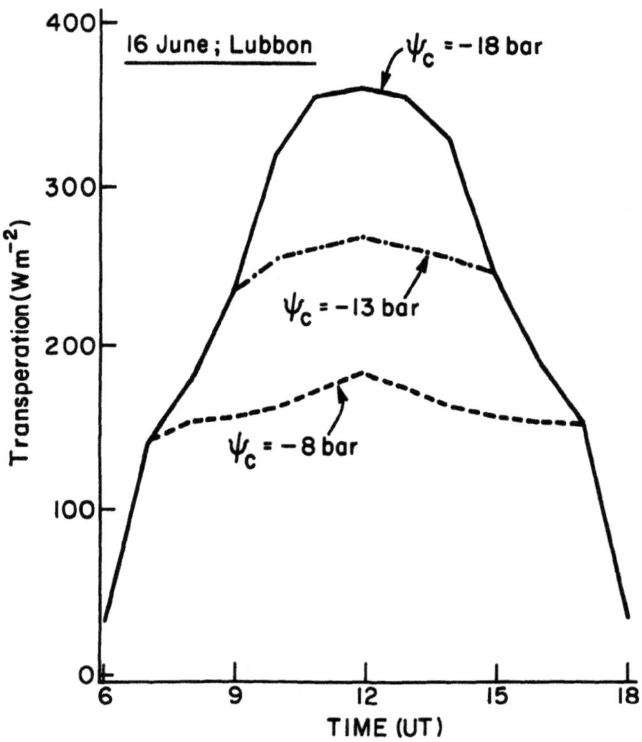

B

FIGURE 21.5. (A) Radiometric surface temperature (T_c, °C) versus time simulated for three values of critical leaf water potential ($\psi_c = -8$; $\psi_c = -13$; $\psi_c = -18$ bar). Simulations pertain to corn for HAPEX, initial conditions for 16 June 1986 near Lubbon, France. Soil moisture availability (M) equals 0.65. (B) Same as a but for transpiration (W m^{-2}) versus time.

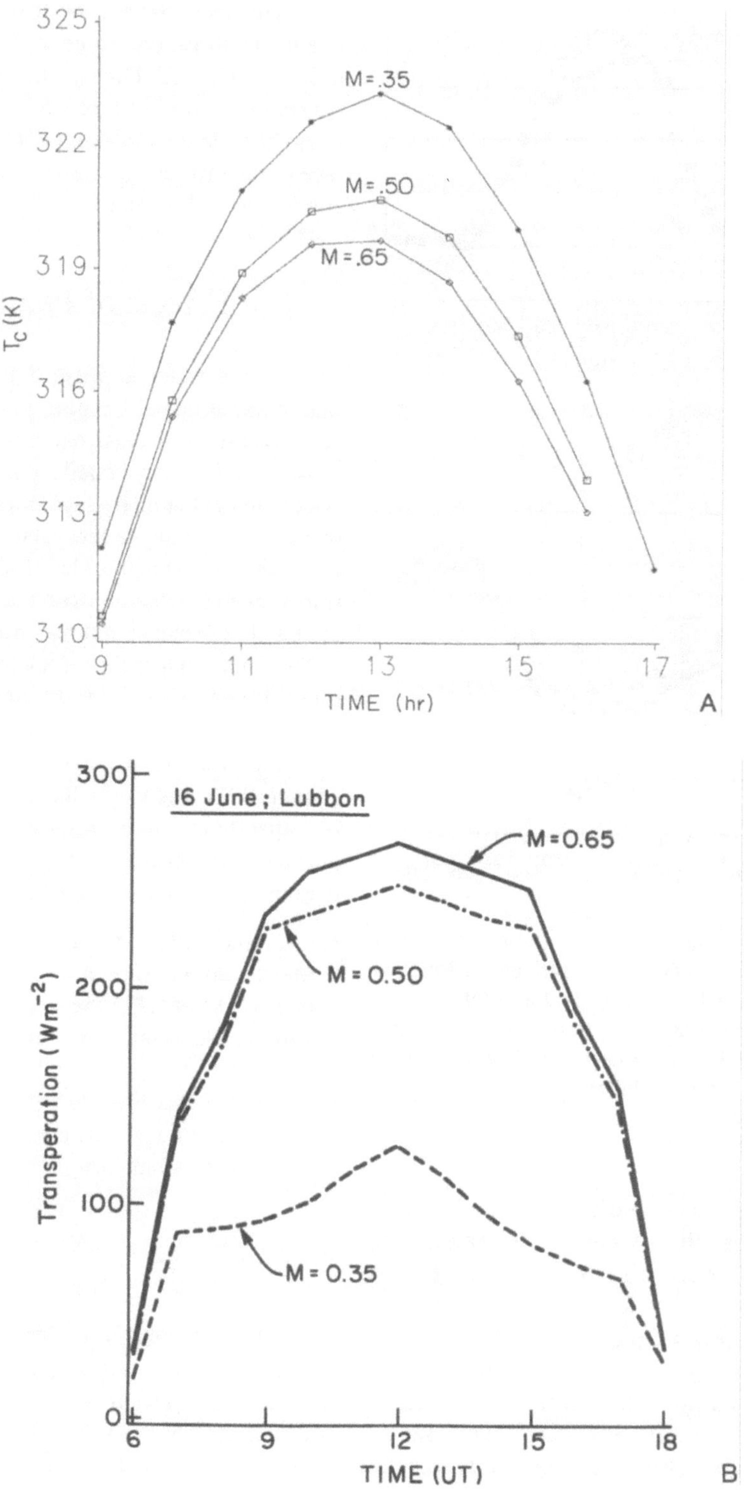

FIGURE 21.6. (A) Radiometric surface temperature (T_c, °C) versus time for three values of soil moisture availability (M = 0.35, 0.5, 0.65) and critical leaf water potential of −13 bars for corn, initial conditions 16 June 1986 near Lubbon, France. (B) Same as (A) but for transpiration as a function of time.

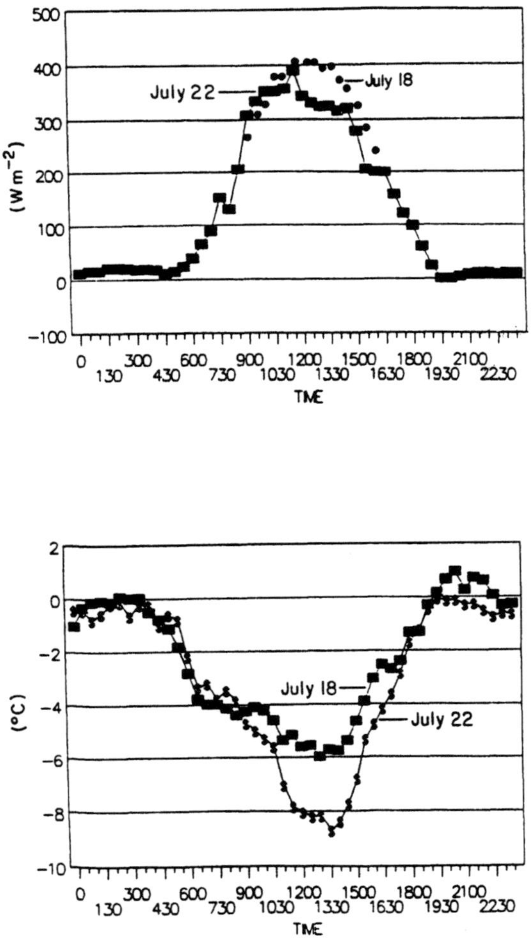

FIGURE 21.7. (top) Latent heat flux (w m^{-2}) measured over corn as a function of time at an agricultural field site (Rock Springs) near Penn State; 18 July (dots) and 22 July 1987 (solid rectangles); (bottom) radiometric surface temperature minus air temperature differences (°C) versus time on 18 July (solid rectangles) and 22 July 1989 (dollar signs).

radiation, substrate sensible heat flux, and the surface sensible heat flux. Figure 21.7 shows these latent heat fluxes over corn at State College, PA, on July 18 and 22, 1987. Note that the latent heat fluxes for the 2 days are identical except when the hat was present. This hat lasted from about 0930 to 1430 local time on July 22. Significant differences in radiometric surface temperature (expressed as differences between surface and air temperatures) between July 18 and 22 occurred only during the period of the hat. Simulations with the model show that the lengthening of the hat between July 18 and 22 resulted from a decrease in soil water content. This decrease enabled the leaf water potential to reach the threshold value of −15 bar for a few hours on July 22. During this period there was a maximum difference of about 50–75 W m^{-2} in latent heat fluxes and about 2°C in radiometric surface temperatures between stressed and unstressed conditions 4 days apart.

4. The Effects of Plant Capacitance

Plants store water and use that water to maintain maximum stomatal opening in the face of limiting soil moisture or excessive atmospheric demand. We have recently incorporated a plant capacitance component in the boundary layer model and have subsequently run extensive sensitivity tests. The sensitivity shows the magnitude of capacitance on the transpiration, radiometric surface temperature, stomatal resistance, and leaf water potentials.

The mathematical framework for the capacitance model follows closely that of Lynn and Carlson (1990), but an additional storage flux term is included. Results are summarized in Carlson and Lynn (1991). Figure 21.8a and b shows the effects of capacitance on transpiration and radiometric surface temperature of a hypothetical forest canopy. Our results suggest the following:

Capacitance seems to be significant in the radiometric temperature of a vegetation canopy under two conditions: (1) the vegetation is large (trees) and (2) the plant experiences water stress (i.e., the hat).

Capacitance has little effect on the transpiration and radiometric surface temperature for wet soil, unless a hat forms.

5. Summary

Variations in vegetation cover, specifically the amount of sunlit, bare soil visible to the radiometer, can profoundly influence the radiometric temperature of the canopy. We suggest a method whereby the spectrum of radiometric surface temperatures as a function of NDVI is inverted to obtain (1) soil moisture estimates appropriate to the surface and the root zone and (2) the fractional

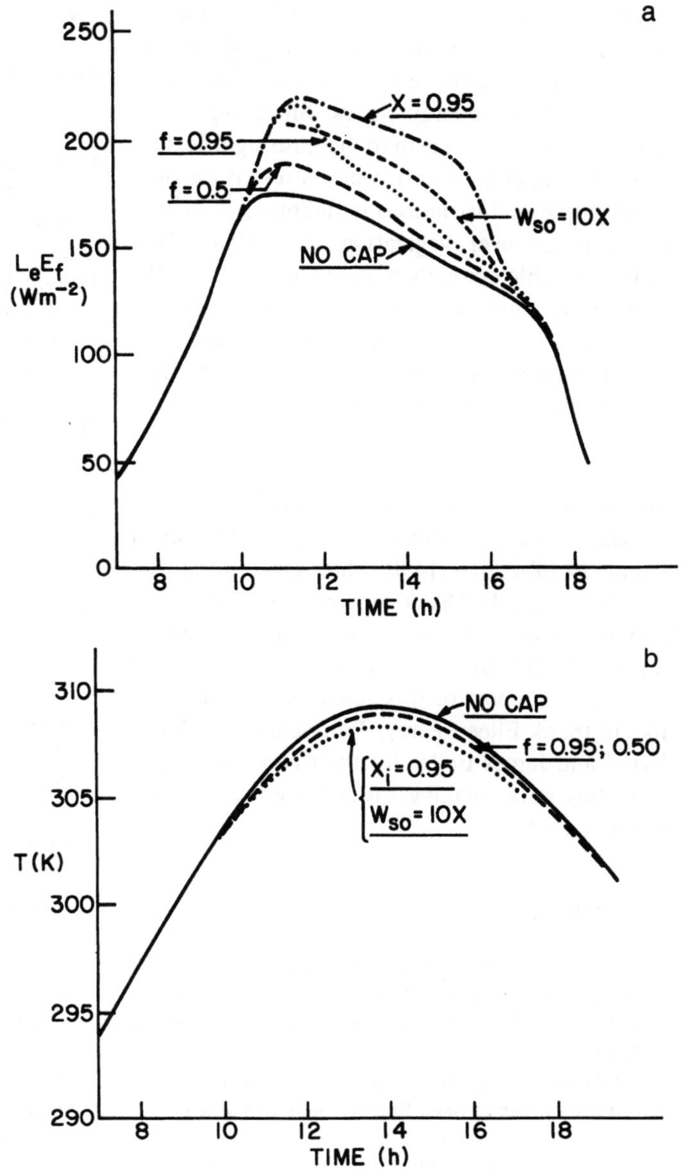

FIGURE 21.8. (a) Transpiration (watts per unit leaf area) versus time for the case of no capacitance (NO CAP), trees with capacitor centered half way between root and leaves (f = 0.5), capacitor centered in leaves (f = 0.95), 10 times the reference water volume in the tree (5×10^{-3} m^3 per unit leaf area; W_{so} = 10×) and 10 times the reference water volume, but with the tree initially 95% of saturation (X = 0.95). Initial soil moisture availability is 0.28. Simulations made for initial conditions 16 June 1986 near Lubbon, France. (b) Same as a but for radiometric surface temperature (T_c, °C).

vegetation cover and surface energy fluxes as a function of NDVI and radiometric surface temperature. The method yields best results when the horizontal variance in soil moisture is small compared to the difference in water content between the surface and root zone.

We show the effects of transient water stress on the radiometric temperature of a plant canopy as a function of soil water content, plant physiology, and atmospheric demand. Radiometric surface temperature during periods of water stress can differ from that for unstressed vegetation by a few degrees centigrade.

The effect of capacitance on radiometric surface temperature is very small except for large vegetation. Trees may store enough water to modify the radiometric surface temperature by about 1 °C, but only during periods of water stress, as manifested by the transpiration plateau. In large trees, the transpiration plateau may be modified substantially by waters from internal sources.

These conclusions signify that infrared surface temperature measurements over vegetation can lead to very misleading estimates of soil water content and surface energy fluxes. Since vegetation temperatures do not seem to increase significantly over air temperature until water stress is encountered, horizontal variations in radiometric surface temperatures over vegetation more than likely reflect variable amounts of bare soil visible to the radiometer than variable soil water content. Horizontal variations in radiometric surface temperatures over vegetation may provide information on the vertical soil water profile.

Acknowledgments. The author would like to thank the United States Geological Survey (USGS) under Grant 14-08-0001-G1490, the Agricultural Research Service, USDA under Grant 58-32U4-8-27, and the National Aeronautics and Space Administration (NASA) under Grant NAG-5-919 for their invaluable support of this research. I would also like to thank Eileen Perry, Tom Schmugge, Barry Lynn, and James Belles for their contributions in providing data, making measurements, and preparing figures.

References

Asrar G, Fuchs M, Kanemasu ET, Hatfield JL (1984) Estimating absorbed photosynthetic radiation and leaf area index from spectral reflectance in wheat. Agron J 76:300–306.

Best GG, Harlan JC (1985) Spectral estimation of green leaf area index of oats. Remote Sens Environ 17:27–36.

Carlson TN, Lynn B (1991) The effect of plant capacitance in a boundary layer model. Ag Forest Meteorol (submitted).

Carlson TN, Perry EM, Schmugge TJ (1990) Remote estimation of soil moisture availability and fractional vegetation cover for agricultural fields. Agric Forest Meteorol 52, 45.

Coakley JA, Bretherton FP (1982) Cloud cover from high resolution scanner data: Detecting and allowing for partially filled fields of view. J Geophys Res 87: 4917–4932.

Cowan IR (1965) Transport of water in the soil-plant-atmosphere system. J Appl Ecol 2:221–239.

Curran PJ (1983) Multispectral remote sensing for the estimation of green leaf area index. Phil Trans R Soc London A 309:259–270.

Deardorff JW (1978) Efficient prediction of ground surface temperature and moisture with inclusion of a layer of vegetation. J Geophys Res 83:1889–1904.

Gallo TF, Daughtry CST, Bauer ME (1985) Spectral estimation of absorbed photosynthetically active radiation in corn canopies. Remote Sens Environ 17:221–232.

Hansen B, Soegaard H (1987) Monitoring vegetation index and biomass production in southern Greenland based on NOAA-AVHRR data. 21st International Symposium on Remote Sensing of Environment, Ann Arbor, MI, 26–30 Oct. 1987.

Holben BN, Tucker CJ, Fan CJ (1980) Spectral assessment of soybean leaf area and leaf biomass. Photogrammet Eng 46:651–656.

Lynn B, Carlson TN (1990) A model illustrating plant versus external control of transpiration. Agric Forest Meteorol 52, 3–43.

Nemani RR, Running SW (1989) Estimation of regional surface resistance to evapotranspiration from NDVI and thermal IR-AVHRR data. J Appl Meteorol 28: 276–284.

Peterson DL, Spanner MA, Running SW, Teuber KB (1987) Relationship of thematic mapper simulator data to leaf area index of temperature coniferous forests. Remote Sens Environ 22:323–341.

Price JC (1983) Estimating surface temperatures from satellite thermal infrared data - a simple formulation for the atmospheric effect. Remote Sens Environ 13: 353-361.

Price J (1989) Estimating evapotranspiration over large areas. IGARSS '89 Vol. H IEEE Proceedings of the 12th Canadian Symposium on Remote Sensing 10–14 July 1989, Vancouver, Canada, pp 2138–2141.

Schmugge TJ, Janssen L (1988) Aircraft remote sensing in HAPEX. Proceedings of the 4th International Colloquium on Spectral Signatures in Objects of Remote Sensing Aussois, France, 18–22 Jan. 1988 (ESA SP-287), pp 463–467.

Taconet O, Carlson TN, Bernard R, Vidal-Madjar D (1986) Evaluation of a surface/vegetation model using satellite infrared surface temperatures. J Climate Appl Meteorol 25:1752–1767.

Tucker C (1979) Red and photographic linear combinations for monitoring vegetation. Remote Sens Environ 8:127–150.

Turner NC (1970) Response of adaxial and abaxial stomata to light. New Phytol 69:647–653.

Turner NC (1974) Stomatal behavior and water status of maize, sorghum and tobacco under field conditions. II at low-soil water potential. Plant Physiol 53:360–365.

22
Computer Simulation of Regional Evapotranspiration by Integrating Landscape Biophysical Attributes with Satellite Data

Steven W. Running

1. Introduction

Evapotranspiration (ET) is an important link between the land-surface energy budget and hydrologic budget. The partitioning of incoming energy into sensible or latent heat substantially influences meteorology from local to global scales. The components of the hydrologic budget that control ET are fairly well understood at local scales (Shuttleworth, 1988). Measurement methods and models exist to describe adequately canopy interception, snowpack dynamics, soil water flux, and ET itself when treated as "point" processes. Even the biological control of ET, primarily manifested through leaf stomatal closure, is now understood reasonably well, and generalized treatment of environmental control functions is available (Jarvis and McNaughton, 1986). However, the challenge is to enlarge our perspective to GCM cell size, macrohydrology as termed by Shuttleworth (1988), which necessitates a new consideration of parameterization of large areas (Eagleson, 1986).

It is this author's contention that calculation of regional ET is not limited by one-dimensional evaporation theory, but by our abilities to describe accurately the highly heterogeneous landscape in a way critical to the calculations. Part of the problem is defining a model with sufficiently powerful, yet simple parameterization requirements so that a landscape can be defined correctly. A corollary to this is having the methodology to parameterize large areas.

In this chapter I will present a modeling system that attempts to solve some of these problems. Our regional ET simulations utilize standard formulas such as the Penman–Monteith equation to calculate ET, but require parameterization of only three site-specific variables. These are (1) leaf area index (LAI) to define the vegetation, (2) soil water holding capacity to define the hydrologic storage, and (3) routine daily meteorological data, extrapolated with a topoclimatological model to drive the model. To facilitate regional applications, LAI can be defined by satellite, topography is defined automatically from digital elevation data, and soils are defined in a geographic information system (GIS). Most significantly, this surface-derived regional simulation system is being used to develop and test a completely satellite-derived surface resistance factor that, when used in conjunction with AVHRR normalized difference vegetation index, may provide dynamic weekly continental-scale calculations of ET.

2. The Ecosystem Model, FOREST-BGC

Our ET calculations are done within the context of a comprehensive ecosystem model developed for coniferous forests of the western United States. FOREST-BGC (*B*io*G*eochemical *C*ycles) is a process-level ecosystem simulation model that calculates the cycling of carbon, water, and nitrogen through forest ecosystems (see Fig. 22.1) (Running and Coughlan, 1988). The model requires daily data for standard meteorological conditions, maximum–minimum temperatures, dew point, incident shortwave radiation and precipitation, and definition of

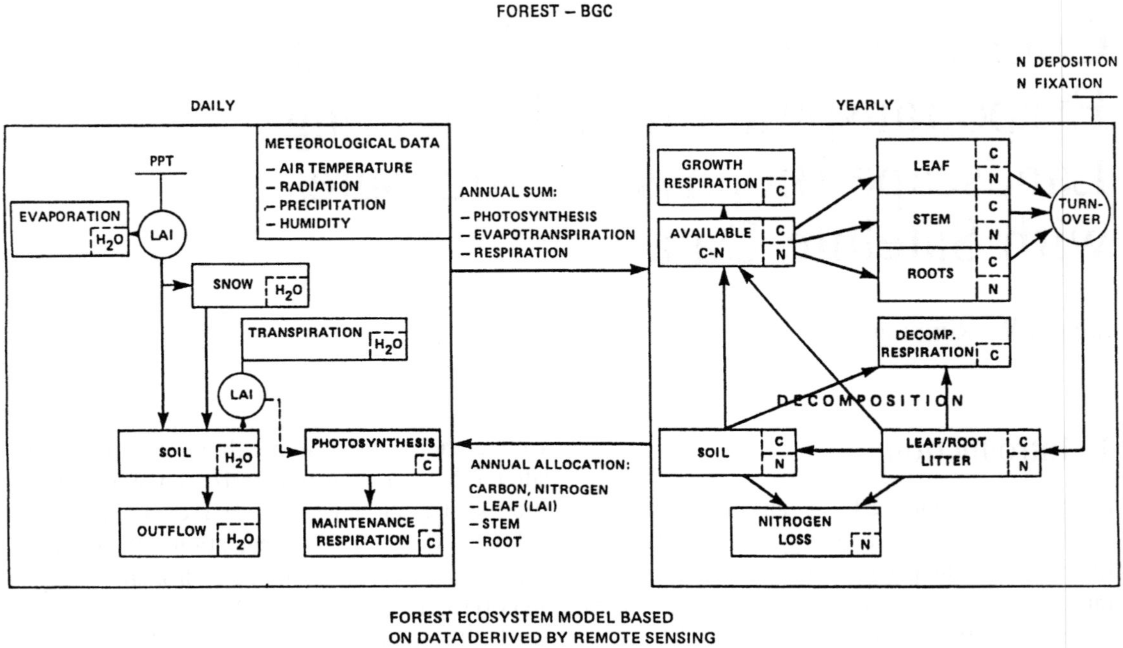

FIGURE 22.1. A compartment flow diagram for FOREST-BGC, an ecosystem process model for carbon (C), water (H_2O) and nitrogen (N) cycles. The daily submodel calculates canopy interception, snowpack dynamics, soil water, evaporation, and transpiration driven by daily meteorological data. Leaf area index (LAI) is typically defined by satellite. (From Running and Coughlan, 1988.)

key site and vegetation variables. The model calculates canopy interception and evaporation, transpiration, soil outflow of water; photosynthesis, growth and maintenance respiration, allocation, litterfall, and decomposition of carbon; and deposition, uptake, litterfall, and mineralization of nitrogen. The model has a mixed time resolution, with hydrologic, photosynthetic, and maintenance respiration processes computed daily, and the other carbon and all nitrogen processes computed yearly. For this study, the important parts of the model are the hydrologic balance, ET, and canopy water stress.

As a brief summary of the model's function, daily precipitation is routed to snowpack or soil dependent on air temperature, a distinction that greatly influences the timing of future hydrologic activity. A canopy interception fraction, effectively a capacitance defined by LAI, is subtracted and evaporated with a Penman equation, and the remaining water goes to a soil compartment. Modeling canopy interception is a critical hydrologic partitioning point because this process determines whether incoming precipitation remains in the system for a few hours and evaporates, or is stored in the soil where it may be transpired many weeks later or routed to streamflow (Lloyd et al., 1988). Forest soils generally have a high soil surface infiltration capacity, so overland flow is rare. Transpiration is calculated with a Penman–Monteith equation.

The canopy conductance term is a complex function of air temperature, vapor pressure deficit, incident radiation, and leaf water potential. Air temperatures below 0°C reduce canopy conductance to cuticular values, and are an important determinant of the active physiological season. Canopy conductance is linearly reduced to a default cuticular value when either average daily vapor pressure deficit exceeds 16 mb or predawn leaf water potential, estimated from soil water availability, decreases below −1.65 MPa. Aerodynamic resistance is assumed as 5.0 s/m in the Penman–Monteith equation, modified slightly by LAI to define surface roughness, because windspeed data are rarely available.

Canopy photosynthesis is calculated by multiplying a CO_2 diffusion gradient by a mesophyll

CO_2 conductance and the canopy water vapor conductance, both of which are controlled by radiation and temperature. The light response surface for mesophyll conductance is asymptotic. A function of $-0.5/\text{LAI}$ is used for Beer's law attenuation of incident radiation to produce canopy average radiation. An inverse parabolic response surface defines high and low temperature compensation points, respectively. Net photosynthesis (PSN) is computed by subtracting a maintenance respiration term, calculated as an exponential function of air temperature with a $Q_{10} = 2.3$ for all stem and root biomass, and then subtracting 35% of the remaining gross photosynthesis for growth respiration. Because the primary interest in this chapter is seasonal evapotranspiration, only the daily half of the model was used. Although species-specific variability occurs in these physiological responses, we have found their effect on ET computations to be rather modest compared to the overriding importance of our three defined variables of LAI, soil water holding capacity, and microclimate (Running and Coughlan, 1988).

3. LAI as the Critical Vegetation Parameter

Critical to our logic for regional implementation is a simple definition of the hydrologically significant vegetation properties. If the basic biome type (forest, grassland, crop etc.) can be defined and general physiological controls assumed for a biome, then the most important general vegetation parameter needed is LAI. For example, multiple variables in the Penman–Monteith equation are influenced by LAI.

$$T = \left[\frac{\Delta Q_n + \rho C_p (VPD/R_a)}{(\Delta + \gamma (1 + R_c/R_a))\lambda} \right] \times \text{LAI} \times \text{DAYL} \tag{22.1}$$

when T is transpiration (or evaporation where R_c is set to 0.0), Δ is the slope of saturation vapor pressure versus temperature, Q_n is net radiation, ρ is air density, C_p is specific heat of air, VPD is vapor pressure deficit, R_a is aerodynamic resistance, R_c is canopy resistance (the reciprocal of canopy conductance), γ is the psychrometric constant, λ is the latent heat of vaporization, LAI is leaf area index, and DAYL is the daylength.

This equation is often referred to as a "big leaf" model because it defines the canopy homogeneously for the area considered. We use LAI to define vertical dimension variability of the vegetation across the landscape. LAI defines the canopy available to absorb incoming shortwave radiation, influencing both the net radiation driver, Q_n, and the canopy resistance terms, R_c, in Equation (22.1). LAI influences the extrapolated surface temperature and vapor pressure deficit in MT-CLIM, the meteorological drivers for evaporation in Equation (22.1). Finally, our formulation defines R_c for a unit LAI of 1.0 to allow parameterization from leaf gas exchange data, so the final P-M equation is multiplied by total LAI of the surface. Consequently ET is proportional to LAI, although not linearly because of the computed attenuation of Q_n through LAI and influence on R_c.

Figure 22.2 illustrates how simulated hydrologic partitioning between evaporation, transpiration, and soil outflow changed with LAI from 3 to 9 in seven contrasting climates of North America. Note how increasing LAI in a wet climate like Jacksonville, FL reduces outflow and increases ET, but in an arid climate like Missoula, MT, additional LAI does little to change seasonal hydrologic partitioning, although within season activity changes. Clearly, varying LAI substantially influences model calculations, primarily because canopy interception and ET are both direct functions of LAI in the model.

A practical reason for choosing LAI as the key vegetation parameter is our ability to estimate LAI by satellite. Figure 22.3 shows LAI estimated by the NOAA AVHRR satellite using NDVI for 53 conifer stands in Montana (from Nemani and Running, 1989a). We have attained similar results using the higher spatial resolution TM sensor, where the NIR/Red ratio of Channels 4/3 was correlated with measured LAI of coniferous forests, $R^2 = 0.91$ (Peterson et al., 1987). Other investigators have also found good correlations between LAI and NIR and Red reflectances of vegetation (Asrar et al., 1989).

4. Microclimate Extrapolation

Even if proper formulas are used, and accurate definitions of vegetation and soils are available, regional ET estimates require accurate meteorological drivers at daily time resolution or greater.

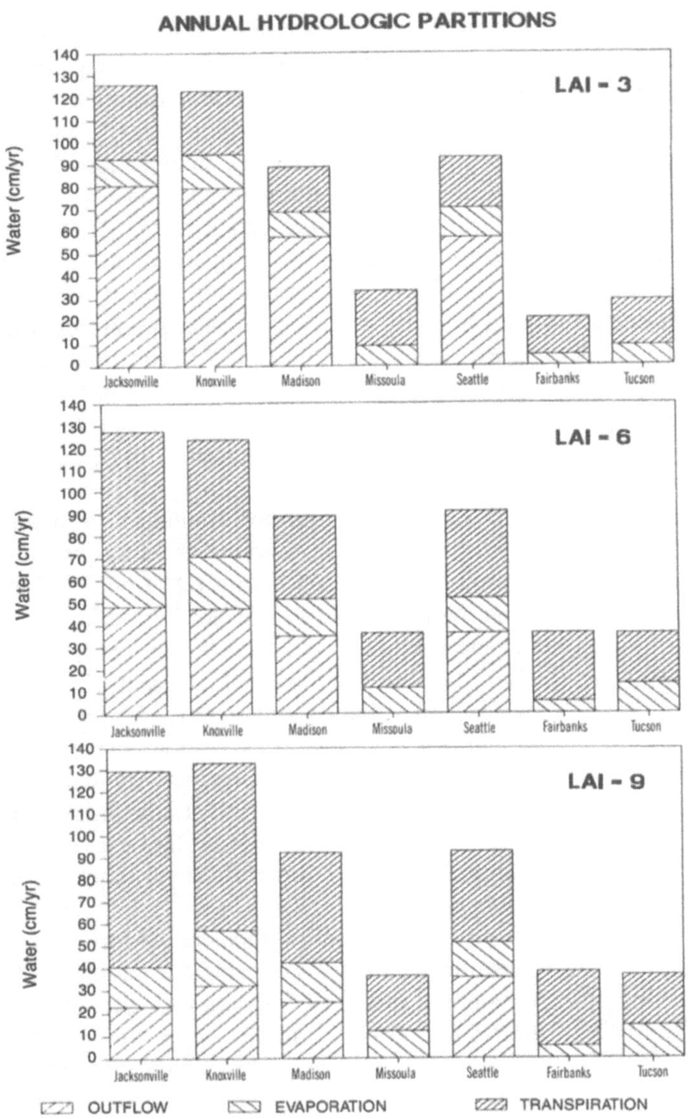

FIGURE 22.2. A comparison of the hydrologic partitioning of annual precipitation between evaporation, transpiration, and soil outflow simulated by FOREST-BGC for seven sites of contrasting climate in North America. Each site was simulated with a hypothetical forest of LAI = 3, 6, 9 to study the interaction of LAI and climate to hydrologic partitioning. (From Running and Coughlan, 1988.)

Although many government agencies collect meteorological data for specific purposes, only the NOAA Environmental Data Service reports data that are consistent and widely available. Because air temperatures, vapor pressure deficit, incoming shortwave radiation, and precipitation are required every day for simulation, standard data sources are necessary. However, multiple problems have been encountered in acquiring these data for regional simulations. First, the "standard" daily data collected by the U.S. National Weather Service and cooperating stations are maximum–minimum temperature and precipitation. The density of primary stations recording humidity and solar radiation in any form is less than 1/100,000 km^2 throughout the western United States. Second, standard meteorological data are not available for diverse mountain sites, as most regular weather stations are in populated valleys. Although one can monitor a few mountain sites with research instrumentation, rou-

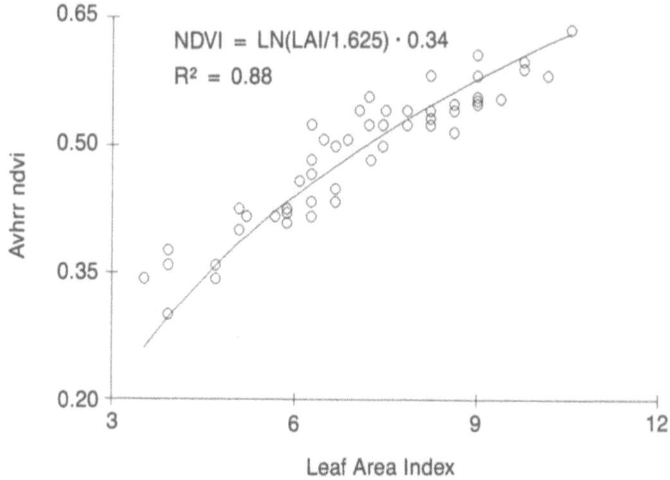

FIGURE 22.3. The correlation found between the NOAA AVHRR normalized difference vegetation index, (NIR−Red)/(NIR+Red) and estimated LAI of 53 coniferous forests in Montana. (From Nemani and Running, 1989a.)

tine monitoring of entire regions is not feasible. Satellite observations can partially solve this problem, but the spatial resolution is very coarse, 8 × 8 km for GOES, and cloudcover problems restrict dependable daily observations. Third, meteorological data specific to ecological processes, such as a photosynthetically active definition of daylength, are never available from routine sources.

To solve these problems, we developed MT-CLIM, a model that allows us to build a daily meteorological map of our study region based on a few direct NWS measurements extrapolated by climatological principles to the surrounding terrain (Fig. 22.4) (Running et al., 1987). The centerpiece of MT-CLIM is a submodel that calculates daily potential incoming shortwave radiation for any elevation, slope, aspect, and latitude based on the original algorithm of Garnier and Ohmura (1968). The observed maximum–minimum diurnal temperature amplitude is then used to calculate an atmospheric transmissivity (which incorporates cloudcover), a surprisingly robust logic developed by Bristow and Campbell (1984) that eliminates the need for using the inaccurate visual cloudcover estimates provided by the NWS. Elevational adiabatic lapse rates are used to correct temperatures, and the differential radiation loading to different slopes, attenuated by LAI, also adjusts final estimates of slope surface temperatures. If humidity data are not available, night minimum temperature substitutes for dew point, then allowing estimation of daytime vapor pressure deficits. Mountain precipitation is adjusted by the ratio of annual isohyets of the base station relative to the study site. Our validations of this model on three north–south mountain slope pairs in Montana found these simple climatological calculations to produce meteorological data of sufficient accuracy for simulation of seasonal evapotranspiration and photosynthesis of our forests (Running et al., 1987).

5. Modeling Regional ET

5.1 Raster Based

A prerequisite to calculating ET of a region is georeferenced definition of our three key parameters. For our current 28 × 55 km study area, we developed a GIS to integrate data for the simulations as shown in Figure 22.5. The NOAA AVHRR sensor was used to define LAI, and defines our minimum cell size of 1.1 km² for the results shown here. However this 1.1-km cell size is limited only by the spatial resolution of the datastream, in this case the AVHRR sensor, and computer power to make calculations for each cell. For some future simulations we plan to use the 30-m resolution Landsat TM data. Daily microclimate of each cell was generated by interpolating with MT-CLIM. Average elevation, slope, and aspect of each cell is determined from digital elevation data, and is essential as part of the microclimate calculation in

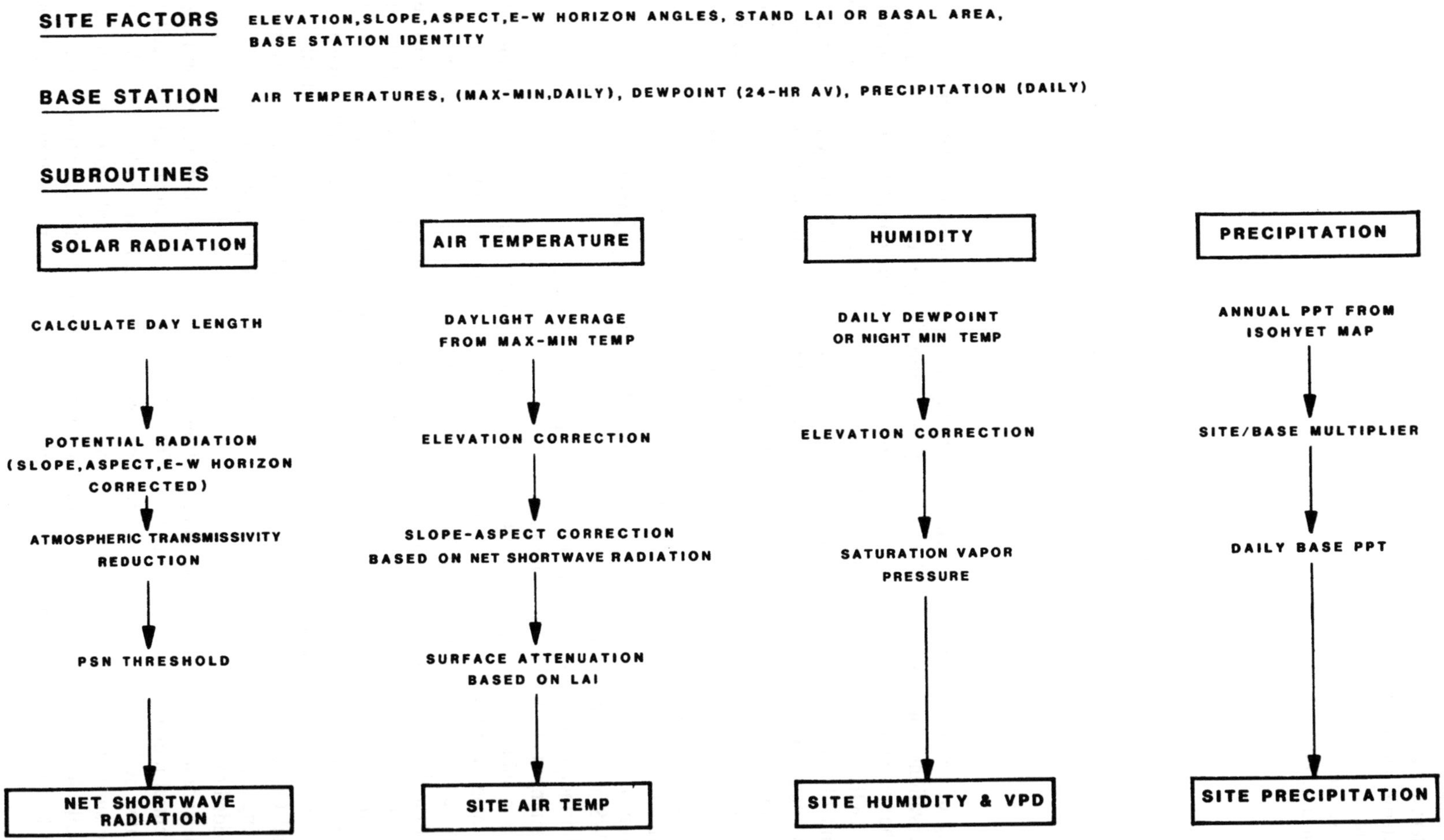

FIGURE 22.4. Flowchart of the MT-CLIM model for estimating daily microclimate conditions in mountainous terrain. The site factors and daily base station meteorological observations shown are required inputs for the model. MT-CLIM is designed primarily to provide meteorological driving variables for ecosystem process models. (From Running et al., 1987.)

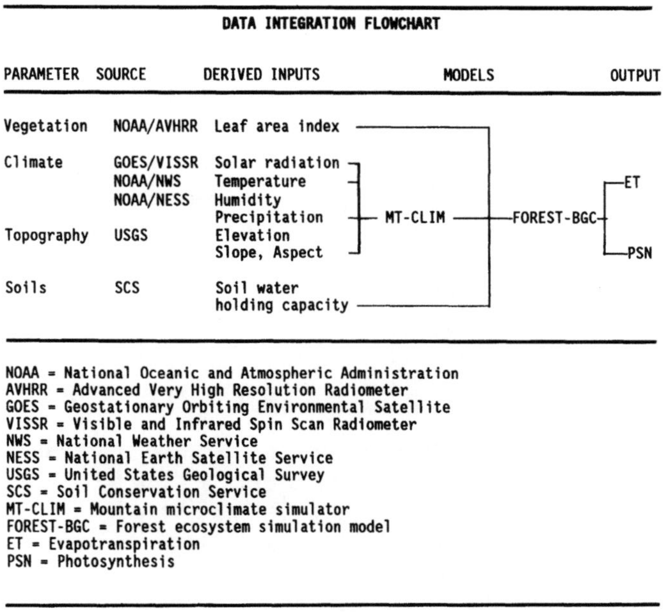

FIGURE 22.5. An organizational diagram showing the sources of raw climatic and biophysical data, the derived variables produced, and their incorporation in the topoclimate (MT-CLIM) and ecosystem simulation model (FOREST-BGC). (From Running et al., 1989.)

MT-CLIM. Finally, soil water holding capacity is inferred by first entering soil classification maps, then translating soil textural characteristics to water holding capacities. Most soils maps provide a taxonomic definition of the soils that is only partially useful for defining the hydrologically important physical characteristics needed; however, this is the only soils data base available for regional scales.

The constant physical variables, topography and soils, the vegetation variable LAI, and the daily microclimate files for each 1.1-km cell produce a file matrix that then drives FOREST-BGC, where daily ET is simulated for the entire year. Figure 22.6 shows a schematic of the 1200-km^2 study area, and the resulting raster map of annual ET for the mountainous conifer forest study site in Montana (Running et al., 1989). Annual ET computed over the region ranged from 35 to 60 cm, with highest values on the mountain slopes. Valley bottoms have lower precipitation, smaller snowpacks, and carry low LAI because of summer plant water stress. The high mountains receive 200 cm/year of precipitation, of which 70% falls as snow, but low evaporative demand, low soil water holding capacities, and temperature limited vegetation and LAI produce ET of only 40–50 cm. Two gauged watersheds within this region where hydrologic balance estimates of ET were possible both measured ET within 10% of our calculated ET.

5.2 Polygon Based

We have recently completed our first effort to define a subset of the Montana study area directly by topographically defined polygons, rather than square grid cells. The polygons are automatically defined from digital elevation data using the terrain model of Band (1986) and Band et al. (1990). This procedure allows us to define the landscape to varying levels of complexity depending on objectives, so one can effectively telescope in to high detail of a watershed, or conversely define only relatively coarse broad regional topography. This procedure also allows definition of a hydrologic drainage network, and reduction of the number of similar cells used in defining a landscape. For example, the 1200 cell map in Figure 22.6 could be represented by as few as eight polygons for some purposes. This simplification reduces computer time for simulation and increases the within cell homogeneity of important biophysical attributes and microclimate. Figure 22.7 shows output of annual ET for the eastern half of the study area simulated in Figure 22.6. With explicit topography in these calculations MT-CLIM calculated more

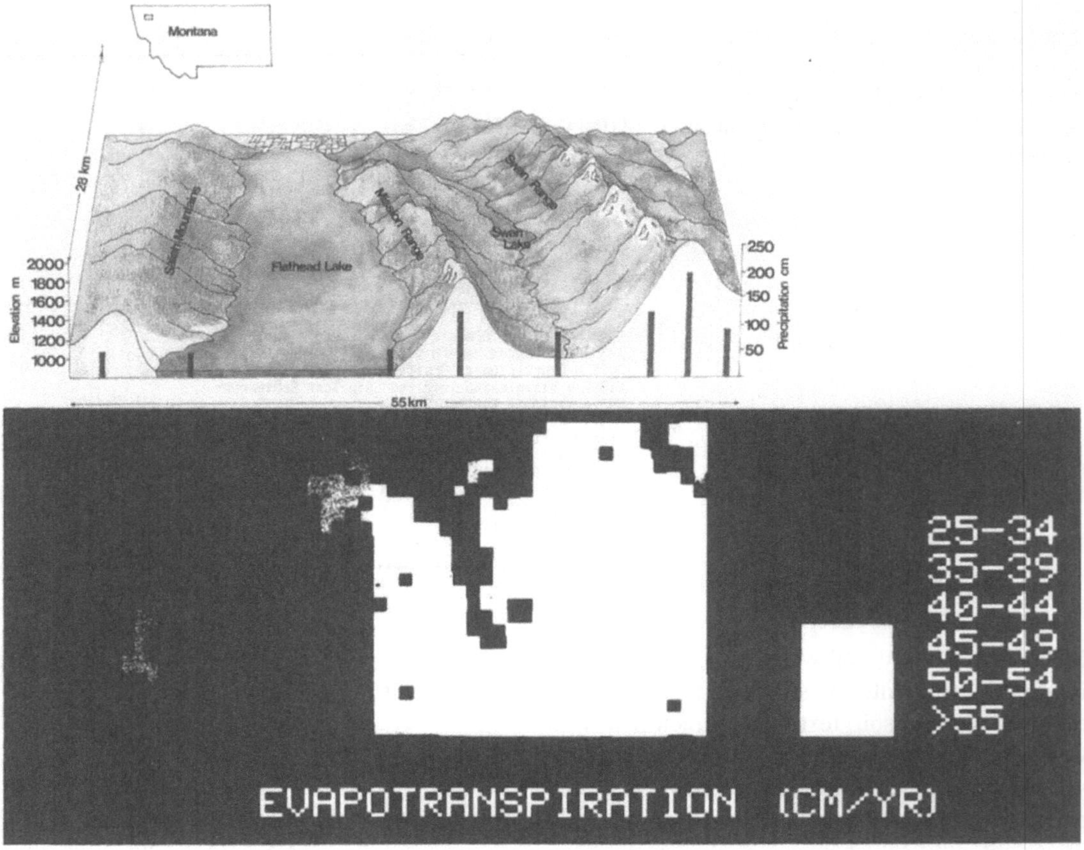

FIGURE 22.6. A schematic diagram of the 28 × 55-km area in Montana, defined by 1.1-km AVHRR pixels, and calculated annual evapotranspiration using the simulation system from Figure 22.4.

accurate slope related microclimates influencing snowpack duration greatly. The dry south-facing slopes of this arid region showed ET as low as 25 cm, whereas opposite north-facing slopes were 10–20% higher in ET for low elevation cells. For higher elevation, energy limited cells, south-facing slopes showed higher ET than north facing slopes, as would be expected at 46°N latitude. (Running et al., 1987).

6. Satellite-Derived ET Estimates

For continental to global scale applications even the above methodology is too data intensive. As an alternative, we have recently tested a completely satellite-derived estimate of surface evaporation that could, if successful, allow much more accurate continental ET estimates. The ET of actively transpiring vegetation, sometimes called potential ET, can be calculated reasonably well from physical meteorological formulas such as the Penman equation. In principle, the meteorological satellites could provide global potential ET estimates now. However, actual ET is rigorously controlled by the physiological condition of the vegetation. Stomatal closure due to temperature, water, or even nutritional stress can reduce actual ET a factor of 10 times below potential ET. Nemani and Running (1989b) found that the slope of the scatterplot of NDVI vs T_s (surface temperature) for a 500-km^2 forested area correlated highly with drought

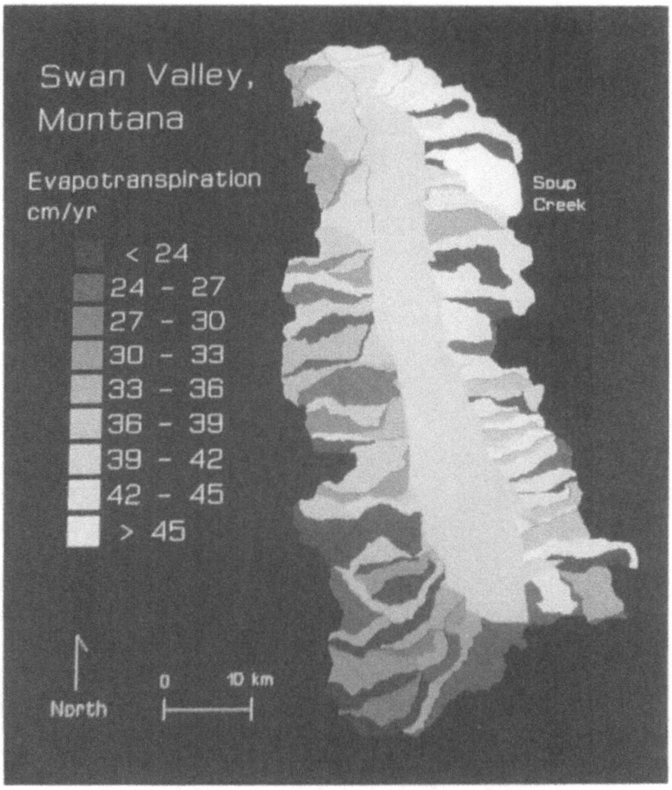

FIGURE 22.7. Simulation of annual ET for the Seeley–Swan basin section of the study area in 6, but defining topographically homogeneous polygons. Landscape definition was produced from topographic simulations by Band (1990) using digital elevation data.

induced surface resistance changes simulated using FOREST-BGC methodology (Fig. 22.8a,b). We hypothesized that the satellite derived ratio, which we designated σ, was inferring seasonal changes of a Bowen ratio partitioning of energy. During wet periods, the surface temperature for a given NDVI (estimated LAI) was lower as more energy was used for latent heat exchange. Conversely as the vegetation became water stressed during drought, higher surface temperatures were sensed as more sensible heat was produced.

Recent derivations (Sellers, 1985, 1987) have illustrated that NDVI is directly proportional to APAR (absorbed photosynthetically active radiation), which suggests that to a first approximation, NDVI should be related to an absorbed energy-defined potential ET. The seasonal integration of NDVI has been found to be highly correlated to annual NPP (Goward, 1988) and simulated transpiration (Fig. 22.9) across continental climatic scales. In another paper, (Running, 1990), I have suggested a combination of this surface resistance σ with weekly time-composited NDVI in the following form

$$ET = t\Sigma T\,(\sigma \times \text{NDVI}) \times \varepsilon_w$$

where

$t\Sigma T$ = seasonal NDVI time integration truncated by surface temperatures < 0°C, to eliminate periods of vegetation inactivity
σ = a surface resistance defined as the slope of the T_s/NDVI as per Figure 22.8b
ε_w = a conversion efficiency term of mm H_2O/NDVI

could be developed to generate dynamic ET maps that are sensitive to water deficits and suboptimal temperature without requiring soil physical data or separate meteorological data. Further development of this logic could produce weekly continental scale actual ET maps defined by satellite.

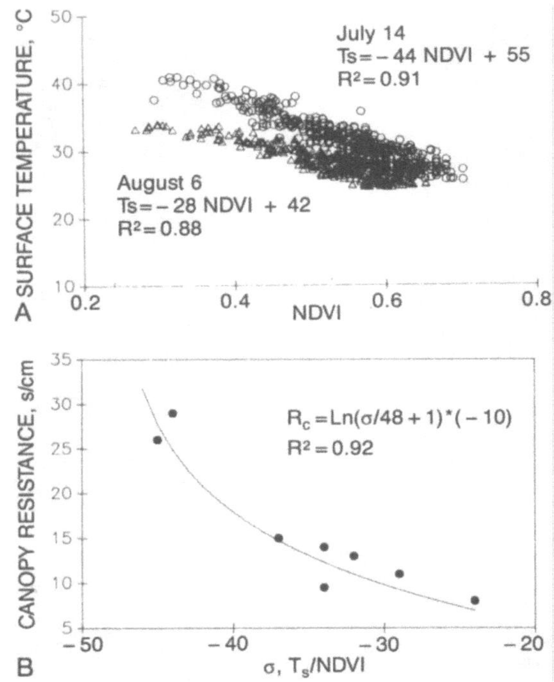

FIGURE 22.8. (a) A scattergram of NDVI and surface temperature, T_s, from the NOAA/AVHRR for a 20 × 25-km study area of conifer forest in Montana on July 14 after 5 weeks without rain, and August 6, after 3.2 cm of rain. (b) The regression relationship found between the slope of T_s/NDVI from (a) and surface resistance simulated by FOREST-BGC for 8 days during the summer of 1985. We hypothesize that the higher surface temperatures on July 14 result from higher Bowen ratio, and that the slope of the T_s/NDVI relationship can be used as a satellite-derived estimate of surface energy partitioning. (From Nemani and Running, 1989b.) NDVI is defined as (NIR−Red)/(NIR+Red) from channels 1 and 2 of AVHRR. T_s was estimated using a split window technique using channels 4 and 5 of AVHRR. (Reprinted with permission from Nemani and Running, *Journal of Applied Meteorology*, vol. 28, no. 4, © 1989 American Meteorological Society.)

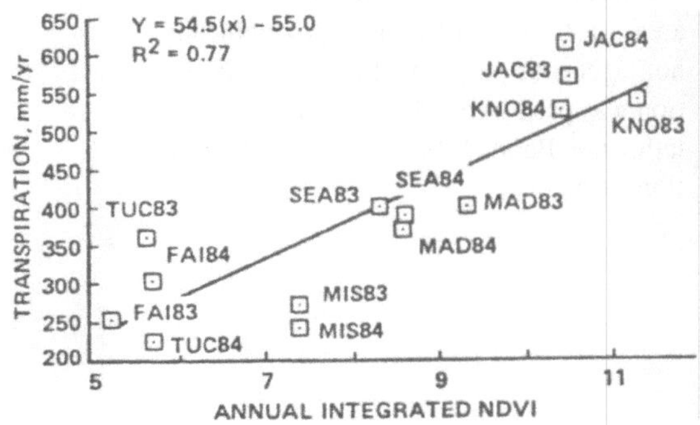

FIGURE 22.9. The correlation between annual integrated NDVI and annual transpiration calculated with the FOREST-BGC model for seven sites across a climatic gradient of North America. The sites are the same as in Figure 22.2. NDVI are from weekly composited AVHRR data. (Reprinted by permission of the publisher from "Relating seasonal patterns of the AVHRR vegetation," by Running and Nemani, Remote Sensing of Environment, vol. 24, pp. 347–67. Copyright 1988 by Elsevier Science Publishing Co., Inc.)

7. Conclusions

I suggest that regional ET research must concentrate on elegantly simple means of defining the heterogeneity of the biophysical and climatic elements of the landscape. Most regional-scale variation in ET can be described with just four parameters: LAI, soil water holding capacity, climate, and canopy resistance. A major challenge is the paucity of regional-scale validation data of ET over short time periods. Although historical watershed mass balance estimates of ET are widely available, they only produce an average ET value over a long time period, often just annually. For purposes such as GCM models, at least daily time scales are required. Satellites must provide the simultaneous regional coverage and high repeat times required for regional ET research (Becker et al., 1988). LAI appears to be a pivotal vegetation parameter that influences many ET system variables, provides a quantitative measure of landscape heterogeneity, and can be monitored by satellite. For maximum predictive capacity, we ultimately must replace our reliance on measured surface meteorological data by interfacing with mesoscale climate models (Segal et al., 1988; Dickinson et al., 1989). Additionally, explicit representation of key ecological processes such as stomatal control of surface resistance is necessary. Major field experiments combining a hierarchy of field measurements, aircraft monitoring, and satellite remote sensing, such as HAPEX (André et al., 1988) and FIFE (Sellers et al., 1988), are important to build a system of both regional land surface parameterization and spatially integrated validation measurements at scales much larger than classical instrumentation can provide. Only when comprehensive, dynamic computations can routinely be done can we explore socially important projection scenarios like the influence of future global change on ET and water resources.

Acknowledgments. Primary funding for this research has been from the National Aeronautics and Space Administration, #NAGW-952, and McIntire-Stennis funding to the University of Montana. Colleagues critical to the development of these ideas are David L. Peterson, Dr. Ramakrishna R. Nemani, Dr. Larry E. Band, and Joseph C. Coughlan.

References

André JC et al (1988) Evaporation over land surfaces: First results from HAPEX-MOBILHY special observing period. Ann Geophys 6:477–492.

Asrar G, Myneni RB, Kanemasu ET (1989) Estimation of plant-canopy attributes from spectral reflectance measurements. In "Theory and Applications of Optical Remote Sensing" (G. Asrar, ed.), pp 252–296. Wiley, New York.

Band LE (1986) Topographic partition of watersheds with digital elevation models. Water Res Res 22:15–24.

Band LE, Peterson DL, Running SW, Dungan J, Lathrop R, Coughlan J, Lammers R, Pierce L (1990) Forest ecosystem processes at the watershed scale: 1. Basis for distributed simulation. Ecol Model (in press).

Becker F, Bolle H-J, Rowntree PR (1988) The International Satellite Land-Surface Climatology Project. United Nations Env. Prog. ISLSCP-Report #10.

Bristow KL, Campbell GS (1984) On the relationship between incoming solar radiation and daily maximum and minimum temperature. Agric Forest Meteorol 31:159–166.

Dickinson RE, Errico RE, Giorgi F, Bates GT (1989) A regional climate model for the western United States. Climatic Change 15:383–422.

Eagleson PS (1986) The emergence of global-scale hydrology. Water Res Res 22:6S–14S.

Garnier BJ, Ohmura A (1968) A method of calculating the direct shortwave radiation income of slopes. J Appl Meteorol 7:796–800.

Goward SN (1988) Satellite bioclimatology. J Climate 7:710–720.

Jarvis PG, McNaughton KG (1986) Stomatal control of transpiration: Scaling up from leaf to region. Adv Ecol Res 15:1–49.

Lloyd CR, Gash JHC, Shuttlesworth WJ, De OMarques AF (1988) The measurement and modelling of rainfall interception by Amazonian rain forest. Agric Forest Meteorol 43:277–294.

Nemani R, Running SW (1989a) Testing a theoretical climate-soil-leaf area hydrologic equilibrium of forests using satellite data and ecosystem simulation. Agric Forest Meteorol 44:245–260.

Nemani R, Running SW (1989b) Estimating regional surface resistance to evapotranspiration from NDVI and Thermal-IR AVHRR data. J Appl Meteorol 28:276–284.

Peterson DL, Spanner MA, Running SW, Teuber KB (1987) Relationship of Thematic Mapper Simulator data to leaf area index of temperate coniferous forests. Remote Sens Environ 22:323–341.

Running SW (1990) Estimating terrestrial primary productivity by combining remote sensing and ecosystem simulation. In "Remote Sensing of Biosphere Functioning" (R. Hobbs, H. Mooney eds.). Springer-Verlag, New York p 65–86.

Running SW, Coughlan JC (1988) A general model of forest ecosystem processes for regional applications. I. Hydrologic balance, canopy gas exchange and primary production processes. Ecol Model 41:125–154.

Running SW, Nemani RR (1988) Relating seasonal patterns of the AVHRR vegetation index to simulated photosynthesis and transpiration of forests in different climates. Remote Sens Environ 24:347–367.

Running SW, Nemani RR, Hungerford RD (1987) Extrapolation of synoptic meteorological data in mountainous terrain, and its use for simulating forest evapotranspiration and photosynthesis. Can J Forest Res 17:472–483.

Running SW, Nemani RR, Peterson DL, Band LE, Potts DF, Pierce LL, Spanner MA (1989) Mapping regional forest evapotranspiration and photosynthesis by coupling satellite data with ecosystem simulation. Ecology 70:1090–1101.

Segal M, Avissar R, McCumber MC, Pielke RA (1988) Evaluation of vegetation effects on the generation and modification of mesoscale circulations. J Atmos Sci 45:2268–2292.

Sellers PJ (1985) Canopy reflectance, photosynthesis and transpiration. Int J Remote Sens 6:1335–1372.

Sellers PJ (1987) Canopy reflectance, photosynthesis and transpiration. II. The role of biophysics in the linearity of their interdependence. Remote Sens Environ 21:143–183.

Sellers PJ, Hall FG, Asrar G, Strebel DE, Murphy RE (1988) The first ISLSCP field experiment (FIFE). Bull Am Meteorol Soc 69:22–27.

Shuttleworth WJ (1988) Macrohydrology—The new challenge for process hydrology. J Hydrol 100:31–56.

23
Implications for Remote Sensing of Natural Switching from Atmosphere-Controlled to Soil-Controlled Evaporation or Infiltration

J. Philip O'Kane

1. Introduction

Fluxes of water from atmosphere to soil, falling as rain, or from soil to atmosphere, rising as evaporated water vapor, are approximately controlled at the pedon scale (≈ 1 m) either by the atmosphere or by the soil. The control may switch suddenly from one to the other, particularly in the case of bare soil.

Consequently, the switching is expressed spatially as patches of soil. Within each patch the switchover to soil control has occurred. During rain, soil patches are covered with ponded water. Under dry conditions, soil patches are evaporating at less than the potential rate.

The patches define two dynamic Boolean fields: one for ponded rain and the other for falling-rate evaporation. Inside each patch the field value is one; outside it is zero. The moving boundaries of the patches as they appear, grow, coalesce, contract, and disappear are the sites where strong nonlinear switching is occurring. The geometry of the patches may be fractal.

Remote imaging may provide high-resolution spatiotemporal data on the dynamics of these patches. Repeated measurements at the same sites, both during rain and evaporation, will provide the data on which to build new physical theory, since the standard physically based models do not apply to land–atmosphere processes at larger scales.

Section 2 contains a derivation of the standard Richards/Philip equation governing the movement of water, in both the liquid and vapor phase, in a column of soil at the pedon scale (≈ 1 m). The gradient in scalar potential energy drives the flow. The thermodynamic equation is the means for partitioning the water content of the soil into its liquid and vapor fractions. The isothermal theory of evaporation is presented in detail since Philip's derivation is dispersed over several papers.

In Section 3 we present the boundary conditions at the bottom and top of the soil column for both infiltration and evaporation at the pedon scale. Infiltration requires a switch in the surface boundary condition after ponding of water on the soil surface. This separates the infiltration process into two stages. The boundary condition for evaporation is nonlinear and of mixed type.

Nonlinear switching in the control of water exchange between soil and atmosphere is examined in Section 4. The nonlinear boundary condition for evaporation from bare soil behaves like a switch and two distinct stages may also occur in the case of evaporation. References are given to both laboratory and field experiments that confirm the switching for evaporation. The complications due to the presence of vegetation are discussed briefly.

Section 5 presents the implications of these nonlinearities for remote sensing and concludes with two recommendations for the design of future HAPEX- and FIFE-type experiments.

2. Water Movement in a Soil Column

2.1 Equations of Motion and of Continuity

The two equations that govern the vertical movement of water in an unsaturated rigid column of bare soil at the pedon scale (≈ 1 m) are

1. a water balance, which states that water mass is conserved, and
2. a dynamic equation, with liquid and vapor water flux proportional to the gradient in soil water potential.

If z is a position coordinate (m) measured vertically downward from the soil surface and t is time (sec), we may write these equations as follows:

$$\rho_w \frac{\partial\theta}{\partial z} = -\frac{\partial(f_l + f_v)}{\partial z}, \qquad 0 \leq \theta \leq \theta_s < 1 \qquad (23.1)$$

$$f_l = -k_l \frac{\partial\psi}{\partial z} \qquad (23.2a)$$

$$f_v = -k_v \frac{\partial\psi}{\partial z} \qquad (23.2b)$$

where θ is the total volumetric content of mobile water, in both liquid and vapor phases (m^3 of water per m^3 of soil). $\theta = \theta_s$ is the saturation or maximum water content of the soil matrix, which we assume to be rigid. ρ_w is the density of water (kg m^{-3}). f_l is the flux density of liquid water (kg m^{-2} s^{-1}) in the direction of positive z, and is proportional to the gradient in the soil water potential $\partial\psi/\partial z$ (Buckingham, 1907). f_v is the flux density of water vapor (kg m^{-2} s^{-1}), and is also proportional to the gradient in the soil water potential $\partial\psi/\partial z$ (Philip, 1955). The liquid and vapor conductivities k_l and k_v are functions of moisture content.

Different authors define the potential energy of soil water at the pedon scale, in one of three equivalent ways, as energy per unit mass, as energy per unit weight, or as energy per unit volume. The gradient of ψ has units of force per unit mass of water, N kg^{-1}, when ψ is measured as J kg^{-1}.

The potential ψ is assumed to be the sum of several component potentials with respect to a common reference state. The most important of these are the gravitational potential ψ_g and the matric potential ψ_m. Hence

$$\psi = \psi_m + \psi_g = \psi_m - gz, \qquad \psi_m \leq 0 \qquad (23.3)$$

g is the acceleration due to gravity. The gravitational potential $\psi_g = -gz$ (J kg^{-1}) is negative because work must be done to raise water from any position z in the soil column to $z = 0$.

The matric potential of soil water ψ_m (J kg^{-1}) is also negative and is a function of moisture content. Thermomechanical work must be done to return the soil water to the reference state, by removing it from the soil, whether by suction, drainage, or evaporation.

The relationship between moisture content and matric potential, the soil moisture characteristic $\psi_m = \psi_m(\theta)$, is a constitutive property of the soil–air–water system and is determined by measurements on soil samples at the pedon scale. The inverse function $\theta = \theta(\psi_m)$ and its derivative, the specific water capacity $C(\psi_m) = d\theta/d\psi_m$, may be used to eliminate θ from Equation (23.1), k_l and k_v, yielding ψ_m as the dependent variable. θ may also be chosen as the dependent variable. The choice of ψ_m is preferable, since ψ_m may be a continuous function of z when θ is not.

Combining Equations (23.1), (23.2), and (23.3) and eliminating the fluxes, we find, for the unknown matric potential $\psi_m(z, t)$

$$\rho_w C \frac{\partial\psi_m}{\partial t} = \frac{\partial}{\partial z}\left[k \frac{\partial\psi_m}{\partial z} - kg\right] \qquad (23.4)$$

where $k = k_l + k_v$. When $k_v = 0$ this is known as Richards' equation (1931) of soil physics; the extension to include vapor transport is due to Philip (1955). This is a nonlinear parabolic equation of the diffusion type with "concentration"-dependent coefficients $C(\psi_m)$ and $k(\psi_m)$.

Extensions of Richards' equation to include two or three space dimensions, hysteresis, the swelling matrices of clay soils, the presence of electrolytes, and the effect of temperature gradients may be found in the literature of soil physics. The above presentation is sufficient for the points that we wish to make in the following paragraphs.

2.2 Potential Energy and the Phase Partitioning of Water

The potential energy, ψ, of water in a state S is equal to the notional work that must be performed by a given process, to convert water from a reference state S_0, to the state S. The reference state is free pure liquid water at $z = 0$ in isothermal equilibrium with its vapor at saturation in a standard atmosphere.

Liquid water in an unsaturated soil is not free, since it is bound to the soil matrix by capillary and adhesive forces. Furthermore the concentration of water vapor in the soil–air, which is in equilibrium with liquid soil–water, has a value that is less than that in the reference atmosphere.

The work done, $\psi_h < 0$, in expanding isothermally the water vapor in the reference state S_0, to achieve a lower humidity is

$$\psi_h = \frac{RT}{M_w} \ln h, \qquad 0<h\leq 1 \qquad (23.5a)$$

assuming that atmospheric water vapor and its partial pressure obey the ideal gas law. [M_w is the molar mass of water (0.018 kg/mol), R the molar gas constant (8.314 J/mol K), and T the constant temperature in degrees Kelvin (293 K at 20°C)].

ψ_m (<0) is the work done in transferring isothermally the liquid water in state S_0 into an inert rigid porous soil. We may equate ψ_m and ψ_h when the liquid and vapor phases of soil water are in thermodynamic equilibrium. Inverting the logarithm we find the relative humidity of soil air h as a "thermodynamic function" of soil water potential ψ_m at any z (Edlefsen and Anderson, 1943; Penman, 1958; Rose, 1966; Campbell, 1977)

$$h(z) = \exp\left[\frac{M_w \psi_m (z)}{RT}\right] \qquad (23.5b)$$

Given $\psi_m(z, t)$, relationship (23.5b) instantaneously partitions θ (ψ_m) between the liquid and vapor phases, and allows us to use one rather than two state variables for soil–water content.

In terms of thermodynamics, ψ, is the partial specific Gibbs free energy of the water in the soil–air–water system (Edlefsen and Anderson, 1943). The work performed by the process in converting water from state S_0 to S must be reversible, independent of path, and depend solely on the initial and final states. Necessary conditions for this are usually given as (1) the process must be so slow that the water passes through equilibrium states only, (2) no friction, and (3) constant temperature (Rose, 1966).

Condition (1) contradicts Equation (23.4) since this parabolic equation has the property that changes in ψ_m are propagated everywhere with infinite speed. For a penetrating critique of the problematic foundations of thermodynamics and its rigorous development, see Truesdell (1981), Truesdell et al. (1984), and Serrin et al. (1986).

2.3 Liquid and Vapor Conductivities

k_ℓ and k_v are the conductivities (kg s m^{-3}) for liquid and water vapor in soil at the pedon scale. They depend on the properties of the soil matrix and on water content θ. The inverse function $\theta = \theta(\psi_m)$ of the soil moisture characteristic may be used to express $k_\ell[\theta(\psi_m)]$ and $k_v[\theta(\psi_m)]$ as functions of ψ_m.

k_ℓ is zero at zero water content, and increases monotonically to a maximum at saturation. k_ℓ is determined by measurement on soil samples at the pedon scale. In contrast to k_ℓ, k_v is zero for both saturated and completely dry soil, and may rise to a maximum at a potential of roughly -10^4 J kg^{-1} (equivalent to -10^3 m of water, pF=3, when potential is defined as energy per unit weight). In the early stages of drying, the relative humidity of soil air is 1 and k_v is negligible compared with k_ℓ. However, when the water potential in the soil drops below -10^4 J kg^{-1}, mass transfer takes place almost entirely in the vapor phase.

Let c be the water vapor concentration in soil air and c_{sat} its value at saturation for a given temperature T. The vapor conductivity k_v is given by the isothermal theory of vapor diffusion in porous media (Penman, 1940; Philip, 1955; Thorstenson and Pollock, 1989)

$$k_v = -\alpha D_0(\theta_s - \theta) c_{sat} \frac{\partial h}{\partial \psi_m} \qquad (23.6)$$

where $h = c/c_{sat}$ is the relative humidity and D_0 is the binary diffusion coefficient of water vapor in still air, at the prescribed temperature. $\theta_s - \theta(\psi_m)$ is the air-filled porosity of tortuosity α, through which vapor diffusion can occur. From expression (23.5b) we can find $dh/d\psi_m$. Hence the vapor conductivity k_v can be expressed entirely in terms of ψ_m.

3. Boundary Conditions at the Pedon Scale

3.1 The Boundary Condition at the Bottom

The boundary condition at the bottom ($z = B$) of the soil column can be expressed in terms of the air-entry potential ψ_e

$$\psi = \psi_{e'} \qquad z = z_B \qquad (23.7)$$

$\psi_e < 0$ is the potential at which air first enters a saturated ($\theta = \theta_s$) sample of soil during drying, coinciding with the emptying of the largest pores in the soil matrix. This condition fixes the top of the saturated capillary fringe at a depth z_B below the surface.

Water fluxes in either direction are possible across the lower boundary and represent capillary rise from or drainage to the saturated zone in the lower deposits of soil. z_B may be regarded as a constant on a time scale of a few days. On longer time scales z_B varies with time and a moving boundary condition is required. A constant value of z_B is sufficient for this presentation.

3.2 Surface Boundary Condition for Infiltration

The boundary condition for infiltration into bare soil is a prescribed flux at $z = 0$, equal to the rainfall rate Q:

$$f_\ell = -k_\ell \frac{\partial \psi}{\partial z} = \rho_w Q, \qquad \theta < \theta_s \text{ at } z = 0 \qquad (23.8)$$

Vapor flux is ignored during infiltration.

As the solution of Equations (23.4), (23.7), and (23.8) evolves in time, from a given initial condition $\theta(z,0)$, $\theta(0,t)$ increases and may reach saturation. When saturation occurs, the first stage of infiltration at the atmosphere-controlled rate is completed. The boundary condition switches to a prescribed concentration $\theta = \theta_s$ and the soil-controlled stage begins:

$$\theta\,(0,t) = \theta_s \qquad (23.9)$$

As the solution of Equations (23.4), (23.7), and (23.9) evolves in time, the flux into the soil $f_\ell\,(0,t)$ decreases from its initial value of Q. During the second stage of infiltration, the surplus water accumulates on the surface of the ground and flows horizontally to the drainage network.

3.3 Surface Boundary Condition for Isothermal Evaporation

We now consider the soil after rainfall has stopped. The patches of ponded water, if there are any, will gradually disappear due to infiltration and evaporation. The infiltrated water will also be redistributed within the soil due to matric forces and gravity.

To simplify the presentation, we begin with a column of bare soil with an initial water content but no ponded water on the surface. The initial water content is the accumulated effect of past rainfall, evaporation, and drainage, modified by spatially varying topography, atmospheric, and soil properties. The relative humidity of the air in the soil surface is 1 and evaporation begins at the potential rate E_p.

The boundary condition for evaporation at the surface is a flux E of water vapor from soil to atmosphere

$$f_\ell = -k_\ell \frac{\partial \psi}{\partial z} = 0, \qquad \text{at } z=0 \quad (23.10a)$$

$$f_v = -k_v \frac{\partial \psi}{\partial z} = -\rho_w E, \qquad \text{at } z=0 \quad (23.10b)$$

The actual evaporation rate E may be defined as follows (Dalton, 1802; Brutsaert, 1982):

$$E = H(w)[c_0 - c_a] \qquad (23.11)$$

where H is a mass transfer coefficient that depends on wind speed w at a reference height above the ground. c_0 is the water vapor concentration in the air in the soil surface and c_a is the water vapor concentration at the reference height in the atmosphere. To combine this boundary condition with Richards' equation, we must relate c_0 to ψ_m at $z = 0$.

The simplest relationship between c_0 and $\psi_m(0)$ is obtained by assuming an isothermal soil–atmosphere continuum. Hence we may define a potential or maximum evaporation rate E_p as the evaporation rate from a wet soil at the same uniform temperature as the atmosphere:

$$E_p = H(w)[c_{sat} - c_a] \qquad (23.12)$$

where c_{sat} is the saturation water vapor concentration in the soil surface at the given temperature.

Numerical experiments by Van Bavel and Hillel (1976), based on an albedo for bare soil that varies with moisture content, show that potential evaporation is not a precise concept even under isothermal conditions. In general, evaporation depends on both soil and atmosphere. Nevertheless, we shall find it a very useful approximation.

FIGURE 23.1. Variation with time of the ratio of infiltration to constant rainfall. t_p time to ponding.

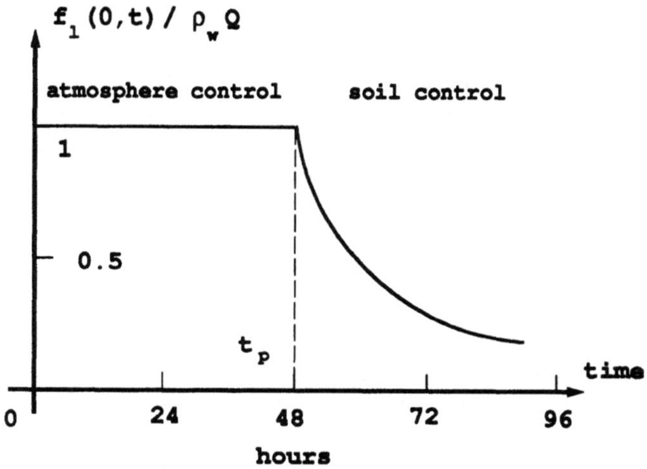

Eliminating H from (23.11) and (23.12) we find (Philip, 1957a)

$$E = E_p\left[\frac{h_0 - h_a}{1 - h_a}\right], \qquad 1 \geq h_0 \geq h_a \qquad (23.13)$$

$h_a = c_a/c_{sat}$ is the relative humidity of the atmosphere at the reference height. $h_0 = c_0/c_{sat}$ is the relative humidity of the air in the soil surface at $z = 0$, and is found by inserting $\psi_m(0)$ into the thermodynamic function

$$h_0 = \exp\left[\frac{M_w\psi_m(0)}{RT}\right] \qquad (23.14)$$

During the isothermal drying of wet soils, the precise value of h_a is not important in Equation (23.13) for small values of t, since $h_0 = 1$, $E/E_p = 1$, and E_p drives the model. Eventually h_0 may decrease from 1 at a rate that depends on soil hydraulic properties that control the upward supply of water. The evaporating surface, which is initially located in the soil surface at $z = 0$, starts to move downward and transport of soil–water in the vapor phase appears. Eventually thermal gradients become important when the soil is dry.

Daily values of E_p may be estimated from Penman's (1948) equation or from corrected open-water evaporation rates. A diurnal cycle may be introduced into E_p in proportion to the elevation of the sun (Van Bavel and Hillel, 1976; Hillel, 1977).

Combining Equations (23.10b), (23.13), and (23.14), we find a nonlinear inhomogeneous boundary condition of the mixed type at $z = 0$:

$$k_v\frac{\partial\psi_m}{\partial z} - \frac{E_p}{1-h_a}\exp\left[\frac{M_w\psi_m}{RT}\right] = k_v g - \frac{E_p h_a}{1-h_a} \qquad (23.15)$$

for isothermal evaporation from wet bare soil.

4. Nonlinear Switching

The most important feature of the solution of Richards' equation for both infiltration and evaporation at the pedon scale is nonlinear switching. This section describes the phenomenon and refers to both laboratory and field experiments that confirm it in the case of evaporation.

4.1 Infiltration

Richards' Equation (23.4) subject to (1) the water table boundary condition (23.7), and (2) the switched boundary conditions (23.8 and 23.9) at the surface of the soil, is the standard model in soil physics for infiltration of rain at the pedon scale. Figure 23.1 shows the nonlinear switch acting at t_p the time to ponding of surface water. At times prior to t_p the ratio of infiltration to rainfall is unity, and no water accumulates or "ponds" on the surface. The infiltration rate is controlled by the atmosphere. After t_p, the rate of infiltration falls below the rainfall rate and water accumulates on the surface of the soil. Now infiltration is controlled by the soil. The ratio of infiltration to rainfall decreases suddenly on a concave curve.

Analytical solutions for both linear and nonlinear simplifications of Equations (23.4), (23.7), and the switched boundary conditions (23.8 and 23.9) are summarized in Kuehnel (1989). Approximate expressions for the time to ponding t_p of soils that are initially air dry and subjected to constant rain are given in Kuehnel et al. (1990a). t_p may vary by orders of magnitude depending on soil properties and rainfall. Ponding will not occur when the rainfall rate is less than the saturated hydraulic conductivity of the soil.

The one-dimensional Richards' equation does not apply at scales larger than the pedon scale, because both it and the infiltration boundary condition are nonlinear. Only linear differential operators have the property that they commute with averaging operators. It is incorrect to drive the pedon-scale infiltration model with spatially averaged values of (1) rainfall and (2) soil properties, because

1. Spatial averaging reduces local peaks in rainfall intensity and underestimates the temporal frequency of ponding.
2. When ponding occurs in a 1-D model, it happens everywhere simultaneously.

The larger the area over which the averages are taken, the larger these discrepancies become.

4.2 The First Two Stages of Evaporation from Bare Soil

Switching of control from the atmosphere to the soil also occurs during evaporation. In this case the switch is a property of the solution of Richards' equation when solved with the boundary condition for isothermal evaporation.

Philip (1957b, 1967) concluded for isothermal steady-state evaporation from water tables at various depths, that the rate of evaporation is virtually independent of the surface relative humidity—and therefore of moisture content also—in the range $0 \leq h_0 \leq 0.99$.

The reason for this may be seen by examining the Fokker–Planck form of Richards' Equation (23.4) where θ is the dependent variable and $D = k/C = k\,[\psi_m(\theta)] \cdot d\psi_m/d\theta$:

$$\rho_w \frac{\partial \theta}{\partial t} = \frac{\partial}{\partial z}\left[D \frac{\partial \theta}{\partial z} - gk \right] \qquad (23.16)$$

Philip (1957b, 1967) noted that the soil–moisture diffusivity function $D(\theta)$ decreases through several orders of magnitude as relative humidity decreases from 1 by as little as 1%. This corresponds to a downward movement of the evaporating surface and the emergence of a vapor bottleneck in the top of the soil.

Once the relative humidity of the soil surface has fallen below about 0.99, the dominant influence on the evaporation process is the resistance to water movement in the soil and not conditions in the atmosphere. Other papers that discuss this very important feature of evaporation are Rose (1963), Jackson (1964), and Sasamori (1970).

Consequently we may distinguish between two stages of evaporation: (1) atmosphere-controlled or potential evaporation when $h_0 = 1$, $E/E_p = 1$ and (2) soil-controlled or actual evaporation when $1 > h_0 > h_a$, $E/E_p < 1$.

Under constant potential evaporation E_p, the switching from stage 1 to stage 2 evaporation, when it occurs, is sudden and precipitate at the pedon scale. The ratio of actual to potential evaporation E/E_p falls rapidly during several hours. This can be verified by numerical solution of Equation (23.4) subject to boundary conditions (23.7) and (23.15) using, for example, Program 9.1 in Campbell (1985). [Consult Scotter et al. (1988) for a discussion of these programs.] A typical solution is shown in Figure 23.2, which shows two stages of evaporation: the constant or potential stage and the falling-rate stage. These features have also been observed in many laboratory experiments and references may be found in Philip (1957b) and Idso et al. (1974).

To find approximate analytical solutions for evaporation, we may imitate the switched boundary conditions for infiltration, and replace the single mixed non-linear boundary condition (23.15) with (1) a prescribed flux E_p during the first stage of evaporation, and (2) a prescribed surface potential $\psi_m(0,t)$ during the second stage.

The switching condition for evaporation may be set as a relative humidity of 0.99 or as the corresponding potential or moisture content at $z = 0$.

Analytical solutions for both linear and nonlinear simplifications of Equations (23.4) and (23.7) and these switched boundary conditions are summarized in Kuehnel (1989). Approximate expressions for the time to switching for soils that are initially

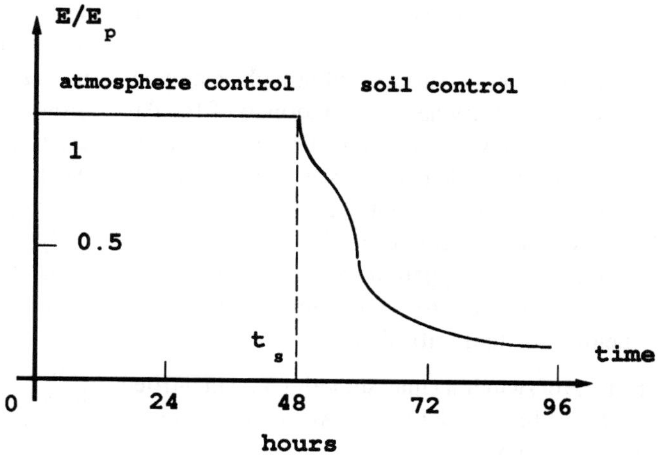

FIGURE 23.2. Variation with time of the ratio of actual to potential evaporation. t_s is the time to switching from stage 1 to stage 2 drying.

saturated are given in Kuehnel et al. (1990b). This time may vary by orders of magnitude depending on soil properties and potential evaporation. Soils with a shallow water table and sufficiently small values of E_p may never enter the second stage of drying.

The concepts of the atmosphere-determined (Stage 1) and soil-determined (Stage 2) stages of isothermal drying are good approximations, but are not exact. Simulations by Sasamori (1970) of coupled heat, momentum, and moisture flux in a combined soil–atmosphere model show that the approximations remain good when a more complete treatment of the physics is considered.

Switching may be most rapid for sand and least for clay soils. Philip (1974) has remarked that "evaporation from an initially wet swelling soil will not exhibit the sharp transition between constant-rate and falling-rate stages which is characteristic of non-swelling soils." Clay soils also crack as they dry. These phenomena are not included in the model presented here. In addition vegetated surfaces are likely to exhibit less pronounced switching because the roots draw water from a volume of soil, whereas in the case of bare soil, evaporation must leave through the soil surface.

The switching from potential (Stage 1) to actual (Stage 2) evaporation is associated with a matching increase in the ratio of sensible to latent heat flux and a consequent rise in surface temperature. When evaporative cooling is switched off, the average daily temperature of the surface of bare soil may rise by as much as 10°. Numerical simulations with diurnally varying E_p also show that the amplitude of the surface temperature may double due to the switch in control (Van Bavel and Hillel, 1976; Hillel, 1977). Two- or three-dimensional models are required to examine the feedback at different scales between the atmosphere and relatively dry fractal patches growing on a wet surface: a "negative" oasis effect. The technique of Partial Analysis may be a fruitful approach in analyzing this problem (Kuehnel et al., 1990c).

The influence of thermal gradients on water flux becomes important as the second stage of evaporation proceeds. Finally evaporation may reach a constant low value: Stage 3 evaporation. This demands a nonisothermal theory of evaporation (Philip and de Vries, 1957).

4.3 Field Evidence

Idso et al. (1974) carried out six major drying experiments on a smooth bare field (72 × 90 m) of clay loam in Arizona, during six periods of the year. The field was irrigated with approximately 10 cm of water and then allowed to dry. Intensive measurements were made of evaporation, soil–water content, soil temperature, soil heat flux, albedo, and net radiation. Their conclusions were as follows:

1. The three classical stages of soil drying may occur in naturally varying field conditions.
2. Simple albedo measurements can often successfully distinguish the transition points between the different stages.
3. Stage 3 drying appears to be initiated at a surface water content that corresponds to a retention of

two molecular layers of water about the soil particles at that level.
4. Rewetting of the surface soil at night may reinstate Stage 2 drying on the morning of the day following Stage 3 initiation. During seasons of low evaporative demand, this oscillation between Stages 2 and 3 may continue for several days.
5. Transitions between Stages 2 and 3 of soil drying may not be apparent from evaporation data if the plane of zero total water flux is located below a certain critical level.

The time of switching may sometimes be identified to within 1 hr. Oscillations between Stages 1 and 2 may also be possible.

5. Implications for Remote Sensing

5.1 The Spatial Expression of Switching

Fluxes of water from atmosphere to soil, falling as rain, or from soil to atmosphere, rising as evaporated water vapor, are approximately controlled either by the atmosphere or by the soil. The control may switch suddenly from one to the other, particularly in the case of bare soil. Consequently, the switching is expressed spatially as patches of soil. Within each patch the switchover to soil control has occurred. During rain, soil patches are covered with ponded water. Under dry conditions, soil patches are evaporating at less than the potential rate.

The patches define two dynamic Boolean fields: one for ponded rain and the other for falling-rate evaporation. Inside each patch the field value is one; outside it is zero. The moving boundaries of the patches as they appear, grow, coalesce, contract, and disappear are the sites where strong nonlinear switching is occurring. The geometry of the patches may be fractal.

5.2 Remote Sensing of Infiltration

The most easily observed feature of infiltration at the pedon scale is the accumulation of rain on the surface of the ground, which coincides with the nonlinear switching of control from atmosphere to soil. We may envisage water starting to pond at points on a large flat area, and growing into fractal patches which may coalesce. Large horizontal areas of bare soil are exceptional. Consequently, spatially varying topography and vegetation will redistribute the ponded water to areas where ponding has not yet occurred. Patches may eventually grow to include parts of the permanent surface-drainage network. After rain has stopped, the patches disintegrate and shrink, due to surface-drainage, infiltration, and evaporation.

By repeatedly observing the Boolean field for ponded soil over selected areas, during rain events of different intensity and duration, we may learn how spatially varying topography, vegetation, and soil properties control infiltration. This is very important for the formulation of hydrological models at scales ranging from 1 m to that of a GCM grid element.

Imaging systems that resolve the Boolean field at the pedon scale over large test areas are required initially. Recent work indicates that remote imaging from towers, aircraft, or helicopters may now be feasible. The pixel size of the imaging system determines the smallest patch that can be resolved. Satellite systems have too big a footprint to resolve the nonlinear switching.

Radar backscatter from soil depends strongly on moisture content. This is due to the large contrast between the dielectric constant of liquid water and that of dry soil at the microwave wavelength. Using data from a field platform, Camillo and Schmugge (1984) demonstrated that passive microwave radiometry may be used to estimate soil wetness in the surface layer of the soil both during and after rain. Choudhury (1990) used the 37 GHz brightness temperature measurements from the SMMR radiometer on the Nimbus-7 satellite to estimate successfully the monthly variation in the width and flow of the Amazon. He included an attenuating layer of vegetation in his radiative transfer model.

Consequently we make the first of two recommendations to the remote-sensing community:

Future HAPEX-MOBILHY and FIFE-type experiments should make provision for (1) the remote imaging of the Boolean field of patches of ponded soil during and after rain, using a range of overlapping pixel sizes starting at 1 m, on bare and vegetated soils and (2) the collection of associated atmospheric and ground truth.

5.3 Remote Sensing of Evaporation

We may learn how spatial variability influences evaporation by repeatedly observing the Boolean field for the evaporation switch over selected areas, under different atmospheric conditions, topography, vegetation, and soil properties. Since the switching is more pronounced for bare soil, it should be considered initially. These data are very important for the formulation of physically based models of evaporation at scales ranging from 1 m to that of a GCM grid element.

Remote imaging of evaporation from towers, aircraft, or helicopters is feasible. Many studies have been carried out on the remote sensing of evaporation from bare soil and from vegetation, for example, Rosema (1975), Hechinger (1979), Recan (1982), Lascano and van Bavel (1983), Camillo et al. (1983), Camillo and Schmugge (1983), Jackson and Schmugge (1986), Gurney and Camillo (1984), and Passerat de Silans (1986).

Typically, these studies begin with the development of a model, which is then applied to the interpretation of remotely sensed data on evaporation. The model is often built on the nonisothermal theory of evaporation (Philip and de Vries, 1957), extended to include the exchange of heat, momentum, and moisture with the lower boundary layer of the atmosphere, and completed with submodels for vegetation and radiative transfer.

A number of general remarks may be made about these pioneering efforts:

1. The nonlinear switching in the surface balances of water, radiation, and energy is the most important feature of the physics at the pedon scale. Frequently this is not identified in the model output.
2. Calibration of the model using data from Stage 1 evaporation only is insufficient. The switching to the second and third stages must also be calibrated.
3. The models apply only at the pedon scale and are unlikely to be correct at larger scales (see Section 4.1).
4. Monte Carlo experiments to extend these physically based models to larger scales are premature, since we have no data on the spatial variability of the nonlinear switching.

The switching from potential (Stage 1) to actual (Stage 2) evaporation in the case of bare soil is associated with a matching increase in the ratio of sensible to latent heat flux and a consequent rise in surface temperature. When evaporative cooling is switched off, the daily-average surface-temperature may rise by as much as 10°C and the amplitude of the daily variation in surface temperature may double (Van Bavel and Hillel, 1976; Hillel, 1977). Idso et al. (1974) have shown that simple albedo measurements can often distinguish the transition points between the different stages of evaporation from bare soil. The time of switching may sometimes be identified to within 1 hr.

Commercially available CCD (charged couple device) imaging systems can probably resolve (1) the emergence of patches where switching is taking place, (2) the growth and coalescence of patches, and (3) the modification of the patches by wind, humidity, radiation, the daily cycle in E_p, topography, sparse vegetation, and perhaps water flowing parallel to the ground in the unsaturated zone.

The diurnal cycle must be filtered from the data to observe the switching clearly. Sampling at intervals of 10 min may be necessary to achieve this. The pixel size of the imaging system determines the smallest patch that can be resolved. A range of pixel sizes should be used starting at 1 m.

Consequently we make the second of two recommendations to the remote-sensing community:

> Future HAPEX-MOBILHY and FIFE-type experiments should make provision for (1) the remote imaging of the Boolean field of the evaporation switch, using a range of overlapping pixel sizes starting at 1 m, on bare and vegetated soils and (2) the collection of associated atmospheric and ground truth. The same sites should be used for both infiltration and evaporation imaging studies.

Acknowledgments. This research has been supported by the climatology programme of the European Communities and was funded under the project "Spatial variability of land surface processes." I thank my colleagues Jim Dooge, Renata Romanowicz, and John Turner in the Centre for Water Resources Research for many stimulating discussions. I also thank the anonymous reviewer for critical suggestions. Any remaining mistakes and omissions are mine.

References

Brutsaert W (1982) "Evaporation into the Atmosphere." D. Reidel, Dordrecht.

Buckingham E (1907) Studies on the movement of soil moisture. US Dept Agr Bur Soils Bull 38:28–31.

Camillo PJ, Gurney RJ, Schmugge TJ (1983) A soil and atmospheric boundary layer model for evapotranspiration and soil moisture studies. Water Res Res 19(2):371–380.

Camillo PJ, Schmugge TJ (1983) Estimating soil moisture storage in the root zone from surface measurements. Soil Sci 135(4):245–264.

Camillo PJ, Schmugge TJ (1984) Correlating rainfall with remotely sensed microwave radiation using physically based models. IEEE Trans Geosci Remote Sens GE-22(4):415–423.

Campbell GS (1977) "An introduction to Environmental Biophysics." Springer-Verlag, New York.

Campbell GS (1985) "Soil Physics with Basic - Transport Models for Soil-Plant Systems." Elsevier, Amsterdam.

Choudhury BJ (1990) Passive microwave remote sensing of some hydrologic parameters. In "Land Surface-Atmosphere Interactions: Parameterization and Analysis for Climate Modelling" (E. Wood, ed.) D. Reidel, Dordrecht (in press).

Dalton J (1802) Experimental essays on the constitution of mixed gases; on the force of steam or vapour from water. Manchester Lit Phil Soc Mem 5(2):536–602.

Edlefsen NE, Anderson ABC (1943) The thermodynamics of soil moisture. Hilgardia 16:31–299.

Gurney RJ, Camillo PJ (1984) Modelling daily evapotranspiration using remotely sensed data. J Hydrol 69: 305–324.

Hechinger E (1979) Contribution a l'interprétation de données de télédétection: Etude d'un modèle thermique de sols et de son emploi pour la réalisation d'images de télédétection dans le visible et l'infrarouge thermique. Thèse de Docteur de Spécialité en électronique et instrumentation, Université Louis Pasteur de Strasbourg.

Hillel D (1977) "Computer simulation of soil-water dynamics; a compendium of recent work." International Development Research Centre, Ottawa.

Idso SB, Reginato RJ, Jackson RD, Kimball BA, Nakayama FS (1974). The three stages of drying of a field soil. Soil Sci Soc Proc 38:831–837.

Jackson RD (1964) Water vapour diffusion in relatively dry soil: I. Theoretical considerations and sorption experiments. Soil Sci Soc Am Proc 28:172–176.

Jackson TJ, Schmugge TJ (1986) Passive microwave remote sensing of soil moisture. Adv Hydrosci 14: 123–161.

Kuehnel V (1989) Scale problems in soil moisture flow. Ph.D. Thesis, National University of Ireland.

Kuehnel V, Dooge JCI, Sander GC, O'Kane JP (1990a) Duration of atmosphere-controlled and of soil-controlled phases of infiltration for constant rainfall at a soil surface. Ann Geophys 8:11–20.

Kuehnel V, Dooge JCI, Sander GC, O'Kane JP (1990b) Duration of atmosphere-controlled and of soil-controlled phases of evaporation for constant potential evaporation at a soil surface. Ann Geophys 8: 21–28.

Kuehnel V, Dooge JCI, O'Kane JP, Romanowicz R (1990c) Partial analysis applied to scale problems in surface moisture fluxes. In "Land Surface-Atmosphere Interactions: Parameterization and Analysis for Climate Modelling" (E. Wood, ed.). D. Reidel, Dordrecht (in press).

Lascano RJ, van Bavel CHM (1983) Experimental verification of a model to predict soil moisture and temperature profiles. J Soil Sci Soc Am 47:441–448.

Passerat de Silans A (1986) Transferts de Masse et de Chaleur dans un Sol Stratifié soumis à une excitation atmosphérique naturelle. Comparison: Modèles-expérience. Thèse de Docteur de l'Institut National Polytechnique de Grenoble.

Penman HL (1940) Gas and vapour movement in soil. I and II. J Agr Sci 30:437–462, 570–581.

Penman HL (1948) Natural evaporation from open water, bare soil and grass. Proc R Soc London Ser A 193:120–145.

Penman HL (1958) Humidity. "Monograph of the Institute of Physics." Chapman & Hall, London.

Philip JR (1955) The concept of diffusion applied to soil water. Proc Natl Acad Sci India 24A:93–104.

Philip JR (1957a) The physical properties of soil water movement during the irrigation cycle. Proc Third Int Congress of the Int Commiss Irrigation Drainage 8.125–8.154.

Philip JR (1957b) Evaporation, and moisture and heat fields in the soil. J Meteorol 14:354–366.

Philip JR (1967) The second stage of drying of soil. J Appl Meteorol 6:581–582.

Philip JR (1974) Water movement in soil. In "Heat and Mass Transfer in the Biosphere – Papers from a Forum on Transfer Processes in the Plant Environment," Chapter 2 (D.A de Vries, N.H. Afgan, eds.), Dubrovnik, August 1974. "Advances in Thermal Engineering," Vol. 3, pp. 29–47, 1975. Scripta Book Co., Washington, and Wiley, New York.

Philip JR, de Vries DA (1957) Moisture movement in porous materials under temperature gradients. Trans Am Geophys Union 38:222–232.

Recan M (1982) Simulation numérique du comportement thermique et hydrique d'un sol nu. Application à l'étude de l'évaporation par télédétection. Thèse. IMPT. Referenced in Passerat de Silans (1984).

Richards LA (1931) Capillary conduction of liquids through porous mediums. J Phys 1:318–333.

Rosema A (1975) Simulation of the thermal behaviour of bare soils for remote sensing purposes. In "Heat and

Mass Transfer in the Biosphere – Papers from a Forum on Transfer Processes in the Plant Environment," Chapter 8 (D.A. de Vries, N.H. Afgan, eds.), Dubrovnik, August 1974. "Advances in Thermal Engineering," Vol. 3, pp. 109–123, 1975. Scripta Book Co., Washington, and Wiley, New York.

Rose CW (1966) "Agricultural Physics." Pergamon Press, London.

Rose DA (1963) Water movement in porous materials. Part 1–Isothermal vapour transfer. Br J Appl Phys 14:256–262.

Sasamori T (1970) A numerical study of atmospheric and soil boundary layers. J Atmos Sci 27:1122–1137.

Scotter DR, Clothier BE, Sauer TJ (1988) A critical assessment of the role of measured hydraulic properties in the simulation of absorption, infiltration and redistribution of soil water. Agric Water Manage 15: 73–86.

Serrin J, et al (1986) "New Perspectives in Thermodynamics." Springer-Verlag, Berlin.

Thorstenson DC, Pollock DW (1989) Gas transport in unsaturated porous media: the adequacy of Fick's Law. Rev Geophys 27(1):61–78.

Truesdell C (1981) "Textbook of Classical Thermodynamics." Department of Mechanics, The Johns Hopkins University.

Truesdell C, et al (1984) "Rational Thermodynamics," 2nd ed. Springer-Verlag, Berlin.

Van Bavel CHM, Hillel DI (1976) Calculating potential and actual evaporation from a bare soil surface by simulation of concurrent flow of water and of heat. Agric Meteorol 17:453–476.

24 An Example of Spatial Integration of a Land-Surface Parameterization in a Meso-Beta-Scale Model

P. Bougeault, B. Bret, P. Lacarrere, and Joël Noilhan

1. Introduction

Natural land surfaces are not homogeneous over the resolvable scales of atmospheric models. Therefore, these models must account for the subgrid-scale variability of underlying surfaces to determine the grid-averaged surface energy budget. The most difficult problem concerns the partition of available energy into sensible and latent heat fluxes, hence the modeling of regional evapotranspiration, since the moisture availability, defined in a broad sense (see Bougeault, this volume, for a review on this parameter), will generally depend on numerous factors: the soil type, the soil moisture content, and the type, age, and density of vegetation. Each of these factors has its own subgrid-scale variability. If we assume that we can compute local evapotranspiration when we know local values of the controlling factors, we are still confronted by the necessity of averaging this result over a model grid box, taking into account the statistical properties of the existing distribution of the controlling factors inside the grid box. Generally, this will require not only the knowledge of mean values and variances but also possible correlations of these elements inside this grid box. The response of a parameterization scheme to this averaging process is expected to be quite nonlinear because of the wide range of observed variation of evapotranspiration, and of the nonlinear dependence of local evapotranspiration on the controlling parameters. However, the exact nature and severity of this nonlinearity is, at the present time, not known.

This chapter is based on a quite optimistic point of view. In the authors' opinion, simple, empirical methods should first be evaluated against existing observations, before going on to more complicated ones. The development of this "optimistic" approach is as follows. First, it is based on the calibration of a parameterization scheme against local measurements, which allows for the determination of the most sensitive parameters. These parameters will clearly require the greatest attention, when trying to obtain area-averaged values. Second, it relies on the progressive elaboration of maps of parameters, by various methods, that will be required anyway if more sophisticated approaches are necessary in the future. Thus, the present work must be viewed as a first, empirical attempt to implement a simple grid averaging method. Its prime merit is that it can be done with the existing information, and that it will set a lower limit for the performances of an averaging method.

The chapter will proceed as follows: we present the basic parameterization scheme in Section 2, along with a list of parameters that need specification to be applied in a mesoscale model (Table 24.1). These parameters have been deliberately kept to a minimum number to simplify the further work of gathering information to produce maps. In Section 3, we discuss *one-dimensional calibration runs*. These runs using the field measurements of the HAPEX-MOBILHY experiment (André et al., 1988), which have been done in large number, allow for the determination of optimal values of those parameters that are not well known, thereby enabling the scheme to reproduce local measurements of the fluxes of sensible and latent heat into the atmosphere, and of the heat flux into the ground. Some of the results of these experiments are summarized in Table 24.2, in which a list of

TABLE 24.1. The parameters used in the NP89 scheme.

Parameters	Notations
Primary	
Dominant type of vegetation	
Dominant type of soil texture	
Secondary	
Saturated volumetric moisture content	w_{sat}
Wilting point volumetric water content	w_{wilt}
Slope off the retention curve	b
Soil thermal coefficient at saturation	$C_{G\ sat}$
Value of C_1 at saturation	$C_{1\ sat}$
Value of C_2 for $w_2 = 0.5w_{sat}$	$C_{2\ ref}$
Coefficients of w_{geq} formulation	a,p
Depth of the soil column	d_2
Fraction of vegetation	veg
Minimum surface resistance	$R_{s\ min}$
Leaf area index	LAI
Roughness length	z_0
Albedo	α
Emissivity	ε

optimal values of all the parameters of the scheme is given for some of the cases that have been experimentally documented during the HAPEX-MOBILHY field experiment. Section 4 deals with the way of deriving maps of these parameters, using various information sources. The three-dimensional (3D) numerical model, using the parameterization scheme and the maps of parameters previously described, is then used to simulate 1 day of the HAPEX-MOBILHY experiment: June 16, 1986. The results of this simulation are described in Sections 5 and 6, and compared to available observations.

2. Formulation of the Basic Parameterization Scheme

For the present study we use the parameterization scheme proposed by Noilhan and Planton (1989, hereafter referred to as NP89). This scheme allows for the computation of the different components of the surface energy and moisture budgets over various types of soil and vegetation, and for the interception of rain or dew on the vegetation.

The scheme has five prognostic variables, which are the surface temperature T_s, the deep soil temperature T_2, the amount of moisture retained at the surface of the vegetation W_r, the superficial soil volumetric water content w_g, and finally the vertically integrated soil volumetric water content w_2. The prognostic equations for these variables are as follows:

$$\frac{\partial T_s}{\partial t} = C_T G - 2\pi(T_s - T_2)/\tau \quad (24.1)$$

$$\frac{\partial T_2}{\partial t} = (T_s - T_2)/\tau \quad (24.2)$$

$$\frac{\partial W_r}{\partial t} = \text{veg}\, P - (E_v - E_{tr}) - P_r, \quad 0 \le W_r \le W_{r\,max} \quad (24.3)$$

$$\frac{\partial w_g}{\partial t} = C_1(P_g - E_g)/\rho_w d_1 - C_2(w_g - w_{g\,eq}), \quad 0 \le w_g \le w_{sat} \quad (24.4)$$

$$\frac{\partial w_2}{\partial t} = (P_g - E_g - E_{tr})/\rho_w d_2, \quad 0 \le w_2 \le w_{sat} \quad (24.5)$$

The definitions of the various parameters found in Equations (24.1) through (24.5) are as follows. The most important one, "veg," is the "fraction of vegetation" within the grid. It must be understood as a foliage shielding factor of the ground from solar radiation. A fraction (1−veg) of incoming solar radiation will effectively reach the soil. Also, a fraction (1−veg) of the rainfall or dew will directly reach the ground.

τ is the diurnal time constant (τ = 86,400 sec). C_T is the thermal inertia of the soil–vegetation medium. It is expressed as

$$C_T = (1-\text{veg})/C_G + \text{veg}/C_v)^{-1} \quad (24.6)$$

where $C_v = 10^{-3}$ km^2 J^{-1} is an arbitrarily large value, which means that the vegetation has a very weak thermal inertia, and C_G, the thermal inertia of the ground, is parameterized as

$$C_G = C_{G\,sat}\,(w_{sat}/w_2)^{b/\log 10} \quad (24.7)$$

G, soil heat flux, is equal to the sum of three components:

$$G = R_n - H - LE \quad (24.8)$$

where R_n is the net radiation and H and LE are the sensible and latent heat fluxes, respectively.

The net radiation R_n at the surface is the sum of the absorbed fractions of the incoming solar radiation R_G, and of the atmospheric infrared radiation R_A, reduced by the emitted infrared radiation.

$$R_n = R_G(1 - \alpha) + \varepsilon\,(R_A - \sigma T_s^4), \quad (24.9)$$

where the albedo α and the emissivity ε combine the soil and the vegetation reflectivities. σ is the Stefan–Boltzmann constant. The turbulent fluxes

TABLE 24.2. Optimal values of some parameters used in one-dimensional experiments.

No.	Site	Vegetation	Texture	w_g	w_2	z_0 (m)	α	LAI	z_0 (m)	R_{smin} (s/m)
1	Estampon	Coniferous	Sand	0.14	0.20	1.00	0.10	2.3	100	0.99
2	Caumont	Cro/sojà	Loam	0.26	0.26	0.02	0.24	1.0	40	0.70
3	Castelnau	Crop/maize	Loam	0.18	0.25	0.02	0.25	0.3	40	0.40
4	Lubbon 2	Crop/maize	Sand	0.17	0.17	0.10	0.15	2.0	40	0.80
5	Tieste	Crop/maize	Loam	0.20	0.26	0.05	0.23	0.5	40	0.40
6	Fusterou.	Crop/maize	Loam	0.16	0.27	0.06	0.25	1.0	40	0.50
7	Lubbon 1	Crop/oats	Sand	0.17	0.17	0.10	0.20	3.0	40	0.90
8	Orchards, vineyards		Loam	0.18	0.25	0.15	0.18	1.0	80	0.40
9	Desert		Sand	0.06	0.12	0.10	0.20	0.1	999	0.10
10	Desert		Loam	0.08	0.15	0.10	0.20	0.1	999	0.10

toward the atmosphere are computed by the classical aerodynamic formulas. For the sensible heat flux

$$H = \rho_a C_P C_H V_a (T_s - T_a) \quad (24.10)$$

where C_P is the specific heat, ρ_a, V_a, and T_a the air density, wind speed, and temperature at the atmospheric level Z_a, and C_H the drag coefficient for heat, depending on the stability of the surface layer.

The water vapor flux E is split into evaporation from the soil surface E_g, transpiration from the vegetation E_{tr} and direct evaporation from the water retained at the surface of the vegetation $E_v - E_{tr}$. Hence, the total evapotranspiration is E_v. These various components are parameterized according to

$$E_g = (1-\text{veg})\rho_a C_H V_a [h_u q_{sat}(T_s) - q_a] \quad (24.11)$$

$$E_v = \text{veg}\, \rho_a C_H V_a h_v [q_{sat}(T_s) - q_a] \quad (24.12)$$

where q_a is the atmospheric specific humidity at level Z_a, h_u is the relative humidity of the surface soil

$$h_u = [1 - \cos(w_g \pi/w_{fl})]/2 \quad \text{if } w_g < w_{fl}$$
$$= 1 \quad \text{if } w_g > w_{fl} \quad (24.13)$$

with $w_{fl} = 0.75\, w_{sat}$ the field capacity.

The Halstead coefficient h_v takes into account the direct evaporation from the fraction δ of the foliage surface covered by water, and the transpiration from the fraction (1 − δ):

$$h_v = (1 - \delta) R_a/(R_a + R_s) + \delta \quad (24.14)$$

with $\delta = (W_r/W_{rmax})^{2/3}$, $R_a = (C_H V_a)^{-1}$ the atmospheric resistance, and R_s is the surface resistance. The expression of R_s, derived from Jarvis (1976), incorporates the various influences of the leaf area index (LAI), the solar radiation, the atmospheric water vapor deficit, the air temperature, and the soil moisture content under the empirical form

$$R_s = R_{smin} (\text{LAI})^{-1} R_1 F_2^{-1} F_3^{-1} F_4^{-4} \quad (24.15)$$

$R_{s\,min}$ is the minimum stomatal resistance of a single leaf, which is assumed to be species dependent, and also may vary according to the degree of maturity for a given crop. The exact formulation of F_1, F_2, F_3, F_4, can be found in Bougeault (this volume). It is also, in principle, species dependent and involves a lot of empiricism.

In Equation (24.3) P is the total precipitation rate reaching the surface before interception by the vegetation and P_r is the runoff of the interception reservoir. This runoff occurs when W_r exceeds $W_{r\,max}$, which depends on the density of the canopy according to an empirical formula

$$W_{r\,max} = 0.2 \text{ veg LAI (mm)} \quad (24.16)$$

In Equation (24.4), P_g is the flux of liquid water reaching the surface after interception. E_g is the evaporation from the soil surface and w_{geq} is an equilibrium value of the superficial moisture content that has been parameterized as a function of w_2 and of the type of soil. d_1 is an arbitrary normalization depth of 10 cm. C_1 and C_2 are two coefficients that depend on the type of soil and on the soil moisture content (see Bougeault, this volume, for the formulation).

Finally in Equation (24.5) d_2 is the total depth of soil active in the water budget.

The parameters that need to be specified to implement the NP89 scheme on a local basis are summarized in Table 24.1. Among these, the ones that describe the thermal and hydraulic properties of the ground itself can be assigned a reasonable value from the knowledge of the soil texture alone, thanks to former work by Cosby et al. (1984), and by the calibrations presented by NP89. This is the case for w_{sat}, w_{wilt}, b, C_{Gsat}, C_{1sat}, C_{2ref}, a, and p. The depth of the active soil column d_2, on the other hand, needs an independent specification. The remaining parameters either concern the type and density of the vegetation (veg, $R_{s\ min}$, LAI) or are global concepts that incorporate information from several sources (Z_0, α, ε). Those parameters are the most difficult to determine for the use in a 3D model. The strategy we have developed consists of obtaining representative values of these parameters for each type of vegetation cover that has been sampled during the field experiment. This is done by a series of trial-and-error one-dimensional experiments. Then, these optimal values are empirically combined to obtain values representative of the grid box.

3. One-Dimensional Calibration Experiments of the Scheme

The main originality of the calibration experiments in this program was to use a 1D version of the physical package of the mesoscale model that was to be later used for the integration itself. This means that the surface scheme was run interactively with the remainder of the model physics (boundary layer and radiation subschemes) for 24-hr periods starting and finishing at midnight. In particular, the atmospheric parameters V_a, q_a, T_a at level Z_a were not imposed, but predicted by the model. As the atmosphere is not 1D, it was necessary to implement a method to constrain the 1D model to follow that part of the evolution that is not related to the surface processes. Therefore, the geostrophic wind as well as the horizontal and vertical advection terms were deduced from mesoscale analyses (Mercusot et al., 1986) to force the 1D model. In all cases, the model was initialized with the observed sounding at midnight and with the soil moisture content derived from the neutron probe observations. The soil parameters were set to the value derived from Clapp and Hornberger (1978), in agreement with the results of the textural analysis at each site.

All parameters that could be inferred from observations have been specified in this way. This was the case for the albedo, and the LAI, which were known for each site. The roughness length has been taken equal to one-tenth of the vegetation height. The minimum surface resistance $R_{s\ min}$ has been deduced following Monteith (1976), and the emissivity was given the constant value of 0.95. Finally, in most cases, only the fraction of vegetation "veg" was left unknown. This parameter was adjusted so as to reproduce as accurately as possible the soil heat flux.

The quality of the ensemble of parameters was checked by comparing the 24-hr predicted fluxes of latent and sensible heat with the observations of the SAMER and HYDRA stations (André et al., 1988).

The present paper will show only one example of such comparison. This is the case number 3 discussed in NP89. It illustrates an example with a significant fraction of bare ground (veg = 0.4). The measurements are taken in Castelnau in a field of corn (0.2 m high) over a loamy soil. The initial values of the soil moisture are $w_g = 0.18$, $w_2 = 0.25$, and the other parameters are set as $Z_0 = 2$ cm, $\alpha = 0.25$, LAI = 0.3, $R_{smin} = 40$ s/m. The results are shown in Figure 24.1. We observe high values of G, reaching some 100 W/m^2 at noon, since the amount of solar radiation reaching the ground is important. The model predicts the observed daily variations of G reasonably well. The magnitudes and diurnal variations of LE and H are also correctly simulated.

As a general result of the 1D calibration/test runs, we have been able to assemble a general table of the optimal parameters for every situation which had been experimentally documented. This is reproduced in Table 24.2. Note that most values are susceptible to vary slowly with time, as the crop changes, and are representative only for a period extending to a few days before and after the test.

4. Mapping the Parameters for the Three-Dimensional Model

The domain of integration of the 3D atmospheric model is shown in Figure 24.2. It is much wider than the 100 × 100 km^2 HAPEX-MOBILHY study area to eliminate the influence of possible bound-

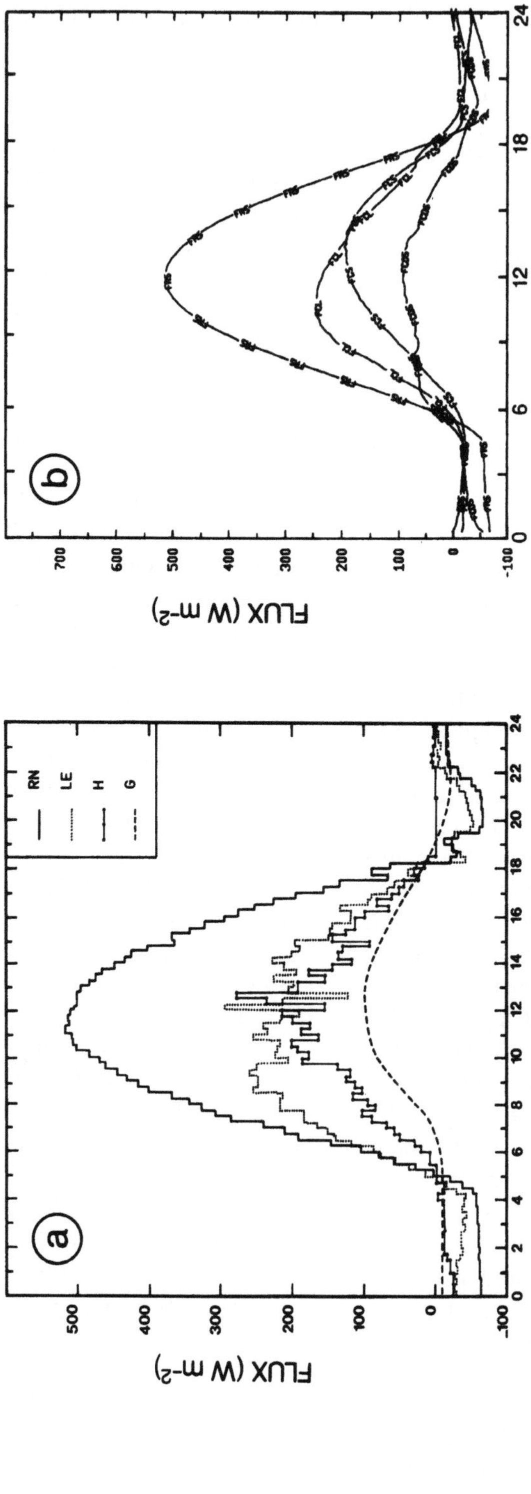

FIGURE 24.1. Diurnal variations of observed (a) and simulated (b) surface fluxes (in W/m^2) for Case 3 of NP89. Net radiation R_n, latent heat flux LE, sensible heat flux H, and soil heat flux G.

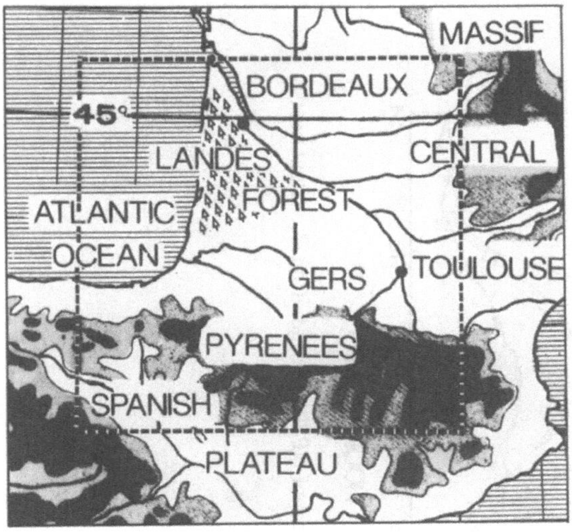

FIGURE 24.2. The domain of integration of the mesoscale model used in the present study.

ary reflections. The domain encompasses large portions of the Massif Central and Pyrénées mountains, part of the Atlantic ocean, and extends to the South as far as the high Spanish Plateau. The grid size is 10 km, and the total area comprises about 400 × 400 km².

Three main sources of information are presently available on this domain. The first one consists of several existing maps of soil types. These maps have been processed by Mascart et al. (1988) to produce a synthetic result for each grid box of the mesoscale model. First, the dominant textural class has been identified, according to a classification compatible with the one of Clapp and Hornberger (1978), used by the NP89 scheme. Second, the depth d_2 of the active soil layer has been mapped. Over the study area, its value ranges from 40 to 80 cm. On the other hand, soils less than 20 cm deep are to be found in the mountainous areas and over the Spanish plateau.

The second source of information is given by the series of Meteosat satellite infrared and visible images that have been archived during the Special Observing Period (SOP) of HAPEX. These images were processed by R.M. Thépenier at the CNRM, following the method of Dedieu et al. (1987) to produce maps of the mean albedo for each month of the SOP, on the grid of the mesoscale model.

Finally, the third source of information lies in the NOAA AVHRR data series, which are available between two and four times a day during the SOP. These data have been processed to obtain NDVI images (Phulpin and Jullien, 1988). The ultimate goal of this research program is to derive from these NDVI maps as much information related to the type of vegetation, and the parameters veg, R_{smin}, and LAI as possible. This goal, however, has not yet been reached. For the present study, we have used only a map of seven dominant vegetation types that has been obtained by a supervised classification method applied to a couple of cloud-free images (27 June and 10 July 1986). The details of this algorithm are given by Phulpin and Jullien (1988). This map is available with the initial AVHRR pixel resolution, i.e., close to 1 km (Fig. 24.3). It has been further processed to obtain the dominant vegetation type (according to the same seven classes) in each grid box of the mesoscale model (Fig. 24.4). These different maps have been used together with the information given in Table 24.2 to compute grid-scale values of the four remaining unknown parameters: veg, R_{smin}, LAI, and Z_0. Several algorithms have been tried, and are currently being tested.

The simplest method was to assign each parameter the value corresponding to the dominant type of vegetation at the scale of the model grid box. Since the NDVI-derived classification does not differentiate between different types of crops, a further simplification of Table 24.2 was necessary, and made possible by the fact that most crops have the same value of $R_{s\ min}$. Thus, a set of simplified correspondence between the type of vegetation and the scheme parameters was developed, as shown in Table 24.3. This most simple approach was tested and resulted in predictions of the mesoscale model including a high degree of noise. A more refined approach was therefore worked out. Taking advantage of the fact that the map of vegetation classes was available at the AVHRR pixel definition, the set of correspondence of Table 24.3 was used at this definition, and representative values of the parameters for each model grid box were computed by averaging the values corresponding to the pixels inside each grid box. This provides maps with smoother, and hopefully, more realistic, spatial variations of the parameters. Several combinations of the two former approaches were tried, and it was

FIGURE 24.3. A map of the vegetation type at the AVHRR pixel-scale resolution for the domain of simulation. (From Phulpin and Jullien, 1988.)

found that the most important effect was on the parameter "veg." Therefore, we will show in the next section results from a simulation where the pixel-averaging method has been used only for "veg" (the other parameters are still set to the value corresponding to the dominant type of vegetation inside the grid box). The field of the parameter "veg" obtained by the explained method is shown in Figure 24.5.

The last quantities requiring a mapping are the initial values of the soil moisture contents W_r, w_g, and w_2. W_r has been set to zero identically for the sake of simplicity. For w_g and w_2 an empirical procedure was defined, to take the best account of the existing information. Semidesertic areas present in the southern part of the domain (discriminated by the condition veg < 0.1) were assigned the values $w_g = 0.15\ w_{sat}$, and $w_2 = 0.3\ w_{sat}$. For all the other regions, the soil moisture content was assumed to depend mainly on the soil type. Average values of w_g/w_{sat} and w_2/w_{sat} were estimated from the neutron soundings for each of the four textural classes that had been experimentally documented within the HAPEX domain.

5. General Results of the June 16, 1986 Simulation

We have chosen June 16 as a convenient day for the first test of the method, since most participants to the HAPEX-MOBILHY program agreed on the fact that this day produced measurements of excellent quality. It should also be mentioned that a preliminary study (Bret and Bougeault, 1988) has shown that the atmospheric mesoscale simulation of June 16 was one of the best among the fair weather days selected for numerical experimentations. This is due to the quality of the mesoscale

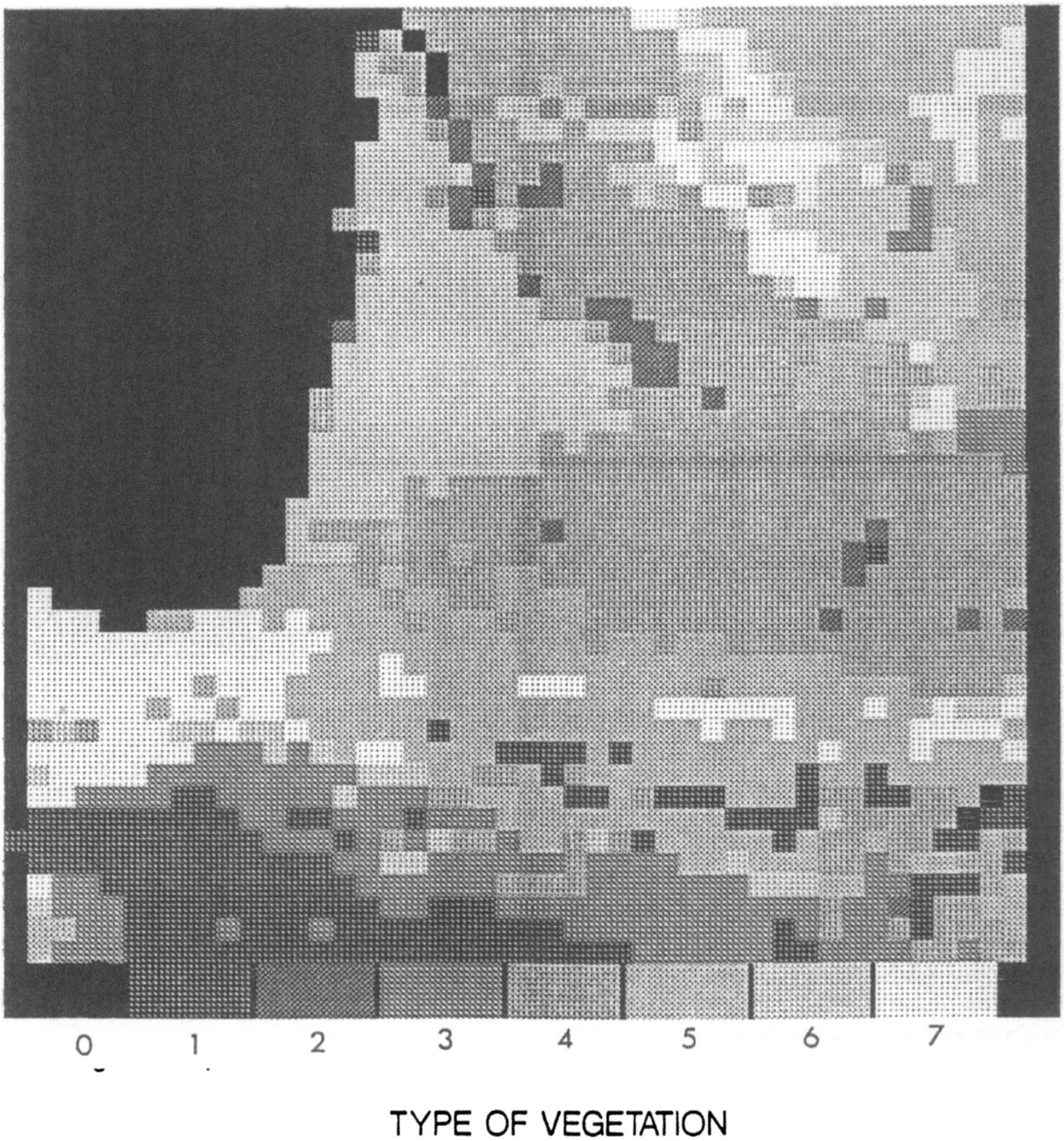

FIGURE 24.4. The map of dominant vegetation type inside each grid box of the mesoscale model. (From Phulpin and Jullien, 1988.)

atmospheric analyses provided by the operational data assimilation algorithms of the French Weather Service (Durand and Bougeault, 1987).

On the area of interest, the low level flow was characterized by weak to moderate southeastern winds during the first part of the day, associated with zero cloud cover, followed by an entry from the Atlantic Ocean of westerly maritime air, starting near 1500 Z, and reaching progressively the entire study area by 0000 Z the following day. The cloud cover was insignificant, except on mountain areas on the border of the study area (upslope circulations) and inside the moist, maritime air, where scattered low level cumuli developed near

1500 Z, followed by more extensive cloud sheets after 1800 Z.

To assess the realism of the dynamic fields predicted by the model, we first compare the predicted anemometer level winds at 1200 Z and 1800 Z with observations. At 1200 Z (Fig. 24.6), the winds are predominantly southeasterly on the study area. We note the influence of the upslope circulation toward the Pyrénées in the southern portion of the study area, where weak, northerly winds are observed. Also, on the coast line, weak westerly flow may be interpreted as an early manifestation of the maritime entry. All these elements are correctly predicted by the model, which additionally depicts a well-organized upslope circulation over the Pyrénées. At 1800 Z (Fig. 24.7),

TABLE 24.3. The simplified set of correspondence between the types of vegetation and the scheme parameters to be used for the mesoscale model.

Dominant surface type	No.	veg	LAI	Z_0 (m)	R_{smin} (s/m)
Water	0	0	0.001	0.001	999
Desert	1	0.10	0.1	0.15	999
Orchards, vineyards	2	0.40	1.0	0.15	80
Mediterranean vegetation	3	0.40	1.0	0.15	80
Crops	4	0.50	1.0	0.15	40
Grasslands	5	0.60	1.5	0.15	40
Coniferous forests	6	0.99	3.0	1.0	100
Deciduous forests	7	0.90	3.0	1.0	100

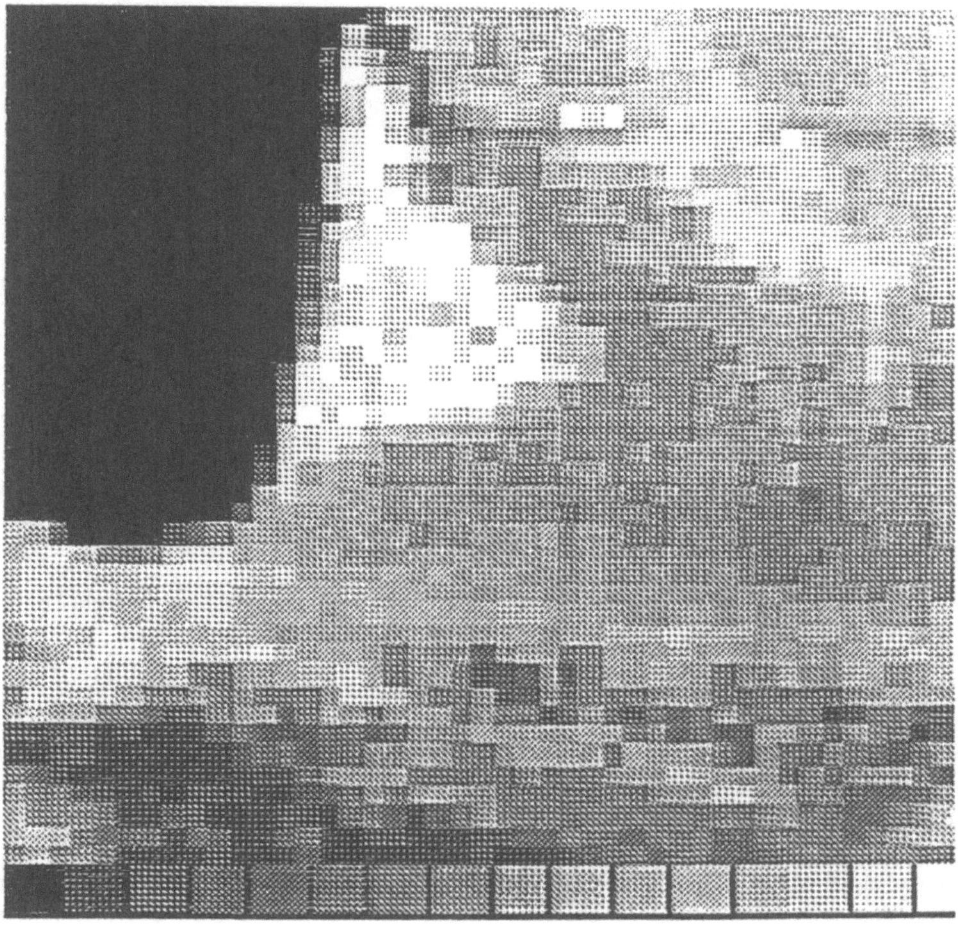

FIGURE 24.5. Map of the fraction of vegetation in each grid box of the mesoscale model, derived as explained in the text.

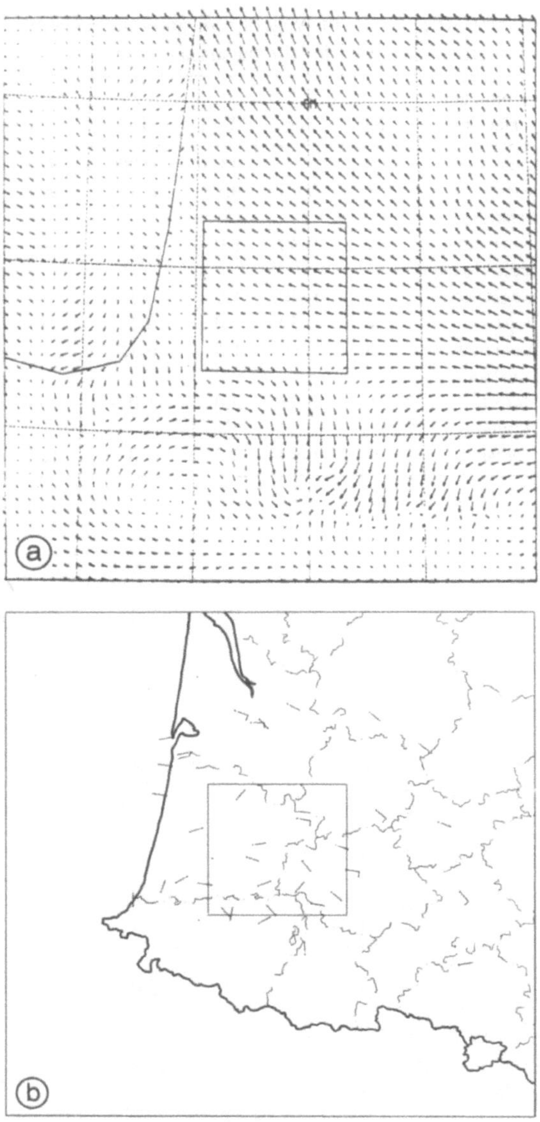

FIGURE 24.6. Computed (a) and observed (b) anemometer level winds at 1200 Z.

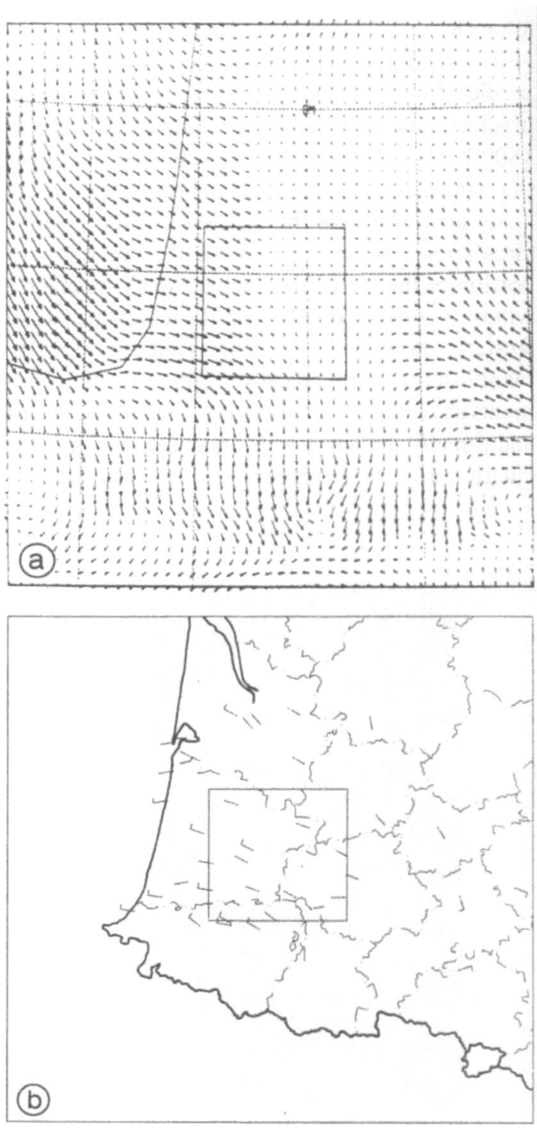

FIGURE 24.7. Computed (a) and observed (b) anemometer level winds at 1800 Z.

westerly winds have reached the middle of the study area and there is a clear separation between the two air masses, since weak to moderate southeasterly winds persist over the eastern part of the simulation domain. The model thus correctly predicts the advance of the maritime air mass.

Similar comparisons can be made for the temperature at screen level. At 1200 Z (Fig. 24.8) the most striking pattern is that the Landes forest has higher temperatures than the surrounding areas, by about 2 K. This is confirmed by the observations. However, the model prediction appears to underestimate by 2–3 K the observations as much over the forest as elsewhere.

The predicted sensible and latent heat fluxes at 1200 Z, shown in Figure 24.9, exhibit several interesting regional differences. First, the Landes forest appears well characterized by higher values

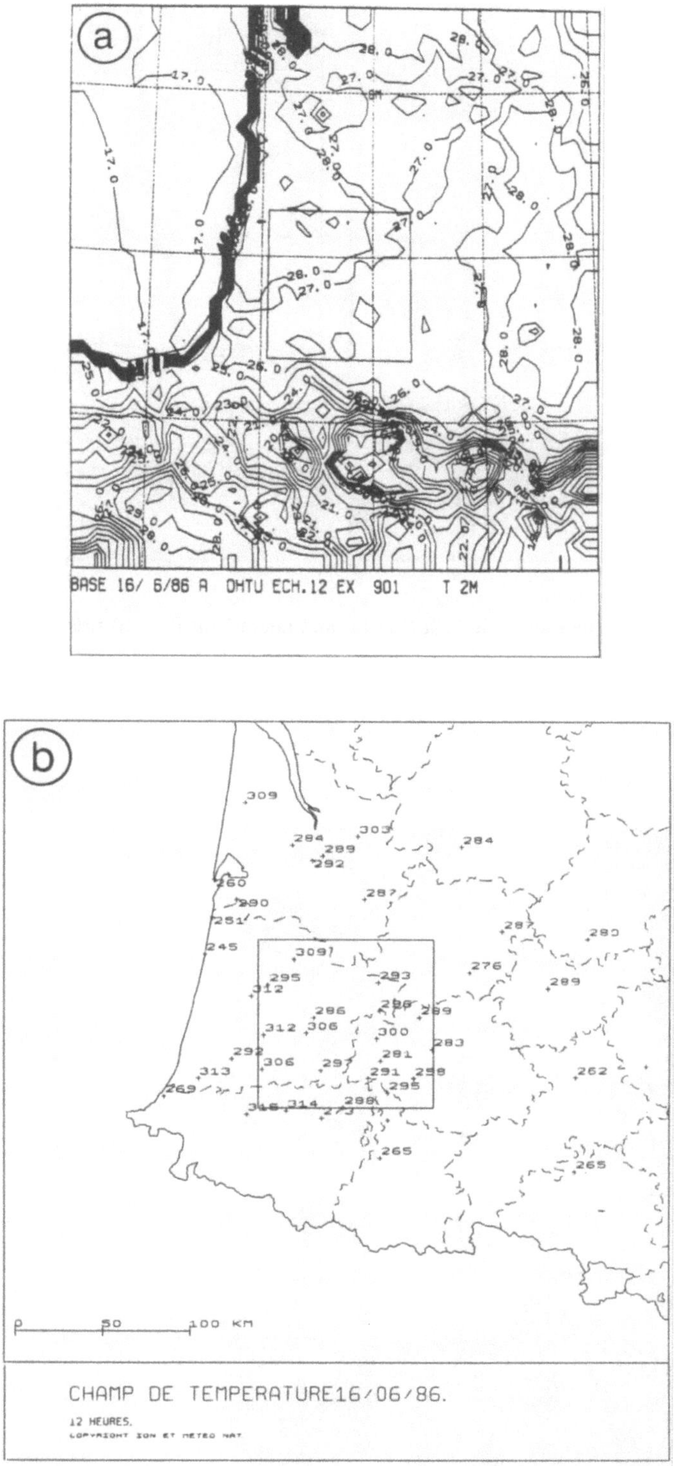

FIGURE 24.8. Computed (a) and observed (b) screen level temperature at 1200 Z.

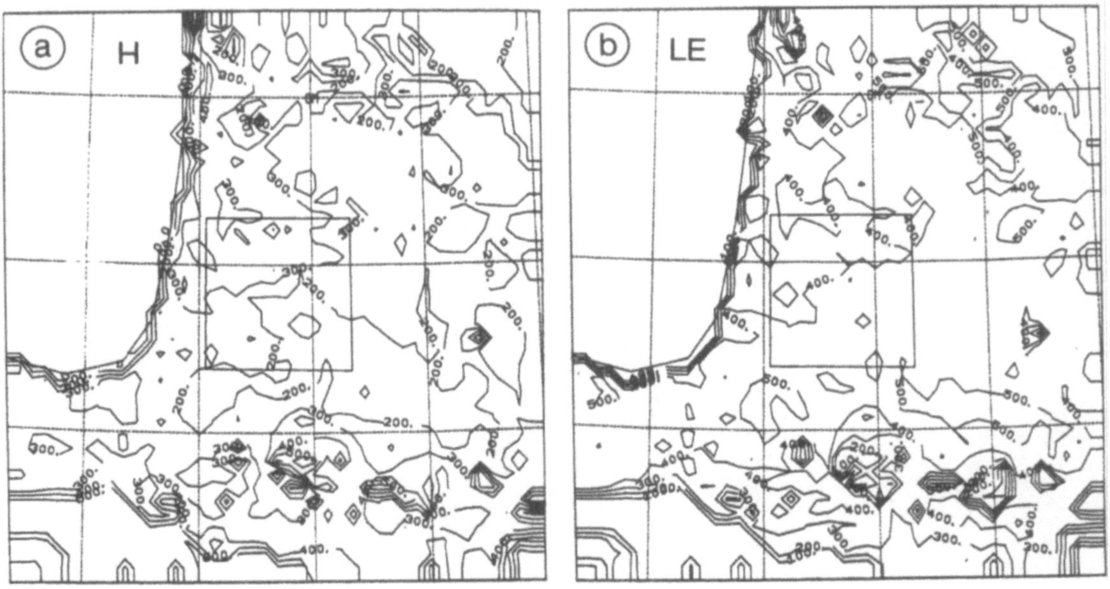

FIGURE 24.9. Surface sensible heat flux (a) and latent heat flux (b) predicted at 1200 Z.

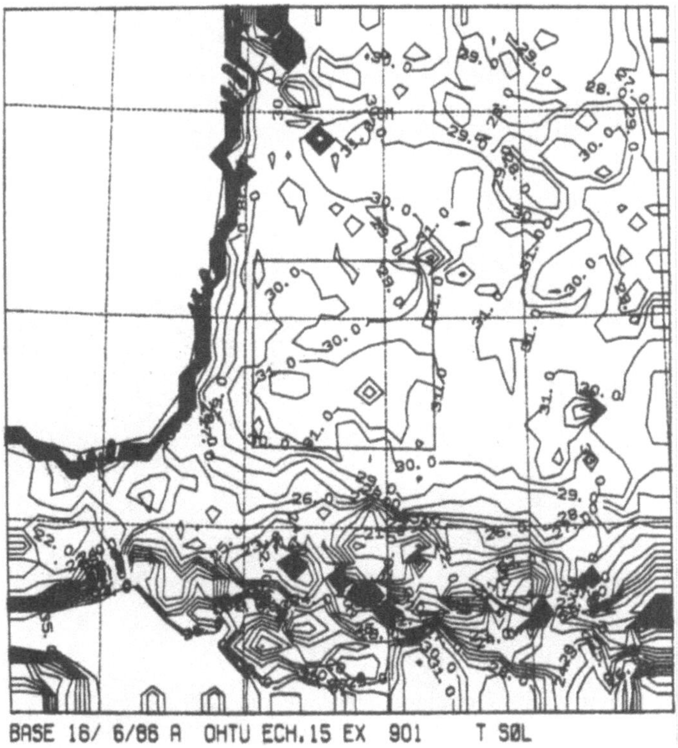

FIGURE 24.10. The surface temperature (T_s) predicted at 1500 Z.

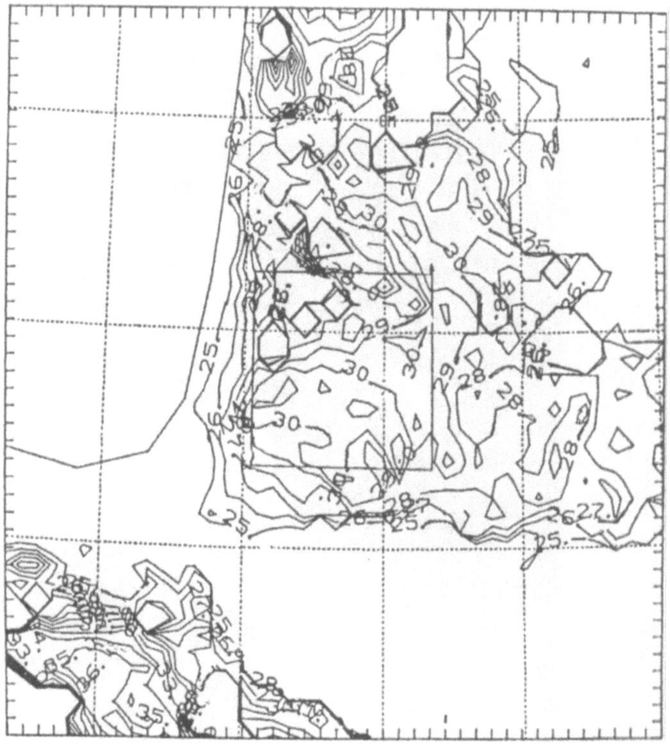

FIGURE 24.11. NOAA6 AVHRR channel 4 picture (infrared) for June 16, 1439 Z.

of the sensible heat flux. Forested areas have generally more than 300 W/m^2, whereas the surrounding areas have usually less. This is in agreement with the aircraft observations (see Section 6). The forest is also visible on the latent heat flux maps, with somewhat smaller values of this flux. The second striking pattern is the very high sensible heat flux predicted over the spanish plateau in the southwestern corner of the domain. This area is classified as semidesertic in the vegetation map used in the model, and the specified soil moisture content is quite low. The superficial soil dries out during the day, leading to these very high values of the Bowen ratio. Finally, we note the relative maximum of the latent heat flux in the grasslands areas which forms the northern border of the Pyrénées mountains.

Next, we will show the predicted surface temperatures at 1500 Z (Fig. 24.10). This time is chosen because it allows direct comparison with the infrared satellite picture shown in Figure 24.11. There is an excellent correspondence between the prediction and the observation if we ignore the very cold spots of the satellite image. They correspond to low level-clouds, which have not been filtered out. It is of special interest to note that the Landes forest has colder surface temperatures than the surrounding areas where crops prevail both in the observations and in the model prediction. This is due to the higher value of the roughness height over the forest. We have noted above that the forest is characterized by higher screen level temperatures than the surrounding areas. Here, the pattern is reversed. This is explained by a totally different value of the vertical gradient of temperature in the surface layer. The mixed crop areas, with low roughness height (Z_0 = 10 cm), need to reach higher surface temperatures than the forest to transfer a comparable amount of heat toward the atmosphere. Maximum values of surface temperatures are noted in the mixed crop area of Gers in the southern part of the study area, as well as in the urban areas of Bordeaux and Toulouse (readily visible isolated spots in Fig. 24.10), and in the Spanish semidesertic areas. All these elements are well supported by the satellite picture. Similar comparison is possible

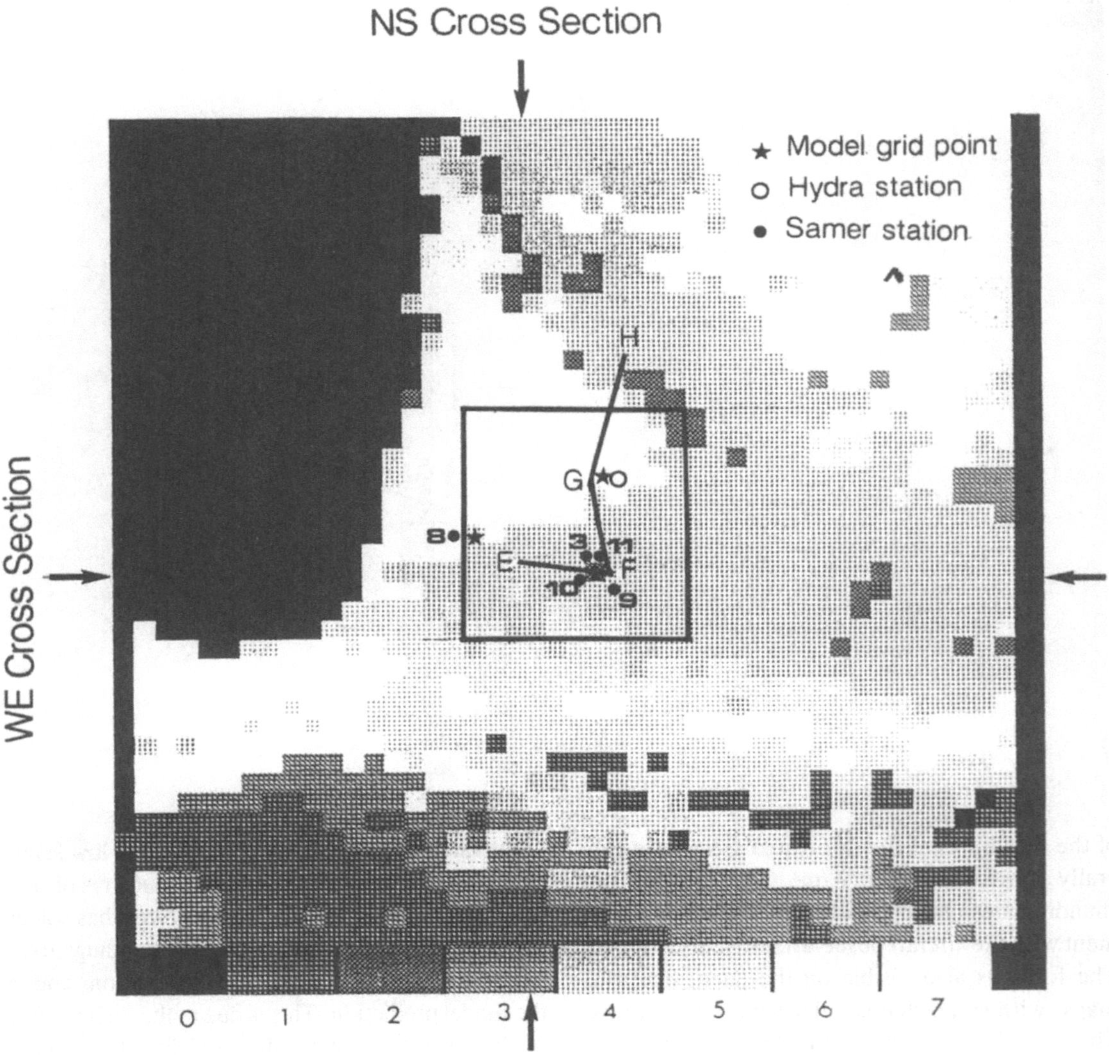

FIGURE 24.12. A map for the localization of the model/observations comparison. (*) Model grid points discussed in Section 6. Numbers stand for SAMER stations. (○) HYDRA station. EFGH, King Air aircraft trajectory.

for the 0300 Z surface temperature, where the Landes forest again appears colder than the surrounding areas.

6. Detailed Comparisons with Available Observations (SAMER, HYDRA, and King Air)

In this section, we will go further into the comparison between existing observations and model-predicted parameters. We will first attempt a direct comparison between the diurnal cycles of temperature, wind, humidity, and sensible and latent heat fluxes at specific locations and the corresponding model-predicted quantities. This is done for three specific locations: one over the forest, in the northern part of the study area, with the HYDRA station; one in the southeastern part of the study area, where four SAMER stations located nearby provide a good estimate of the evaporation over crops; and the last one in the southwestern part of the study area, to illustrate the effect of non-representativeness of a SAMER station. The location of the reference measurements and of the model grid points that are used for comparison is shown in Figure 24.12.

The comparison of observed and predicted values of the screen level temperature and specific humidity over the forest is shown on Figure 24.13. The diurnal cycle of both quantities if correctly predicted, with an amplitude of about 15 K for the temperature, and 2 g/kg for the specific humidity. One possible discrepancy is the surface temperature at 0300 *Z*, where the model reaches the very low value of 10°C. There is no reference measurement for this quantity. However, this drop seems to be overestimated, and may be due either to a too low value of the drag coefficient for heat under strong static stability conditions or to a too low value of the surface thermal inertia assumed in the model in conditions of high vegetation cover. Work is in progress to improve this aspect of the simulation. Another discrepancy with observations is the evolution of specific humidity after 2100 *Z*. The model drops rapidly, while in the observations, q increases to 11 g/kg. This is probably more related to an underestimation of the moisture in the maritime air mass, which is advected over the domain at that time, than to a wrong behavior of the surface scheme. Indeed, the error in the forecast of the specific humidity is seen at various places in the model, always starting with the arrival of the maritime airmass. Rapid variations in the observed specific humidity during the afternoon are not reproduced by the model. The origin of these variations is not known to the authors.

The diurnal evolution of the fluxes of sensible and latent heat is compared to the observations over the forest in Figure 24.14. The agreement can be considered as excellent, especially if we take into account the large changes of the Bowen ratio during the day. The maximum values observed and predicted are about 300 W/m² for the latent heat flux and 400 W/m² for the sensible heat flux. However, during the morning and evening hours, the latent heat flux is larger than the sensible heat flux. The success of the model in reproducing this behavior is due to the F_3 factor in Equation (24.15), which introduces the dependence of the surface resistance on the atmosphere moisture deficit, as first proposed by Jarvis (1976), and verified on the HAPEX-MOBILHY data by Gash et al. (1989).

We next compare the model results for a grid point situated in the southeastern corner of the

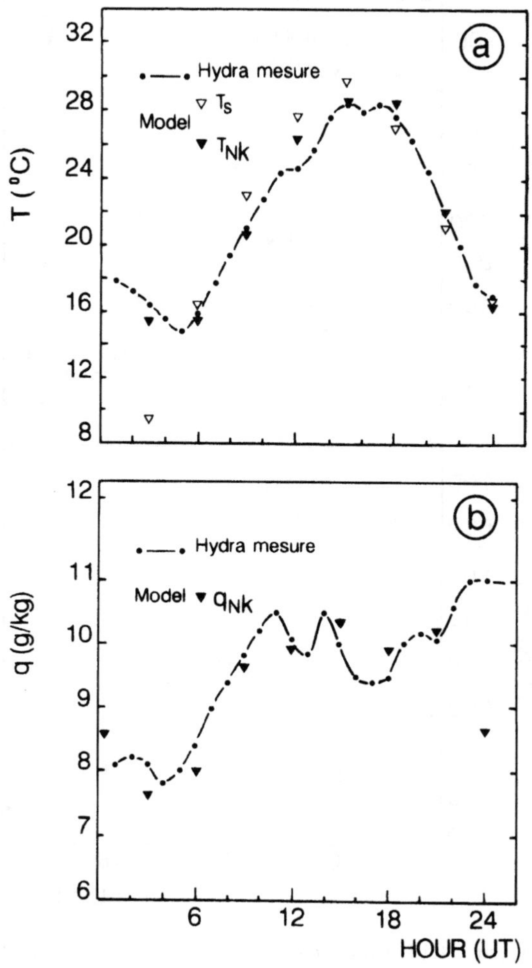

FIGURE 24.13. Comparison of predicted and measured temperature (a) and humidity (b) over the forest.

study area with a set of four SAMER stations describing various conditions present in this area. These stations are Caumont (A), over a field of soja, and Tieste (B), Castelnau (C), and Fusterouau (D) in fields of corn with little development. The fraction of vegetation for the reference grid box of the model has the quite low value of 0.48. This allows for a significant heat flux into the ground. The comparison of the sensible heat flux is next shown in Figure 24.15a. The model predicted values fall well within the observed range for this quantity. The differences between the four stations come from several factors: the different development of the culture, slightly different atmospheric conditions, and different soil moisture contents. Here, it is rewarding to see that the model is able to

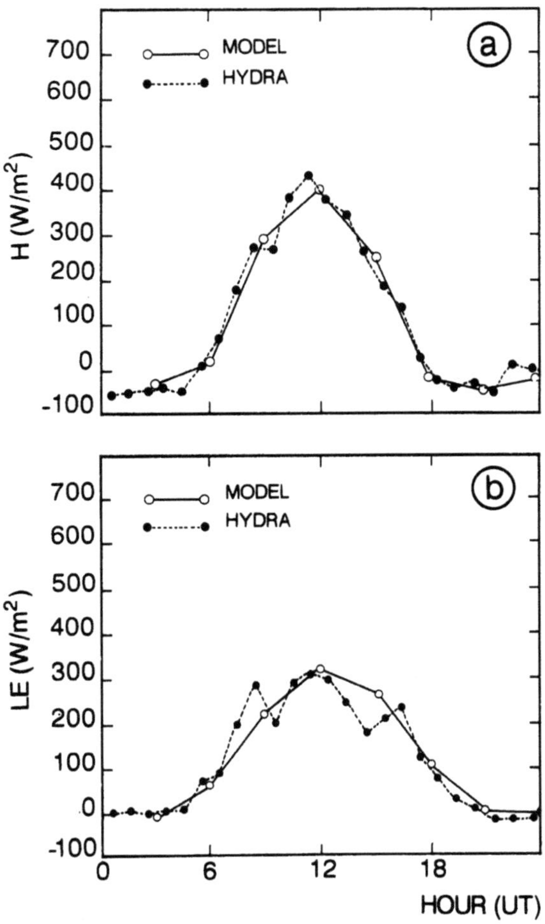

FIGURE 24.14. Comparison of predicted and measured sensible heat flux (a) and latent heat flux (b) over the forest.

predict a kind of average. The agreement is not so good for the latent heat flux, shown in Figure 24.15b. Here, the model predicts a value higher than the average between the four stations, but close to the Caumont observations. We have found several possible origins for this discrepancy. First, the model-predicted global solar radiation is larger than the average observations. This could be due either to measurement errors, since the two different measuring systems are not in agreement (see, e.g., Goutorbe and Marre, 1988), or to the underestimation of possible aerosol effects by the model. There are also differences between the albedo observed in the different sites, and the mean albedo determined by remote sensing used in the model, (see above). Further work is needed to establish the most probable reason. Second, the model-predicted Bowen ratio is wrong, since the sensible heat flux is approximately correct and the latent heat flux is underestimated. This calls for an error on the mean grid box value of veg. In fact, the procedure described above is not designed to provide grid box average values of veg in agreement with the local observations. Here, we must remember that there is no reason why the model-predicted values should coincide with the means of the four SAMER stations!

Finally, we will compare the model results from a grid point in the southwestern corner of the study area with a single SAMER station (Vicq). This station is placed in a field of corn situated in a clearing inside the Landes forest. For a more significant comparison, we have chosen a model grid point with characteristics close to those of a crop area.

The net radiation is again overestimated by the model (see discussion above). As a consequence, the latent and sensible heat fluxes (Fig. 24.16) are overestimated but in very different proportions. The latent heat flux is overestimated by 10%, and the sensible heat flux by more than 100%! The origin of this error on the Bowen ratio can be ascribed to an oasis effect. Looking at the prediction of the wind speed, temperature and specific humidity at screen level, it is clear that the model underestimates the temperature. We interpret this in relation with the fact that the station of Vicq is located in a clearing. The air temperature at Vicq is therefore very much influenced by the nearby forest, which is supposed to be warmer than the clearings. On the other hand, we have selected a "nonforest" grid point for comparison and the air temperature at this grid point is less influenced by the forest. In conclusion, these three comparisons between model-predicted quantities and local observations illustrate the complexity of this exercise. On the one hand, it is absolutely necessary to assess the general behavior of the model, since local observations are at the present time the only sources that provide continuous information throughout the diurnal cycle of all quantities of interest. On the other hand, it should not be expected that the model exactly reproduce the measurements of a local station, or even of a group of stations, which are not, except on very special cases, representative of the regional scale.

One never knows whenever discrepancies are noted between model results and local observa-

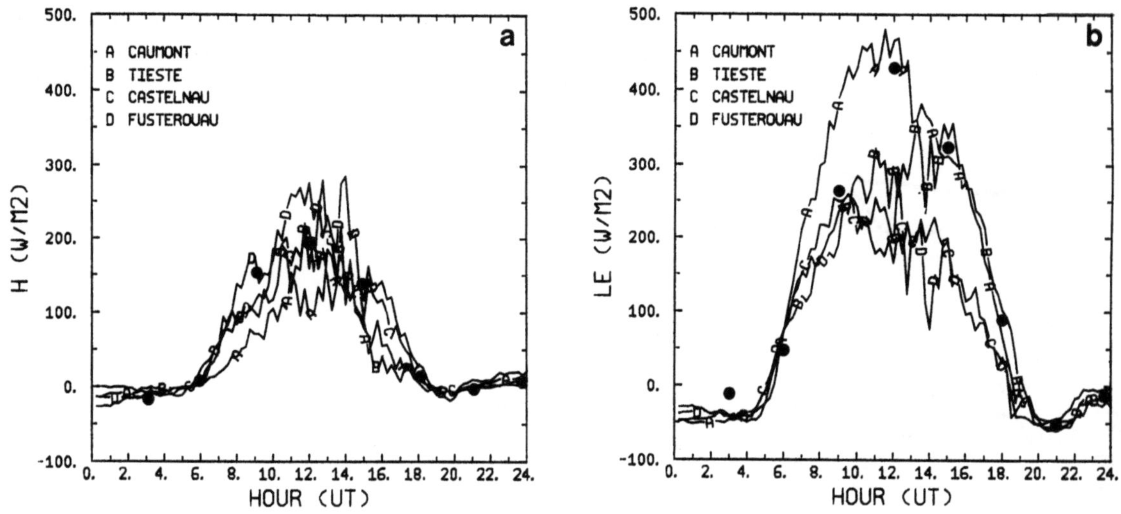

FIGURE 24.15. Comparison of sensible heat flux (a) and latent heat flux (b) observed and predicted by the model.

tions what part of these discrepancies should be attributed to model deficiencies, and what part to the nonrepresentativity of the measurements.

One way to escape this problem is to use aircraft measurements. This has been the major axis of the HAPEX-MOBILHY program strategy, and much is expected from the comparison between model results and aircraft flux measurements. Detailed comparison with the data reported by Hildebrand (1988) are underway. Here we will only show a preliminary result. To make this comparison, the aircraft trajectory (see Fig. 24.12) has been computed on the model horizontal and vertical grid. For each 20-km segment for which measurements are available, we have interpolated the model result at 1200 Z. The comparison is made with the "synthetic" 1200 Z aircraft measurement, which is obtained by combining three actual measurements taken at different times. These measurements are taken at an altitude of 100 m above ground level.

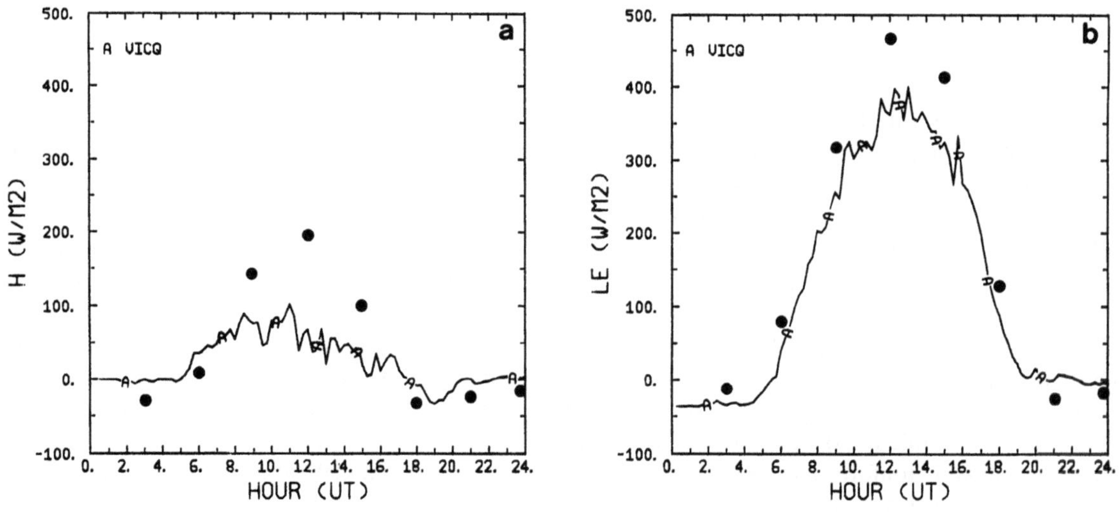

FIGURE 24.16. Sensible heat flux (a) and latent heat flux (b) observed at Vicq and predicted by the model.

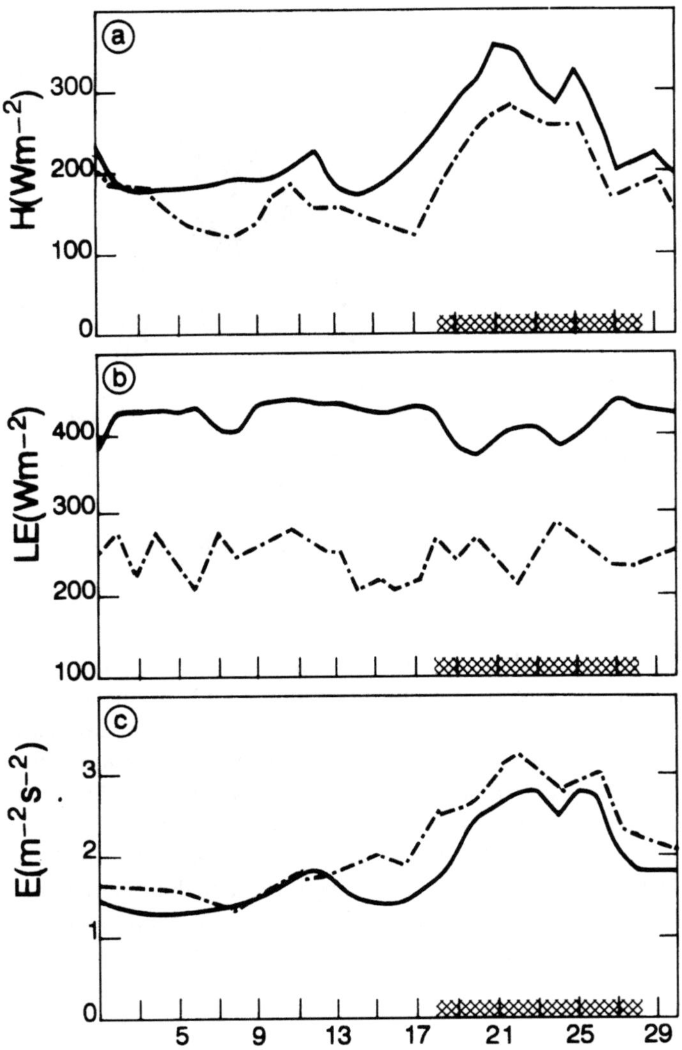

FIGURE 24.17. Sensible heat flux (a), latent heat flux (b), and turbulence kinetic energy (c) from aircraft measurements and model.

Comparison is shown in Figure 24.17 for the turbulent kinetic energy, the sensible heat flux, and the latent heat flux. It should be noted that the turbulent kinetic energy from the model is also interpolated at 100 m above ground level (the model ground), but the sensible and latent heat fluxes are *surface fluxes* (we intend to use the fluxes computed at the altitude of 100 m for further work). There is clearly a satisfactory correspondence for both turbulent kinetic energy and sensible heat flux. The only visible regional contrast is between the forested and nonforested areas. The model reproduces the observed increase of sensible heat flux (from 150 to 300 W/m^2) and turbulent kinetic energy (from 1.5 to 3 $m^2\ s^{-2}$) over the Landes forest. This result may be considered as one of the main successes of the mesoscale modeling program for HAPEX-MOBILHY.

Turning now to the comparison of the latent heat flux, it can be seen that the aircraft observations do not support the model results. The difference is as large as a factor of 2, the observations being smaller. This seems difficult to understand at the time this chapter is written. One may of course question the empirical procedure that has been used to set the model parameters, and this will be done in the near future. However, it is puzzling to note that the disagreement is also present over the forest, where horizontal homogeneity is a good approximation.

7. Conclusion

It is obvious from the previous sections that this chapter is an attempt to discuss some results from work that is still underway. We have built a new method of parameterization for land-surface properties that is simpler than most previous methods, but incorporates the essential physics of the processes under study. We have shown by a series of one-dimensional experiments that this scheme has the capacity to reproduce correctly the surface energy budget at the local scale. For inclusion inside the mesoscale model, the problem of subgrid scale variability has been treated in a deliberately simplified manner, as a first attempt to investigate the difficulty, and to remain consistent with the amount of existing information. The subgrid-scale variability of the soil texture and water content was ignored. The subgrid-scale variability of the vegetation cover was treated by linearly averaging the fraction of vegetation deduced at the AVHRR pixel scale by a combination of a supervised classification method on two clear NDVI images and a table of correspondence deduced from the one-dimensional experiments. This method allows for the derivation of a map of the fraction of vegetation that has been used in the mesoscale model, in lieu of the local value of this parameter, whereas the other parameters were specified at the value corresponding to the *dominant* vegetation type inside the grid box. The results of the mesoscale model show a high degree of realism in its simulation of atmospheric parameters. The apparent surface temperatures predicted by the model are in fair agreement with infrared satellite pictures. For the regional surface energy budget, and the prediction of the sensible and latent heat flux, the limitations of local measurements to assess the quality of the model results have been discussed, and some comparisons with available aircraft measurements have been shown.

Those measurements support the model results for the sensible heat flux and turbulence kinetic energy. On the other hand, the aircraft measurements of latent heat flux are not in agreement with the model results, and the screen level temperatures of the model seem to be underestimated by 2–3 K as compared to conventional measurements. This calls for further tuning of the parameters of the scheme.

If this problem can be solved and in view of the general results, we feel that given the degree of quality of the mesoscale analysis and the amount of information available to qualify the model results, there is no basis at the present time to increase the complexity of the basic parameterization scheme. The method to account for subgrid scale variability may need some refinement. It is most useful to try and obtain similar results on other days of the SOP, taking into account the evolution of those parameters that describe the vegetation properties and the soil water content. The present technique to obtain the grid-average value of veg cannot be extrapolated outside the SOP, and our intention is to work out a method to obtain this parameter directly from the NDVI images, using our continuously expanding set of reference values deduced from the 1D calibration experiments. This, however, will be a difficult task, because the NDVI evolution with time is a function not only of veg but also of such parameters as R_{smin} and LAI.

Acknowledgments. We gratefully acknowledge the splendid work of all our colleagues who have been involved in the HAPEX-MOBILHY field phase. In addition, the present work was only made possible through intensive collaboration with P. Mascart and J.P. Pinty (LAMP), who elaborated the map of soil texture and depth, T. Phulpin and J.P. Jullien, who produced the various maps of vegetation type, and R.M. Thépenier, who worked out the albedo maps. Comparison with the aerological and aircraft data were made easier by software developed by J.L. Champeaux and P. Peris.

References

Andre JC, Goutorbe JP, Perrier A, Becker F, Bessemoulin P, Bougeault P, Brunet Y, Brutsaert W, Carlson T, Cuenca R, Gash J, Gelpe J, Hildebrand P, Lagouarde JP, Lloyd C, Mahrt L, Mascart P, Mazaudier C, Noilhan J, Ottlé C, Payen M, Phulpin T, Stull R, Shuttleworth J, Schmugge T, Taconet O, Tarrieu C, Thepenier RM, Valencogne C, Vidal-Madjar D, Weill A (1988) Evaporation over land-surfaces: first results from Hapex-Mobilhy Special Observing Period. Ann Geophys 6:477–492.

Bret B, Bougeault P (1988) Simulation de quatre journées Hapex-Mobilhy à l'aide du modèle Péridot. Note EERM No. 211.

Clapp R, Hornberger G (1978) Empirical equations for some soil hydraulic properties. Water Res Res 14: 601–604.

Cosby BJ, Hornberger GM, Clapp RB, Ginn TR (1984) A statistical exploration of the relationships of soil moisture characteristics to the physical properties of soils. Water Res Res 20:682–690.

Dedieu G, Deschamps PY, Kerr YH (1987) Satellite estimation of solar irradiance at the surface of the earth and of surface albedo using a physical model applied to Meteosat data. J Climate Appl Meteorol 26:79–87.

Durand Y, Bougeault P (1987) L'analyse objective Péridot. Note EERM No. 193.

Gash JHC, Shuttleworth WJ, Lloyd CR, André JC, Goutorbe JP, Gelpe J (1989) Micrometeorological measurements in Les Landes forest during Hapex-Mobilhy. Agric Forest Meteorol 46:131–147.

Goutorbe JP, Marre JL (1988) Analyse du rayonnement–Période d'Observation Intensive. Expérience HAPEX-MOBILHY (available from CNRM).

Hildebrand P (1988) Flux and sounding data from the NCAR King-Air aircraft during Hapex. NCAR Tech. Note TN-319 (Available from NCAR, Boulder, CO 80307-3000 USA).

Jarvis PG (1976) The interpretation of the variation in leaf water potential and stomatal conductance found in canopies in the field. Phil Trans R Soc London Ser B 273:593–610.

Mascart P, Gelpe J, Pinty JP (1988) Etude des caractéristiques texturales des sols dans la zone Hapex-Mobilhy 86. Note OPGC No. 95.

Mercusot C, Bougeault P, Durand Y (1986) Programme Hapex-Mobilhy. Atlas des analyses Péridot.

Monteith JL (1976) "Vegetation and the Atmosphere. Vol. 2: Case Studies." Academic Press, New York.

Noilhan J, Planton S (1989) A simple parameterization of land surface processes for meteorological models. Mon Weather Rev 117:536–549.

Phulpin T, Jullien JP (1988) A study of the vegetation cover with AVHRR during Hapex-Mobilhy. Proceedings of the 4th International Colloquium spectral signatures of objects in remote sensing. Aussois, France, 18–22 January, 1988.

25
HAPEX-MOBILHY Data Base

J.-P. Goutorbe and C. Tarrieu

1. Introduction

The principal focus of the HAPEX-MOBILHY program (André et al., 1986, 1988) is on the hydrological budget and evaporation fluxes at the scale of a general circulation model grid square, i.e., of the order of 100 × 100 km. The field program was conducted for 1 year and provided a data set including hydrological, pedological, surface, and atmospheric parameters against which parameterization schemes for hydrological processes and evaporation can be tested or developed. The observational program took place in southwestern France (Fig. 25.1) and extended over most of year 1986. It included a special observing period (SOP) from May 7 to July 15, during which additional observations were made. The northwestern third of the square extends over the Landes pine forest; the remaining part is mostly agricultural land with a 25% wood cover. The main crop is maize. Other cereals, meadows, orchards, and vineyards also cover important surfaces.

The many parameters that control the water budget and evaporation flux have very different time and space scales. On the one hand surface hydrology and upper-soil properties have a typical horizontal scale of the order of 100 m, while their time scale is relatively large, i.e., around 10 days. For this reason, soil moisture has been monitored on a weekly basis at 14 representative locations for 2 years. At each site, four to six access tubes were installed to deal with small-scale variability. On the other hand, surface values of atmospheric variables and fluxes vary on a fairly large spatial scale, close to 10 km, but with a short time scale. Accordingly a sampling of these parameters, more limited in time and spanning the vegetative season of 1986, was organized at 12 of these locations (Fig. 25.1). Other, more costly measurements have been performed during the special observing period. This includes the monitoring of the boundary layer development and improved analysis of the weather patterns using radiosoundings. It also includes direct measurements of eddy fluxes by aircraft. This is needed because the various point measurements of sensible and latent heat fluxes have to be aggregated to provide an area-integrated value. This has been done by flying an instrumented airplane over the surface network during the special observing period.

1.2 Description of the Data Set

The various items available are listed in Table 25.1. All data are on magnetic tape ASCII code, 1600 bpi. The measurements performed are briefly described in this section. A list of the teams responsible for the various subprograms is given at the end of the text. The measurements are ordered according to their duration, starting by the long-term monitoring.

2. Description of the Area

2.1 Soil Map

Essentially for modeling purposes a map of the dominant soil types for large grid elements (10 × 10 km) has been established by Mascart et al. (1988).

TABLE 25.1. List of contents of the HAPEX-MOBILHY data base.

Data	Number of tapes	Agency(ies)
Radiosoundings (P,T,U,V), Lubbon, Toulouse, Bordeaux	1	SCEM, CNRM
SAMER (12 stations)	6	CNRM
PATAC (29 stations)	5	SCEM/SMIR SO
Met. network (14 stations)	1	SCEM
Neutron soundings (13 stations)	1	CNRM, INRA, Compagnie des coteaux de Gascogne
Forest fluxes	1	Institut of hydrology, Wallingford
Forest met. data	1	Institut of hydrology, Wallingford
Aerodynamic mast	1	INRA
Airplane fluxes (analyzed data only)	1	National Center for Atmospheric Research
Acoustic sounder	1	CNET/CRPE
Satellite and Radar (data METEOTEL)	17	SCEM
Remote sensing from aircraft	21	NASA–USDA
Weather analysis	3	CNRM, SCEM
Hydrological modeling	1	Ecole des Mines
Vegetation classification (Landsat data)	1	GSTS

2.2 Vegetation Classification

Data from the Landsat Thematic Mapper on June 23, 1986 have been used by J. Hirsch (GSTS) to classify land use in the area. Twelve vegetation types have been classified.

2.3 Climate

The climate of the area, based on data from 1951 to 1980, has been studied in Choisnel et al. (1987). The study also includes simulations of the water balance for selected points in the square.

3. Permanent and Semipermanent Networks

3.1 Stream Gauges

For the needs of hydrological studies 29 stream gauges have been used. An inventory of available hydrological data can be found in Girard and Boukerma (1985). This information as such is not included in the data base (see Section 5.9).

3.2 Weather Stations

Data from 14 of the weather stations operated by the French Weather Service (Météorologie Nationale) have been archived from April 1985 to December 1986. These stations are Agen, Auch, Biarritz, Biscarosse, Bordeaux, Cazaux, Dax, Mont de Marsan, Pau, Saint-Girons, Tarbes, Toulouse, Cap-Ferret, and Captieux. For the first 12 stations, pressure, air temperature, dewpoint temperature, wind speed and direction, and precipitation are recorded for 6-hr intervals. For the 2 last stations the same parameters are recorded only once a day. Only Mont de Marsan and Captieux are actually inside the test area. The remainders are just outside.

3.3 Automatic Weather Stations

The network of PATAC (Prevision Ameliorée, Technique d'Affinement de la Climatologie) stations is also operated by Météorologie Nationale. The data are centralized by telephone line on a computer. Each PATAC station records rainfall every 5 mins, air temperature, relative humidity, wind speed, and direction at 10 m, and solar radiation every 30 min. The station also measures temperature 10 cm above the surface and temperature in the soil at 10 and 50 cm. The station also records for every 30-min interval, maximum and minimum temperature and maximum wind speed.

From December 1985 to February 1987, 29 PATAC stations (Fig. 25.2) have been archived in the HAPEX-MOBILHY data base.

FIGURE 25.1. SAMER network.

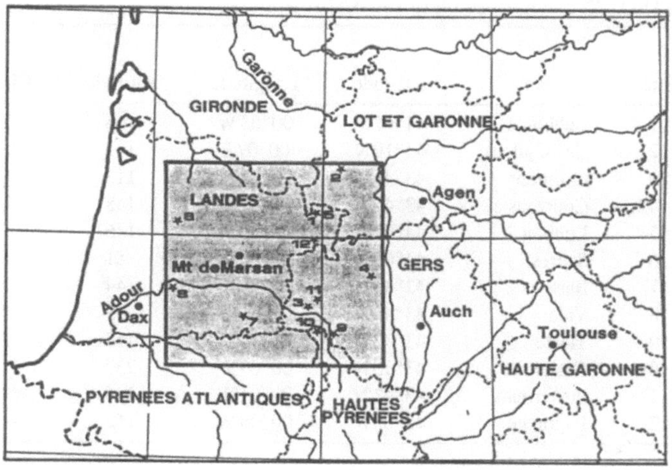

4. Year-Round Monitoring

4.1 Soil Moisture

Measurements were performed using the neutron scattering method. The probe used was a Solo 25 from Nardeux. The measurements were performed for the years 1985 and 1986. At each site, four to six access tubes were installed, separated by 10 m. The depth of the tubes ranged from 0.7 to 2.0 m. For each tube, measurements were performed at intervals of 10 cm, starting at 5 cm from the soil surface, and later averaged for all tubes. The raw data are neutron counts that have to converted into volumetric water content. The calibration procedures, together with an illustration of the data set, can be found in Goutorbe and Tarrieu or Cuenca (this volume).

4.2 Forest Climate

Two automatic weather stations (Didcot Instruments) were mounted on top of a 25-m-high tower in the Landes pine forest. They record solar and

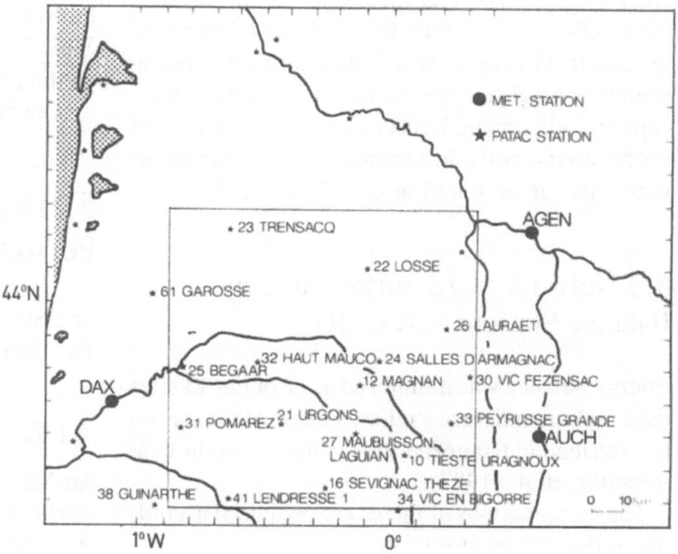

FIGURE 25.2. PATAC network.

TABLE 25.2. Description of the SAMER sites.

Site	Name	Latitude	Longitude	Altitude	Type of soil (%) Clay	Silt	Sand	Crop	Irrigation
1	Lubbon 1	44°07'N	00°03'W	146	2	5	93	Oat	No
2	Casteljaloux	44°19'N	00°07'E	131	35	49	16	Corn	Yes
3	Caumont	43°41'N	00°06'W	113	17	46	37	Soja	No
4	Courrensan	43°49'N	00°16'E	148	–	–	–	Wheat	No
5	Lubbon 2	44°07'N	00°03'W	146	1	2	97	Corn	Yes
6	Sabres	44°05'N	00°50'W	81	3	1	96	Corn	Yes
7	Bats	43°38'N	00°26'W	144	18	51	31	Corn	No
8	Vicq	43°46'N	00°51'W	15	10	16	74	Corn	Yes
9	Tieste	43°32'N	00°02'E	145	12	61	27	Corn	Yes
10	Castelnau	43°35'N	00°03'W	239	11	66	23	Corn	No
11	Fusterouau	43°42'N	00°01'W	146	20	34	46	Corn	Yes
12	Lagrange	43°58'N	00°03'W	152	–	–	–	Corn	Yes

net radiation, dry and wet bulb temperature in an aspirated screen, and wind direction at 5-min intervals. Accumulated wind run and rainfall are also recorded every 5 min. One-hour averages are then computed and data from the two stations, also averaged to produce a single set. Measurements extend from February 9, 1986 to January 3, 1987.

During this period, throughfall was measured in a 30 × 30 m plot situated close to the tower. This area was marked out as a grid at 1-m intervals and was kept clear of all ground vegetation. Twenty-two bottle gauges sampled throughfall within this grid, each gauge being randomly relocated every 2 weeks. Stem flow was measured on six adjacent trees within the throughfall grid. In addition 10 ground level gauges were installed under a representative bracken cover to monitor bracken interception. All tipping bucket gauges were recorded for 5-min intervals. Information about these measurements can be found in Gash et al. (1989).

4.3 Network of 12 Surface Energy Balance Stations (SAMER)

Energy balance was monitored at 12 of the 13 sites used for soil moisture studies. The SAMER station is described in Goutorbe (this volume) and in Bessemoulin et al. (1987).

The computations of fluxes are organized according to the following scheme:

Wind speed vertical gradient Temperature vertical gradient Height of probe	Sensible heat flux
Height of vegetation Net radiation Ground heat flux	Latent heat flux

The other measured parameters are rainfall, wind speed, air temperature, relative humidity, and shortwave and total radiation upward and downward. Data are recorded at a 15 min rate and sent by phone to a central computer in Toulouse. Figure 25.1 is a map of the network; stations 1, 5, 6, and 8 were located in clearings in the Landes forest; the others were in the agricultural area. Information on the soils and crop types is given in Table 25.2. SAMER operations were concentrated on the growing season of 1986. The data availability can be found in Table 25.3.

5. The Special Observation Period Data Set

The SOP extended from May 7 to July 15, 1986. The data set includes the following.

5.1 Eddy Fluxes over the Forest

An Mk2 Hydra eddy correlation device (Shuttleworth et al., 1988) was mounted on the forest mast at a height of 29.0 m. Measurements were

TABLE 25.3. HAPEX-MOBILHY, SAMER stations status.[a]

SAMER site	1986												1987	
	J	F	M	A	M	J	J	A	S	O	N	D	J	F
1. Lubbon 1	+	+	+	+	*	*	+	–	–	+	*	+	+	*
2. Casteljaloux	+	+	+	+	+	*	+	*	–	–	–	+	+	–
3. Caumont	+	+	+	+	–	*	+	+	*	+	+	*	+	–
4. Courrensan	–	–	–	+	+	*	+	*	*	+	–	–	–	–
5. Lubbon 2	–	–	–	–	+	+	–	–	–	–	–	–	–	–
6. Sabres	–	–	–	–	+	+	–	–	–	–	–	–	–	–
7. Bats	–	–	–	–	–	*	+	+	–	–	–	–	–	–
8. Vicq	–	–	–	–	–	*	+	*	+	–	–	+	+	–
9. Tieste	–	–	–	–	–	*	*	+	–	–	–	*	+	+
10. Castelnau	–	–	–	–	–	*	+	–	–	–	–	–	–	–
11. Fusterouau	–	–	–	–	–	*	+	+	–	–	–	–	–	–
12. Lagrange	–	–	–	–	–	*	*	*	*	+	*	+	+	–

*,[a] Continuous or almost continuous recording (between 90 and 100%); +, recording between 40 and 90% of time; –, very sparse or no data (less than 40%).

taken between June 10 and July 22. The Hydra uses a vertical sonic anemometer, an infrared absorption hygrometer, a fine wire thermocouple, and a fast response cup anemometer to give hourly average values of evaporation, sensible heat flux, and friction velocity. The calculations are performed in real time using a battery-powered microprocessor.

5.2 A Mast-Born Aerodynamic System

This mast was located in a large maize field close to SAMER station 10 (Fig. 25.1). Air temperature and wind speed are measured at six levels, from 0.50 to 12 m. Incoming and reflected solar radiation, net radiation, and a brightness temperature are measured. There are seven levels of temperature measurements in the ground, down to 1 m, and heat flux plates. Humidity is measured at screen height. The data set also includes computation of fluxes. All data are averaged over 30 min. The experiment lasted from May 27 to July 7, 1986.

5.3 Aircraft Fluxes

During the special observing period a King Air aircraft from the National Center for Atmospheric Research flew over the research area. The flight pattern consisted of a repetitive low-level flight path designed to sample the various surface conditions. The forest tower and seven SAMER stations were flown over. The pattern was flown every 3 days. Data from these flights are more fully described in Hildebrand (1988a).

The King Air aircraft measured wind, temperature, humidity, pressure, altitude, solar and terrestrial radiation, and surface radiative temperature. Using this instrumentation the latent and sensible heat flux, the stress, the surface temperature, and the radiative fluxes have been measured along the flight track. For this program special attention has been paid to humidity measurements. Two NCAR-designed Lyman-alpha hygrometers and two EG&G 137 hygrometers were used. Information on humidity measurements and more generally on data processing can be found in Hildebrand (this volume).

5.4 Radiosoundings

During the intensive period, radiosoundings were launched from the central site of the experiment, Lubbon (site 1, Fig. 25.1) and from Toulouse. At Bordeaux, which is a regular sounding station, the sounding frequency was increased. As a result, four soundings a day, at 0, 6, 12, and 18 GMT, were launched every day from May 7 to July 15, 1986. In addition the boundary layer development was monitored on special days with up to seven launches.

Each sounding extends from the surface to an altitude of 16 km. The vertical increment is 60 m in

TABLE 25.4. Available NS001 (N) and TIMS (T) data.[a]

Day	Altitude	Hour (GMT)	Sensor	R_g	SAMER sites
May 29	MA		N		9,1,15,10,1
June 6	MA		T		9,10
			N		1,5,9,10,12
June 14	MA 900	11.49	T	470	5
			N		12,10
June 16	MA 1500	11.34	T	880	1,5,12,10,9
			N		1,5,12,10,9
	HA 6000	15.11	T	650	1,5,12,3,10,9,2,4
			N		1,5,12,3,10,9,2,4
June 19	HA 6000	14.19	T	740	1,4,2,5,12,3,10
			N		1,4,2,5,12,3,10,9
June 20	MA 1400	12.01	T	850	1,5,10,9
			N		1,5,12,10,9
June 23	HA 6000	9.54	T	830	1,4,5,12,3,10,2,9
			N		1,4,5,12,3,10,2,9
	MA 1200	12.09	T	610	1,5,12,10,9
			N		1,5,12
June 25	MA 1500	11.54	T	890	1,5,12,10,9
			N		9
June 27	MA 1500	11.38	T	820	1,5,12,10,9
			N		1,5,12,10,9
	HA 6000	14.37	T	670	1,5,12,3,10,9,2
			N		1,5,12,3,10,9
July 2	HA 6000	10.16	T	660	4,2,9,10,3,12,1,5
			N		4,2,9,10,3,12,1,5
	MA 1500	12.15	T	830	1,5,12,10,9
			N		1,5,12,10,9

[a] Altitude, altitude (m) above the site; R_g, global radiation (W/m²·μm); MA, middle altitude; HA, height altitude. The SAMER sites are listed in chronological order. Coverage for site 5 implies that the adjacent site, No. 1, is also covered.

Lubbon and approximately 600 m in Toulouse and Bordeaux. The parameters recorded are pressure, temperature, humidity, and wind. The data set consists of 914 soundings. A more complete description of soundings and of their potential use is given by Parlange and Brutsaert (1989).

5.5 Acoustic Sounders

Two acoustic sounders operated at the central site of the experiment, one in a small clearing in the forest and the other over cultivated land in Lubbon. The data set consists of reflectivity, horizontal and vertical wind, and standard deviation for horizontal and vertical wind. The range of measurements extends from 20 to 350 m with a height resolution of 17 m and a time resolution of 15 min. The study extended from May 22 to June 30, 1986. Information about doppler SODAR methodology can be found in Weill (this volume).

5.6 Meteotel Imagery

This set of 17 "METEOTEL" tapes includes Radar echoes maps from Toulouse and from Bordeaux and a Radar composite over France, for selected periods (an image every 15 min). It also includes visible and infrared imagery for meteosat (an image every 30 min). Days for which information is available (but not necessarily the whole set) are May: 10, 16, 17, 18, 22, 23, 25, 26; June: 5, 15, 16, 17, 18, 19, 20, 23, 25, 26, 27, 28, 30; and July: 1, 2, 3, 11.

5.7 Aircraft Remote Sensing Data

The C-130 from NASA Ames Research Center operated over the HAPEX-MOBILHY area from May 29 to July 3. The aircraft was fully equipped for visible and near-infrared imagery (NS001 multispectral scanner), thermal infrared (TIMS: thermal infrared multispectral scanner), and microwaves (PBMR: push-broom microwave radiometer). Detailed information about the objectives of this experiment and first results can be found in Schmugge and Janssen (1988). The large data collection is not included as such in the HAPEX-MOBILHY data set. However, some thermal infrared multispectral images from SAMER sites 1, 2, 3, 4, 5, 9, 10, and 12 are in the data set. The chapter by Carlson (this volume) gives an example of the potential use of these data. These scenes are extracted from flight lines at 6000 m at 1500 m altitude (Table 25.4).

5.8 Weather Analysis

The standard analysis procedure of the French Weather Service (Durand and Bougeault, 1987) has been improved from May 7 to July 12, 1986 by incorporating the extra soundings in Bordeaux, Toulouse, and Lubbon (Mercusot et al., 1986). The weather analysis was performed every 6 hr. The archived data consist of temperature, humidity, and wind fields at 15 levels on a 300 × 300 km area centered on the HAPEX box.

5.9 Hydrological Modeling

The tape contains the hydrological model used. Input for years 1974–1986 includes evaporation, soil moisture, and runoff for 10 × 10 km grid elements over the square. Additional computations can easily be performed.

6. Additional Data

6.1 Leaf Potential (Lubbon)

INRA-CNRM-CRPE study. Measurements available on magnetic tape at CNRM.

6.2 Tensiometric Measurements (Castelnau)

INRA-CRPE-CNRM study. Available in paper form at CNRM.

6.3 Weather Description (SOP)

Available in paper form (74 pp) at CNRM: "Atlas des situations synoptiques du 7 Mai au 15 Juillet 1986" by R. Cazalens and B. Séjourné.

7. Agencies and Teams

1. Direction Météorologie Nationale (France)
   - Establissement d'étude et de recherches météorologiques (EERM)
   - Centre National de la Recherche Météorologique (CNRM–Toulouse)
   - Service Central d'Exploitation de la Météorologie (SCEM–Paris)
   - Service Météorologique Inter-Régional Sud-Ouest (SMIR-SO Bordeaux)
2. Institut National de la Recherche Agronomique (INRA, France)
3. Centre National d'études des télécommunications (France)
   - Centre de Recherche en Physique de l'Environnement (CNET/CRPE)
4. Ecole des Mines (France)
   - Centre d'Hydrologie Mathématique (Fontainebleau)
5. Groupement Scientifique de Télédétection Spatiale (GSTS–Strasbourg, France)
6. Laboratoire Associé Météorologie Physique (LAMP–Clermont-Ferrand, France)
7. Institute of Hydrology (IH) (Wallingford, GB)
8. National Center of Atmospheric Research (NCAR) (Boulder, Colorado, USA)
9. NASA/Ames Research Center (California, USA)
10. Hydrology laboratory, U.S. Department of Agriculture (USDA) (Beltsville, Maryland, USA)

References

André J-C, Goutorbe J-P, Perrier A (1986) HAPEX-MOBILHY: A hydrologic atmospheric experiment for the study of water budget and evaporation flux at the climatic scale. Bull Am Meteorol Soc 67:138–144.

André JC, Goutorbe JP, Perrier A, Becker F, Bessemoulin P, Bougeault P, Brunet Y, Brutsaert W, Carlson T, Cuenca R, Gash J, Gelpe J, Hildebrand P, Lagouarde JP, Lloyd C, Mahrt L, Mascart P, Mazaudier C, Noilhan J, Ottlé C, Payen M, Phulpin T, Stull R, Shuttleworth J, Schmugge T, Taconnet O, Tarrieu C, Thepenier RM, Valencogne C, Vidal-Madjar D, Weill A (1988) Evaporation over land-surfaces–first results from HAPEX-MOBILHY special observing period. Ann Geophys 6:477–492.

Bessemoulin P, Desroziers G, Payen M, Tarrieu C (1987) Atlas des données SAMER, 256 pp. Internal report available at CNRM, avenue G. Coriolis, 31057 Toulouse Cedex. France.

Choisnel E, Payen D, Lamarque P (1987) Climatologie de la zone du project HAPEX-MOBILHY. Edité par la direction de la Météorologie Nationale, Paris.

Durand Y, Bougeault P (1987) L'analyse objective PERIDOT. Note de travail de l'Establissement d'Etude et de Recherches Météorologiques No. 193. (Available at Direction de la Météorologie Nationale, 73-77 rue de Sèvres, 92 Boulogne Billancourt, France.)

Gash JHC, Shuttleworth JW, Lloyd CR, André JC, Goutorbe JP, Gelpe J (1989) Micrometeorological measurements in Les Landes forest during HAPEX-MOBILHY. Agric Forest Meteorol 46:131–147.

Girard G, Boukerma B (1985) "Projet HAPEX-MOBILHY, calage du modèle hydrologique, exploitations spécifiques et propositions d'utilisation." Rapport LHM/RD/85/110, Centre d'Informatique Géologique. (Available from CIG, 35 rue Saint-Honoré, 77305 Fontainebleau Cedex, France.)

Goutorbe JP, Noilhan J, Valancogne C, Cuenca RH (1989) Soil moisture variations during HAPEX-MOBILHY. Ann Geophys 7:415–426.

Hildebrand P (1988a) Flux and Sounding Data From the NCAR King Air Aircraft During HAPEX. NCAR Tech. Note, NCAR/TN-319+STR.

Mascart P, Gelpe J, Pinty JP (1988) Etude des caractéristiques texturales des sols dans la zone HAPEX-MOBILHY. Note O.P.G.C., 95, Université Blaise Pascal, Clermont-Ferrand, France.

Mercusot C, Bougeault P, Durand Y (1986) Programme HAPEX-MOBILHY, Atlas des analyses PERIDOT. (Available from Centre National de Recherches Météorologiques, 31057 Toulouse Cedex, France.)

Parlange MB, Brutsaert W (1989) Regional roughness of the Landes forest and surface shear stress under neutral conditions. Bound Layer Meteorol 48:69–81.

Schmugge T, Janssen L (1988) Aircraft remote sensing in HAPEX. Proceedings 4th International Colloquium on Spectral Signatures in Remote Sensing, Aussois, France, 18–22 January 1988 (ESA SP-287), pp. 463–467.

Shuttleworth JW, Gash JHC, Lloyd CR, Moore CJ, Wallace JS (1988) An integrated micrometeorological system for evaporation measurement. Agric Forest Meteorol 43:295–317.

26
The FIFE Data

Donald E. Strebel, P.J. Sellers, and F.G. Hall

1. Introduction

The FIFE [First ISLSCP (International Satellite Land Surface Climatology Project) Field Experiment] collected data on and around the Konza Prairie Research Natural Area near Manhattan, Kansas during 1987–1989. FIFE is one of the most complex interdisciplinary research efforts yet undertaken in earth science. The objective is to combine the skills of remote sensing scientists, atmospheric physicists, meteorologists, biologists, and modelers to provide not only more detailed biophysical understanding of the land-surface–atmosphere interactions, but also a means of monitoring these interactions using satellite remote sensing. The principal focus of the project is the exchange of energy, heat, mass (CO_2 and H_2O), and momentum between the (vegetated) land surface and the atmosphere.

The first of the two goals of the experiment was to test the soil–plant–atmosphere models developed by biometeorologists for small-scale applications (millimeters to meters) and to develop methods to apply them at the larger scales (kilometers) appropriate to atmospheric models and satellite remote sensing. Second, the experiment was targeted at exploring methods for using satellite data to quantify important biophysical states and rates (at these larger scales) for input to models. To achieve these goals, the field phase of the experiment was designed for the simultaneous acquisition of satellite, atmospheric, and surface data over a wide range of temporal and spatial scales.

As a consequence of these ambitious goals, FIFE required an information system that could support the collection, distribution, and analysis of more than 100 gigabytes of data. The data and the data system problems encountered are likely to be common to the pursuit of other interdisciplinary earth science studies throughout the 1990s, including those associated with the Earth Observing System (EOS; NASA, 1984).

2. FIFE Experiment Design

The FIFE site consists of the Konza Prairie Research Natural Area and surrounding privately owned pastures, located a few kilometers southeast of Manhattan, Kansas. The site selection criteria included a requirement for a grassland of moderate topography (measuring fluxes over forests or steep slopes is difficult), strong seasonal climatic forcing (i.e., a mid-continental climate), proximity to a research organization for logistical support, and airfields for housing the research aircraft.

The size of the experiment site was constrained by the practicality of meshing surface observations with satellite data and airborne eddy correlation measurements. That is, the site had to be small enough to achieve a reasonable sampling density with the available surface instrumentation, but large enough so that the turbulent structure of the planetary boundary layer could be adequately sampled with aircraft and a reasonable number of satellite pixels would lie within the site boundary. The resulting compromise was a 15 × 15 km site in which the surface instrumentation was distributed using a stratified sampling allocation based on land management treatment and topography (see Fig. 26.1).

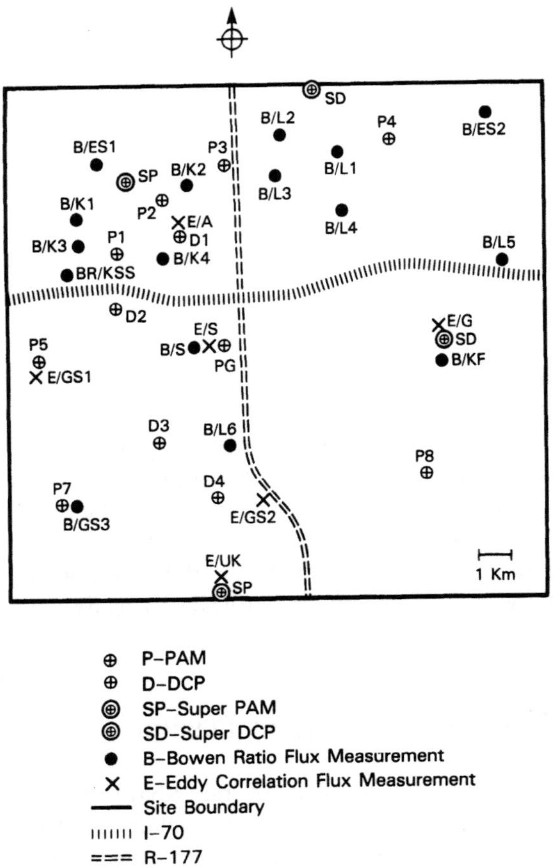

FIGURE 26.1. Arrangement of automatic meteorological stations and flux measurement instruments within the FIFE site. (PAM, Portable Automated Mesonet station from NCAR; DCP, Data Collection Platform from Army Corpse of Engineers.) The PAM instruments reported surface meteorological conditions every 5 min throughout the 3-year data collection phase of the experiment.

The duration of FIFE was determined by the need for data covering a range of surface conditions in terms of soil moisture, vegetative phenology, incoming radiation, and the constraints on the measurement resources that were available. Therefore, a monitoring program of continuous data acquisition—satellite, automatic meteorological station, hydrological and biometric data—was initiated in early 1987 and extended through October 1989. In 1987, the monitoring program was punctuated by four intensive field campaigns (IFCs), during which teams of experimenters, specialized equipment, and research aircraft were committed to the field for periods of 12 to 20 days, timed to allow coverage of the different phenological stages of the vegetation. An additional IFC was conducted in 1989 to obtain data on the critical "dry down" conditions that were poorly sampled in 1987.

3. Measurements

The design of the field experiment itself was complex and the data produced by it are voluminous and diverse. A brief description of the monitoring and field campaign activities follows; for more details see Sellers et al. (1988, 1989).

3.1 Monitoring

Satellite data from all available sensors were acquired for the site from January 1987. Additional data prior to this date were assembled from agency archives. Ten (later increased to 12) Portable Automated Mesonet Stations (PAM, supplied and operated by NCAR) were installed in May 1987 and provided a variety of micrometeorological and radiation measurements at 5-min intervals through October 1989. A time series of biophysical (leaf area, biomass, etc.), hydrological (soil moisture, streamflow), and atmospheric (optical depth) data were acquired for the same period by cooperating investigators at Kansas State University.

3.2 Intensive Field Campaigns

The IFCs were dedicated to collecting the detailed, coincident data necessary for exact comparison of satellite observations with surface and atmospheric processes. The scientists involved were organized into six "disciplinary" teams:

Atmospheric Boundary Layer: This group used three flux aircraft, radiosondes, LIDAR and SODAR to investigate atmospheric profiles and turbulent fluxes over the site.

Surface Fluxes: Twenty-one flux measurement stations, including six eddy correlation rigs, were distributed about the site to measure near-surface fluxes of radiation, heat, moisture, and in some cases momentum and CO_2.

Correction and Calibration: Sun photometer and other observations were made to provide data for the atmospheric correction of the satellite and airborne radiometric data. Additionally, a coordinated effort was made to provide calibration information for all of the radiometers used in the study.

Surface Radiation and Biology: A range of studies was performed to relate observed radiances with surface biophysical and physical states.

Soil Moisture: A variety of airborne instruments and surface instruments were used to characterize the temporal and spatial variation of soil moisture over the site.

Staff Science: A staff science team, consisting largely of NASA and KSU personnel, supported the above efforts by assuming responsibility for ancillary measurements, logistic support, and experiment coordination, including an integrated information system and an extensive airborne remote sensing program using radiometers and scanning instruments operating in the visible, near-infrared, thermal infrared, and microwave regions.

4. The FIFE Information System (FIS)

An integral part of the FIFE Science Plan (Schmugge and Sellers, 1986) was a data system that could serve the FIFE investigators as a tool for designing the experiment as well as for organizing and manipulating the complex data set during and after the data collection effort. A dedicated, remotely accessible data system was developed at NASA's Goddard Space Flight Center to meet this requirement. For a full description of the evolution and accomplishments of FIS, see Strebel et al. (1990a,b).

The volume of digital data exceeds 100 gigabytes. Most of these data are the "raw" (Level 0) digital imagery from aircraft platforms and satellite sensors such as AVHRR, SPOT, and Landsat TM. A reduction factor of about 20 is achieved as the digital imagery is screened for suitability for further analysis during processing to Level 1 (reformatted data, with location information added) or Level 2 (geometrically and/or radiometrically corrected data). In addition, there are over 100 online tables, occupying approximately 250 megabytes of disk storage, which contain a complete inventory, reference information, and all of the nonimage digital data mentioned in the previous section. An archive of analog data, primarily maps, photography, and video tapes, is also part of the data system.

TABLE 26.1. Data dimension vs. reference method.

Dimension	Example	Method
0	Soil moisture at a given point	Site ID number
1	Trace gases over transect	Site number and distance
2	Multispectral imagery	Latitude and longitude files
3	Aircraft flux measurements	*X*, *Y*, *Z* coordinates

The data set diversity and volume could be handled effectively only by establishing standards for data input, storage, and output. At the same time, it was desired to avoid a rigidity that would discourage submission of data or act as a barrier to access. In defining working standards, we looked for common or near-universal capabilities in three areas: formats, measurement units, and coordinate systems.

The time and three-dimensional spatial location of each measurement are particularly crucial to the derivation of the energy and mass fluxes through the FIFE site. In the FIS, this coordinate information is associated with each data set as it is assimilated into the system. The methods range from storing explicit values for point source data to appending georeference data files as additional layers to image products. The method used depends largely on the physical dimensionality of the measurement technique (Table 26.1).

5. Utilization

Active distribution of data by FIS began during the preliminary experiment design phase of the experiment. Soon after the October 1987 field campaign, user logins climbed rapidly to a level of 100–200 queries per month. During this period, the primary use of the online portion of FIS has been to obtain supporting data sets interactively and to obtain information about processed image tapes, which are then distributed by mail. The interactive usage shows increasingly intense periods, and more frequent use of the analog data and documentation is being experienced in the archive located at GSFC.

At the most basic level, FIFE acquired a data set that will allow direct comparison between satellite observations and (near-) surface parameters and processes associated with surface–atmosphere exchanges of energy and mass. As the integrated analysis and modeling phase of the experiment begins, these data are being used to test various methods of integrating our understanding of small-scale processes (e.g., photosynthesis, transpiration, scattering of light by leaves) up to the scale of satellite pixels and atmospheric turbulent transport, that is, several kilometers. In addition, plans are being completed to open FIS for access by investigators outside of the FIFE project, to promote a wider range of relevant studies.

6. Conclusion

FIFE is a trailblazing effort in terms of both the data collected and the information system that handles that data. The FIFE Information System may represent the first practical tool supporting a new way of doing science. Through computer technology and common sense organization, specialists from various fields are succeeding in collaborating in the collection and analysis of the data required to answer a significant interdisciplinary research question. Many similar questions are becoming pressing as the seriousness of possibilities such as global climate change are recognized. The FIS model can serve as a tested starting point for organizing the data activities of future interdisciplinary research projects that will be required to address these questions.

Preliminary results of the FIFE analysis are reviewed in Hall et al. (1989). Individuals wishing more information about FIS or the FIFE data should contact the authors.

Acknowledgments. Peter Camillo, to whom this volume is dedicated, provided valuable guidance to FIS as a member of the FIFE Information System Science Working Group.

The Konza Prairie Research Natural Area, which was included in the FIFE site, is owned by the Nature Conservancy and operated by the Division of Biology at Kansas State University. The work reported in this paper was partially performed under contract to the National Aeronautics and Space Administration by employees of Versar, Inc.

References

Hall FG, Sellers PJ, MacPherson I, Kelly RD, Verma S, Markham B, Blad B, Wang J, Strebel DE (1989) FIFE: Analysis and results–A review. Adv Space Res 9:7:275–293.

NASA (1984) Earth Observing System. Vol I: Science and Mission Requirements. Working Group Report, National Aeronautics and Space Administration, Technical Memorandum 86129.

Schmugge T, Sellers PJ (1986) FIFE Science Plan. NASA Technical Report.

Sellers PJ, Hall FG, Asrar G, Strebel DE, Murphy RE (1988) The First ISLSCP Field Experiment (FIFE). Bull Am Meteorol Soc 69:1:22–27.

Sellers PJ, Dorman JL, Walthall C, Hall FG, Strebel DE, Markham BL, Wang JR, Kelly RD, Verma SB, Blad BL, Schimel DS, Grossman R, Brutsaert W, Desjardins R, Fritschen L, Frouin R, Kanemasu E, MacPherson I (1989). First ISLSCP Field Experiment: Experiment execution and preliminary analyses. Remote Sensing and Large-Scale Global Processes. Proceedings of the IAHS Third International Assembly, Baltimore, MD, May 1989. IAHS Publ. No. 186.

Strebel DE, Ungar SG, Ormsby JP, Newcomer JA, Landis D, Goetz SJ, Huemmrich KF, van Elburg-Obler D, Sellers PJ, Hall FG (1990a). "The FIFE Information System: Support of Interdisciplinary Science." Preprint volume, Symposium on the First ISLSCP Field Experiment, pp. 140–147. American Meteorological Society, Boston.

Strebel DE, Newcomer JA, Ormsby JP, Hall FG, Sellers PJ (1990b). The FIFE information system. IEEE Trans Geosci Remote Sens 28:4:703–710.

Appendix: Acronyms and Abbreviations

ABL	Atmospheric Boundary Layer
APAR	Absorbed Photosynthetically Active Radiation
AVHRR	Advanced Very-High Resolution Radiometer
BATS	Biosphere–Atmosphere Transfer Scheme (Model)
CCM	Community Climate Model (NCAR)
CDT	Central Daylight Time
CFL	Constant Flux Layer
CNES	Centre National d'Etudes Spatiales (France)
CNRM	Centre National de la Recherche Météorologique (Toulouse)
CNRS	Centre National de la Recherche Scientifique (France)
CRPE	Centre de Recherche en Physique de l'Environnement
ECMWF	European Centre for Medium Range Weather Forecasts Model (Reading, UK)
EERM	Establissement d'Etude et de Recherches Météorologiques
EML	Elevated Well-Mixed Layer
ENSMP	Ecole Nationale Superièure des Mines de Paris
EOS	Earth Observing System
ET	Evapotranspiration
FFT	Fast Fourier Transforms
FIFE	First Islscp Field Experiment
FIS	Fife Information System
GCM	General Circulation Model
GIS	Geographic Information System
GOES	Geostationary Operational Environmental Satellite
GPS	Global Positioning System
GSFC	Goddard Space Flight Center (NASA)
GSTS	Groupement Scientifique de Télédétection Spatiale (Strasbourg, France)
HAPEX-MOBILHY	Hydrologic Atmospheric Pilot Experiment-Modélisation du Bilan Hydrique
IBL	Internal Boundary Layer
ISCCP	International Satellite Cloud Climatology Project (World Meteorological Organization)
IFC	Intensive Field Campaign

IH	Institute of Hydrology (Wallingford, UK)
INRA	Institut National de la Recherche Agronomique (France)
ISLSCP	International Satellite Land Surface Climatology Project
LAI	Leaf Area Index
LAMP	Laboratoire Associé Météorologie Physique (Clermont–Ferrand, France)
LIDAR	Light Direction and Ranging
MO	Meteorological Office Model (UK)
MPP	Multiple Parallel Processor
NASA	National Aeronautics and Space Administration
NCAR	National Center for Atmospheric Research (Boulder, Colorado)
NDVI	Normalized Difference Vegetation Index
NIST	National Institute of Standards and Technology (formerly, the Bureau of Standards)
NMC	National Meteorological Center
NOAA	National Oceanic and Atmospheric Administration
NWPMS	Numerical Weather Prediction Model
NWS	National Weather Service (USA)
ORSTOM	Institut Français de Recherche Scientifique pour le Développement en Coopération
PAM	Portable Automated Mesonet Stations
PAR	Photosynthetic Active Radiation
PBL	Planetary Boundary Layer
SAMER	Systeme Automatique de Mesure de l'Evapotranspiration Re'ele
SCEM	Service Central d'Exploitation de la Météorologie
SHE	Systems Hydrologique European
SiB	Simple Biosphere Model
SODAR	Sound Detecting and Ranging
SOP	Special Observing Period
TIMS	Thermal Infrared Multispectral Scanner
TMS	Thematic Mapper Simulator
TOPMODEL	Beven and Kirkby Catchment Model
USDA	U.S. Department of Agriculture
UT	Universal Time
WCRP	World Climate Research Program
WML	Well-Mixed Layer
WMO	World Meteorological Organization

Index